# COURSES IN
# MINING GEOLOGY

## FOURTH EDITION

# COURSES IN
# MINING GEOLOGY

## FOURTH EDITION

### R.N.P. AROGYASWAMY

*Formerly, Chief (Oil & Minerals)*
*Planning Commission, New Delhi*
*Emeritus Scientist, C.S.I.R.*
*Visiting Professor (University of Madras)*

## Oxford & IBH Publishing Co. Pvt. Ltd.
New Delhi
*( A Unit of* CBS Publishers & Distributors Pvt Ltd *)*

CBSPD

## CBS Publishers & Distributors Pvt Ltd

New Delhi • Bengaluru • Chennai • Kochi • Kolkata • Lucknow • Mumbai
Hyderabad • Jharkhand • Nagpur • Patna • Pune • Uttarakhand

# COURSES IN
# MINING GEOLOGY
## FOURTH EDITION

© 1973, 1980, 1988, 1995, RNP Arogyaswamy
**Reprint:** 1996, 2017, 2020, **2024**

ISBN 978-81-204-0937-X

## OXFORD & IBH
New Delhi
( A Unit of CBS Publishers & Distributors Pvt Ltd )

Published by **Satish Kumar Jain** and produced by **Varun Jain** for

### CBS Publishers & Distributors Pvt Ltd
4819/XI Prahlad Street, 24 Ansari Road, Daryaganj, New Delhi 110 002, India.
Ph: 011-23289259, 23266861
Website: www.cbspd.com
Fax: 011-23243014
e-mail: delhi@cbspd.com

*Corporate Office:* 204 FIE, Industrial Area, Patparganj, Delhi 110 092
Ph: 011-4934 4934
Fax: 011-4934 4935
e-mail: publishing@cbspd.com; publicity@cbspd.com

### Branches

- **Bengaluru:** Seema House 2975, 17th Cross, KR Road, Banasankari 2nd Stage, Bengaluru 560 070, Karnataka, India
  Ph: +91-80-26771678/79          Fax: +91-80-26771680          e-mail: bangalore@cbspd.com
- **Chennai:** 7, Subbaraya Street, Shenoy Nagar, Chennai 600 030, Tamil Nadu, India
  Ph: +91-44-26680620, 26681266          Fax: +91-44-42032115          e-mail: chennai@cbspd.com
- **Kochi:** 42/1325, 1326, Power House Road, Opp KSEB, Power House, Ernakulum Kochi 682 018, Kerala, India
  Ph: +91-484-4059061-65,67          Fax: +91-484-4059065          e-mail: kochi@cbspd.com
- **Kolkata:** 147, Hind Ceramics Compound, 1st Floor, Nilgunj Road, Belghoria, Kolkata-700056, West Bengal, India
  Ph: +033-25633055, 033-25633056          e-mail: kolkata@cbspd.com
- **Lucknow:** Basement, Khushnuma Complex, 7 Meerabai Marg (Behina Jawahar Bhawan), Lucknow-226001, UP, India
  Ph: +0522-4000032          e-mail: tiwari.lucknow@cbspd.com
- **Mumbai:** PWD Shed, Gala no 25/26, Ramchandra Bhatt Marg, Next to JJ Hospital Gate no. 2, Opp. Union Bank of India,
  Noorbaug, Mumbai-400009, Maharashtra, India
  Ph: 022-66661880/89          e-mail: mumbai@cbspd.com

### Representatives

| | | | | | |
|---|---|---|---|---|---|
| • Hyderabad | 0-9885175004 | • Jharkhand | 0-9811541605 | • Nagpur | 0-8692091830 |
| • Patna | 0-9334159340 | • Pune | 0-9664372571 | • Uttarakhand | 0-9716462459 |

*Printed at* Mudrak, Noida, UP, India

## PREFACE TO THE FOURTH EDITION

It is gratifying to note that the "Courses in Mining Geology" has been accepted not only by the teachers but also by the students, as anticipated by the late Dr. D.N. Wadia, F.R.S., in his foreword to the first edition. The present edition takes into account the recent changes in the Mineral Concession Rules and also in the Mineral Conservation and Development Rules as a result of which geologists, with the requisite experience, have certain specific advantages when appearing for the Examinations held by the Department of Mines Safety, in regard to the "certificate of competency" for mine managers. It is also possible for geologists, with 5 years of field experience, to qualify as "Recognised Qualified Persons" as required by the Indian Bureau of Mines.

Hence, with these additional avenues of employment being open, it is necessary that geologists equip themselves with the requisite knowledge, especially with regard to open cast mining, including the deployment and utilisation of heavy earth moving machines, in addition to the techniques of blasting as applied to open cast mines. The other additional material now included provides an insight into aspects of open cast mining such as deployment and utilisation of heavy earth moving machines and the techniques of blasting as applied to open cast mining.

It is, however, to be pointed out that in addition to technical knowledge, it is essential that the mining geologist gets acquainted with legislation as applicable to mines, since this is also an integral part. Technical ability contributes to improving efficiency and economy, while legislation takes care of the third aspect "safety". Hence familarity with the Mines Act, as well as the Regulations and Rules framed under the Act cannot be neglected.

R.N.P. AROGYASWAMY

# FOREWORD

During the last three decades or so, geological education in India has made considerable progress and from four or five geology teaching institutions, today we have more than 20 universities with fairly equipped geological faculties imparting postgraduate instruction in various branches of earth sciences. On the whole, this development has been on more or less correct lines and now we have a body of over 2,500 young working geologists in the country, the majority of whom, in spite of the bias on theoretical and academic aspects in their training, are giving creditable account of work even when placed in unfamiliar or difficult environments.

With the expansion of mining and mineral-based industries in the country, there is bound to be a call for larger numbers of well-trained practical geologists in the near future and our universities and technical institutions have to adjust their syllabii more and more to the new applied branches of earth sciences—aerial and photo geological surveys, geophysics and geochemistry, mineral technology, oil geology, ore geology, improved mining practices, etc.

Mr. Arogyaswamy, as a senior officer of the G.S.I., from his long experience of teaching and field geological survey and mapping has produced a valuable compendium which provides within a small compass a wide range of useful information on the above mentioned subjects that all working geologists today should possess. This book will stimulate in students an interest in the practical applications of geology, which the orthodox textbooks do not give and lead them to seek further knowledge in more specialised works. In this way, the author has done a great service to both the teachers and the students of geology in our universities and he deserves our thanks for this.

D.N. WADIA

## PREFACE TO THE FIRST EDITION

Geology from a subject of mere amateurish, popular interest crystallised into a science in the days of Sir Alfred Lyell. With the rapid industrial expansion, during the present century, the applied aspect has gained importance. Thus, mining geology, engineering geology, photo geology etc. have emerged and so have geomorphology, ground water geology, geochemistry etc. attained a status of their own. Geology now demands the use of other disciplines, inclusive of the social sciences, in keeping with its growing importance. "The stone, which the builders refused, has become the headstone of the corner" (Psalms 118, 22). How true is this of geology one can easily see.

Portions of the book formed the basis of the lectures which the author delivered at the Presidency College Madras in 1942, and later at the University of Patna in 1963-64. Some parts of the book, e.g., notes on survey methods, were written for use in the "Training Camps" held by the Geological Survey of India between the years 1948 and 1950 while other portions formed the articles contributed to the Tamil encyclopaedia.

In drawing up the contents of the book, the author has been guided mainly by his experience as a field geologist for over 25 years, and has thus been able to sort out the grain from the chaff. Familiarity with the syllabus, in mining geology, prescribed by certain Indian universities, has been of advantage in setting the broad limits to the extent of the courses. Thus, the general pattern of the book may be said "to fill up the gaps" and remedy the deficiencies, so that, with the additional material found herein, students taking pure geology courses may acquire the adequate applied bias. Consequently, subjects like physical geology, geomorphology, mineralogy, economic geology and structural geology, forming part of the pure geology course, have not been included. It is however, not the intention of the author to make a mining engineer out of a geologist, but to make the young geologist alive to the problems which he will have to face, in the field, and give him the necessary self-reliance, especially, when he is left alone, to develop a small prospect and be responsible for its running. With this object in view, only the principles of the important machines, employed in mining operations, and their uses have been indicated, while the technical details have been strictly excluded. If the geologist attached to a larger mining concern, after going through these pages, is able to understand the language of the mining engineer, and make the mining engineer understand him, the purpose of this book will also have been served to a large extent.

It is common experience that no industrial undertaking or project can survive unless it is based on sound economics. Thus, more recently, the tendency for laying greater emphasis, on the study of humanities by students of technology, has been growing. Hence, courses in industrial management and economics have also sought to be included. In consonance with this view, subjects of practical utility, such as Book-keeping, Accounting, Cost Accounting, Economics and Store-keeping have been given adequate emphasis.

Incidentally the courses included herein, to a large extent, cover the syllabus in mining geology and elements of mining etc. for the competitive examinations of the Union Public Service Commission for the recruitment of geologists in the Geological Survey of India.

Finally, as a note of caution, it is to be pointed out, that this book is by no means a compendium of all knowledge. It is just a humble initiation to a vast field of study and it is well for the student to remember the wise Arabian proverb:

"There are four sorts of men.
He who knows not, and knows not he knows not;
he is a fool—shun him.
He who knows not, and knows he knows not;
he is simple—teach him.
He who knows and knows not he knows;
he is asleep—wake him.
He who knows and knows he knows;
he is wise—follow him."

—Lady Burton

The present work is the outcome of the encouragement the author received from the older generation of geologists of the Geological Survey of India. The indefatigable geologist, late Sir Cyril S. Fox, was a source of inspiration and it was under his guidance and direction that mining geology came to its own in India. Late Dr. John A. Dunn, whose book "Indian Mining" was long used as a guide by students of Mining Geology, set the standard, and late Dr. M.S. Krishnan, whose studious habits were contagious, was an example. The author recollects with pleasure his early initiation to geophysics by Shri M.B. Ramachandra Rao (Retd. Member, O.N.G.C.) and his eagerness to help and encourage.

It is superfluous to point out that in a work of this nature information is necessarily to be drawn from various sources, such as those listed in the references, in addition to one's experience. It gives me utmost pleasure to acknowledge the valuable information which has been supplied most willingly and almost spontaneously by the foremost mining undertakings in the country. My sincere gratitude is due to these concerns, viz., the Kolar Gold Mining Undertaking, Oorgaum; the Indian Copper Corporation, Ghatsila; the Shivrajpur Syndicate (Killick Industries Ltd., Bombay) the Tata Iron & Steel Company, Jamshedpur; and the Metal Corporation of India Ltd., Calcutta. The author is thankful to Shri Ashok Mitra, Secretary, Planning Commission, who has been a source of help and encouragement.

Advice has been sought from various quarters, and the author wishes to place on record the thoughtful guidance he received from the late Prof. C. Mahadevan of the Andhra University. Suggestions and contributions were received from late S.V. Subramanyan, General Manager, Jaduguda Mines, Shri S. Narayanaswamy, Shri B. Arogyaswamy and Shri G.S. Lodha, geophysicist. The amendments and improvements of Shri M.N. Saxena and Shri M.L. Kapur of the Planning Commission are also gratefully acknowledged. The author is, however, solely responsible for any lapses.

To (late) Dr. D.N. Wadia, F.R.S., who has been the source of inspiration to several geologists in this country and abroad, I express my thanks for so willingly accepting to write the foreword to this humble work and thus launching it on its maiden voyage.

Last but not the least it is my pleasure to thank Messrs. Oxford & IBH Publishing Co. Pvt Ltd., who have spared no pains in arranging the manuscript and expediting the publication, without sacrificing the high quality in execution.

It is proposed to undertake the revision of the book very shortly. Any suggestions for its improvement will be gratefully received by the author.

June 1973                                          R.N.P. AROGYASWAMY

# CONTENTS

## Part I—Topographical Surveying for Geological Work

## Part II—Photogrammetry

## Part III—Photogeology

## Part IV—Geochemical Prospecting

## Part V—Geophysical Methods in Mineral Explorations

## Part VI—Drilling

## Part VII—Ancilliary Operation in Prospecting

## Part VIII—Mining Methods

## Part IX—Coal Mining Methods

## Part X

PART I


TOPOGRAPHICAL SURVEYING
FOR GEOLOGICAL WORK

# INTRODUCTION

Besides routine geological mapping, on standard topographical maps, on the usual scale of 1 inch=1 mile, the geologist in India is often required to undertake various other types of work which call for a fair knowledge of surveying methods. Instruments used in surveying, their construction and detailed descriptions have been given in several standard text books on the subject. For carrying out his normal work, the field geologist requires only a few practical hints as to the handling and the care of these instruments. Thus the purpose here is to show how the survey instruments are employed in carrying out surveys for geological work.

Simple methods of surveying, such as tape and compass survey, chain survey, and plane table survey have been given their due importance. Sufficient emphasis on the use of the Brunton Compass has also been laid. In the methods of survey, often used by the geologist, for mapping various types of mineral deposits, the Brunton Compass is indispensible. It is also handy for underground work, such as stope surveys and detailed geological mapping of mine workings, and in the examination of excavations for large scale projects.

Considerable improvements have been affected to surveying instruments within the last two decades. The introduction of the microptic instruments, such as the microptic telescopic alidade, and the microptic theodolite have gone a long way in relieving the burden of the surveyor. These instruments are sufficiently robust and can be used with considerable advantage in the survey of mineral deposits. The construction and use of such instruments has been indicated under the relevant chapters. The later developments in this line, are the introduction of microptic stadia instruments known as self reducing tacheometers. These instruments have further contributed to easing the task of the field geologist. But at the present time these instruments are rather expensive and may be beyond the pocket of the private practising geologist. These instruments are obtainable from manufactures in the USSR and W. Germany etc., but it is expected that similar instruments of indigenous manufacture will be more easily available in the near future.

3

Since World War II application of air photography to surveying has increased manyfold; and air survey has become a routine for surveying virgin terrain, since air photos, especially verticals provide additional information, besides data for the construction of maps. Air photos are used in soil conservation, agriculture, irrigation projects, engineering projects, forestry etc., and often contain a fund of geological information. Thus the subject of photogeology has come into existence.

Photogrammetry, which includes, preparation of maps from air photos, together with measurement of heights from photos, by stereoscopic methods, and contouring, is an essential requisite for the photogeologist. An endeavour has been made to compress all essential information required for preparing simple maps from air photos, both verticals (stereopairs) as well as from obliques.

## PREPARATION FOR SURVEY WORK

1. Assembling of instruments, accessories and surveyor's kit, required for the particular method to be adopted, is primary.

2. Reconnaissance or preliminary examination of the area by traverses is carried out, so as to obtain a thorough knowledge of the lay out, with a view to decide what details are to be included, depending on the method proposed to be adopted and the amount of time available. A pair of binoculars is useful for purposes of reconnoitering.

3. Fixing of stations—The stations should be so located such that there is no obstruction to direct visibility between any two adjacent (at least) stations, unless, such an obstruction is unavoidable.

4. Referencing stations—All important stations should be located with reference to three permanent objects such as masonry work or large trees etc., and a sketch plan showing distances and direction from such permanent objects should be made out in the field.

5. The sketch plan of the whole area should be prepared showing the stations and other salient features. The north line and approximate scale should also be indicated.

## CHAIN SURVEY

The equipment consists of chain (30 m) or 100 ft. long, 10 arrows, (15 m) or 50 ft. metallic tape, ranging rods, required number of station flags, plumb bob, field book, pencil, and eraser.

In the ideal chain survey, no angular measurements are made. The stations are so located that the area, to be surveyed, is divided up into triangles and all the sides of these triangles are measured. The triangle is chosen, as it is the only plane figure whose shape can be fixed merely by linear measurements. However, for the survey to be accurate, the triangles should be "well conditioned", i.e., the angles should be as near to 60° as possible. This condition is essential for obtaining clear intersections when the survey is plotted. Acute intersections cannot be pinpointed, and the ill conditioned triangle or triangles in which any angle is less than 30° are to be strengthened by "*tie lines*", which may either be perpendiculars from any angle to the opposite sides, or any line which connects any angle to any known point on the opposite side. Such tie-lines will serve as a check on the accuracy of the plotting of the survey.

Three men are normally required for chain survey : (*a*) Leader, (*b*) Recorder (*c*) Follower. (The work of follower and recorder can also be combined if necessary.)

Procedure : Flags preferably of conspicuous colours, attached to poles are planted at the various stations. The follower takes his stand at the starting station and opens the chain in the manner described later and gives one of the handles to the leader, who also takes the plumb bob, arrows and a ranging rod.

In order to open the chain, a firm grip is taken of both the handles with the left hand and a few links are paid off, while the rest of the folded chain is gripped in the right hand. Now the unopened portion of the chain in the right hand is thrown forward, in the direction of the line to be measured, while the left hand is slightly jerked backwards. The chain opens out but remains folded at the 50th link i.e. the mid link. The follower now throws one of the handles to the leader who takes it up, moves in line with the forward station, and stretches the chain to its full length. The follower aligns the leader's ranging rod with the forward station. The chain is now stretched in line with the rod and the arrow planted alongside the handle. The follower leaves the chain and accompanies the recorder, who walks along astride the chain noting the measurements, while the follower assists him in taking offsets to objects or other features seen on either side of the chain line with the metallic tape. The measurements are recorded in the manner given in Fig. 1.1.

The follower, in all subsequent operations, picks up the arrows so that by counting the number of arrows, both with the leader and follower, a mutual check, on the distance measured, is obtained.

Note : (1) The various shapes of the tags on the chain—the 10 ft. and 90 ft. tags have one point. The 20 and 80 ft. tags have two points. The 30 and 70 ft. tags have three points. The 40 and 60 ft. tags have four points while the 50th feet tag is circular.

(2) The double line in centre of the field book (Fig. 1.1) represents the chain line, and the figures noted between the double line, denote chain distances. Figures noted outside indicate the offset distances of respective points from the corresponding points on the chain line.

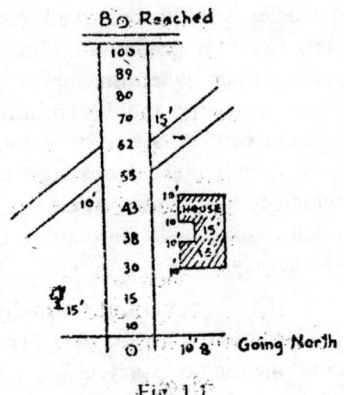

Fig. 1.1

(3) In recording the distance of a point to be 'off-set' the follower places the end of the tape at the point. The leader pays of the tape and then holding it taut and level over the chain describes an arc and obtains the minimum tape reading as well as the chain reading (at the tangential point).

(4) The interior filling in or the location of details, of the survey is done either with the help of extra tie lines or by means of subsidiary compass and tape surveys.

Method of closing the chain is as follows : The chain is pulled by the middle tag (i.e. the 50 ft. tag) so that the two handles come together. The two links on either side of the 50th tag, are held together in the left hand, while the alternate-pair of links are picked up two together at a time with the right hand folded upwards and held in the left hand.

Note : It is a good and convenient practice for the follower to align the leader with the forward stations by a show of hands. Shouting is tiresome and leads to confusion. The right hand is used for indicating movement towards the right side and the left hand for movements towards the left. In case, the same hand is employed to indicate movement in both directions confusion will arise. While winding the metallic tape with the right hand, the tape is made to pass between the forefinger and the second finger of the left hand in which the case is held, so that any twists in the tape automatically straighten out.

## COMPASS AND TAPE (OR CHAIN) SURVEY

This is the method most frequently used by geologists in mapping and in preliminary mineral investigations, in small areas or for interior filling.

**Instruments required** : (1) Compass, which may be (a) Prismatic-com pass, (b) Clino compass, (c) Miner's dial, (d) Brunton compass; (2) Chain and arrows or tape; and (3) Ranging rods etc. as in Chain Survey.

**Procedure** : The angular measurements are made with the compass and recorded either as whole circle bearings or as quadrant bearings and the distances are measured, with the chain or tape. It is always advisable to close such traverses, so as to come back to the starting point. This will enable the surveyor to judge the magnitude of errors, if any, involved in the survey. If, on plotting, the errors are small, then the closing error could be distributed in the manner described later.

*Note :* The compasses are of two types :

**A.** Those in which the needle moves (relatively) over a fixed graduated scale, *e.g.*, Clino compass, Brunton compass, and Miners' dial. In these instruments there is an interchange of the east and west points so as to compensate the relative movement of the needle to the stationary scale so as to enable quadrant bearing to be obtained directly. Quadrant bearings are expressed in the following manner, *e.g.* NW, NNW, WNW, or ENE, NE, NNE, etc. or N 30° E, N 20° W, S 10° E, S 5° W, *i.e.* always in relation to the N or S direction.

The whole circle forward bearing is the reading of the north end of the needle on the 360° scale. For example in (Fig. 1.2), the forward bearing of AB is 90°, *i.e.*, bearing from A to B, while the back bearing from B to A is 270°. If during the course of mapping it is required to locate one position A, with reference to a known point B, then the back bearing from the unknown point A to the known point B is measured and the distance plotted on the map from the known point B or if on the other hand the position of A is known and it is required to locate further points, then the forward bearings are determined and the distances plotted. Frequently in practice, intersection of 2 or more plotted directions are used for a location of greater accuracy.

**B.** Those instruments, in which the graduated scale moves along with the needle, the bearing, with reference to the magnetic north is obtained directly by reading the graduations, *e.g.*, prismatic compass. These are the whole circle bearings and these are recorded in the following manner—1° to 360°. *Note :*—The instrument must be level or else inaccuracies will result. In instruments that have liquid immersion for damping the movement of the disc, air-bubbles when formed in the liquid chamber disturb the accuracy. Compass surveys are normally not recommended in areas susceptible to magnetic influence.

### Compass traverse (alternative method)

It is easier to complete the traverse by observing both the bearings of the forward line and the backward line from each station as indicated earlier. The difference between the values of the bearings of the forward and the backward stations gives the included angle at the instrument station. Thus it is possible to plot the traverse with the value of the included angles.*

The closing error in a compass traverse should not exceed 1 in 300.

## DISTRIBUTION OF CLOSING ERROR IN A COMPASS TRAVERSE

If on plotting a compass traverse, a small closing error AA' is obtained (Fig. 1.2), it may be reasonably assumed that this is due to discrepancies in both angular and linear measurements, and if the error is not large, e.g., 1 : 1000, it may be distributed proportionally between the distances AB, BC, CD, DE, EA without seriously affecting the angles of the traverse individually.

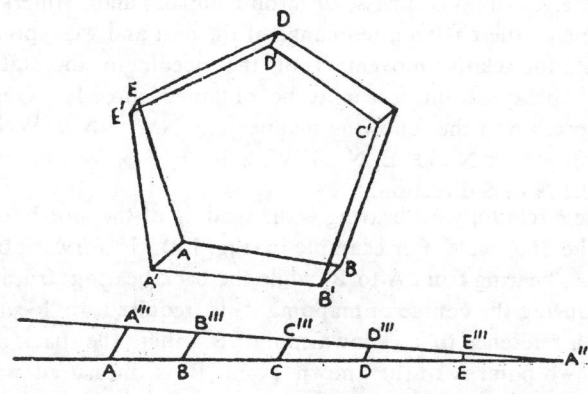

Fig. 1.2

Let AA' (Fig. 1.2) be the closing error of the traverse which is to be distributed equitably. Draw any line AA'' and on it mark of lengths AB, BC, CD, DE, EA'', proportional or equal to AB, BC, CD, DE, EA', respectively of the traverse. From A draw A A''' equal and parallel to AA'. Join A''' to A'''. Similarly draw BB''', CC''' DD'', EE''', parallel to AA''' meeting A'' A''' in B''', C''', D'', E'''. Draw lines parallel to AA, i.e. the closing error from B, C, D, and E of the traverse plan and plot lengths, BB'', CC'', DD'', EE'', respectively on these. Join A'B'C' D'E'. The figure A'B'C'D'E' represents the traverse after distributing the error.

----

*See addendum

## COMPASS SURVEY IN THE PRESENCE OF ROCKS
## CONTAINING MAGNETIC MINERALS

In the case of compass survey of an area where there is magnetic attraction *e.g.*, in the vicinity of large basic intrusions, magnetic ore bodies or in a mine where there is a large amount of iron work, like rails, pipes etc. or in an electrified part of a mine, the following procedure may be adopted :

1. The traverse is started from a point where there is no magnetic attraction, *i.e.* where the back and fore readings on a straight line are in agreement or show no deflection, the difference being 180°.

2. The survey is continued and back sight and fore sight readings are taken at each station. The following example (Fig. 1.3) will serve as a guide to computations. It will be seen that where there is a difference between the back and fore sight readings, the extent of the difference is the amount of correction to be applied. The same procedure is applied for the rest of the traverse.

Where the bearings (whole circle) have been recorded, it is possible to calculate the included angles in a traverse as can be seen from the example given below :

If $f$=the forward bearing at the station (2) and $b$=the bearing (forward) at the previous station (1) then the included angles can be computed from the formula :

$$180° - f + b \text{ or,}$$

where $f > 180°$ then, $360° - (f + b)$.

It may be pointed out that if there is a difference between the back sight and the foresight, on the same straight line, the amount of the difference, is the amount of the correction to be applied to the succeeding foresight reading got at the station from which the back sight was taken. It should also be noted whether the correction is to be applied in the clockwise direction or reverse. In the following example, the computation has been done in terms of whole circle bearings. (fig. 1.3).

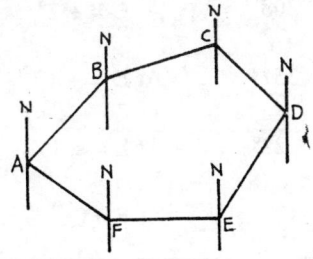

Fig. 1.3

In case wholè circle bearings are used, the computations are easier, as it will
be noticed from the examples given below:

| Line | Observed bearing | | Correction | Corrected bearing | |
|------|------|------|------------|------|------|
|      | FB | BB |            | FB | BB |
| AB | 15°20′ | 195°20′ | At B = 0 | 15°20′ | 195°20′ |
| BC | 19°33′ | 199°33′ | At C = 0 | 19°33′ | 199°33′ |
| CD | 108°09′ | 282°25′ | At D = +5°44′ | 108°09′ | 280°09′ |
| DE | 273°11′ | 88°35′ | At E = + 4°36′ | 278°55′ | 98°55′ |
| EF | 92°14′ | 276°50′ | At F = + 4°36′ | 96°55′ | 276°50′ |
| FG | 350°11′ | 170°11′ | At G = 0 | 354°47′ | 174°47′ |

| Line | Observed bearing | | Correction | Corrected bearing | |
|------|------|------|------------|------|------|
|      | FB | BB |            | FB | BB |
| AB | 50° | 230° | At B = 0 | 50° | 230° |
| BC | 130° | 308° | At C = + 2° | 130° | 310° |
| CD | 178° | 357° | At D = + 1° | 180° | 360° |
| DE | 303° | 122° | At E = + 1 | 304° | 124° |
| EF | 306° | 126° | At F = 0 | 307° | 727° |

## BRUNTON COMPASS

The Brunton compass is one of the most popular survey instruments for geological work. It can be used for the following purpsses :

1. As a compass or clino compass, where (a) the line of sight is horizontal or slightly depressed, (b) where the line of sight is steeply inclined or depressed.
2. As an Abney level.
3. As a prismatic compass.

1. (a) For measuring angles in azimuth or horizontal angles, Fig 1.3a., the body (B) of the Brunton is held horizontally while the sight vane (V) is made vertical and the lid (L) with the mirror is opened out at say 135° or 140°.

The instrument is held in this position at waist level, and the object aligned or sighted by turning round and not by moving the hands. The intersection is perfect when the centre line of the image of the slot in the sight vane (V), conicides with the black centre line of the mirror on the inside of the lid (L). Now with the observer's eye (E) overlooking the mirror, horizontality is checked up by the bulls' eye level (circular). The bearing in azimuth is obtained either as whole circle or quadrant bearing by reading the north end of the needle.

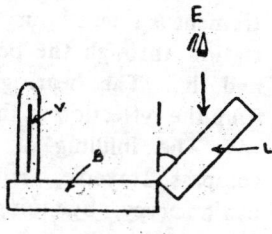

Fig. 1.3a

The above method is followed where the line of sight does not make an angle of more than 45° above the horizontal and not more than a few degress below the horizontal.

(b) Where the line of sight is inclined at steep angles, 15° from the horizontal, the instrument is held in the reverse position as shown in Fig. 1.3b, with the sight van (V) towards the observer and the lid (L) with the mirror away.

Observations are made by sighting the object through the sight vane (V) and the hole (h) in the lid (L) Fig. 1.3b.

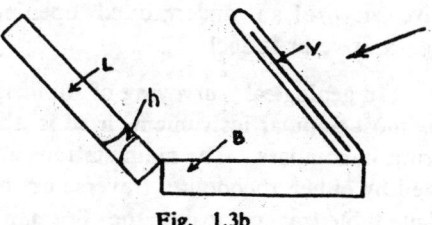

Fig. 1.3b

2. For obtaining the general angle of dip of a formation the mirror lid (L) is opened out fully and the upper flat edge of the instrument is aligned with bedding plane or contact plane of a conformable series. Then the bubble tube is brought to the centre of its run by operating the lever at the back of the instrument. The reading of the vernier gives the angle of dip.

Where it is intended to us the Brunton compass as an Abney level, to measure vertical angles, the sight vane (V) is opened out fully and made

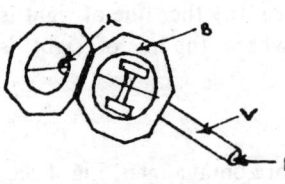

level with the body (B). The upper digit of the sight vane carrying the peep hole (P) is bent at right angles to the sight vane. The lid (L) is kept at an angle of 45° (Fig. 1.3c). The peep hole (P) on the sight vane (V) is aligned with hole (h) in the lid (L). The instrument is now held in the vertical position (edgewise) as shown in Fig. 1.3c and the object is sighed through the peep hole (P) and the hole (h). At the same time the long bubble attached to the vernier is brought to the centre of its run. The instrument is now read at the vernier to obtain the angle of elevation or depression.

Fig. 1.3c

3. For using the Brunton compass as a prismatic compass, the instrument is opened out as shown in Fig. 1.3d. The observer sights the station through the peep holes 'P' and 'h'. The bearing is obtained from the reflection in the mirror.

The infilling of details, in a compass traverse, is done, as in chain survey, by taping off-set distances from the traverse line and recording the details in the same manner.

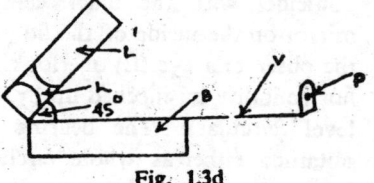

Fig. 1.3d

## TUNNEL SURVEY WITH BRUNTON COMPASS

During the course of geological work it sometimes becomes necessary to survey tunnels driven in connection with engineering projects, e.g., railway, electricity, irrigation, or road projects etc. In addition, surveys of inclined shafts, adits, levels, drives, crosscuts, raises and winzes etc. and other openings made in connection with mining and mineral exploratory operations are to be surveyed geologically.

Figure 1.4 gives the perspective view of an underground opening e.g., a drive or tunnel.

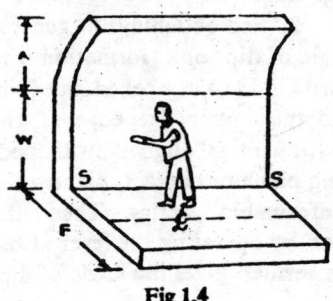

In geological surveying of tunnels, the most popular instrument used is the Brunton compass. The main stations are fixed by either theodolite traverse or by plane table traverse (using the Beaman's stadia arc). After this the geological details, such as mapping of rock types, location of structures, mineralised zones

Fig 1.4

seepage zones etc. are done by means of Brunton compass.

There are two methods adopted for mapping tunnels. For purposes of normal underground geological surveying, all details are recorded at breast height. The surveyor (geologist) moves along the centre line of the tunnel holding the Brunton at breast height, and notes the bearing of the strike of all linear or planar features. The direction and amount of plunge and dip of such features are also recorded simultaneously.

Thus a planar (dyke, fault, joint etc.) feature oblique to the centre line of the tunnel will be recorded at the point where it crosses the centre line at breast height. The direction and amount of dip are also recorded.

While carrying out geological mapping of tunnels (Fig 1.4a) for engineering purposes, a modified practice is adopted. In general, the features on the floor of the tunnel F are covered largely by debris and dirt and hence all features visible on the walls W and in the arch are recorded. The mapping is carried out at spring level S. In this method, three separate plans will result, one each for the two walls below the spring line, and one for the arch.

Where a planar feature, e.g., bedding plane, joint plane, foliation plane, fault plane, is encountered in the tunnel : (Fig. 1.4a).

(a) When the planar feature is vertical and is oriented at right angles to the traverse line, e.g., ADECB then its projection on the floor of the tunnel AB and on the walls AD & BC lies in a straight line AB. (Fig. 1.4b(a).

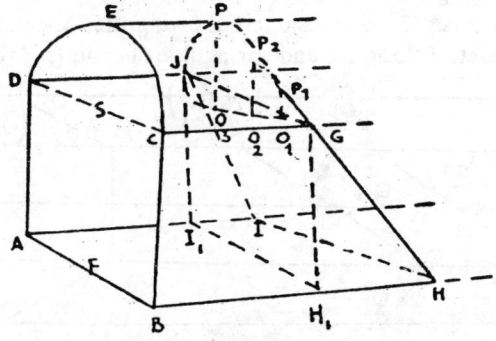

Fig. 1.4a

(b) When the planar feature is inclined, but the strike is normal to the traverse line, e.g. HPJI the projections on the walls are inclined and the angle of inclination of the trace in respect to the floor corners will be equal to the dip. The traces of the planes on either wall are symmetrical with reference to the traverse line (Fig. 1.4b) but will be off set from HI to $H_1I_1$.

(c) When the planar feature is vertical but it cuts the traverse line at an angle (*i.e.* oblique to it) then the trace on the walls bear the same angular relation to the corners and the trace line is off set. (Fig.1.4b (*b*)).

(d) When the planar feature is oblique to the traverse line and is at the same time dipping at an angle $<90°$ the traces on the walls make an angle Q equal to the dip and show an off set. (Fig.1.4b(*c*))

The recording of the features on the walls is relatively simple, but the recording of certain plunging linear features on the arch or dipping planar features on the arch, requires careful judgement. Fig. 1.4*a* illustrates the procedure with respect to plotting dipping planar structure. The planar feature is HGPJI, and the portion GPJ is above the spring level. When projected to a horizontal plane at spring level the curve $GO_1$, $O_2$ $O_3J$ will be obtained.

The following are some of the important points to be noted in connection with the geological examinations of tunnels made for mineral exploration and engineering projects, where the 3 dimensional mapping is undertaken.

(1) The dip or plunge of the planar features, is to be recorded. (2) The rake of a linear feature is to be shown. (3) In recording the features present on the tunnel walls imagine that the walls are hinged to the tunnel floor and that they open out on either side. Thus all linear traces on the walls will rotate about the hinge as axis, when the walls are opened, maintaining the same angle as the dip.

Plan of Tunnel (3—dimensional) showing the orientation of planar features in respect of floor 2-2 and spring line (1) and (1), (Fig. 1.4b).

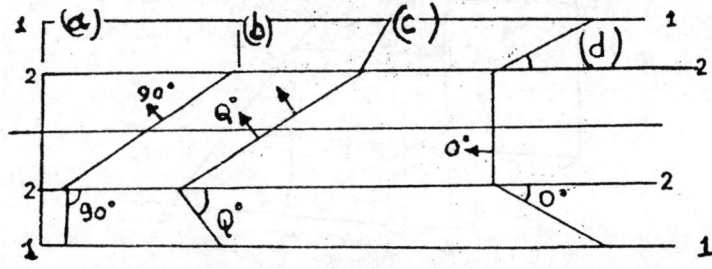

Fig. 1.4b

Figure 1.4*b* (*a*) shows a trace of a vertical linear feature, which is at right angles to the traverse line, as projected on the floor of the tunnel.

Figure 1.4*b* (*b*) shows the trace of a vertical linear feature, which is at an angle to the traverse line, as projected on the floor of the tunnel.

Figure 1.4 (*c*) Shows the trace of inclined linear feature, which is dipping at an angle Q as shown by the arrow, as projected on the floor of the tunnel.

Figure 1.4d shows the trace of an inclined linear feature, which is dipping at an angle O in the direction of the arrow, as projected on the floor of the tunnel.

In general, it may be stated that the procedure adopted is to record the geological features by projecting these on a plane on the floor level of the arch. It involves three stages :

(1) Projecting the features present on the roof of the arch, to a plane at ground or floor level.

(2) Projecting any feature present on the wall (say left side) at ground or floor level. This operation may be achieved by imagining the spring lines of the arch 2—2, from where the wall starts on Fig. 1.4b to form a hinge with the floor of the tunnel. Next, imagine the wall to open outwards, so that all linear features rotate about the hinge line, but yet remain attached to their respective points of intersection with the hinge line.

(3) Projecting of the features present on the right wall of the tunnel is done by adopting a similar procedure in respect of these, as discribed in (2) above.

## SOME HINTS ON TRAVERSING

### A. Ranging lines

(a) For measuring off-set distances the follower takes the end of the tape and holds it on the point to be off-set, while the recorder standing astride the chain stretches the tape and swings it over the chain, circumscribing a circle so as to find out the minimum reading, which gives the perpendicular, tangential or off-set distances.

(b) In order to range a line at right angles to the traverse line from any given point on it, any one of the following methods may be employed.

(1) The optical square or surveyor's cross.

(2) An angle measuring instrument like the theodolite or dial.

or  (3) Advantage may be taken of simple geometrical relationship such as those given later.

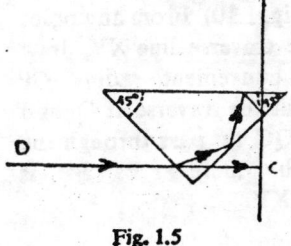

Fig. 1.5

(1) Optical square is an instrument in which a right angled prism with a mirrored side is employed to obtain reflections of objects at right angles in any one direction. Figure 1.5 shows the principle involved in the construction of the instrument.

There are two models of the surveyor's cross. One is a very elemenary device in which there are two sight varies at right angles or two

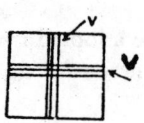

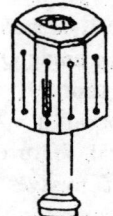

Fig. 1.6               Fig. 1.7

V-shaped grooves at right angles to each other, on a flat square piece of wood (Fig. 1.6). The other type of cross consists of an octagonal hollow right prism made of brass, with pairs of sighting slits at right angles to each other (Fig. 1.7).

(2) The theodolite and dial are costly angle measuring instruments and these are not employed in preliminary work.

(3) Geometrical methods—Case (i) (Fig. 1.8). If P is the point from which the perpendicular is required, measure AP=30 ft. with the chain.

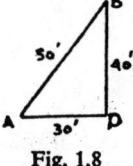

Then take a length AB=50 ft. on the chain, or the helper takes hold of the 50th tag and a follower measures a further length of 40 ft., *i.e.*, the end of the 90th link, which is brought to rest on P. BP is perpendicular to AP at P.

Fig. 1.8

Case (ii) (Fig. 1.9). When using a 100 ft. chain, measure off PA=PB=25 ft. on either side of the point P on the traverse. Then take hold of C the 50 ft. tag and rest the ends (handles) of the chain one at A and the other at B. Stretch the chain at point C. The point C lies on the prependicular to AB at P.

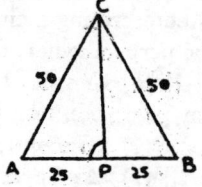

Fig. 1.9

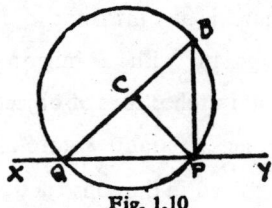

Fig. 1.10

Case (iii) (Fig. 1.10). From any point C away from the traverse line XY, draw a circle with any convenient radius CP (15 or 20 ft.) to cut the traverse at Q and P. Now draw QC to pass through the centre and meet the circle at B. BP is perpendicular to XY.

It is to be pointed out that methods (i) and (ii) may also be employed in the construction of angles of 60° or 30°.

**B. Avoiding obstacles**

(*a*) By ranging a line CD (Fig. 1.11) parallel to the traverse line AB.

Make     AO=OD and

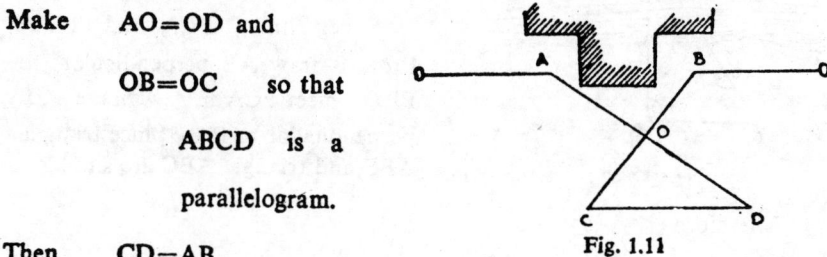

OB=OC     so that

ABCD     is a

parallelogram.

Then     CD=AB.

Fig. 1.11

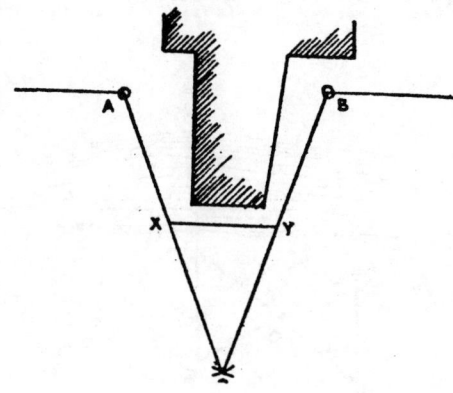

(*b*) Construct triangle ACB to be isoceles, so that AC=CB. Make AX=BY so as to avoid the obstracle ( Fig. 1.12 ). Further $\frac{AC}{CX} = \frac{AB}{XY}$, and since values of AC, CX and XY can be measured, the value of AB can be calculated from the above equation.

Fig. 1.12

(*c*) Range line AX at right angles to OA at A by any one of the methods already explained.

| | | | | | | |
|---|---|---|---|---|---|---|
| similarly | range | ,, | XY ,, | ,, ,, | AX at X | ,, |
| | range | ,, | YB ,, | ,, ,, | XY at Y | ,, |
| and | range | ,, | BP ,, | ,, ,, | YB at B | ,, |

and continue the traverse (Fig. 1.13).

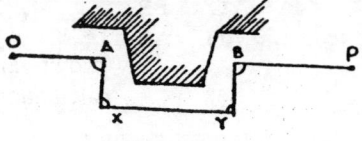

Fig. 1.13

### C.  Obtaining distances of inaccessible points

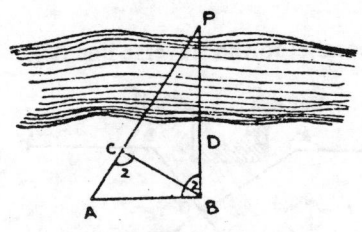

Fig. 1.14

1.  Let P be the inaccessible point (Fig 1.14).

Peg lines PCA and PDB. From B draw AB perpendicular to PB, to meet PCA in A.  Make CB perpendicular to PA.  Since triangle ABP and triangle ABC are similar.

Therefore

$$\frac{CA}{CB} = \frac{AB}{BP}$$ (where CA, CB and AB are known and

BP is got by calculation).

2.  Range any line PQ. Peg any point B, and make an aligenment PBC.  Range any other line parallel to PQ.  (Fig. 1.15), *e.g.* by one of the methods given earlier.

Let line QB meet RC in R.

RCQP   is   a   parallelogram

$$RC = PQ.$$

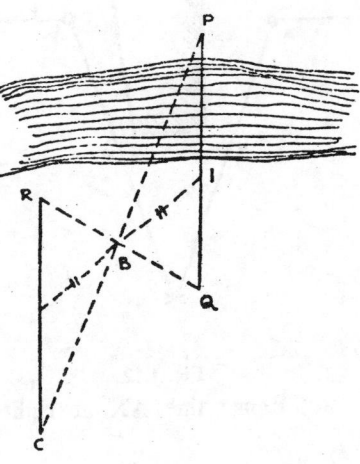

Fig. 1.15

3.  The method suggested in (Fig. 1.16) may also be employed.

$$\frac{YO}{AO} = \frac{PB}{AB}$$ where YO, AO and AB are known and PB is got by calculation.

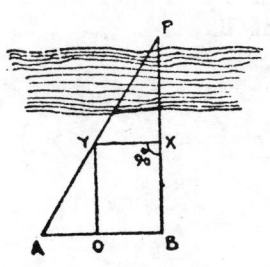

Fig. 1.16

### D. Double ranging

In case where a high ground intervenes and one of the stations B is not visible from station A (Fig. 1.17) the following procedure will be useful. Two persons P and Q get to the top of the high ground, so that both can see each other. P can see station A and Q can see station B. P aligns Q with station B. This operation is continued until P and Q are in a line with A and B.

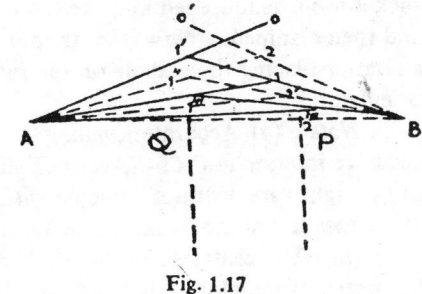

Fig. 1.17

There Q shifts successively from O′ to 1′, 1″ and 1‴ consequently .P shifts from position O to 2, 1,2″ and 2‴, until both P and Q are finally in line with A and B.

## PLANE TABLE SURVEY

**Equipment :** Plane table, tripod-stand (telescopic legs preferable), alidade, spirit level, trough conpass, drawing paper (non-shrinking), drawing pins (or adhesive tape), two or three ordinary pins, and a plumb bob, if available; chain with arrows, plotting scale.

**Theory :** The plane table is considered to be a point or at least of negligible dimensions, when compared to the area to be surveyed. Thus no accurate centering or placing over stations, is necessary, and in practice (scale 100′ = 1 inch) rough centering is sufficient, except where the area to be surveyed is comparable in respect of that of the plane table.

**Procedure :**

1. Open out the stand and adjust the legs to convenient length. Plant the legs firmly in the ground.

2. Attach the table board to the stand by means of the fly nut, but do not tighten. Fix the drawing paper, by means of drawing pins, or better with strips of adhesive tape, to the top of the board.

3. Align one edge of board parallel to any two legs of tripod by slight rotation.

4. Place bubble on the board parallel to the edge just aligned with the pair of legs in operation (3) above.

5. Move the third leg back and forth, or left and right (side-ways) *i.e.* in two directions at right angles, till the bubble comes to rest in the middle.

6. Check by placing the bubble in a position at right angle to the original. Fix this third leg also. Repeat operation (5) above if necessary.

If the stand is provided with a ball and socket head or (Johnson head) the levelling is simpler. The capstan nut for clamping the ball and socket joint is loosened and the table is levelled with the bubble tube, and then clamped. Now the fly nut at the bottom is loosened slightly and after placing the alidade on the ray for back sighting, the table can be oriented.

*Note* : (*a*) Accurate levelling of the board is essential so as to obtain accurate intersection of the various stations. If levelling is bad the wire in the sight vane will not coincide with a vertical rod at a station but will cut across it, and no definite intersection can be got.

(*b*) The centering of the table over the station point may be done by means of the plumb bob or by dropping a stone from beneath the plotted position of that station (on the board) on to the station point (on the ground). A discrepancy of 2″ or 3′ is allowable depending on the extent of area. This depends on the scale of the survey *i.e.* this error should not be of a magnitude which can be plotted, *e.g.* one can not plot 4″ on a scale of 1 inch=100ft. It also depends on the length of sights. In case of short sights a large error in centering will cause a variation in angle.

(c) Mark the north line after orienting the table (see procedure below).

(d) In the case of light wooden alidades, a pin planted at the station point is used as a pivot for turning the alidade and sighting the next station.

**Plane table traverse** : If ABCDE be the traverse (Fig. 1.18) the table is fixed at A, centered, and levelled by means of the bubble tube placed in two directions at right angles, and clamped in such a manner so as to accomodate the plan of the area, on the table when plotted to the scale required. It is a good practice to draw the north line by means of the trough compass (see adjustments above).

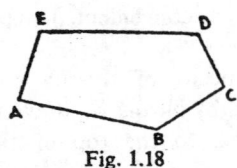

Fig. 1.18

Now station B is sighted with the alidade, and the ray drawn from A to B by using a fairly hard pencil. The distance AB is measured and the position of B plotted on this ray to scale. The table is now removed and set up at station B. After the necessary adjustments, explained earlier (i.e. centering, levelling etc.), A is sighted back along BA. This is done by loosening the fly nut at the bottom of the table, so as to release the board and by placing the alidade along BA, so that the peep-hole is on the side of B. Looking through the alidade the station A is sighted back, and the board is tightened up in the position.

Thus the table is oriented. Station C is now sighted and ray BC drawn and the distance BC measured and plotted to scale. The traverse is similarly continued onwards, to station C, D, and E.

### Location of Points by Intersection and Resection

**Triangle of error**

(a) The location of points on the traverse is accomplished by means of intersecting rays generally two or sometimes three drawn from two or three stations respectively depending on the importance. This is known as intersection. If the table orientation has been perfect the intersection of the rays is at the same point. Very often the third ray when drawn gives rise to a small triangle and the object will have to be located in respect of this "triangle of error".

(b) Quite accurate location can be obtained by taking a ray from a station to the object measuring the distance and then plotting the same. This location is then checked up by a similar measurement from another station. This procedure is however likely to be more time consuming.

Sometimes, it becomes necessary to locate one's position on the plan, during the course of a plane-table survey. This can be accomplished only when three stations are visible. This method of location is called "resection". The plane table is set up at the desired point, and by inspection a point is selected on the plan so as to correspond as closely as possible to the instrument station. Three rays are then drawn passing through the three visible stations. Most often the three rays will not meet at a point, and a triangle known as, "the triangle of error" will be formed (Fig. 1.19). The following rules are helpful in fixing the position of the new station. (a) If the new station point lies on or near the great circle, passing through the three stations in the field, its position becomes indeterminate. (b) If the new station lies within the triangle formed by the field stations, the strength of determination increases

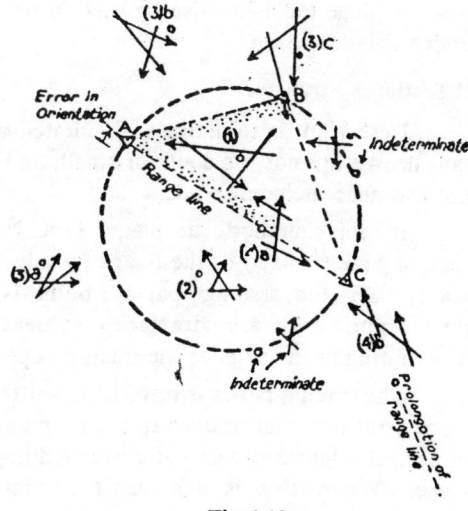

Fig. 1.19

as its position approaches the centre of gravity of the great triangle *i.e.*, formed by the field stations. (*c*) If the new station lies outside the great circle, its strength of determination will be weak (Fig. 1.19).

The following rules are handy in solving the triangle or error.

1. When the new station is within the great triangle formed by the field stations, then its location on the table is within the triangle of error, and its approximate position will be represented by a point whose perpendicular distances from the three sides are proportionate to the respective distances of the three stations sighted, from the new location.

2. When the new station is outside the triangle, then its position in the plan will be outside the triangle of error, but then its location on the plan will be, to the same side of the ray to the most distant station, as well as to that of the rays to the other two stations, *e.g.*, if the point is to the left of the ray to the most distant station, then it will also be found to the left of the other two rays.

3. When the new station lies within any of the three segments of the great circle, passing through the three stations, then its location would be as far distant from the median through one of the points (of the great triangle) as the point of intersection of the other two rays.

*Note* : A plane table in combination with a dumpy level is very useful for contouring in a moderately hilly terrain. In a rugged country a plane table with telescopic alidade equipped with Beaman stadia arc is to be preferred (for description of Beaman stadia arc see under stadia survey and tacheometric levelling).

A plane table can also be used in triangulation. The method is given under Triangulation.

### Resection by transfer

Resection is sometimes facilitated when (*a*) the angles between the rays drawn are not too acute or small, and (*b*) when the possible location is not too near the great circle.

In this method the plane table is set up and levelled. A piece of tracing paper is fixed to the board by adhesive tape, at any convenient place. On the tracing paper a point is marked. With this point as the origin, rays are drawn to at least three triangulation stations, terminating at the edge of the tracing paper, and propertly indexed.

The tracing paper is now detached from the board placed over the place (station) and moved in such a manner that the rays drawn on it are in perfect alignment with the corresponding points alreay plotted cn the place. When this is achieved the point of origin (on the tracing paper) is pricked through with a fine needle and transferred to the plan below.

The point thus obtained is checked up by orienting the table in the usual manner *i.e.*, (a) by loosening the fly nut, (b) aligning a ray and back sighting, (3) locking the fly nut, and (4) checking up correspondence between other stations.

## THEODOLITE AND MINER'S DIAL SURVEYS

*Permanent adjustment* : It is taken that the instrument is in perfect adjustment and hence this has been dealt with in the appendix.

1. Open the stand and spread the legs of the tripod keeping the head centrally over the station point. Further centring over the station may be done by dropping a small stone through the centre of the opening at the top of the head. See that the head is level and remove any shake or "headache" by tightening the fly nuts, and also see that shoes are firmly planted. Fix the cap to the leg or place it in the box.

2. Note the lie of the instrument in the box.
    (a) See which is the tribrach end and which is the objective end.
    (b) Note the position of the vernier clamps and the vertical circle.

3. Hold the uprights (supporting Y) firmly with the left hand, take hold of the tribrach with the right hand and lift the instrument gently out of the box.

4. Place the instrument on the stand, so that the opening in the tribrach fits on the screw in the head of the stand. Rotate the tribrach in an anticlockwise direction, until a slight click indicates that the screws are engaged. Now tighten up in the clockwise direction.

5. Remove plumb bob from box and suspend it to the hook below central spindle of the instrument.

6. Bring the three foot screws to the centre of the run.

7. Roughly centre the instrument over the station by manipulating the sliding mechanism in the tribrach.

8. Clamp the tribrach by tightening the large fly nut at the bottom.

9. Bring the bubble on the horizontal plate over any two foot screws and level by rotating both the foot screws inwards or outwards to an equal amount, simultaneously.

10. Turn the plate through 90°, so that the bubble is now over the third food screw, now operate the single foot screw and bring the bubble to the centre.

Repeat operations 9 and 10 until the bubble remains in the centre in all positions of the plate.

11. Remove any parallax by focussing the cross hairs of the telescope. Also check up by moving the eye from side to side.

*Further manipulations for normal transits :* Note on which side the vertical circle lies, *i.e.*, to the right or left of the line of sight.

(a) Clamp the body (*i.e.* lower clamp). (b) Loosen the vernier clamp. (c) Looking into the vernier 'A' rotate the plate and bring the vernier to read as near zero as possible. (d) Clamp vernier and then with the help of the vernier tangent screw, make the reading exactly zero. (e) Loosen body clamp and loosen telescope vertical arc clamp and sight the back station, so that the horizontal plate yet reads zero, when sighting the back station. (f) Clamp body, and intersect back station accurately by means of body tagent. (g) Release vernier clamp and rotate the telescope in a clockwise direction and sight the forward station. Clamp vernier and make an accurate intersection with the help of the vernier tangent. Read horizontal angle recorded by the vernier.

*Note :* (a) Always intersect the lowest point of the flag placed at stations, as the upper portions are liable to deflections.

(b) While recording vertical angles, always make a note whether it is positive or negative.

In the case of the modern "microptic" instruments the reading of both the horizontal and vertical circles are projected on to an auxiliary telescope and can be read simultaneously.

(Specimen book—see appendix).

## VERNIERS

There are two types of verniers. (a) The verniers in which the smallest vernier division or subsidiary scale division is less than the smallest main scale divison, are known as forward reading verniers. (b) Those verniers in which the least vernier scale division or subsidiary scale division is greater than the main scale division, are called the retrograde or backward reading verniers. All modern instruments, however, employ the forward reading vernier so that the vernier is read in the same direction as that of the main scale. When handling any instrument, always determine the least count of the vernier or the value of the smallest division of the vernier.

In verniers for angle measuring instruments, the least count is calculated in the following manner. Suppose a degree of the main scale is divided into three equal parts or 20 minute divisions and if 59 of these are taken and again divided into 60 parts, the least count of the vernier will be 1/60×20 min. or 20 seconds. The graduations on the verniers are marked minutewise, i.e., 0 to 20 minutes. Now to read the vernier the following example may be cited.

Suppose in a certain reading of the instrument the arrow indicating the zero is located between 152° to 153°. Suppose the arrow is beyond the 2nd small main scale division from 152°. This shows that the value of the angle is greater than 152° 40'. If now it is found that the 1st small vernier division after the graduation marked "15" on the vernier, (i.e. 16th division) coincides with a main scale division, then the value of the angle is read as 152° 55' 20", i.e. , 152° on the main scale, plus 40' on the main scale plus 15' 20" on the vernier scale.

**Definition of terms used :** In a theodolite, when the vertical circle is to the left hand side of the operator or the line of sight, the position is called *face left*, when it is on the right hand side, the position is known as *face right*. *Transisting* or *changing* face consists in rotating the telescope through a vertical angle of 180° and then rotating the horizontal plate through 180° so that the telescope points in the same direction as before. For accurate work, repeat measurements of all angles made, first on face left and then on face right or vice versa. The readings so obtained should be recorded in a suitable tabular form. For further computation and calculation (see appendix 3).

**The microptic tacheometers :** are instruments similar to (Watts type) theodolite. These instruments differ from normal theodolites in that they have an arrangement for measuring distances.

The instrument is kept either in a box or sometimes in an all-steel shell. In the later case, the clamps holding down the shell to the base are loosened and the shell removed. Next the holding down clamps fixing the instrument to the base plate are loosened and the instrument set free.

The stand can be levelled with its own bull's eye bubble after placing it over the station point. The instruments are attached by means of the hollow nut, found in the head of the tripod. In some instruments it is possible to do preliminary centring by means of a prism arrangement attached to the tribrach. With the help of this device it is possible to see the ground stations and thus do the centring. After this the plump bob is attached to the other end of the hollow nut in the head of the tripod.

The other temporary adjustments are similar to those for normal tacheometers and theodolites.

The small circular plane mirror at the side is for reflecting light for illuminating the various scales, viz. vertical and horizontal. These scales in the case of these microptic instruments, are totally enclosed and are not visible. The images of these two scales are projected for convenient viewing by means of reflecting prisms, and can be seen in the auxiliary telescope placed on the main telescope. The image· of the micrometer vernier scale can also be seen in the auxiliary telescope. This vernier is of the forward reading type and is common to both the horizontal and vertical scales (and designated H & V respectively) which are graduated in 20 minutes. The vernier is graduated in the same manner as in normal transits or theodolites and has a least count of 20 seconds; but it is used in a slightly modified fashion.

The vernier is operated by rotating the knurled nut on the right hand side of which the word "micrometer" is engraved. It will also be noticed that movements of the vernier from 1 to 20, after clamping the horizontal and vertical plates, produces a translation of 20 minutes on other circles. This shows that the least count of the instrument is 20 seconds. It is a good practice to set the vernier at zero at the start of each setting of the theodolite.

In replacing the instrument, the telescope is clamped in the vertical position and the base plate is placed in a position to fit on to the clamps provided.

In the case of the microptic theodolite (Watt's type) the least division (least count) of the main scale is 20', and that of the vernier is 20". There is a micrometer vernier which is common to both the circles, i. e. horizontal and vertical. After completing the temporary adjustments the micrometer vernier is made to read 'zero' by rotating the micrometer milled nut on the upright. After intersecting the station (either in azimuth or in the vertical), the auxiliary telescope readings are noted correct to the nearest 20' on the lower side, e. g., if while reading the horizontal scale the index point lies between 189° 20' and 189° 40', it is moved towards 189° 20'. Now the micromer is rotated so that the micrometer vernier reading which was initially at zero now increases, whereas the index point moves towards 189° 20', where further movement is stopped. If now the micrometer vernier reads 12' 20", this is added to the initial reading of the horizontal scale that is 189° 20' + 12' 20" = 189° 32' 20". The same manipulations are carried out with respect to the vertical scale readings also.

### FIXING OF STATIONS AND TRAVERSING

(a) Traverse stations for surface surveys are usually indicated by a small cement posts one foot long and eight inches square which are planted flush with the ground level. A cross made on the top (near the centre) of the exposed portion denotes the actual station point. Sometimes when permanent marks are not required it is convenient to drive a ½ inch pipe or peg to mark stations. The station flag is placed on the station point. When not in use a cairn or a stone heap is built up over the station points.

(b) In underground surveys, as stations are likely to be disturbed when placed along roadways, the station pegs are nailed on to the roof. A plumb bob is suspended from a hook or eye in the peg and the string or line is illuminated from behind and sighted though the telescope (of the theodolite etc).

### Methods used in traverse surveys

If bearings are obtained with the Miner's dial the readings can be recorded as :

1. Whole circle bearings calculated with reference to the north. The values are recorded as 35°, 197°, 352°.

2. Quadrant bearings are calculated either with reference to the north or south and further indicated by east or west sector e. g. the values are recorded as N 20°E, S 30°W etc.

In the case of both theodolite and dial traverses, all angles are measured by the following methods :

1. Back and fore sight.

2. Double fore sight.

3. Continuous bearing (not recommended at the earlier stages).

### 1. Back and fore sight

The instrument is set up at stations 1, 2, 3, 4 etc. and the angular measurments made in the manner indicated by the arrows (Fig. 1.20) presuming the instrument to be right handed. The angles are read on both verniers, and recorded separately. In the Fig. 1.20, X is the eyepiece end and O is the objective end. The bearing. of any one of the lines is also measured.

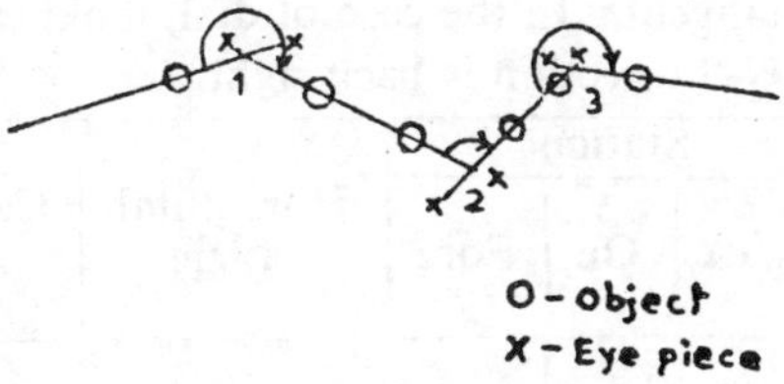

O – object
X – Eye piece

Fig. 1.20

The azimuth of all the lines is calculated with reference to that of any one line, by adopting the following procedure.

Add the measured angle to the measured or assumed azimuth of the preceding line, if the sum exceeds 360° subtract 360° and this will not affect the azimuth.

**Example**

| Vernier   Readings | measured or<br>Assumed   Azimuth |
|---|---|
| 187° 15' | 187° 15' |
| 98° 20' | 285° 35' |
| 256° 09' | 181° 44' |
| 160° 12' | 341° 56' |

**2. Double foresight method** (This is the method commonly adopted).

This instrument is set up at station, 1, 2, 3 etc. and angular measurements taken as indicated by arrows, assuming the instrument to be right handed (Fig. 1.21).

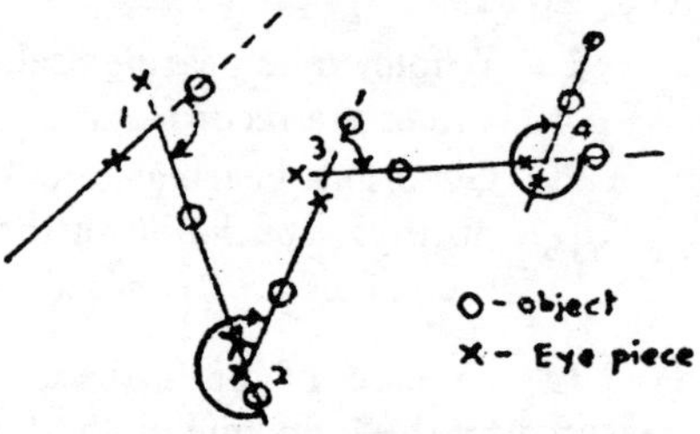

Calculations of azimuths when the bearing of any line of the traverse is known, is done as follows :

Add the value of the forward bearing of the *preceding line* to the measured value of the included angle, deduct 180° if the sum is more than 180°. If on subtraction the balance is more than 360° further subtract 360°. When the sum is less than 180° add 180°.

O - object
X - Eye piece

Fig. 1.21

*Note* : Fore sights are taken by using the plate or vernier clamp and tangent screw, and back sights are taken by handling the body clamp, and tangent. In the case of dial, looking in the S-N direction is fore sight, and N-S direction is back sight.

| Station | | | Horizontal<br>angle | Calculated bearings | Remarks |
|---|---|---|---|---|---|
| Back | On | Fore | | | |
| A | B | C | 127° 14' | 92° 00' | Bearing of AB=92° |
| B | C | D | 220° 27' | 39° 14' | as measured in the |
| C | D | E | 92° 34' | 79° 41' | field. |
| D | E | F | 98° 12' | 352° 15' | |
| E | F | G | 130° 46' | 270° 27' | |
| F | G | H | 215° 09' | 221° 13' | |
| G | H | A | 115° 17' | 256° 22' | |
| H | A | B | 80° 21' | 191° 39' | |

Total 1080° = (180 × 8) − 360

The following is an example showing the method of calculation:

## Method of Calculating Bearings

| | |
|---|---|
| Bearing of HA | 92° 00′ |
| ∠B | 127° 14′ |
| | 219° 14′ |
| | 180° 00′ |
| Bearing of AB | 39° 14′ |
| ∠C | 55° 27′ |
| Bearing of BC | 94° 41′ |
| ∠D | 77° 34′ |
| | 172° 15′ |
| | 180° 00′ |
| Bearing of CD | 352° 15′ |
| ∠E | 108° 12′ |
| | 450° 27′ |
| | 180° 00′ |
| Bearing of DE | 270° 27′ |
| ∠F | 39° 46′ |
| | 310° 13′ |
| | 180° 00′ |
| Bearing of EF | 121° 13′ |
| ∠G | 315° 09′ |
| | 436° 22′ |
| | 180° 00′ |
| Bearing of FG | 256° 22′ |
| ∠H | 115° 17′ |
| | 371° 39′ |
| | 180° 00′ |
| Bearing of GH | 191° 39′ |
| ∠A | 80° 21′ |
| | 272° 00′ |
| | −180° 00′ |
| Bearing of HA | 92° 00′ |

## 3. Continuous bearing method

In this method (Fig. 1.22) the instrument is set up at sation A. The vernier is brought to read 0°00′ 00″. The trough compass is attached to

Y support of the telescope
by the studs provided. The
body clamp is released and the
instrument rotated until the
centre point marked on the
compass coincides with the
vertical mark at the end of the
trough. Now the telescope is
pointing north.

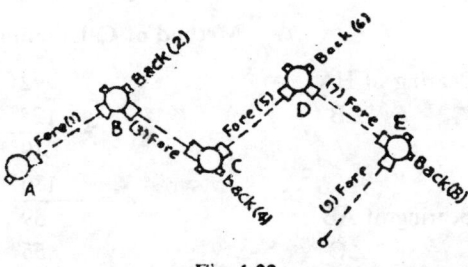

Fig. 1.22

The body is clamped, then the vernier clamp is released and the
instrument is turned in a clockwise direction to sight the fore station B.
The station is accurately intersected by vernier tangent. The instrument
now records the bearing of AB.

Next the instrument is carried carefully and set up at station B. The
reading on the vernier is the value of the fore bearing of AB. Body clamp
is released and station A is intersected by using body clamp and tangent
The telescope is now transitted, plate clamp released and station C is
intersected using the plate tangent screw. The instrument now records
the bearing of BC. Thus, when similar operations are repeated at each
station, the forward bearings are automatically recorded by the instrument. This method is, however, not recommended as it requires additional
care. Any accidental shifting of the reading of the instrument at any stage
especially when transporting the instrument from station to station will
effect the readings recorded.

## STADIA OR TACHEOMETER SURVEY

A tacheometer is a distance measuring telescope in which besides
the normal vertical and horizontal wires, there are two extra horizontal
cross wires (generally short) on the vertical cross wire, one of which is
placed above and the other symmetrically below with respect to the horizontal wire. These short cross wires are known as the stadia wires.

### Principles and theory

In this method no actual distances or heights are measured but
calculated from readings obtained from the tacheometer and graduated
staff. Hence this method is handy when a survey of moderate accuracy
(say 100 ft. = 1 inch) 1 : 1200 is required in a rugged country. The principle
involved is illustrated by Figure 1.21 where A is a pin hole and no optical
device is employed.

This is the basic principle of the subtence bar commonly used by
surveyor's for obtaining the distances to inaccessible objects.

In triangle XAY (Fig. 1.23)

XY $=i$, AZ $=a$

angle XAY $=\theta$

$$\text{Tan}\frac{\theta}{2}=\frac{XZ}{AZ}=\frac{i}{2}\times\frac{1}{a}$$

$$=\frac{i}{2a} \quad ...(1)$$

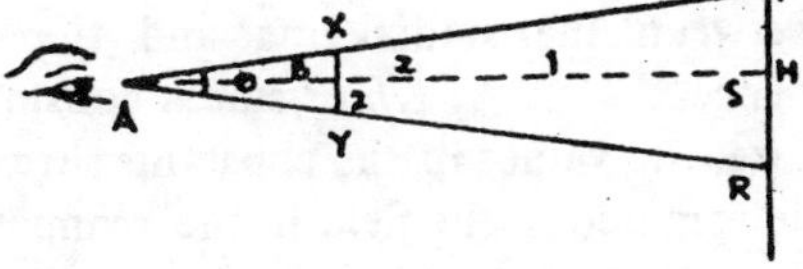

In triangle PAR

PR $=h$, angle PAR $=\theta$

Fig. 1.23

$$\tan\frac{\theta}{2}=\frac{PB}{AS}=\frac{h}{2}\times\frac{1}{AS}, \quad 2AS=\frac{h}{\tan\theta},$$

$$AS=\frac{h}{2\tan\frac{\theta}{2}}.$$

Substituting the value of $\tan\frac{\theta}{2}$ from (1)

we have.

$$AS=\frac{h}{2}\times\frac{2a}{i}=\frac{ha}{i} \qquad ...(2)$$

$\theta$ is known as anallatic angle, $h$ is known as generating number, $\frac{a}{i}$ is stadia coefficient $=k$ (constant).

Case (i) = When a telescope is used instead of a pin hole, but the sights are horizontal (Fig. 1.24). As $fh/i$ AS$/h=f/i$ where $f=$ focal length of the telescope, K is known, as the stadia coefficient Fig. 1.24.

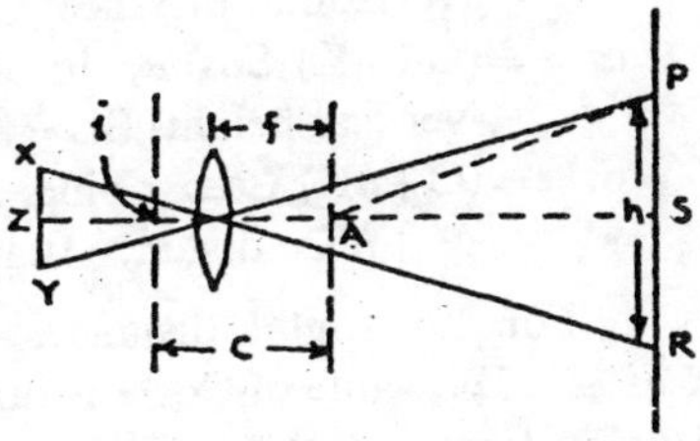

$f$, focal length by construction;

C, is known as the instrument constant.

Since distances are large when compared to the focal length

$$\frac{AS}{h}=\frac{f}{i}$$

$$AS=\frac{hf}{i}$$

where $\frac{f}{i}=$K or stadia coefficient.

$$AS=Kh \qquad ...(3)$$

But the horizontal distance

$$=C+AS$$
$$=C+Kh \qquad ...(4)$$

(K$h$+C) is known as the distance function, K stadia coefficient, and C instrument constant and $h$ the intercept or generating number. Thus the distance MC can be determined. In most instruments the distance, between the stadia ring and the object, is made equal to 0.5 $f$ and hence C=1.5 $f$. K$h$ or stadia constant is generally made equal to 100. In case, the values of the constants are not given by the makers, these can be determined in the field in the manner described later.

Case (*ii*). When a telescope is used and the observations are made with the telescope in the inclined (Fig. 1.25) position.

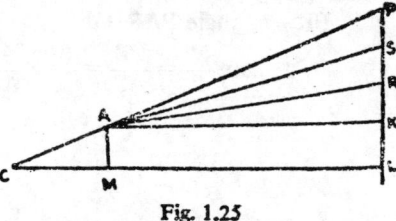

Fig. 1.25

M=Centre of instrument,
CA=C or instrument constant $\angle$ PAK=$\alpha$ or angle of inclination
PR=$h$ or generating number.
$\angle$ PAR=2 $\theta$ or anallatical angle, $\angle$ SAR=$\theta$

(Horizontal distance) CL=CM+ML=C Cos$\alpha$+K$h$ Cos$^2\alpha$......(5)
or    =(C+$h$K) Cos$^2$ $\alpha$, since C Cos $\alpha$ is approximately=C............(6)
(vertical height) SL=KL+KS=C Sin $\alpha$+$h$K Sin$\alpha$ Cos $\alpha$......(7)
or    =(C+$h$K) Cos$\alpha$×Sin$\alpha$, since C Sin$\alpha$ in approx.=C ..............(8)
C+$h$K=distance function.

For horizontal distances equation (6) may be taken as (C+$k$$h$) Cos$^2\alpha$. The value of Cos$^2\alpha$ is obtained from tables. For obtaining heights or elevations, equation (8) is taken as (C+K$h$) Sin$\alpha$ Cos$\alpha$. The value of (Sin$\alpha$ Cos$\alpha$) is obtained from tables. The quantity (C+K$h$) is known as distance function.

The value of C+K$h$ is calculated in the manner described later and the values of Cos$^2\alpha$ and (Sin$\alpha$ Cos$\alpha$) are got either from the tables which have already been computed for various values of $\alpha$, called stadia or tacheometer tables or from graphs. These may also be obtained by means of a special slide rule. The value can be easily got from the scientific calculator.

### Use of stadia or tacheometric tables

(Tacheometric tables by Louis & Caunt have been used in the following calculations.)

**Example**: If the distance function is 658.5 and the vertical angle is 15° 10', then refer to the tables at 15° or 135° down the vertical column at 10′, and horizontally for partial "distances", and "heights" indicated at Dt. at Ht. respectively in the table. The partial distances are given in units from 1 to 10. The corresponding figures will have to be multiplied by 1000, 100, or 10 etc. if the distance function contains units of these orders, e. g., the value for partial distance of 6 is 5·589 hence for 600, it is 558·9. The others are similarly computed.

| Distance function | Dt. | Ht. |
|---|---|---|
| 600 | 558.9 | 151.5 |
| 50 | 37.3 | 10.1 |
| 8 | 7.5 | 2.0 |
| 0.5 | 0.5 | .1 |
| 658.5 | 604.2 | 163.7 |

Distance is 604.2 and the corresponding vertical height in the height column is 163.7.

Thus in the above example, if 100 is the reduced level of the instrument station, and the height of instrument is 4.9 ft. and the reading of the middle cross-wire is 6.4 ft. on the staff intercepted then, the height of the staff station will be $100+4.9+163.7=262.2$ ft. (For derivation of the formula see tacheometric levelling). If the angle of sight exceeds 30° even then the tacheometric table can be employed in a modified form. If D be the distance function, and the angle of inclination the horizontal distance is equal to $D \sin (45°-\alpha) \cos (45°-\alpha)+D/2$ and the corresponding vertical height is equal to $D \cos^2 (45°-\alpha)-D/2$. Therefore the *horizontal distance* is obtained by referring to the stadia tables for $(45°-\alpha)$ and obtaining the height given there and adding D/2 to it. The corresponding *vertical height* is got by obtaining the horizontal distance given in the table for $(45°-\alpha)$ and subtracting D/2 from it.

**Example :** If the distance function (D) be 420 ft. and the angle of inclination 38° 20′ the table is turned to $(45°-38° 20′)=6° 40′$, and the corresponding height and distance is noted as 48.4 and 414.3. Hence the required values are :

Horizontal distance$=48.4+420/2=48.4+210=258.4$ ft. and
Vertical height$=414.3-420/2=414.3-210=204.3$ ft.

This rule in general applies to angles not exceeding 45°. For angles exceeding 45°, the angle corresponding to $(\alpha-45°)$ is referred to in the tables, and the required horizontal distance is obtained by subtracting from D/2, the value obtained from the table for height. The vertical height is obtained by subtracting D/2 from the value of the horizontal distance as given in the table.

The introduction of electronic calculators has not only reduced the time factor involved in the computations but it has provided an additional advantage, in that the calculations required in connection with techeometric/stadia surveys can to a large extent be carried out in the field. Further, one need not buy the telescopic alidade equipped with the Beaman Stadia arc, but for plotting the distances and heights; in the field, an ordinary telescopic alidade with the vertical circle graduated in degrees and fractions of a degree together with a 'scientific' calculator can do the job just as efficiently as its Beamen arc counterpart. With this calculator the values of $\cos^2$ can be found out as well as the value of the tan of the angle. In case, an ordinary calculator is used, then the values of the cos can be xeroxed as well as the values of the tan from a book of mathematical tables and placed in a transparent plastic envelope, back to back for reference in the field. It may be mentioned that the telescopic alidade should have the stadia lines engraved together with the cross hairs. The values of $\cos^2\theta$.*

### Determination of instrument constant and the stadia co-efficient

The instrument is racked up to the solar focus (i. e. focus at infinity) and the distances (1) between the diaphragm and the centre of the objective (2) between the objective and the centre of the theodolite, are measured with a steel scale and the sum of these gives the instrument constant i. e. (1) plus (2).

In order to determine the stadia coefficient K the theodolite is set up at a station. From the station point a length equal to the instrument constant is measured out. From this latter point, distances equal to 50, 100, 150, 200 ft. and so on are measured and pegged out. Keeping the instrument horizontal, stadia intercepts are booked.

$$K = \frac{\text{distance}}{\text{intercept}} = \frac{50}{h} = \frac{100}{h} = \frac{150}{h} \text{ etc. } h = \text{intercept value.}$$

Thus the value of K is got.

Tabular columns for the recording of field reading and office work are shown in Appendix 4 and 4a.

### Self reducing tacheometers

In these instruments instead of the normal stadia lines, certain curves etched on a glass plate are introduced into the field, in the focal plane of the eyepiece. In the Zeiss instrument for example ; the lowest horizontal curve is designated the base curve (1) and it has a companion curve (2) running left to right. The intercept on the staff between curves (1) and (2) gives the factor for horizontal distance. Similarly there are other curves designated −10 or −20 etc. (3) The intercept on the staff between the base curve (1) and curves (3) give the reduction factor for the vertical height.

---

*See addendum.*

For obtaining the horizontal distance the staff intercept between curves (1) and (2) is multiplied by 100, and in the case of vertical height the staff intercept between (1) and (3) is multiplied by the factor inscribed on the curve used, e.g., $\pm$ 10 or $\pm$ 20.

**Example :** If the intercept between curves (1) and (2) is 573 m.m. the horizontal distance $= 0.573 \times 100 = 57.3$ metres.

If the intercept between curves (1) and (3) is 652 m.m. and the number inscribed on the curve is 10, then the vertical height is $0.652 \times 10 = 6.25$ metres.

The formula used in working out the distance curves are given below :

$$a = \frac{F\cos^2\alpha}{K \pm \frac{1}{2}\sin^2\alpha}$$

where a is intercept between the distance curve and base.

$$h = \frac{\frac{1}{2} F\sin^2\alpha}{K \pm \sin^2\alpha}$$

where h is the intercept between the height curve and the base.

$f =$ equivalent focal length of the system
$\alpha =$ the angle of elevation or depression
$K =$ instrument constant.*

## Telescopic alidade with beaman stadia arc

This instrument also utilises the same primary principle as the tacheometer, *i.e.*, measuring distances by stadia intercepts. The telescopic alidade is very convenient when used in conjunction with the plane table for topographic surveying and geological mapping. The telescope is mounted on a scale rule which enables the proper plotting of direction of rod shot stations. The theory and the method of computation of stadia measurements have already been discussed in connection with the teacheometer. The same principles apply here also.

The Beaman stadia arc is a mechanical accessory to the alidade, which obviates the reading of vertical angles and eliminates laborious calculations of stadia and instrumental data, as well as any reference to tacheometric tables.

The accuracy of the method compares favourably with that of the tacheometric survey and for normal topographic-geological survey of mines and mineral deposits on moderately large scale, say upto 100 feet $= 1$ inch, considerable accuracy can be achieved, specially in areas where the topography is moderately rugged.

## Use of telescopic alidade with beaman stadia arc for levelling

The telescopic alidade, with Beaman stadia arc in combination with

---

*See addendum.*

the plane table, is an ideal instrument for topographic contouring. It has an advantage over the tacheometer, as intermediate staff stations can be straightaway plotted on the plane-table sheet, with their reduced levels and the contour lines interpolated and sketched, in the field. The difference in elevation, between the plane-table station and rod station, is added to or subtracted from the reduced level of the plane-table station, depending on whether the shot is a positive (up-slope) or negative (down-slope) one, and the reduced level of the staff station is calculated, as shown in Figs. 1.32 and 1.33.

The Beaman arc is a specially graduated vertical arc attached to the horizontal axis of the telescope. In addition to the scale of degrees in the vertical arc, there are two scales of Beaman factors, one marked "Hor" or "H" scale and the other "Vert" or "V" scale. The principle involved in the construction of these scales is the same as in teacheometer and shown in Figs. 1.26 and 1.27.

The horizontal distance $H = (C + Kh) \cos^2 \alpha$

The vertical height $\quad V = (C + Kh) \sin \alpha \cos \alpha$

As indicated earlier the instrument constant C is very small and in ordinary geological and topographical surveys, it need not be taken into account in the solution of the formulae. The formulae are thus simplified to

H or Horizontal distance $= C + hK \cos^2 \alpha$ is modified to
$$H = Kh \cos^2 \alpha$$

V or Vertical height $= C + hK \cos \alpha \sin \alpha$ is modified to
$V = Kh \sin \alpha \cos \alpha$, where $h$ is the staff intercept and K is the stadia coefficient.

In most telescopic alidades, the telescope is so constructed that the stadia coefficient K is 100.

Thus $H = \text{Rod intercept} \times 100 \times \cos^2 \alpha$

$V = \text{Rod intercept} \times 100 \times \sin \alpha \cos \alpha$

In some types of Beaman arc the values $100 (\cos^2 \alpha)$ and $100 (\sin \alpha \cos \alpha)$ are directly obtained from the scales as "Hor" factor and "Ver" factor respectively. These are described later as 'B' type arcs. In other types, the values are obtained by very simple arithmetical manipulation of the "Vert" and "Hor" scale readings. These are described as 'A' type arcs.

### Adjustment

(1) With the table levelled, and oriented with reference to the back station, or with reference to some triangulation station, clamp the tele-

scope and level it. Measure the height of the instrument over the station point.

(2) Set the Beaman arc with the scales exactly at the zero point (either 0 or 54 as the case may be), by means of the tangent screw. In the American type alidades, the right tangent screw or gradimeter screw is used for this adjustment, and this screw is not disturbed during further operations.

(3) Sight the telescope towards stadia rod, taking care to see that the edge of the alidade rule does not move from the station point on the table. For accurate work, it is better to prick holes at the station position and slide alidade carefully, without the aid of a fixed pin, on the station point, as is done in the case of the open-sight alidade method.

(4) Raise or lower the telescope, by the clamp and tangent screws, until the vertical cross-hair falls on the staff. Fix either the top or bottom stadia wire on a whole number of the staff and quickly read all the three intercepts.

(5) Level the Beaman arc, by observing the bubble attached to it, and by operating its tangent screw; and note the "Hor" and "Vert" scale readings.

(6) Calculate the horizontal distance, difference in elevation and reduced level by following the methods given earlier.

The instrument readings and calculations are recorded as shown in Appendix (5 & 5a).

(7) Without shifting the instrument, draw the rays to staff shot stations. The horizontal distance, to a particular station is plotted and the reduced level noted against the points.

(8) When a number of staff stations are plotted, the contour lines may be interpolated and sketched in accurately, by observing the slope of the ground, and topographic details in the field.

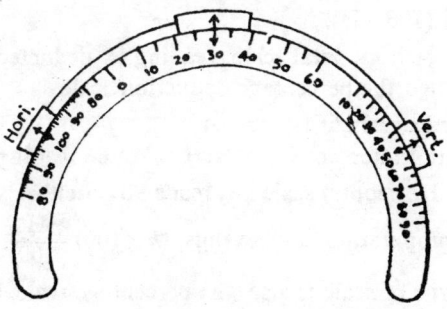

Fig. 1.26

'*A*' *type arc* : The horizontal scale divisions are proportional to 100 (1—Cos²α) and the vertical scale divisions are proportional to 100 (Sin α Cos α). The scales are so constructed that when the index mark of the "Hor" scale coincides with 0 or 100, the index mark of the "Vert" scale coincides with 50.

(1) In some American instruments the divisions on the "Hor" scale either read 90 on either side with 100 in the middle or 10 on either side with 0 in the middle. The "Vert" scale reads from 10 to 90 with 50 in the middle (Fig. 1.26).

*or*

(2) In English (Watts microptic alidade) instruments, the divisions on the "Hor" scale read from 0 to 20 and the "Vert" scale 0 to 40 (Fig. 1.27). The readings on each scale have the sign prefixed, viz. +ive or—ive. In this instrument the indices of both scales read 0 simultaneously.

Fig. 1.27

'*B*' *type arc* : Although the "Hor" and "Vert" scales move as one unit on the same arc, they are in separate quadrants—one on the left and the other on the right. The zero points of these scales are 100 and 50 respectively. The divisions (in some American make alidades used by the U.S.G.S.) are 80 to 100 on the "Hor" scale with 80 in the middle and 10 to 90 on the "Vert" scale as in the case of the 'A' type arcs.

**Calculation of distances and heights (tacheometric levelling )**

1. '**A**'**type arc.**

(*a*) Horizontal distances
=staff intercept (*h*) × 100
—(staff intercept *h* × horizontal scale reading H)
=*h* (100—H).

Since the H or horizontal scale reading is deducted from 100 this type of scale is called the percentage deduction scale.

(*b*) Vertical height (+ or —)
=staff intercept (*h*) × vertical scale reading (V) =*h* × V.

In case, the Horizontal scale (H) reads 90 on either side with 100 in the middle, the computation is done thus, $(h \times 100) \frac{H}{100}$.

Thus this type of scale is also the percentage scale, as the Horizonal scale reading H gives the percentage directly.

*Note.* In the American type alidades, the vertical scale reading is obtained by noting the reading of the scale index—50 (algebraic). In Watts' microptic alidades, this is directly read on the arc. The "Vert" reading may be plus or minus, depending on whether the staff reading is up-slope or down-slope.

2. **'B' type arc.**

(a) Horizontal distance=staff intercept (*h*) × Horizontal scale reading (H)=$h \times H$.

(b) Vertical height (+or—)=staff intercept (*h*)× Vertical scale reading (V)=$h \times V$.

**'*A*' type Arc (*Watts' Microptic alidade*)**

(a) For obtaining reduced level :

(1) Looking through the telescope intersect the staff at any convenient point and bring the bubble to the centre of its run.

(2) Move the telescope up or down so that the middle cross hair reading coincides with a whole number on the "V" scale or Vertical scale either + or —.

Note the reading on the (V) scale and the corresponding staff reading (M), of the middle cross wire at this position.

(3) Note stadia reading (i.e. intercept) (S) between the top and bottom stadia lines.

Thus the R.L. of the middle cross hair point is given by (V)×(S)

The R.L. of the ground point is given by (V) × (S)—(M).

**Example :**

Suppose the middle cross wire reading on the "V" scale is 4.00

The middle cross wire reading (M) on the staff is 7.10

The stadia intercept (S) is 5.20.

The R.L. of the middle wire intersection on the staff is 4.00 × 5.20=20.80

R.L. of the ground point is 20.80—7.10=14.70.

(b) For obtaining horizontal distances :

(1) Read the horizontal scale "H" (H) when reading the "V" scale.

(2) Obtain stadia intercept (S)

(3) The stadia constant (C)=100

Thus the horizontal distance is obtained from
$$(S) \times (C) - (S \times H) = S(C - H)$$

Suppose the "Hor" scale reading is 8.00 ; and S=5·20 as in the previous case, then the horizontal distance,

$$= S(100 - H)$$
$$= 5·20(100 - 8·00)$$
$$= 5·20 \times 92$$
$$= 478·40$$

'B' type Arc :   Staff intercept      5·20
                "Hor" reading       92·00
                "Vert" reading      4·00

Horizontal distance $= 5·20 \times \dfrac{92}{100} = 47·84.$

Vertical height      $= 5·20 \times 4 = 20·80.$

The difference in elevation between the instrument station and the staff station is calculated as below :

Calculation of elevation of a point in relation to the instrument station = staff intercept (S) × Vertical scale reading (V) — Middle cross-hair reading (M) + Height of instrument (HI).

**Example :**   Suppose when sighting a stadia X
           staff intercept is           5·20
           "Vert" reading is         4·00
           middle cross-hair reading is   7·10
           height of instrument is      4·20
           then the elevation of X above the initial station is
$$(5·20 \times 4·00) - (7·10) + 4·20) = 29·82$$
           (See Tacheometric levelling)

## LEVELLING AND CONTOURING*

Levelling is the process of determining the relative heights of points, above or below, an assumed horizontal datum ; when this is applied to surfaces or areas it is known as contouring. Levelling carried out on an alignment is known as profiling. In some cases, the sea level is assumed to be the datum. In practice, the level of any one point is taken to be the standard, and the relative difference in levels of all other points, known as reduced levels, in relation to this are then worked out.

For approximate levelling for use with prismatic compass and Brunton compass surveys (1) an ordinary mason's spirit level or (2) the Abney level or (3) the Bruton alone may be employed with advantage or (4) an aneroid barometer calibrated as an altimeter, can be employed.

However, when a higher degree of precision, in contouring, is

*See addendum.

required, the plane table with telescopic alidade and Beaman's stadia arc, is to be used. The plane table can also be used in conjunction with a levelling instrument or level (to be described later) for this purpose. Tacheometric levelling as described earlier can also be resorted to.

(1) **Mason's level or spirit level** : A makeshift sightvane and peep hole are attached to either end of the bubble as shown in Fig. 1.28, and the instrument is mounted on a light tripod or stand and placed over a starting station. After levelling the bubble, the level may be rotated and radial sights are taken on to a graduate staff or scale held at various spots. Hence, if the level of the instrument station and the height of instrument are known the relative

Fig. 1.28

levels of the other points can be derived. The method of calculating the relative or reduced levels of stations has been given at the end of this chapter.

(2) **Abney level** : This instrument (Fig. 1·29) consists of a sighting tube (1) to which is attached a semicircular scale (5). The line joining the zero of the scale to its centre is at right angles to the axis of the tube. To the vernier (4) of the scale is attached a bubble tube (3). The bubble is in the centre of its run, when the vernier reads zero, and, hence the sighting tube is also horizontal. On the top of the sighting tube is an opening, and below the opening and within the tube is fixed an inclined mirror (6), with a horizontal centre line. The bubble tube (3) is not cased at the top and bottom, and this naked portion is just over the mirror. The instrument is so constructed that

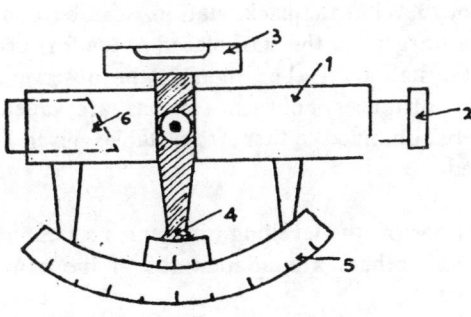

Fig. 1.29

when looking through the sighting tube the image of the bubble can be seen in the mirror.

When sighting the station point, through the tube subsequent adjustment of the vernier scale to set the bubble back on the centre of the mirror, is affected by rotating the vernier. The movement is registered either as an angle of elevation or depression, on the vernier.

The Abney level is handy for rough contouring as the vernier could be set at zero, and sights taken on to a graduated staff held at various points (no stand or sight vanes are required) in the manner described earlier for the use of the mason's level. If the level of the observer's eye above ground level is measured, the relative heights of points can be computed.

In case, only the heights of a few inaccessible points are required, and the distances are great, the method indicated below can be employed (Fig. 1.30).

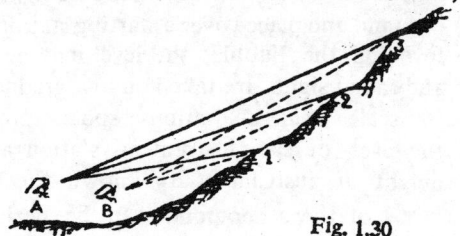

Fig. 1.30

Here the distance AB is measured, A and B are selected either on a level portion of the ground, or they are points whose reduced levels have been determined beforehand. The vertical angles to stations 1, 2 and 3 are measured first from A and then from B. The heights can be obtained either by geometrial constructions or by trigonometrical calculations. An approximate contouring can also be attempted with this method but this will be a tedious process. An alternative method of contouring with an Abney level is to run traverses across the area with a prismatic compass. The stations on the traverse are placed at such intervals, so that the Abney level at the back station, can be conveniently employed to sight a target on the staff placed at the forward station. The targets placed on the staff are at the height of the observer's eye from the ground. The slope distances between stations are taped. The heights or depression can be obtained either from tables given in Appendix (1) or can be calculated.

### Levelling instruments

There are various modifications of the levelling instruments. Some are merely improved models while others are adaptations of the same instrument for a particular purpose.

One of the earliest forms of the levelling instrument is the Y-level. The telescope is supported on short Ys. This instrument is obsolete. The dumpy level is an improved model. The telescope is attached to a vertical spindle which fits into a cup. This instrument is very popular because of its simplicity and robustness of construction.

Further modifications and improvements have resulted in the manufacture of the reversible types—the Cooke reversible or the Zeiss reversible. The instruments are capable of being easily adjusted or corrected (see permanent adjustment) but they are not as robust as the dumpy level.

In addition to these models are the quick set levels in which there are several modifications. In the simpler models, the telescope is supported on a ball and socket arrangement with a single graduated drum screw for levelling the instrument along the axis of the telescope, *i.e.*, line of collimation.

Finally, there are the instruments known as gradiometers, which are particularly useful in projects involving the pegging out of graded alignments —as in laying roads or railroads; on surface or underground, or canals. The instrument is similar to the quick set type, but in addition the drum screw placed below the telescope is graduated in gradients also, *e.g.*, 1 in 100, 1 in 50 etc.

Some levels are also equipped with stadia device, for tacheometric measurements of distances.

**Dumpy level :** The most popular instrument for levelling is the dumpy level. This in combination with the plane table is useful for all ordinary purposes. The level sections can be run, either along radial lines from a known point or run in two sets of intersecting parallel lines at right angles to each other, constituting a grid. The reduced levels of several points are determined along these lines and the contours interpolated. Sometimes, it is more convenient to locate points of known reduced levels by trial and error.

*Note :* As the level is not a distance measuring instrument normally, hence it is not necessary to place it over stations.

**Adjustment (temporary) :** Adjustments required to be carried out for setting up the instrument, for carrying out observations are called temporary adjustments, as opposed to adjustment required to be made for setting right any defects in the instrument. The following are the various steps in setting up the dumpy level.

(1) Open out the stand, place it firmly on the ground and remove the cap.

(2) Note the lie of the instrument in the box, so as to enable its quick replacement.

(3) Support the instrument by the tribrach (as far as possible) and lift it out of the box.

(4) Place the instrument on the stand and screw in, after making sure that the threads have engaged (see Theodolite).

(5) Bring the foot screws to the middle of the run.

(6) Press the legs "home", *i.e.*, plant them firmly.

(7) Place the longer bubble, which is parallel to the axis of the telescope, in alignment with the two legs of the tripod stand.

(8) Now move the third leg back and fore and sideways, so as to bring the bubble to the middle of the run.

(9) Place the bubble next over the third leg, *i e.*, in a direction at right angles to its original position. Repeat as (7) above, by pressing one of the other two legs, so as to move the bubble to the centre.

(10) When the bubble is fairly steady, the free leg is also pressed down and the instrument is finally levelled by means of its three foot screws (see instructions regarding theodolite adjustment given below).

**Theory and procedure**

The process of levelling may be considered analogous to making measurements of the depth of water in a lake by taking soundings. In this process, it is assumed that the surface of water is horizontal and so it can be taken as a plane of reference. Once the levelling instrument is set up, and adjusted, (as stated earlier) the line of collimation describes a horizontal surface, when the instrument is rotated about its axis. This forms the surface of reference or datum for measurements of the relative difference or variations in levels.

*To start with, the staff is placed vertically at the control point or datum point. Then staff is placed over the other points, and readings are taken similarly. An increase in staff readings, over that noted for the control point, indicates a fall in ground level, while a decrease in the staff reading (from that of the control point) signifies a rise in ground level, assuming that the height of the instrument is the same at all stations.

While continuing levelling operations, over long distances, it is often necessary to shift the position of the instrument. When the levelling instrument is shifted to a new place, the staff is fixed at the last point sighted. After completing the temporary adjustments at this point, the staff is again sighted back and the new reading recorded for

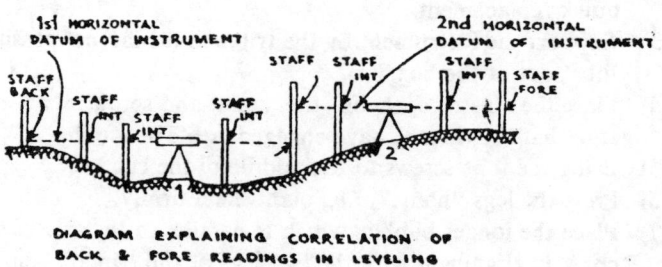

DIAGRAM EXPLAINING CORRELATION OF
BACK & FORE READINGS IN LEVELING

Fig. 1.31

the same point. By these operations, the correlation between the horizontal datum of the instrument at the first setting with the new horizontal datum of the second setting is established. The last reading of the first setting, taken for connecting up is thus recorded as 'fore' and the first reading, at the same point, from the new setting is recorded as 'back', while the other intermediate readings are booked as 'inter' (Fig. 1.31).

### Rise and fall method

Starting with the initial point, the relative, rise or fall in ground level is calculated by applying the principles stated earlier, i.e., increasing staff reading shows fall in ground level and vice versa. Then assuming any one point as reference standard, reduced levels for successive points or obtained by either adding or subtracting the 'rise' and 'fall' value as the case may be.

The levelling data are recorded in the field, in the following tabular form in columns 1 to 7, and the reduced levels of the various points calculated at the office later on in the manner of the example given on **next page**.

In the table given below the same example is worked out using the rise and fall method.

| Station | Sight | | | Rise | Fall | Reduced |
| | Back | Inter | Fore | | | level |
| --- | --- | --- | --- | --- | --- | --- |
| 7.20 | | | | | | 100.00 |
| | | 8.62 | | | 1.42 | 98.58 |
| | | 10.43 | | | 3.23 | 96.77 |
| | | 9.32 | | | 2.12 | 97.88 |
| | | 11.71 | | | 4.51 | 95.49 |
| | | 14.60 | | | 7.40 | 92.60 |
| 6.42 | | | 14.92 | | 7.72 | 92.28 |
| | | 8.92 | | | 2.50 | 89.78 |
| | | 6.44 | | | 0.02 | 92.62 |
| | | 4.11 | | 2.31 | | 94.59 |
| | | 3.20 | | 3.22 | | 95.50 |
| 9.28 | | | 2.10 | 4.32 | | 96.60 |
| | | 8.48 | | 0.78 | | 97.38 |
| | | 6.56 | | 2.72 | | 99.32 |
| | | 5.72 | | 3.56 | | 100.16 |
| | | 4.10 | | 5.18 | | 101.78 |

*Height of Instrument Method* : This method of levelling is also commonly used by surveyors. In this method, the initial back sight is taken to a known Bench Mark (B.M.). The staff reading on the B.M. is noted and this is added to the B.M. height to obtain the instrument height. For example, if the B.M. height is 10 m and the staff reading is 1 m at the point, then the height of instrument is 11 m. If from this value, the reading on the foresight or inter sight, is deducted, the reduced level of the fore or inter sight point, is obtained. For example, if the fore or inter reading is 1.3 m, then the reduced level is 11 m − 1.3 m or 9.7 m. The fig 1.31a illustrates the principle of the procedure.* The method of recording the data is given in the table below.

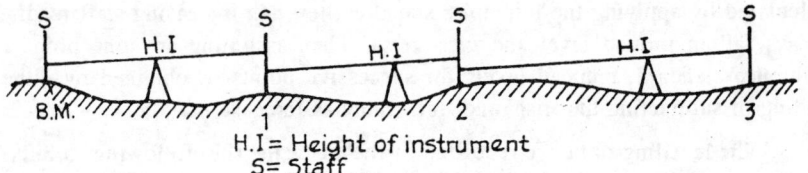

$H.I$ = Height of instrument
$S$ = Staff

Fig. 1.31(a)

| Back | Height of Inst | Inter | Fore | Reduced level |
|------|----------------|-------|------|---------------|
| 7.20 | 107.20 |  |  | 100.00 |
|  |  | 8.62 |  | 99.58 |
|  |  | 10.43 |  | 96.77 |
|  |  | 9.32 |  | 97.88 |
|  |  | 11.71 |  | 95.49 |
|  |  | 14.60 |  | 92.60 |
| 6.42 | 98.70 |  | 14.92 | 92.28 |
|  |  | 8.92 |  | 89.78 |
|  |  | 6.44 |  | 92.26 |
|  |  | 4.11 |  | 94.59 |
|  |  | 3.20 |  | 95.60 |
| 9.28 | 105.88 |  | 2.10 | 96.60 |
|  |  | 8.48 |  | 97.40 |
|  |  | 6.56 |  | 99.32 |
|  |  | 5.72 |  | 100.16 |
|  |  | 4.10 |  | 101.78 |

**Tacheometric levelling :** The tacheometer can also be employed for purposes of levelling and contouring. The principle of the tacheometer has already been explained, while dealing with stadia or tacheometer survey methods Two cases arise, while computing the reduced levels from the field data. These are solved in the following manner.

---

*See addendum.*

*Case* (*i*) — When the vertical angle is positive, *e.g.*, (Fig. 1.32).

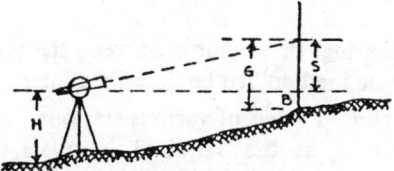

Fig. 1.32

To obtain the reduced level of B with reference to the reduced level of A ; the following formula may be employed :

R.L. of B = (R.L. of A + H + S) − G

where  H = height of instrument at A

S = vertical height as obtained from tacheometric calculation (or from the Beaman arc).

G = is the middle wire reading.

*Case* (*ii*) — Where the vertical angle is negative *e.g.*, (Fig. 1.33).

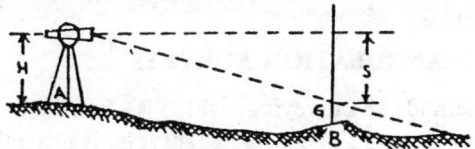

Fig. 1.33

To obtain the reduced level of B with reference to the reduced level of A ; the following formula may be used.

R.L. of      B = (R.L. of A + H − S) − G.

=(R.L. of A + H − S + G).

**Procedure for tacheometric levelling**

1. Set up the tacheometer on the traverse stations or triangulation stations (see adjustments).

2. (*a*) Before sighting the staff stations, adjust the horizontal plate so as to make the verniers read zero with reference to a known field station.

(*b*) Note the height of the instrument axis above ground station.

(*c*) The above procedure should be repeated after setting up the instrument at every station.

3. Direct the telescope on the staff stations and read the (*a*) horizontal angle, (*b*) vertical angle, (*c*) stadia intercept, *i.e.*, top and bottom stadia wire readings, and (*d*) middle wire readings.

4. Calculate the horizontal distance, difference in elevation and

reduced level of the rod stations with reference to the tacheometer station.

5. Plot the bearings or azimuths between the stations following the "co-ordinate method" to be described later.

6. Plot the reduced levels of various stations, and interpolate the contour lines, at the required intervals, by calculation as well as by sketching in the field, where minor details can be brought in.

Note : (a) As the speed in reading the staff is of great importance it is advisable to adjust the vertical arc tangent in such a manner so that either the top or bottom stadia wire coincides with a whole figure on the staff, i.e., 2 feet or 4 feet so that the other intercepts are read quickly. If readings are not taken quickly, the staff is likely to oscillate and cause inconvenience and delay.

(b) Steps described under paragraphs 4, 5 and 6 are carried out in the office.

## TRIANGULATION SURVEYS

Triangulation method of surveying, which is often followed in theodolite, tacheometric surveys and plane table surveys, is generally utilised where considerable accuracy is required. Triangulation method is fast and enables horizontal control over large map areas. It is best adopted in regions, where long sight lines from many points are possible, especially in areas of rugged relief, as in hilly terrain.

Principle : In triangulation, map locations of ground points are obtained by the intersection of two or more rays, taken to the point from two different instrument stations. Briefly, the most simple triangulation method involves the following sequence· of operations :—

(1) Fixing the base line by measuring its length and direction. The base line fixes the scale of the map as well as its orientation with respect to the sheet. The line should be measured on an open level ground and the two ends must be visible from each other as well as from at least three other triangulation stations. If the area is large, the base line should be located somewhere in the central portion, so that the net need not be extended too far in any one direction. If the area is greatly elongated, the base line should be oriented as far as possible in the direction of the larger dimension, and the measurement is accurately done with invar steel tape. A chain may be used in case of triangulation with prismatic compass. Where the country is uneven, stadia measurement may be made as a check on tape measurement. For very accurate theodolite triangulation,

the base line is measured with invar steel band applying corrections to compensate for (*a*) extension, due to tension or pull, (*b*) thermal expansion and (*c*) sag.

This method of measurement is outside the scope of normal geological work, and it is not required for use in connection with the methods of survey used in geological work.

(2)   In the field, the location of any point, which is visible from both base line points, is obtained by sighting the unknown point from the base line points and drawing the two rays. The intersection establishes the relative location of the unknown point in question and horizontal distance from other known points. It is sometimes advisable to check the accuracy of intersections by a third intersecting ray from a third station.

(3)   Establishment of the elevation of the base line station is done by taping up to known bench mark or trigonometrical station of known R. L. and by running a fly level traverse.

(4)   Establishment of a base net is done by sighting two known stations, one on either side of the base line, and sighting back from those to the base line stations, as well as on to each other; as shown in Fig. 1.34.

A triangulation net is established and extended over the area to be mapped, in precisely the same manner as in the case of establishing the base net. From stations in the base net, rays are drawn towards all

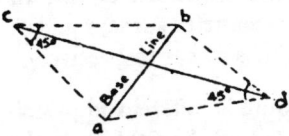

visible stations, and these are located by the intersection of rays (Fig. 1.34). The relative strength of these locations, however, depends on the angles made by the intersecting rays and also on the number of rays drawn to each of these points. The location is strongest,

Fig. 1.34

when this angle is 90° and weakest if the angle less than 30° is greater than 150°.

There are four types of triangulation nets as shown in Fig. 1.35 (*a*) Chain of triangles, (*b*) Chain of quadrilaterals, (*c*) Chain of polygons,

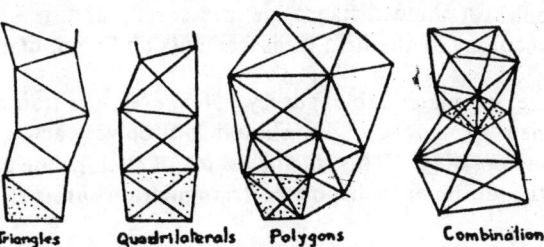

Triangles    Quadrilaterals    Polygons         Combination

Fig. 1.35

and (d) Combination net. The chain of triangles is the weakest construction of the four and the combination net is the strongest.

A check base is established, when the triangulation system has been greatly extended from the base net (as shown in Fig. 1.36), in order to check up if the system is "out of scale" or orientation. The stations, at the two ends of the check base line, are located by intersection from three or more triangulation stations, and the distance measured accurately with the steel tape. If the work has been carried out accurately, the measured

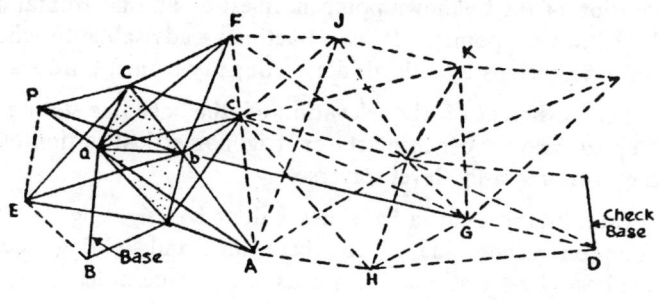

Fig. 1.36

distance and the scaled distance, as obtained from the map, should agree.

The recording of instrumental data and calculations are shown in Appendix 3 for theodolite surveys, Appendix 4 tacheometer surveys and 5, 5a, for telescope alidade Beaman arc, and plane table surveys respectively.

*Note* : Triangulation survey with a theodolite is different in principle, since the distances are not measured or plotted in the field, as in the plane table surveys. In the case of theodolite triangulation, only a base line is measured, with the greatest precision and all angles determined accurately with a theodolite. The area is divided into triangles. Starting with the base line, the lengths of all the sides of all traingles are mathematically computed, A check base, forming the side of one of the computed triangles, at some distance, is measured, and this serves as a check on the accuracy of the field measurements and computations.

In the case of plane table surveys, however, the traingles must be "well-conditioned", i. e., the angles should not be very acute. and should be as near 60° as possible. This factor is not of such prime consequence, in the case of theodolite or tacheometer triangulation surveys.

### Interior filling of details in surveys

The filling in of interior details—topographic, geologic etc.—in most surveys, except in accurate theodolite triangulation, or traverse, is generally done either by compass and tape survey, or by compass and pace survey. The compass used, may by prismatic compass, Brunton compass or clinometer compass.

The measurement and recording of details in chain and compass traverse is shown in Figs. 1.37 and 1.38. Offsets, taken on either side of the chain or compass traverse, are measured with a tape. Subsidiary traverses shown in dotted line may also be made from the main traverse

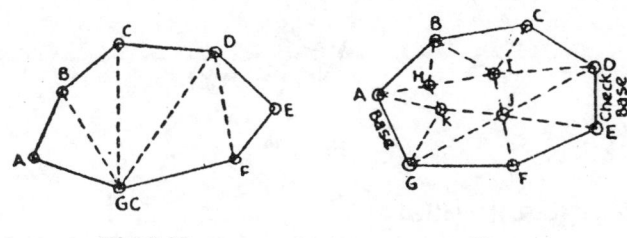

Fig. 1.37            Fig. 1.38

shown in firm lines for filling in further details. The same procedure is adopted again for the purpose as explained under chain survey.

### Theodolite traverse and plane-table surveys

In the case of theodolite and plane-table surveys, subsidiary traverses are made to cover the areas within easy reach of each rod-station, and such details are generally obtained by again employing the compass and pacing method or by compass and tape method. For such subsidiary traverse initial stations are radially located at short distances from each rod station. In a rugged country, requiring accurate work, it is advisable to have a larger number of such staff stations or rod shot stations, as these form, known points at close intervals, distributed all over the area. If carefully executed and plotted, pace and Brunton compass survey, (Fig. 1.39 & 1.40) or pace and clino-compass survey, for such work is as accurate as compass and tape measurement for map scales which are most commonly used, say upto 200 ft. to an inch. A specimen of the staff man's field notes, is shown below :

### From station A—Station 1

"At contact of granite and manganese ore. Ore 5 ft. thick low grade consisting mainly quartz, subordinate manganese ore.

(Specimen 3/Rs/64).

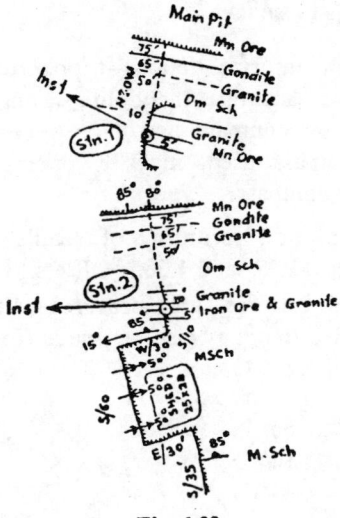

Fig. 1.39

### From station A—Staff station 2

On the  same contact as at  stn. 1 and at .the  top of a vertical cutting (Fig. 1.39).

### Staff station 51

On top of qzt. hillock  due  NNE  of  stn. A.  Qzt. bed  55 ft. thick-finely bedded with a core of massive wide recrystallised quartz".
(Specimen 2/RS/64) (Fig. 1.40).

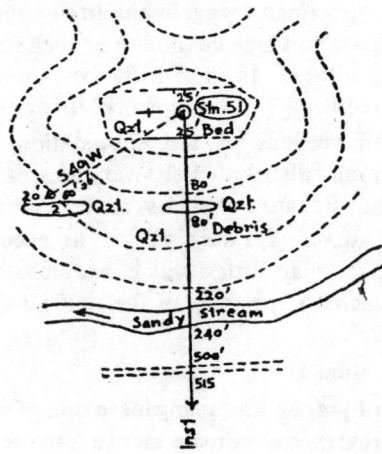

Fig. 1.40

# DETERMINING TRUE NORTH

**Shadow method**—is sufficient for most purposes and the following is the procedure. The principle involved depends, on the symmetry in the length and direction of the shadow cast by a vertical linear object placed in the sun, during the course of the day. The sun is a bright star, moving round the celestial pole and, hence transits the celestial meridian once each day like every other star.

The traverse, after being plotted, is fixed to a level table (or plane table after levelling), and placed over a point P, the location of which is known, and which is free from interfering or extraneous shadows. The traverse is oriented with reference to the field.

Concentric circles of varying radii (Fig. 1.41) are drawn round this point P, at invervals of 1/4th inch upto 3 inch radius, and at intervals of ½ inch upto 6 inches radius. A straight steel pin, about 1/8 inch thick and 5 inches long is planted firmly and verti-

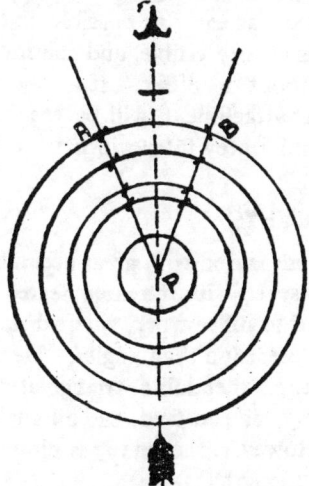

cally at the plotted point P. The of the pin may be checked, in four directions at right angles to each other, either with the right angle of a setsquare or by a plumb bob. The measurements are started in the morning and completed in the evening, say between 8 A. M. and 4 P. M. As the tip of the shadow cast by the pin, just touches a particular concentric circle, the point of contact is marked on the circle by a cross. Thus, at the end of the day, when all such crosses are joined, two lines AP and BP meeting at the point P are obtained. The magnitude of the angle between the two lines, depends on the latitude of the place,

Fig. 1.41

and also on the season of the year. The bisector of this angle gives the direction of the true north.

**Equal altitude method :** When a theodolite is available, this method can be employed and greater precision obtained. This is also simple in operation. In this method, advantage is taken of the fact that when a star (or the sun) is sighted at a certain altitude when it is rising, and again at the same altitude when it is setting, and if directions of these two positions in the horizontal plane i. e. azimuths are known, the medial position gives the celestial meridian or the north-south line passing through that point. The theodolite is set up and temporary adjustments done, on any station in the traverse. The horizontal plate, say vernier A, is made to read zero. By releasing the body clamp A station B is sighted. At night a frosted lamp or beacon is required to be held behind the rod, planted at the station B for sighting.

Now the horizontal plate, vernier clamp and the vertical circle clamp, are released, and the star is sighted. The clamps are tightened, and finer intersection completed with the respective vernier tangents. The readings of both, the horizontal and the vertical verniers are recorded. Both the verniers are continuously read and the readings recorded, at suitable time intervals. It will be noticed, that the value of the vertical angles reach a maximum (until the azimuth of celestial meridian is crossed) and then decrease on the other side, where vertical angles, of equal value to those measured earlier, will be obtained. If the horizontal angle, subtended by two equal altitude positions, is bisected; the direction of the bisector, gives the position of the meridian, thus fixing the true north. The observations, on a suitable star, should be started, as early as possible, in the evening after dusk when the particular star is clearly visible, and continued until the early hours of the morning without a break. If the sun is observed, special equipment, known as solar attachment, will be required for reading vertical angles (in our latitude) and a grey filter will have to be employed to protect the eyes.

## PLOTTING OF SURVEYS

It may be observed that angles can be measured with great accuracy, in the field, whereas distances cannot be measured with the same degree of precision. But reverse is the case with regard to office work, where distances can be plotted with a greater degree of precision than angles. Hence in the case of the most accurate surveys e. g., theodolite triangulation, only a few distances are measured accurately, in the field, and all angles are measured with a precision instrument. However, the survey is plotted, in the office, by a method which involves only lengths.

In general, the accuracy of the method of plotting employed, should be commensurate with the degree of accuracy of the survey. Thus for a compass survey, the plotting may be done by means of plotting scales and protractor. For chain surveys, the plotting scale may be employed.

### Coordinate Method

In the case of a theodolite traverse, the plotting should invariably be done, by the co-ordinate method described below :

    (a) The bearings or azimuths of various lines, in the traverse are calculated, as indicated earlier; vide, theodolite and Miner's dial traverses, or the bearings may be obtained directly, by the use of the continuous bearing method.

(b) The whole-circle bearings thus obtained are then converted to quadrant bearings.

(c) If the length of any line and its quadrant bearing are known, taking one end to be the initial point, the co-ordinates are determined for the other end. The co-ordinates so obtained are called the partial co ordinates. Similarly partial co-ordinates may be calculated for any number of points, in the traverse. When the same initial point is taken for calculating the coordinates of the next and the succeeding points, then the co-ordinates so obtained will be known as total co-ordinates. The

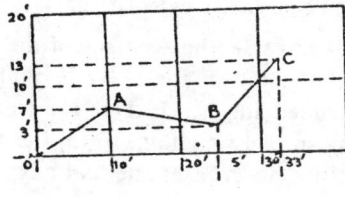 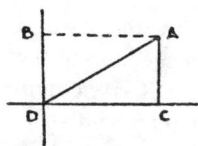

Fig. 1.42                    Fig. 1.43

principle involved in the calculation of partial and total coordinates is explained in Figs. 1.42 and 1.43.

The partial co-ordinates of A are + 10′ and + 7′,

for B, are +15′ and +4′,

for C, are + 8′ and +10′ ;

whereas the total co-ordinates are

for A, +10′ and +7′,

for B, +25′ and +3′, and

for C, +33′ and +13′.

The co-ordinates on the Y axis or vertical axis are called latitude, and the co-ordinates on the X axis or horizontal axis are called departure. Both latitude and departure can be easily calculated as shown below (Fig. 1.43).

Partial latitude of A

   =Cos θ×AD

Partial departure of A

   =Sin θ×AD

where $\theta$ is the quadrant bearing of the line AD.

The partial co-ordinates of all the points or stations in a traverse are calculated in a similar manner.

The total co-ordinates are thus computed from the partial co-ordinates by mere algebraic summations, i. e., paying heed to sign ; whether positive or negative in respect of the point of origin. An example of the procedure involved is given in table on page 58*.

The total coordinates have been computed by reference to tables known as traverse tables, in which the values of partial latitude and departure for various values of quadrant bearings have been calculated for various distances from 1 to 1000 ft.

### Checks on angular measurements and coordinate calculations

1. In the case of a closed traverse, the accuracy of the field work can be checked, since the sum of the included angles=2 X right angles—360°, where X is the number of included angles. If there is a small angular error, this may be distributed giving special weightage to (a) the amount or magnitude of the angle, (b) difficulties in measurement at any particular point etc.

2. In a closed traverse, the calculations of the latitude and departure may be arithmetically checked. The total latitude and departure should be equal to zero, since the traverse terminates at the point of origin. Further, the sum of northings=sum of southings and the sum of eastings=sum of westings. It should be noted, that this check is only in respect of the computations, and not in regard to the accuracy of the field work.

### The scale of chords

This method is at times employed for plotting angles. In order to construct an angle of required magnitude at a point on straight line, to start with, a circle, of radius equal to the length of the 60° chord of the scale, is constructed (Fig. 1.44). Then from the intersection point of the

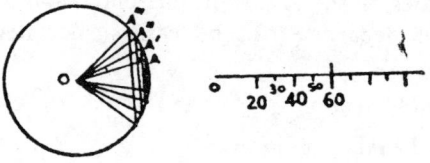

Fig. 1.44

---

*See addendum.*

circle and the straight line, a length of chord equal in value to that of the desired angle is marked of on the circle (on required side of the straight line). This point on the circle, when joined to the centre gives the angle required.

**Principle of the scale of chords :** In any circle, the length of the chord subtended by an angle of 60° is equal to the radius (Fig. 1.44).

In the scale of chords, length of the chords subtended by various angles at the centre of the circle, are utilised to constitute a scale. Thus AB is the length of a chord which subtends an angle of 30° at the centre, AB' is also the length of a chord subtending angle of 40° at the centre of the circle.

These lengths are marked as 30°, 40°, 50°, and 60° on the scale.

Thus with the scale of chords, since the values of the angles are plotted as linear measurements some amount of accuracy can be achieved.

**Preparation of Mine Plans:** The underground mine plan is a map which shows the position of the workings projected on to the horizontal plane. Since coal seams generally dip at low angles the common practice is to prepare the plan according to the above convention. This is mostly the case in regard to the coal mines of India. In the case of the metalliferous lodes, however, where the dips are often very high, being of the order of 70 to 80 degrees, and where the dip and strike are also inconsistent, plans for each level are drawn separately, placed one below the other, in the correct relative orientation and traced out, taking into account the dip of the ore body. In case the dips are very steep, the offset is increased so that the plan of each level is well separated, as shown in figure (1.45A). Sometimes combining the level plans in their correct orthographic position, in order to obtain the composite plan, is very useful in the identification of geological structures, which otherwise will be diffused in the individual level plans. This aspect is brought out in the figure (1.45B).

As opposed to the mine plan the practice in all underground metal mines, is to prepare longitudinal sections of the ore body, in which the levels, raises, winzes shafts etc., are projected on to the vertical plane at right angles to the dip. Thus the contortions seen in the level plans, due to the variation in strike, are eliminated and the levels appear as straight lines, as shown in figure (1.46). The longitudinal section is best suited for depicting structural features, such as faults, joints, bedding, wall rock contacts etc. It is, however, to be noted that the true dips of such features is to be converted to the apparent dip corresponding to that of the ore body, while

Tables known as traverse tables have been worked out for various values ∠BDA = 0 (bearing) and for various lengths of DA BD. Thus at a glance the partial latitude and partial departure of any given point for any given bearing can be determined by referring to the table

| Sl. No. | Line | Distance in feet | Horizontal angle | Calculated whole circle bearing | Quadrant bearing | Partial latitude | | Partial departure | |
|---|---|---|---|---|---|---|---|---|---|
| | | | | | | N | S | E | W |
| A | HA | – | – | 92° 0' | S 88° E | – | 16.3 | 522.7 | – |
| A | AB | 523 | 127° 14' | 39° 14 | N 39° 14 E | 271.1 | | 221.3 | |
| B | BC | 350 | 55° 27' | 94° 41 | S 85° 19' E | | 61.9 | 754.5 | |
| C | CD | 757 | 77° 34' | 352° 15' | N 7° 45 W | 768.7 | | | 104.6 |
| D | DE | 776.7 | 108° 12' | 270° 27' | N 89° 33 W | 4.7 | | | 595.1 |
| E | EF | 595.1 | 39° 46' | 121° 13' | S 41° 13 W | | 246.7 | | 226.2 |
| F | FG | 328.0 | 315° 09' | 256° 22' | S 76° 22' W | | 108.4 | | 447.0 |
| G | GH | 460.0 | 115° 17' | 191° 39' | S 11° 39' W | | 609.2 | | 125.6 |
| H | HA | 622.0 | 98° 39' | 92° 0' | S 38° E | 1044.5 | 1042.5 | 1498.5 | 1498.5 |

The following calculations will have to be done before computing the co-ordinates.

plotting. A variant of the longitudinal section is what is known as the inclined longitudinal section. This is in reality the plan of the ore body in the plane containing the dip and strike. (See Underground Mining Plan and Opencast mine Plan.)

## SURVEY EQUIPMENT WITH REFERENCE TO THE PROBLEM

In general, the instruments chosen will be in keeping with the accuracy required and the time available for completion of the survey :

| Instrument employed and method | Problem involved or conditions |
| --- | --- |
| 1. Chain Survey. | It is employed when no angle measuring instrument is available. It can be carried out with the help of untrained hands. The country should be more or less flat. The method can also be employed for the interior filling of theodolite surveys. Scale of plotting, 1 inch = 100 ft. |
| 2. Compass Survey. | It is good for ordinary preliminary work. Distances are, in general, obtained by taping or pacing. As such it cannot be employed for accurate work. Scale of plotting is generally 4″ = 1 mile. It can be used for interior filling of chain surveys and for normal geological mapping, on large scale. It can also be used in stope surveys underground (with taping). |
| 3. Plane table Survey. | It is useful in flat or gently undulating topography. It is as accurate as chain surveys and can be usefully employed for interior filling of theodolite surveys. When telescopic alidade, equipped with Beaman stadia arc is available, the method can be applied even in fairly rugged terrain. It is recommended for. detailed mapping and proving of mineral deposits—underground mine surveys, geochemical surveys, and geophysical surveys. |

| | |
|---|---|
| 4. Theodolite Survey. | Traverses are required for accurate work—when extreme accuracy is required, then theodolite triangulation is essential. Tacheometric surveys are as accurate as chain surveys but the accuracy is relatively higher in rugged country. The method is useful in contouring hilly country. It is useful in proving mineral deposits, mines surveys, and correlation surveys, etc. |
| 5. Abney Level. | Useful in contouring in conjunction with compass surveys or in the interpolation of heights in such surveys in rugged country. It can be used also with chain surveys. |
| 6. Dumpy Level. | Very useful in accurate contouring of theodolite and plane table surveys in a flat or gently undulating topography. |

**Instruments and mineral deposits**

General scheme giving a few examples of mineral deposits and the methods of survey.

| Type or nature of deposit | Instruments or methods to be employed |
|---|---|
| 1. Limestone and clay in a flat country. | If no high order of accuracy is required the compass and pacing with abney level will do. If a higher order of accuracy is wanted, then plane table with dumpy level, will be necessary. |
| 2. Limestone or coal in flat country. | Theodolite traverse with dumpy level will be required. |
| 3. Metalliferous lode in plane country (Gold or copper or manganese etc.). | Theodolite traverse, or plane table or chain with dumpy level will be required. |
| 4. Metalliferous lode in a hilly country. | Tacheometer survey (or Telescopic alidade with Beaman stadia arc) may be employed. |

| | |
|---|---|
| 5. Iron ore occurring as a capping on a flat topped hill. | Theodolite. Normally no compass survey should be employed. Traverse with tacheometer and levelling by stadia or dumpy level will be satisfactory. |
| 6. Mine survey for coal or metal. | Theodolite traversing or Dial survey may be done. Underground traverses may also be done with plane table with Beaman stadia arc attachment. |
| 7. Stope survey. | Compass and tape survey (Brunton compass is convenient) will be best suited. |

*Note* :—(1)  In general the more precious or valuable a metal is, the more accurate should be the method of survey chosen.

(2)  And the nature of the work should also be considered *i.e.*, whether preliminary or detailed investigation.

**Appendix 1**

**Horizontal Distance in Feet with Corresponding Slope-Distance**

Slope in degrees

| Slope distance | 1 | 2 | 3 | 4 | 5 | 6 | 7 | 8 | 9 | 10 | 20 | 30 | 40 | 50 | 100 |
|---|---|---|---|---|---|---|---|---|---|---|---|---|---|---|---|
| 2 | 0.99 | 1.99 | 2.99 | 3.99 | 4.99 | 5.99 | 6.99 | 7.99 | 8.99 | 9.99 | 19.99 | 29.98 | 39.98 | 49.97 | 99.94 |
| 4 | 0.99 | 1.99 | 2.99 | 3.99 | 4.98 | 5.98 | 6.98 | 7.98 | 8.98 | 9.98 | 19.95 | 29.93 | 39.90 | 49.88 | 99.76 |
| 5 | 0.99 | 1.99 | 2.99 | 3.98 | 4.98 | 5.98 | 6.97 | 7.97 | 8.97 | 9.96 | 19.92 | 29.89 | 39.85 | 49.81 | 99.62 |
| 6 | 0.99 | 1.99 | 2.98 | 3.98 | 4.97 | 5.97 | 6.96 | 7.96 | 8.95 | 9.95 | 19.89 | 29.84 | 39.78 | 49.73 | 99.45 |
| 8 | 0.99 | 1.98 | 2.97 | 3.96 | 4.95 | 5.94 | 6.93 | 7.92 | 8.91 | 9.90 | 19.81 | 29.71 | 39.61 | 49.51 | 99.03 |
| 10 | 0.98 | 1.97 | 2.95 | 3.94 | 4.92 | 5.91 | 6.89 | 7.88 | 8.86 | 9.85 | 19.70 | 29.54 | 39.39 | 49.24 | 98.48 |
| 12 | 0.98 | 1.96 | 2.93 | 3.91 | 4.89 | 5.87 | 6.85 | 7.82 | 8.80 | 9.78 | 19.56 | 29.34 | 39.13 | 48.91 | 97.81 |
| 14 | 0.97 | 1.94 | 2.91 | 3.88 | 4.85 | 5.82 | 6.79 | 7.76 | 8.73' | 9.70 | 19.41 | 29.11 | 38.81 | 48.51 | 97.03 |
| 15 | 0.97 | 1.93 | 2.90 | 3.86 | 4.83 | 5.79 | 6.76 | 7.73 | 8.69 | 9.66 | 19.32 | 28.98 | 38.64 | 48.29 | 96.59 |
| 16 | 0.96 | 1.92 | 2.88 | 3.84 | 4.81 | 5.77 | 6.73 | 7.69 | 8.65 | 9.61 | 19.23 | 28.84 | 38.45 | 48.06 | 96.13 |
| 18 | 0.95 | 1.90 | 2.85 | 3.80 | 4.76 | 5.71 | 6.66 | 7.61 | 8.56 | 9.51 | 19.02 | 28.53 | 38.04 | 47.55 | 95.11 |
| 20 | 0.94 | 1.88 | 2.82 | 3.76 | 4.70 | 5.64 | 6.58 | 7.52 | 8.46 | 9.40 | 18.79 | 28.19 | 37.59 | 46.98 | 93.97 |
| 22 | 0.93 | 1.85 | 2.78 | 3.71 | 4.64 | 5.56 | 6.49 | 7.42 | 8.34 | 9.27 | 18.54 | 27.82 | 37.09 | 46.36 | 92.72 |
| 24 | 0.91 | 1.83 | 2.74 | 3.65 | 4.57 | 5.48 | 6.39 | 7.31 | 8.22 | 9.14 | 18.27 | 27.41 | 36.54 | 45.68 | 91.35 |
| 25 | 0.91 | 1.81 | 2.72 | 3.62 | 4.53 | 5.44 | 6.34 | 7.25 | 8.16 | 9.06 | 18.13 | 27.19 | 36.25 | 45.31 | 90.63 |

Horizontal distance

## Horizontal Distance in Feet with Corresponding Slope Distance—Continued

| Slope in degrees | Horizontal distance | | | | | | | | | | | | | | |
|---|---|---|---|---|---|---|---|---|---|---|---|---|---|---|---|
| | 1 | 2 | 3 | 4 | 5 | 6 | 7 | 8 | 9 | 10 | 20 | 30 | 40 | 50 | 100 |
| 26 | 0.90 | 1.80 | 2.70 | 3.59 | 4.49 | 5.39 | 6.29 | 7.19 | 8.09 | 8.99 | 17.98 | 26.96 | 35.95 | 44.94 | 89.88 |
| 28 | 0.88 | 1.77 | 2.65 | 3.53 | 4.41 | 5.30 | 6.18 | 7.06 | 7.95 | 8.83 | 17.66 | 26.49 | 35.32 | 44.15 | 88.29 |
| 30 | 0.87 | 1.73 | 2.60 | 3.46 | 4.33 | 5.20 | 6.06 | 6.93 | 7.79 | 8.66 | 17.32 | 25.98 | 34.64 | 43.30 | 86.60 |
| 32 | 0.85 | 1.70 | 2.54 | 3.39 | 4.24 | 5.09 | 5.94 | 6.78 | 7.63 | 8.48 | 16.96 | 25.44 | 33.92 | 42.49 | 84.80 |
| 34 | 0.83 | 1.66 | 2.49 | 3.32 | 4.15 | 4.97 | 5.80 | 6.63 | 7.46 | 8.29 | 16.58 | 24.87 | 33.16 | 41.45 | 82.90 |
| 35 | 0.82 | 1.64 | 2.46 | 3.28 | 4.10 | 4.91 | 5.73 | 6.55 | 7.37 | 8.19 | 16.38 | 24.58 | 32.77 | 40.96 | 81.92 |
| 36 | 0.81 | 1.62 | 2.43 | 3.24 | 4.05 | 4.85 | 5.66 | 6.47 | 7.28 | 8.09 | 16.18 | 24.27 | 32.36 | 40.45 | 80.90 |
| 38 | 0.79 | 1.58 | 2.36 | 3.15 | 3.94 | 4.73 | 5.52 | 6.30 | 7.09 | 7.88 | 15.76 | 23.64 | 31.52 | 39.40 | 78.80 |
| 40 | 0.77 | 1.53 | 2.30 | 3.06 | 3.83 | 4.60 | 5.36 | 6.13 | 6.89 | 7.66 | 15.32 | 22.92 | 30.64 | 38.30 | 76.60 |
| 42 | 0.74 | 1.49 | 2.23 | 2.97 | 3.72 | 4.46 | 5.20 | 5.94 | 6.69 | 7.43 | 14.86 | 22.29 | 29.73 | 37.16 | 74.31 |
| 44 | 0.72 | 1.44 | 2.16 | 2.88 | 3.60 | 4.32 | 5.03 | 5.75 | 6.47 | 7.19 | 14.39 | 21.58 | 28.77 | 35.97 | 71.93 |
| 45 | 0.71 | 1.41 | 2.21 | 2.83 | 3.53 | 4.24 | 4.95 | 5.66 | 6.36 | 7.07 | 14.14 | 21.21 | 28.28 | 35.35 | 70.71 |
| Slope distance | 1 | 2 | 3 | 4 | 5 | 6 | 7 | 8 | 9 | 10 | 20 | 30 | 40 | 50 | 100 |

**Appendix 2**

## Vertical Height in Feet with Corresponding Slope Distance

| Slope in degrees | Vertical height | | | | | | | | | | | | | | |
|---|---|---|---|---|---|---|---|---|---|---|---|---|---|---|---|
| 2 | 0.03 | 0.07 | 0.10 | 0.14 | 0.17 | 0.21 | 0.24 | 0.28 | 0.31 | 0.35 | 0.70 | 1.05 | 1.40 | 1 74 | 3.49 |
| 4 | 0.07 | 0.14 | 0.21 | 0.28 | 0.35 | 0.42 | 0.49 | 0.56 | 0.63 | 0.70 | 1.40 | 2.09 | 2.79 | 3.49 | 6.98 |
| 5 | 0.09 | 0.17 | 0.26 | 0.35 | 0.44 | 0.52 | 0.61 | 0.70 | 0.78 | 0.87 | 1.74 | 2.62 | 3.49 | 4.36 | 8.72 |
| 6 | 0.10 | 0.21 | 0.31 | 0.42 | 0.52 | 0.63 | 0.73 | 0.84 | 0.94 | 1.05 | 2.09 | 3.14 | 4.18 | 5.23 | 10.45 |
| 8 | 0.14 | 0.28 | 0.42 | 0.56 | 0.70 | 0.83 | 0.97 | 1.11 | 1.25 | 1.39 | 2.78 | 4.18 | 5.57 | 6.96 | 13.92 |
| 10 | 0.17 | 0.35 | 0.52 | 0.69 | 0.87 | 1.04 | 1.21 | 1.39 | 1.56 | 1.74 | 3.47 | 5.21 | 6.95 | 8.68 | 17.36 |
| 12 | 0.21 | 0.42 | 0.62 | 0.83 | 1.04 | 1.25 | 1.45 | 1.65 | 1.87 | 2.08 | 4.16 | 6.24 | 8 32 | 10.40 | 20.79 |
| 14 | 0.24 | 0.48 | 0.73 | 0.97 | 1.21 | 1.45 | 1.69 | 1.93 | 2.18 | 2.42 | 4.84 | 7.26 | 9.68 | 12.10 | 24.19 |
| 15 | 0.26 | 0.52 | 0.78 | 1.03 | 1.29 | 1.55 | 1.81 | 2.07 | 2.33 | 2.59 | 5.18 | 7.76 | 10.35 | 12.94 | 25.88 |
| 16 | 0.28 | 0.55 | 0.83 | 1.10 | 1.38 | 1.65 | 1.93 | 2.20 | 2.48 | 2.76 | 5.51 | 8.27 | 11.03 | 13.78 | 27.56 |
| 18 | 0.31 | 0.62 | 0.93 | 1.24 | 1.55 | 1.85 | 2.16 | 2.47 | 2.78 | 3.09 | 6.18 | 9.27 | 12.36 | 15.45 | 30.90 |
| 20 | 0.34 | 0.68 | 1.03 | 1.37 | 1.71 | 2.05 | 2.39 | 2.74 | 3.08 | 3.42 | 6.84 | 10.26 | 13.68 | 17.10 | 34.20 |
| 22 | 0.37 | 0.75 | 1.12 | 1.50 | 1.87 | 2.25 | 2.62 | 3.00 | 3.37 | 3.75 | 7.49 | 11.24 | 14.98 | 18.73 | 37.46 |
| 24 | 0.41 | 0.81 | 1.22 | 1.63 | 2.03 | 2.44 | 2.85 | 3.25 | 3.66 | 4.07 | 8.13 | 12.20 | 16.27 | 20.34 | 40.67 |
| 25 | 0.42 | 0.84 | 1.27 | 1.69 | 2.11 | 2.54 | 2.96 | 3.38 | 3.80 | 4.23 | 8.45 | 12.68 | 16.90 | 21.13 | 42.26 |
| Slope distance | 1 | 2 | 3 | 4 | 5 | 6 | 7 | 8 | 9 | 10 | 20 | 30 | 40 | 50 | 100 |

## Vertical Height in Feet for Corresponding Slope Distance—Continued.

| Slope in degrees | 1 | 2 | 3 | 4 | 5 | 6 | 7 | 8 | 9 | 10 | 20 | 30 | 40 | 50 | 100 |
|---|---|---|---|---|---|---|---|---|---|---|---|---|---|---|---|
| 26 | 0.44 | 0.88 | 1.31 | 1.75 | 2.19 | 2.63 | 3.07 | 3.51 | 3.95 | 4.38 | 8.77 | 13.15 | 17.53 | 21.92 | 43.84 |
| 28 | 0.47 | 0.94 | 1.41 | 1.88 | 2.35 | 2.82 | 3.29 | 3.76 | 4.22 | 4.69 | 9.39 | 14.08 | 18.78 | 23.47 | 46.95 |
| 30 | 0.50 | 1.00 | 1.50 | 2.00 | 2.50 | 3.00 | 3.50 | 4.00 | 4.50 | 5.00 | 10.00 | 15.00 | 20.00 | 25.00 | 50.00 |
| 32 | 0.53 | 1.06 | 1.59 | 2.12 | 2.65 | 3.18 | 3.71 | 4.24 | 4.77 | 5.30 | 10.60 | 15.90 | 21.20 | 26.49 | 52.99 |
| 34 | 0.56 | 1.12 | 1.68 | 2.24 | 2.80 | 3.35 | 3.91 | 4.47 | 5.03 | 5.59 | 11.18 | 16.78 | 22.37 | 27.96 | 55.92 |
| 35 | 0.57 | 1.15 | 1.72 | 2.29 | 2.87 | 3.44 | 4.01 | 4.59 | 5.16 | 5.74 | 11.47 | 17.21 | 22.94 | 28.68 | 57.36 |
| 36 | 0.59 | 1.18 | 1.76 | 2.35 | 2.94 | 3.53 | 4.11 | 4.70 | 5.29 | 5.88 | 11.76 | 17.63 | 23.51 | 29.39 | 58.78 |
| 38 | 0.62 | 1.23 | 1.85 | 2.46 | 3.08 | 3.69 | 4.31 | 4.93 | 5.54 | 6.16 | 12.31 | 18.47 | 24.63 | 30.78 | 61.57 |
| 40 | 0.64 | 1.29 | 1.93 | 2.57 | 3.21 | 3.86 | 4.50 | 5.14 | 5.78 | 6.43 | 12.86 | 19.28 | 25.71 | 32.14 | 64.28 |
| 42 | 0.67 | 1.34 | 2.01 | 2.68 | 3.35 | 4.01 | 4.68 | 5.35 | 6.02 | 6.69 | 13.38 | 20.07 | 26.77 | 33.46 | 66.91 |
| 44 | 0.69 | 1.39 | 2.08 | 2.78 | 3.47 | 4.17 | 4.86 | 5.56 | 6.25 | 6.95 | 13.89 | 20.84 | 27.79 | 34.73 | 69.47 |
| 45 | 0.71 | 1.41 | 2.12 | 2.83 | 3.53 | 4.24 | 4.95 | 5.66 | 6.36 | 7.07 | 14.14 | 21.21 | 28.28 | 35.35 | 70.71 |

Slope distance

**Appendix 3**

Table for Recording Theodolite Traverse Field Data

| Station from | Station to | Face right | | Face left | | Mean value of angles | Distance | Remarks |
|---|---|---|---|---|---|---|---|---|
| | | Vernier A | Vernier B | Vernier A | Vernier B | | | |

**Appendix 4**    Tabular Form for Recording Field Data in Tacheometer or Stadia Surveys

| I Back sation | II On station | III Fore station | IV Height of Inst. | V Horizontal angles | VI Vertical angles | VII Staff readings on | | VIII Remarks regarding Geology or Topography |
|---|---|---|---|---|---|---|---|---|
| | | | | | | Stadia lines | Centre line | |

**Appendix 4a**    Tabular Form for Calculating Horizontal Distances and Reduced Levels in Tacheometry

| I Station Point | II Azimuth (Horizantal) angle | III Generating Number $h$ = staff intercept | IV Distance function $(Kh + c)$ | V Horizontal distance $D = (kh + c) \cos^2\theta$ | VI Vertical heigh $(D \times \cos\theta \sin\theta)$ | Reduced levels | |
|---|---|---|---|---|---|---|---|
| | | | | | | VII of Inst. | VIII of Point |

## APPENDIX 5

### Tabular Form for Recording Plane-table Telescopic Alidade Survey Data (with Beaman arc).

| Sl. No. | Inst. Station | Ht. of Inst. (HI) | Rod Station | Rod Inter (RI) | Hor. Beam (HR) | Hor. Distance (HR×RI) | Vert. Beam (VR) | Cal. Diff. in Elev. | Net Difference in Elevation | R. L. of Instrument | R.L. of Rod Station |
|---|---|---|---|---|---|---|---|---|---|---|---|
| (1) | (2) | (3) | (4) | (5) | (6) | (7) | (8) | (9) | (10) | (10) | (12) |
| 1 | | | | | | | | | | | |
| 2 | | | | | | | | | | | |
| 3 | | | | | | | | | | | |

## APPENDIX 5a

### Tabular Form for Reporting Triangulation Data (Telescopic Alidade).

| Sl. No. | Inst. Station. | Ht. of Inst. | Field Station Int. | Rod Int. | Rod Read | Vert. Angle | Measured Hor. Dist. | Cal. Diff. in Elevation | Net Diff. in Elevation. | R.L. of Inst. Station | R.L. of Field Station |
|---|---|---|---|---|---|---|---|---|---|---|---|
| (1) | (2) | (3) | (4) | (5) | (6) | (7) | (8) | (9) | (10) | (11) | (12) |
| 1 | | | | | | | | | | | |
| 2 | | | | | | | | | | | |
| 3 | | | | | | | | | | | |

## BORE HOLE SURVEYING METHODS

**Introduction :** Apart from logging bore holes by the various techniques, such as (a) lithological logs obtained by inspection and examination of cores and cuttings, (b) Caliper logs, (c) geophysical data e.g. S.P. and resistivity, (d) γ-ray data, (e) dip meter reading etc., the geologist is often required to know some essentials of the methods used in bore hole surveying.

Deep bore holes, and sometimes holes exceeding 1000 ft. are liable to deflection, either due to faulty drilling techniques or due to certain geological factors. Some of the chief causes for deviation of the diamond drill bore holes are, (a) the presence of rocks of differential hardness dipping a steep angles. In such a case, a steeply inclined bore hole, after passing through softer rocks, is deflected on touching a relatively very hard formation, and tends to follow the dip of that formation, *e.g.*, an inclined hole after passing through shales or phyllites may be deflected if it touches a steeply dipping band or hard quartzite; (b) jointing and fracturing present in the rocks, also cause deviation; (c) sudden variations and excessive pressures, applied during drilling, either due to increased rate of feeding or increased rock resistance, are another common cause. In general calyx holes, using shot bit, are more liable to deflection than diamond drill holes. A deflection, in the case of a shot drill hole, causes greater abrasion of the walls and dilution of the sample since the shot, leave the bottom of the hole and comes to the sides. The same is true of diamond drill holes to a lesser extent. A deviation will cause greater abrasion of the walls when the reamer is used; and the bit and reamer tend to cut downwards more than forwards. The result will be that the hole is more elliptical than circular; when the downward cutting is considerable. It may also be necessary, at times, to deflect a bore hole from a certain depth. In both cases, the survey of the bore hole becomes essential so as to know what exactly is happening at depth. Under these circumstances, the driller may not be aware of the deflection. Sometimes deflection, of a bore hole, is necessary when additional geological data is to be obtained. In such cases, the bore hole is deflected with a purpose. Bore holes, when once deflected may take the form of large arcs, or spirals with a large pitch. The deflection may be to either side, upwards or downwards or a combination of all these possibilities. Surveying of bore holes, thus becomes essential if the geological data obtained by drilling is to be assembled in the correct perspective, or correlated with data obtained by drilling in the adjacent areas. It is in fact, a method of underground surveying, under special circumstances, when the geologist has no direct access, to the traverse.

In the methods of bore hole surveying, so devised, distances (i.e. depths either vertical or inclined distances) are measured by means of the drill rods. The inclination of the bore hole, at the point of measurement, is indicated by a plumb bob arrangement, either directly or indirectly. A magnetic compass or a gyro compass is employed for obtaining the bearing of the bore hole at the point of measurement. All data, i.e., the inclination and bearing of the bore holes are recorded, either directly or photographically.

## Inclinometer

The inclinometer is a simple apparatus for measuring the angle of the hole, at various depths. This consists essentially of a glass cylinder, protected by a metallic case. The cylinder is partially filled with a solution of HF of suitable concentration, generally 5%–10% and the inclinometer assembly is lowered into the bore hole to the desired depth, where it is kept sufficiently long to enable the HF to etch the glass container. The inclination of the bore hole is obtained by measuring the angle, which the itched ring left by meniscus of HF, makes with the axis of the glass cylinder.

It is safer to use the instrument in which the principle of electrolytic deposition of copper on iron is utilised. If a copper sulphate solution is placed in a iron cylinder and lowered into the drill hole, copper is deposited on the iron, and the reading of the angle, made by the plane of the meniscus, to the axis of the cylinder gives the dip. Instead of a cylinder of iron two iron sheets, welded at right angles to each other, can be fitted into a suitable glass cylinder, filled with $CuSO_4$ solution and lowered into the drill hole. These fins of iron plates show the deposition of copper. From the inclination of marks left by the copper deposit, the dip or angle of the hole can be worked out.

In the earlier methods, the inclination of the bore hole was obtained by freezing a gel contained in a glass cylinder, in which a plumb bob was suspended.

## Photographic method

In the photographic methods, the apparatus (Fig. 1.45) used comprises a metallic cylinder, (C) which is attached to the end of the drill rods (T). At the bottom of the cylinder, is a spring loaded marking device or scriber (S) which makes a deep linear mark on the top of the stub at bottom of the hole (before drilling). Within the cylindrical housing, as a sensitised disc of photographic paper (P), over which is a free moving magnetic needle attache to a graduated transparent scale (D). Immediately above the centre of the ransparent disc is a plumb bob (B). The centre of the transparent disc a d the point of suspension of the plumb bob are in alignment when the e uipment is vertical. On the same alignment is, a collimation hole, (H) above which is a small battery operated electric lamp (L). The shadows of ae magnetic needle, and the plumb line fall on the photographic paper, aud thus the direction of the deviation and amount are recorded. A ontinuous record is possible, where the disc of photographic paper is replaced by a roll of paper moved by a electric motor.

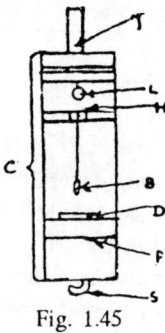

Fig. 1.45

In modern bore hold surveying instruments, a gyro-compass is employed instead of a magnetic needle. This makes the equipment suitable for work in areas where magnetic influence is very pronounced.

The gyroscope has been in use, for well over 60 decades, inmarine navigation and is also used for direction finding in aircrafts. The gyro compass is essentially gyroscope or spinning disc. In effect, the gyro is very much like the spinning top. Just as the upper end of the top's axis, while spinning steadily and after standing vertically, shows a circular movement about of the vertical axis, the gyro also behaves similarly. This movement of the axis is akin to that of the earth's axis and is known as precession.

### Gyro Compass

Latter modifications of the equipment have utilised the photographic devices increasingly. The gyro compass was first available for use, only in the large diameter bore holes, but now models for use in holes upto about 2 diameter are available. The instrument is housed in a cylindrical shell with a diameter of $1\frac{3}{4}$". Once the gyro compass is set in motion, and corrected for precession, the axis maintains a constant direction irrespective of the movement of the housing. Thus the gyro device is proof, against effects of deflection as well as that caused by extraneous magnetism, and provides the basis of directional correlation in azimuth. A plumb bob can be included in the assemblage, for indicating the inclination of the hole. Continuous photographic recording, of the position of the gyro axis and the plumb bob simultaneously, provides a very clear picture of the deflections and deviations (Fig. 1.46)

Gyros can be classified under two main types:

(a) Free gyro, and

(b) Compensated or counter balanced gyro.

(a) The free gyro Fig 1.46 comprise a free spinning wheel (W) mounted on the axis 3–3. The arrangement is mounted on a universal joint as shown in Fig 1.46 or gimbals or in a fluo-liquid suspension. The last principle of suspension resembles the mounting of the human eye ball, within the

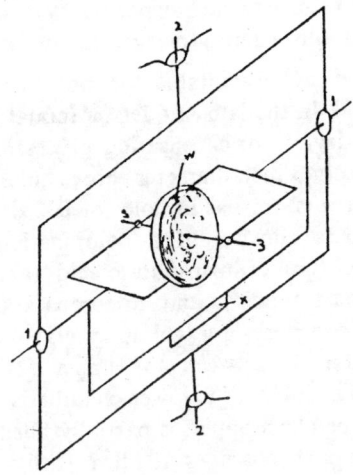

Fig. 1.46

eye socket. Thus the spinning wheel (W) can move in two mutually perpendicular directions 1–1 and 2–2.

When the wheel (W) is made to spin on 3–3 axis, at a high velocity, the entire set up shows a slow rotation about 2–2 axis, called the gyroscopic precession of the axis, and the movement has been attributed to conservation of angular momentum.

(b) In the counter balanced gyroscopic set up, this precession is either arrested or compensated either by applying a torque, or by a small electric motor which induces a controlling torque, on the rotating spinning wheel (W). This type of gyro is also called the north seeking type. as it compensates the rotation and precession, in the manner described below :

If the gyro axle 3—3 is oriented in a random position, making an angle with the true north, then the torque is impressed on the spinning wheel (W) about 3—3. This is mainly caused by the rotation of the earth from west to east, which makes the wheel (W) tilt on 1—1. This tilt initiates a precession. If, however, the axle 3—3 can be maintained in a horizontal position, by the use of a pendulum (X) attached to the 1—1 axis, then precession is about the 2—2 axis. The precession stops once the 3—3 axis aligns itself with the N—S (north-south) direction.

In the earlier bore hole surveying instruments, a free gyroscope, together with a plumb bob type angle (inclination) measuring unit, were employed, to give the direction and inclination of the bore hole, at various points. The measurements were, made continuously on a 16 m.m. camera film. The entire equipment was lowered into the bore hole, by wire line, and the time taken for recording each exposure was only ½ minute. The spool contained enough film length, for even the deepest bore hole.

The above equipment was suitable for bore holes of diameters exceeding 6⅝ inch (OD). In the later models, a miniature gyro, employing an electro-mechanical device, for expensating precession, is utilised. The instrument can be used in 3 inch diameter bore holes, and very soon the manufacture of instruments, for use in holes of BX size can be expected.

The Masas compass consists of a combination of both the inclinometer, using a dilute solution of hydrofluoric acid (about 12 parts of water to 1 part of acid), and a compass needle. The instrument comprises a 6 inch long glass tube, with a median partition (1) between the upper (2) and a lower portion (3) Fig.1.47. The lower chamber which is fitted with a gutta-percha stopper, is partially filled with H.F. solution. The upper chamber is filled with a solution of 1 2/3% gelatine. In the chamber is a combination which consists of a small magnetic compass (b) fitted with a guard ring, (c) suspended from a float (a). Tha combination is placed in a bronze casing, which is attached to the string of rods and lowered into the bore hole. Sufficient time is allowed, for the gelatine to set, and for the acid to etch the glass. The setting time for the gelatine varies from 20 minutes to about 50 minutes. The relation between the level (surface) of the galatine, to the axis of the compass tube, gives the angle of inclination. The orientation of the tube axis, in relation to the compass

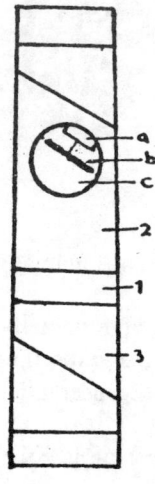

Fig. i 47

(magnetic) needle, gives the bearing of the inclined hole.

The Briggs Clinophone Fig.1.48 is an instrument which measures deflections with a high order of accuracy. The instrument consists of two portions, I and II, the former is lowered into the bore hole while the latter remains on the surface. The first portion Fig.1.48 I$a$ consists of a brass (cylindrical) body (1) with a bakelite cup at the bottom (2). In the cup are four terminals $N_1W_1S_1E_1$ Fig.1.48 I. A plumb bob (3) in suspended from a insulate plug (5) by means of the wire (4) which terminates in point $C_1$. The terminals in the cup are connected to connectors N, W, S, and E in the insulator plug. (5), and $C_1$ to $C_2$.

The portion II (on the surface) Fig. 1.48 II consists of an insulator cup in which there are four corresponding terminals $N_2$, $W_2$, $S_2$ and $E_2$ immersed in an electrolyte as well as additional point $C_2$. A 5-wire lead connects the corresponding terminals of II to I (at the connectors attached to the insulator (5) $N_1$ and $N_J$, $S_1$ and $S_2$, $W_1$ and $W_2$, $E_1$ and $E_2$, and $C_1$ and $C_2$).

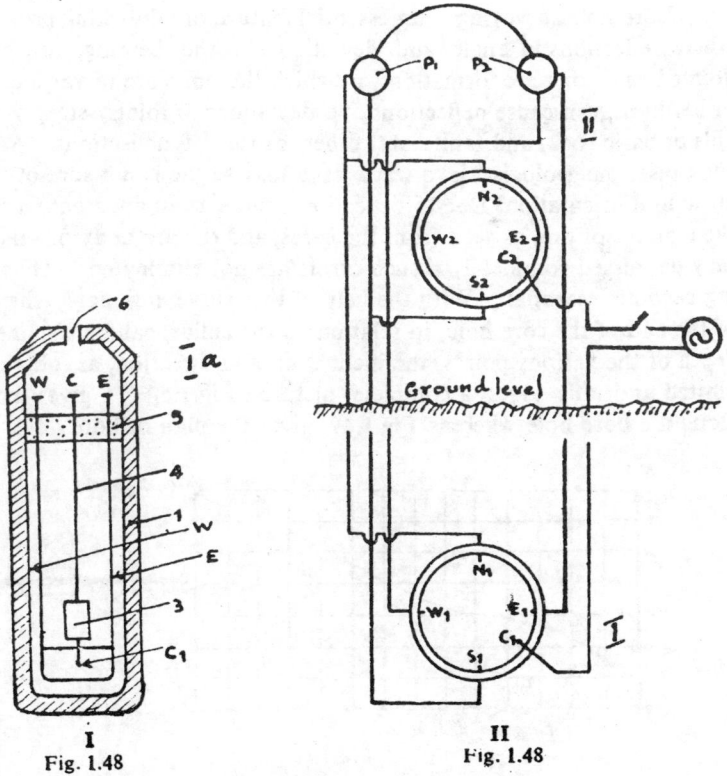

I
Fig. 1.48

II
Fig. 1.48

$N_1$, $N_2$, $S_1$, $S_2$ are connected to the ear phones $P_1$ while $E_1$, $E_2$, $W_1$, $W_2$ are connected to the ear phone $P_2$. The moving terminals $C_1$ and $C_2$ are connected to an alternating frequency generator (of audio frequency range).

Since the N—S and E—W terminals are separately connected to the ear phones $P_1$ and $P_2$, they form, as it were, reference coordinate for the location of the moving terminal $C_1$ attached the plumb bob (3). In operating the instrument, sufficient time is allowed for the plumb bob to come to rest. The alternating frequency generator is started up, and the switch P is closed when the operator wears the ear phones. The terminal $C_2$ in Fig. 1.48 II *i.e.*, upper portion of the instrument, is moved from place to place, within the cup, and the intensity of the tone in the ear phone is noted. It is only at one point that the volume of the note, generated by both ear phones, is identical. The location of $C_2$ in respect of $N_2W_2S_2E_2$, in portion (II) will be identical to that of $C_1$ in respect of $N_1W_1S_1E_1$ in the portion (I) of the instrument. Thus the inclination and the bearing of the hole can be made out.

Bore hole surveying is an essential feature of all drilling programmes, where deflections in angle and deviation, in the bearing, are expected. Jointed rock, or rock formations, in which the units are of varying degrees of hardness, can cause deflections and deviations. Joints, steeply dipping sills of basic rock, and faults are other causes of deflections. Whatever, the cause, the geologist is always at a loss as he is not sure of the exact structural orientation. Due to deflections, there is an apparent variation in the amount of dip, as seen from the cores, and the ore body or oil reservoir may be missed completely if such errors are not eliminated. Thus surveying becomes essential. With the help of the surveying data, the location of the end of the core hole, in relation to the collar, can be obtained. The depth of the various points, the inclination and direction, as obtained, are plotted and this gives a complete picture. Fig. 1.50    gives the section along the bore hole whereas Fig. 1.49 gives the plan of bore hole.

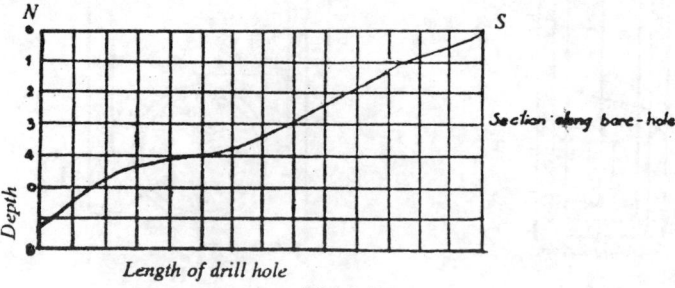

Fig. 1.49

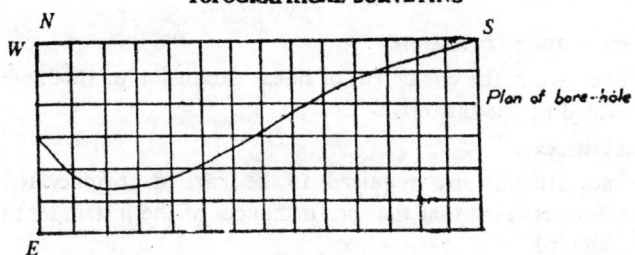

Fig. 1.50

## MISCELLANEOUS CALCULATIONS AND ADJUSTMENTS

Corrections to be applied to distance measurements with steel tape for base line measurements.

(a) Slope correction.

If Sc = Slope correction

L = Slope length

h = Difference in level between the two ends

D = Horizontal distance

the $Sc = D - L + - \dfrac{h^2}{2L} - \dfrac{h^4}{8L^3} - \dfrac{h^6}{16L^5} - \dfrac{5h^8}{128L^7}$

(Tables are available for this).

(b) Tension correction.

If the additional tension applied is T lbs.

S = Cross sectional area in sq. in.

L = Length of tape in ft.

C = Increase of length due to extension of tape in ft.

E = Young's modulus

$C = \dfrac{L \times T}{E \times S}$; Young's mod. of steel = 28,000,000.

(c) Sag correction.

Sa = Sag correction

C = Actual length

L = Sagged length

h = Sag

t = Tangential tension at the bottom of the sag.

W = Weight per foot lengh of tape.

$Sa = - \dfrac{W^2 L^2 C}{24t^2} = \pm \dfrac{W^2 L^3}{24t^2}$

Fig 1.51

(d) Temperature correction.

This depends on the coefficient of linear expansion of steel, which is normally provided by the makers.

**Permanent adjustments**

These adjustments are required to be carried at intervals, after a period of use for verifying that the performance of the instrument is upto the desired standard.

**A. Dumpy level :** After carrying out the temporary adjustments, it is essential to check ;

1. whether the axis of the bubble tubes, lies in a plane, perpendicular to the vertical axis of the instrument,

2. whether the horizontal cross hair is really horizontal,

and 3. whether the line of sight is parallel to the axis of the bubble tube.

For the purpose of carrying out the various adjustments,

1. The instrument is set up firmly on a ground which is nearly level, and temporary adjustment is carried out. If the long bubble, on the telescope, does not remain in the centre of its run, on a rotation of 180°, between two foot screws, then half the error is corrected by raising or lowering the capstan nut, at one end of the bubble tube and the remaining half adjusted by moving the foot screws. This procedure is repeated until the bubble does not move from the centre.

The telescope is now turned through 90° and brought over the third foot screw. If the bubble leaves the centre of the tube, half the discrepancy is adjusted by the capstan nut, at the end of the tube, and the remainder corrected by moving the single foot screw. It may be necessary to repeat the operations, until horizontality is achieved, when the bubble remains in the central position during a complete rotation through 360° of the telescope.

2. In order to make the horizontal cross wire actually horizontal, any point, within the field of view and lying on the cross hair is selected. The telescope is moved on the horizontal axis and the movement of the point, in relation to the cross hair, is observed. In case, the point is found to move away from the cross hair during rotation, the cross hair is not horizontal. The four set screws holding the ring and containing the cross hairs are loosened and the ring is rotated in the required direction, so as to eleminate the observed "tilt", and the point under observation moves on the cross wire, when the telescope is moved.

3. In this adjustment, the line of sight is made parallel to the axis of the bubble tube, so that when the instrument is levelled, the line of sight will lie on a horizontal plane.

For this purpose, two stations A and B are pegged down on a nearly level stretch, about 300 ft. apart. The instrument is set up in a central position, between A and B and temporary adjustments completed. A levelling staff is set at A and the reading taken. Similarly, the reading on a staff placed at B is also noted. Since A and B are equidistant, from the instrument and exactly 180° apart, the difference in level can be readily obtained as seen from the Fig.1.52. Thus the readings obtained indicate the relative levels of A and B.

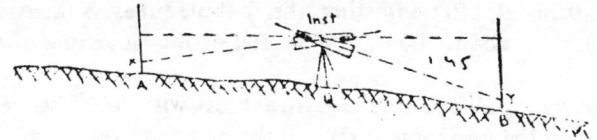

Fig. 1.52

Now the instrument is moved to a point P in alignment with A and B, Fig.1.52 (a) The staff at A is sighted, and its reading ($X_1$) noted. Since the difference in level between A and B is known the corresponding reading on the staff at ($Y_1$) can be calculated.

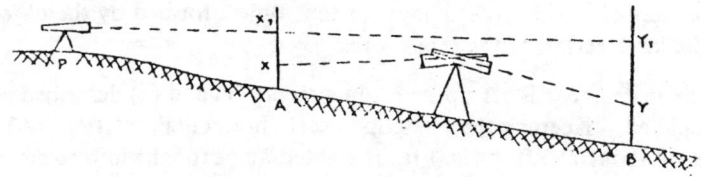

Fig. 1.52(a)

The cross hair ring is, now, so adjusted that the intersection coincides with $X_1$ on the staff at A, and $Y_1$ on the staff at B.

**Example**

Instrument on station Q.
Readings on the staff at A = 5.32
Readings on the staff at B = 7.54
Difference in elevation between A and B = 2.22
Instrument at station P.
If the reading on the staff at A = 8.42
Then the reading on the staff at B should be 8.42+2.22 = 10.64.

**B. Theodolite** : The accuracy of the instrument depends on its condition also, hence regular checking is necessary for rectifying any defects.

1. To make the axis of the plate bubbles horizontal, *i. e.*, lie on a plane normal to the vertical axis of the instrument.

The adjustment is identical, to that described for the dumpy level (adjustment 1) ; except that the adjustment is carried out in respect of both the plate bubble tubes, which are arranged at right angles to each other.

2. (a) To make the vertical cross hair vertical to the horizontal axis of the instrument. This is an important adjustment since the theodolite is an angle measuring instrument.

(b) To make the horizontal cross hair really horizontal. This adjustment is also identical to that carried out on the dumpy level (adjustment 2), (Figs.1.51 and 1.52)

The instrument, after adjusting the bubble tubes, is focussed on any convenient mark at about 100 or 200 ft. away, for adjusting the vertical cross hair.

After focussing a point on the vertical cross wire the telescope is moved up and down, on the horizontal axis. If the point moves away from the vertical cross wire, then the ring holding the cross wire is adjusted, so that coincidence is maintained between the point sighted, and the cross wire at all positions of the telescope.

3. To make the axis of collimation, coincident with the optical axis of the telescope, so that the image of the object formed by the objective lies at the intersection of the cross wires.

The instrument is set up and adjustments (1) and (2) described above are completed. Keeping the telescope nearly horizontal, station (A), at a distance of about 250 ft. or 200 ft., is sighted and brought into focus, without parallax. Both body and vernier plates are clamped. The telescope is transitted and made more or less horizontal. Any station (B), at a distance of about 200 ft., is fixed in the alignment of the cross wires. The body clamp is released and station A sighted back with body clamp and tangent. The telescope is again transitted. If the instrument is in adjustment, station B should again lie on the cross wires, if not, another station (C) is fixed at about the same distance from the instrument, say about 200 ft. away. The distance between the two stations B and C is measured and at $\frac{1}{4}$ this distance station $C_1$ is located ($CC_1 = \frac{1}{4} BC$). The screws, holding the cross wire ring, are loosened and the vertical cross wire is moved to coincide with C. The procedure is repeated, and if on transitting station B falls on the cross hair, the instrument is in adjustment.

4. To make the horizontal axis of the telescope truly horizontal. The instrument is set on a nearly plain ground and adjusted as described above.

A point (Q) at a height of about 50 ft. above the instrument is sighted say e. g., the top of a building. The telescope is lowered and the

Fig. 1.53

point is transferred to the ground $(P_1)$. The instrument is transitted and the point Q is sighted again,· and the telescope lowered to intersect the ground. If the point $(P_2)$ is not coincident with $P_1$ the instrument requires adjustment.

The adjustment is affected by means of the screws, found on the standard, just below the bearing of the horizontal axis. The screw is raised or lowered so that the vertical cross wire is moved from $P_2$ to $X_2$.

The horizontal distance $P_1P_2$ is measured and $\frac{1}{4}$ $P_1P_2$ is taken from $P_2$ so as to fix point $X_2$.

The procedure is repeated, so that on transitting the point P, when sighted and transferred to the ground, is constant at Q'.

5. The bubble tube fixed to the vertical circle is now adjusted, as indicated for the adjustment of the longer bubble tube (under dumpy level). Now, when the reading on the vertical circle is "O", then the bubble tube on the telescope should be in the middle of its run.

## NOTES ON INSTRUMENTS AND ACCESSORIES

### (1) The Automatic Level

Sketch showing the principle of an automatic level.

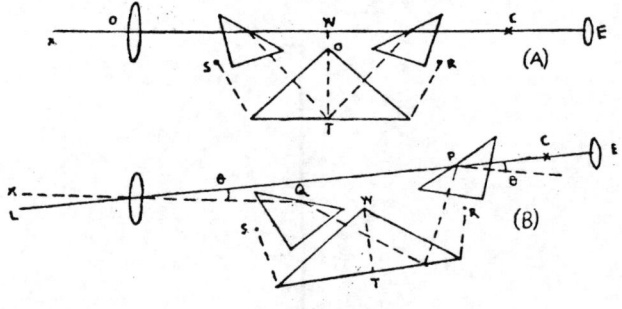

Fig. 1.54

In the self adjusting level, the coarser adjustment is carried out by means of the 3 foot screws and the bull's eye or circular bubble. When this is completed, the line of sight remains horizontal, even though the axis of the telescope is tilted. Figure (1.54) above illustrates the principle.

O is the objective.          W is a pendulum with counter weight.
E is the eye piece.          C is the position of the cross wire.
P, Q and T are reflecting prisms.

Prisms P, Q and T are enclosed in a darkened box, as also the pendulum. The prism T is suspended by wires of equal lengths, from the box at points S and R, while prisms P and Q are adjusted so as to be in the axis of the telescope.

When the telescope is horizontal, the line of sight coincides with the axis of the instrument, as shown in Fig. 1.54 A, and the line connecting the points of suspension is horizontal, i. e., they lie on a line parallel to OE, the line of collimation.

When the telescope is tilted by $\theta°$, Fig. 1.54 B the two prisms P and Q move together through angle $\theta$, while prism T moves relatively but also through $\theta$. Since the reflecting surface of T is parallel to the points of suspension R and S, which are fixed to the dark box lying within the telescope tube, any movement of the telescope produces the same displacement of the reflecting surface T.

Thus in Fig. 1.54 B if the axis of the telescope is tilted by $\theta°$, the reflecting surface T tilts by $\theta°$ so that the line of sight OX remains horizontal. The angle between the line of sight XO, and the axis of the telescope is also $\theta$.

(2) **Levelling :** Sensitivity of the bubble tube is an important factor in levelling.

In order to determine the sensitivity of the bubble the following procedure is adopted.

A levelling staff is held at any convenient distance.

The instrument is levelled such that the bubble is 3 to 4 divisions away from the centre. The reading of this end of the bubble is noted, and the reading on the staff is recorded.

The instrument is again adjusted in such a manner that the bubble is made out of centre by 3 or 4 divisions in the opposite direction. The staff reading is recorded again.

$$\text{Sensitivity} = S = \frac{1}{0.00000485} \frac{d}{Ln}$$

where    d = difference in staff readings.

L = distance between centre of the instrument and staff.

n = total number of divisions through which the bubble has moved.

S = sensitivity in seconds per bubble division.

**Note :**—Difference between the 3 foot screw and 4 foot screw levels is, that for any setting of the stand or tripod the instrument height is not changed by adjustment, in the case of a 4 foot screw instrument. This is not so with the 3 foot screw instrument.

(3) Optical micrometer or telemeter comprises a thick circular disc of glass in which both the surfaces are optically plane and are parallel to each other (*i.e.*, plano-parallel).

The glass plate is mounted diametrically within a cylindrical holder, on which it is capable of being rotated by means of a graduated drum or micrometer (Fig. 1.55 A).

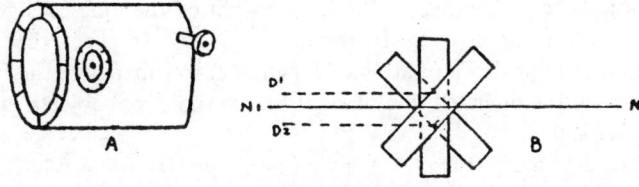

Fig. 1.55

When the micrometer is attached to the telescope, the surfaces of glass plate are normal to and, intercept the line of sight. Since the surfaces of the plate are parallel, the line of sight is, deviated parallel to itself according to the laws of refraction as shown in Fig. 1.55 B.

$N_1N$ is the normal ray, $D_1$ and $D_2$ are deviated rays.

(4) **Horizontal stadia:** In the instrument, so far described the stadia hairs (or wires) are horizontal and are placed across the vertical cross wire. Instruments have been made, where vertical stadia

hairs are placed on the horizontal cross wire. When using such instruments for stadia measurements. (1) The staff is held in a horizontal position. (2) It is made level. (3) It is at right angles to the line of sight. (4)· It is fixed at the same height as that of the instrument above ground level at the point.

It will be seen that the stand for the staff is of a special design. In the tripod stand, the rod carrying the staff forms one of the legs and this is planted over the station point. This vertical leg is also used for intersection and for obtaining the angles in azimuth, as well as vertical angles, as no staff is placed over the station. Levelling of the horizontal staff is carried out by means of the level tube or bubble provided.

(5) **Distance wedge:** The distance wedge is an attachement (Fig. 1.56), which enables horizontal stadia measurements. It consists of two thin wedges of glass, making an achromatic combination. The wedge covers a narrow strip in the field of view of the objective. When viewing

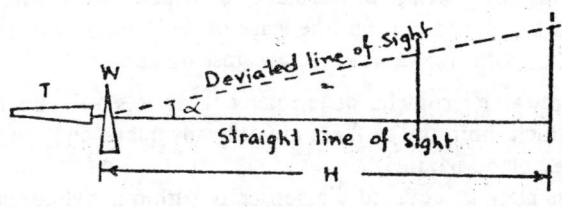

Fig. 1.56

through the telescope it is seen that, due to the presence of the wedge, the images of objects found, either to the left or right of the line of sight, are super imposed on those lying in the field of view. In effect, the wedge serves to (1) deflect or deviate the line of sight, (2) to super-pose the images of the objects on the deviated line of sight on to those of objects on the straight line of sight.

Viewed in azimuth (Fig. 1.51) any point Y appears along side X. The glass wedge is so constructed that, if H is the distance from the wedge to an object X and Y is an object (at distance D from X) the deviated image of Y falls on X.

In most instruments the ratio

$$D/H = \frac{1}{100}$$

$$H = D \times 100.$$

The distance wedge is used with a special staff, which has a vernier attached to it. When sighting, through the telescope on the station (*i.e.*, the vertical leg of the staff stand) and the staff, it will be seen that one end of the staff is superposed on the other end, due to deviation caused by the wedge. Thus for, *e.g.*, 1.00m may appear against 7.00m. or 2.00m may appear against 6.00m etc. etc. In the first case the stadia reading is 6.00m. In the 2nd case the stadia reading is 4.00m. etc. Thus the horizontal distances are 600m and 400m. respectively. The vernier is used to read up 0.01m.

(5a) In some distance wedges a plane parallel glass plate is provided. The plate is placed in front of the wedge and can be rotated about a vertical axis. The effect of this rotation, is to cause the deviated ray (passing through the wedge) to move parallel to itself as shown in Fig. 1.57.

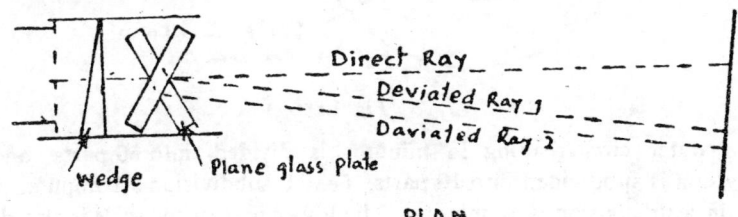

PLAN

Fig. 1.57

The formula for horizontal distance is
$$H = 100S \, Cos\theta$$
The formula for vertical height is
$$V = 100 \, Sin \, \theta$$
where,          S = Stadia reading
                $\theta$ = is the angle of elevation or depression of the telescope (*i.e.* vertical arc reading).

The distance wedge is more accurate than normal stadia measurement as

(1) both ends of the staff are read simultaneously,

(2) neither cosine corrections nor addition constants are needed as the staff is held horizontal.

(6) **Notes on Watts Microptic Theodolite :** The auxiliary tube on side is the reading microscope. The field of view presents, both the horizontal scale reading designated by "H", and the vertical circle reading designated "V".

The upper window, gives directly the main scale graduations which are at 10 minutes intervals. The lower window shows 2 sets of readings, (a) the scale of the top read directly in minutes and can be approximated to the nearest minute while (b) the scale below reads directly in seconds. Thus the reading indicated in Fig. 1.58 is 72°06'30".

The vertical scale is also similarly divided.

The micrometer scale seen in the lower window is calibrated thus :

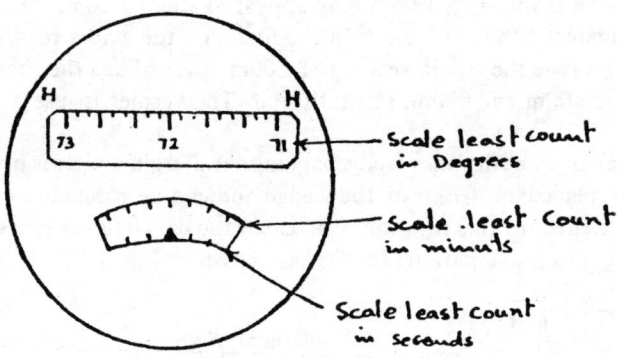

Fig. 1.58

the outer circle reading in minutes is divided into 60 parts, and each division is subdivided into 10 parts, *i e.*, 1 subdivision=1 minute, as the main scale division is 10 minutes. The lower part of the scale is also divided into 60 parts, *i.e.*, reading directly in 1″.

The micrometer is so adjusted, that when the reading of the scales is "0" minutes and seconds, *i.e.*, the readings in the lower arcuate scale, are zero and the reading on the upper main scale is a complete number of 10 minntes, *e.g.*, 28°10′(0″, 28°20′00″, 28°30′00″ etc. Consequently, any reading other than "zero" on the minutes and seconds scales indicates a departure of main scale from coincidence with the indicator mark. The micrometer scale is actuated by a knob, the movement of which causes the main scale reading to be altered so as to coincide with the nearest 10′ reading. This operation also involves a change in the minute and seconds scales from which the angular value can now be read off in minutes and seconds.

(7) **Self reducing tacheometers** : In these instruments the usual stadia wires are replaced by a set of tacheometric curves. When looking through the telescope, the tacheometric curves are seen to occupy one half

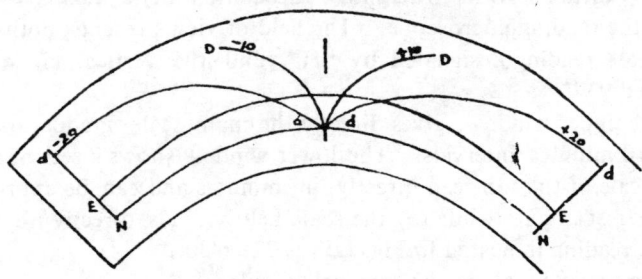

Fig. 1.59

of the field of view. The curves are etched on a transparent glass laminate, housed in the vertical supports, in the same manner as the vertical arc (Fig. 1.59).

The tacheometric curves comprise the reference curve (N-curve), which constitutes the "zero", for both the horizontal measurements and the vertical measurements. Alongside N-curve, at some distance, is the E-curve on which measurements for horizontal distances are made. The curves for measuring vertical height, consist of 2 pairs of companion curves, designated +D or —D for the lower range (*i.e.*,±10 factor) and +d or —d for the higher range (*i.e.* ±20 factor).

The N-curve which is concentrically placed, with reference to the horizontal axis of the telescope, is so adjusted that it represents the horizontal reference line, and forms a tangential contact with the horizontal cross wire.

The curve E can also be identified, (in the field of view) as being nearly concentric to curve N, while the curves ±D and ±d bear a marked cross cutting relationship to both the E and N curves. (Fig. 1.59).

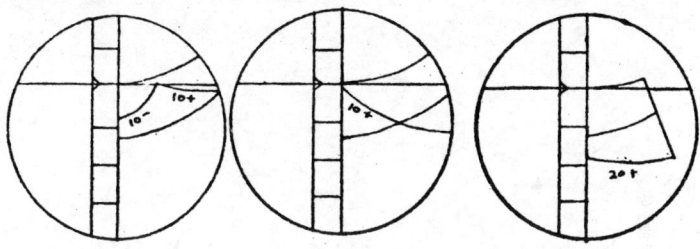

Fig. 1.60

The basis for working out the E, D and d curves in relation to the N curve is as follows :

Intercept between N and E$=i \cos^2\theta$
       ,,       ,,    N and D$=i \sin\theta \cos\theta$
       ,,       ,,    N and d$=i \sin\theta \cos\theta$

where $i$ is the distance of separation of the stadia hairs

when                        $f/i=100$
or                          $i=f/100$
                        $\theta=$ vertical angle.

**Note :** (*a*) the N curve is tangential to the horizontal cross wire,

        (*b*) when an angle of depression is indicated, only the negative counterpart of the d or D curves comes against the staff, (Fig.1.60) .

(c) when an angle of elevation is indicated, the positive counterparts of the same curves abuts against the staff.

(d) The N curve is tangential to the horizontal cross wire.

(e) Intercept on the staff subtended by the N and E curves gives the horizontal distance factor.

(f) The intercept between N curve and D or d curve gives the vertical height.

# PART II

# PHOTOGRAMMETRY

# INTRODUCTION

For details of the various aspects of air surveying, the reader will have to consult more exhaustive treatises on the subject. It is proposed to give only a brief outline of the essential features, which will enable the understanding of the subject with a view to aid photo interpretation and map preparation.

This chapter, includes a short digression in photogeology and the means used in the interpretations of air photos. This aspect has developed into a separate science and it may be stated without any reservations that photo geology is a great boon to the geologist traversing difficult virgin terrain. The texture and grain of the photograph, often reveal the geology, and regional structures are well displayed. With a little practice the estimation of dips can also be made—in spite of exaggeration caused by the eye base, to air base ratios and other factors.

During recent years aerial surveying has been developed into a science. Aerial surveying can be divided into three parts :

1. Air photography—relates to technique and instrumentation in photography, such as—

   (a) The equipment employed, e.g. air craft, cameras, films, control equipment etc. and the limitations.

   (b) Aerial photography— techniques of photographing, verticals, obliques, etc., defects and the limitations.

   (c) Air surveys and purposes of air survey—aircraft control, air craft positioning, laying out of surveys etc.

2. Photogrammetry—relates to the process by which accurate measurements of distance, height or depth are obtained from air photographs, leading to the compilation of contour maps. This includes (a) stereoscopy, (b) methods of planimetric plotting, and (c) methods of contouring.

3. Interpretation of air photographs, relates to the specialised study of the details present on air photographs in relation to any one branch of science, or project, e.g., geology, forestry, or road alignment or irrigation projects etc.

Aerial surveying and photogrammetry are valuable in geological surveying. Vertical photos, where good exposures are available, give a complete picture of the topography, soils, drainage, variation in rock types, and also the nature and type of vegetation. Very often, geological structure is very clear. Hard rock bands stand out prominently, due to differential weathering and with a little experience, and correlation with ground measurements, the dips can be estimated. Dykes and intrusives are often distinctive, due to greater resistance to weathering, characteristic relief and vegetation. Faults can also be located, either by the drainage pattern or by displacement of contacts between formations of rock types. Thus verticals, where exposures are good, can save considerable time and energy for the field geologist, who, with experience, will be in a position to delineate boundries of rock types, dips and strike, structures, such as folds, faults etc. with considerable facility. Ground locations can be made for checking up as and when required. Ground traverses will also be required for the collection of rock specimens, samples for analyses, detailed work on mineral deposits, etc.

## AIR PHOTOGRAPHY

### (a) Equipment

Aerial cameras are of two types, (1) single lens, ordinary or wide angle and (2) multi lens.

1. The single lens camera is very simple and robust in construction, and free from optical defects. It is generally equipped to take roll films, and can accommodate about 200 frames or more. The operation of the camera is generally semi-automatic or automatic. The lens should be free from any type of aberration and the range of focal length is normally between 6″ and 24″. The common apertures are f/4 to f/7, but in certain Zeiss wide angle types, the aperture employed is f/6.3 or as in the Ross ultra wide angle type, f/5.5. The shutter of the camera is so designed as to eliminate, as far as possible, the distortion of the image, on the photograph. The focal plane shutter is in vogue, but the improved cameras, fitted with the rotating disc type shutter, have practically eliminated distortion effects, and this type of shutter is gradually finding favour. Shutter speeds vary between 1/50 to 1/300 sec., while the sizes of air photographs are usually 7″ × 7″ or 9 × 6½″ or 8″ × 8″ or 18 cms. × 18 cms.

The Wild RC9 camera, with the ultra wide angle super Aviogon lens, with 88 mm (about 3.5 in.) focal length, gives 120° coverage along the diagonals of a 9″ × 9″ photo format, is a recent development. This camera has the advantage of greater coverage per photo, this being about three times of the 6″ lens at normal height of flight. Thus there are fewer photos and the cost is also less. By flying at lower altitudes (12,000 ft) the haze of the tropics can be avoided and clearer photos can be obtained. Infra red photography, sometimes lacks ground resolution, because of a lower flying height, but the lack of haze compensates this defect, besides, enhancing the exaggeration in the vertical scale, *i.e.*, increasing the air base to flying height ratio.

The Zeiss Pleogon is also a modern air camera, besides the Hilger and Watts "105".

2. Multi-lens cameras are used for taking vertical photographs as well as oblique photographs. The camera is made of several independent chambers, each equipped with a separate lens, shutter and film magazine, all of which are synchronously operated. There are various models. In the Zeiss, 4-lens type, there are four oblique chambers. In the Fairchild, 5-lens type, there is one central vertical chamber surrounded by 4 oblique chambers. In the Barr and Stroud 7-lens type there is one central vertical section and 6 oblique sections.

There are also slight differences in the mode of attachment of the camera to cock-pit of the aircraft, but whatever the type of fixture, the purpose is to compensate by differential motion any accidental tilt of the aircraft, during the photographic flight and for maintaining horizontality while taking vertical photos.

Panchromatic film is invariably used. Distortion proof nitrate film base in employed, but this is highly inflamable. However, distortion proof nitrate safety film, is also available. When panchromatic films are employed, in conjunction with suitable filters, good contrast and tone differentiation are obtainable. The introduction of the red sensitive emulsion, has widened the range of filters, and has also greatly improved the differentiation of details. Infra-red emulsions have been used to penetrate haze, but these are about four times slower than ordinary emulsions. Considerable research has been done to minimise distortion of sensitised paper and in the production of positives or prints. Material having an uniform differential shrinkage of only 0.015 percent is now available.

**Accessories and Auxilliary Equipment :** (1) Spirit level, (2) View finder, (3) Intervalometer—a divice which enables the taking of photographs automatically at calculated intervals of time, (4) Colour filters employed with various film emulsions for bringing out specific features, (5) Altimeter for recording heights and in the improved cameras statoscopes are attached.

In practice, provision is also made in the aircraft to accommodate geophysical equipment , for airomagnetic surveys or radiometric surveys, and also an air borne profile recorder.

The types of aircraft employed are D.C. 3(Dakota) D.H. Dove, Canso, Airocommander, Mosquito and Survey Prince, The normal flying altitudes are of the order of 2,000 ft. and occasionally 25,000 ft., while the photo scales, of cameras with lens of 6 in. focal, length, vary between 1 : 40,000 and 1 : 50,000.

(b) **Technique** : The pilot is provided with a flight map, on which individual flight lines, and prominent landmarks are indicated. He takes off when the weather conditions are favourable. Photographs, required for mapping purposes, are taken in "strips" along the lines of flight. In each strip, the smaller unit is the individual photograph. Each photograph bears two numbers—one refers to the "strip" and the other refers to the individual photograph. In order to facilitate mapping from vertical photographs, an overlap of about 60% is maintained between adjacent photographs, in the direction of flight and an overlap of about 30% is allowed between adjacent photographs of adjoining strips. This is known as "the side lap" (Fig. 2.1).

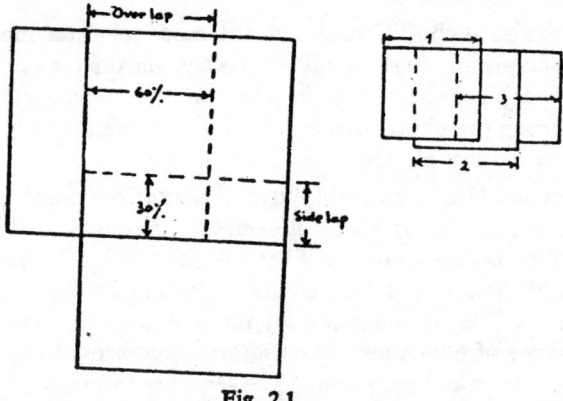

Fig. 2.1

These large overlaps (in vertical photos) are intended to accommodate defects in either, (a) navigation or (b) photography. The navigation defects, which tend to reduce the effective overlaps are mainly (1) "drift", casued when the aircraft is blown sideways by a cross wind, and (2) "crab" which results while attempting to compensate for the drift and the aircraft flies along a course with the nose pointing across the direction of motion. The defects due to photography are, (1) distortion of various types, especially towards the fringes of the photographs and (2) tilts. A tilted vertical photograph will have to be "rectified", but the overlap is reduced during rectification.

The introduction of the "automatic pilot" has, however, minimised tilts to within 1/4°. and navigational defects have also been overcome to a large measure by the introduction of radar ground control systems.

Range finders, using gallium arsenide diode lasers and ruby lasers have been developed with ranges of the order of a mile and an accuracy of

1 in 10⁴. The diode laser has the additional advantage of ease of modulation, compactness, and high pulse repetition rate.

## TYPES OF AIR PHOTOGRAPHS

**1. High Oblique Photo** : When an oblique or tilted photo shows the horizon, it is known as a high oblique. The axis of the camera may be inclined at angles even upto 20°. The oblique, unlike the vertical photo, readily shows up relief and it presents the entire view in a familiar perspective. The definition of the details in the photo, varies from foreground to background, but the high oblique covers much larger ground than the vertical. The scale of the oblique diminishes progressively from the foreground to the horizon. When the topography is flat, the high oblique can be used for determining distances. However, in an area of high relief the slopes towards the camera are exaggerated and the slopes away from the camera are decreased. High peaks, may at times obscure details behind them. Mapping from such obliques is discussed later on.

**2. Low Oblique** : A photo is termed low oblique when the horizon is not seen. The photo resembles the vertical photo, when the tilt is small, Perspective effects produced, are similar to those of high obliques. Photos of this type can be, however, rectified as in the case of tilted verticals.

**3.** Composites, are photographs taken with the multiple lens aerial cameras. They consist of one vertical, with two or more obliques. These are, however, rectified by means of special printers.

**4. Verticals** : Vertical photos are taken with the axis of the camera in the vertical position. These are most commonly used in air survey for map making.

**Mosaic** : A mosaic is an assembly of vertical photos, of an area, which enables the reconstruction of the topographic and other details, and simulates a topographic map. Mosaics may, either be of the controlled type or of the uncontrolled type. In the former type, the individual photos are corrected for distortion due to relief or tilt, or variation in air base (*i.e.* scale). The uncontrolled mosaic is prepared from uncorrected photos.

## SCALES

**1. Large Scales :** 1/1000 to 1/5000 and occasionally 1/500 are normal map fractions, employed for plans used in engineering investigations, and for mineral prospecting. Where accurate height determinations are required, say within 2 or 3 feet, ground surveys are to be employed.

**2. Medium Scales :** 1/5000 to 1/30,000 are used in reconnaissance surveys, engineering projects and mineral surveys. Good contour plans can be prepared with the help of suitable ground control points.

**3. Small Scales :** 1/30,000 to 1/1250,000 are used mainly for reconnaissance work—1 : 80,000 to 1 : 10,0000 may be taken as adequate. Vertical photographs cover less ground per photograph and require reduction to smaller scales. It is a common practice to take high angle obliques or obtain composites with multi-lens cameras, in the case of preliminary surveys.

### USES OF AIR SURVEYS

1. As a means of normal topographical surveys, it is quicker than ground survey.

2. For engineering projects, such as power or irrigation schemes, photographs provide additional data for the selection of suitable sites for dams and reservoirs and also for outlining the extent of submergence.

3. In surveys for power transmission or aerial ropeway alignments.

4. Surveys for road and railway engineering works.

5. Town and country planning.

6. For drainage, flood control, irrigation, soil erosion surveys.

7. Soil surveys.

8. Military purposes.

9. Forest surveys where under favourable conditions, the various types of vegetation can be identified and even heights of trees can be measured.

10. In mineral prospecting—for mapping rocks and earth structures. This aspect has been well developed and has given birth to the science, called photogeology or aerogeology.

### VERTICAL PHOTOS : CHARACTERISTICS

(a) **General :** Normally, little relief can be noticed in a vertical photo, but even so this kind of photograph provides the most complete picture.

The effects of prespective are generally subdued, and it is only when the area photographed has high relief, with tall vertical objects, that perspective effects appear *i.e.*, in the photograph, tall object of uniform dimension such as a tower or chimney will appear to taper towards the top.

(b) **Scale :** Scale of vertical photograph may be expressed in the form

$$S = \frac{f}{H}$$

where S is the representative fraction or scale fraction.

f=focal length of camera lens, and H=altitude of flight. f and H are expressed in the same units.

(c) **Distortion**

(i) *Due to uneven topography* : As a result of parallax, the photographic images of two points lying on the same vertical plane are radially displaced with reference to the centre. This feature is clearly noticed in vertical photographs of tall structures, such as towers or chimneys etc. where the top of the tower etc. is displaced outwards in relation to the base, inducing an apparent slanting of the image (Fig. 2.2). The position of an object in a depression is similarly displaced inwards.

In Fig. 2.2, A and B will be the actual plotted position in plan, while in the photograph, image represents points A″ and B″. Thus B has shifted inwards by D′ and A has shifted outwards by D.

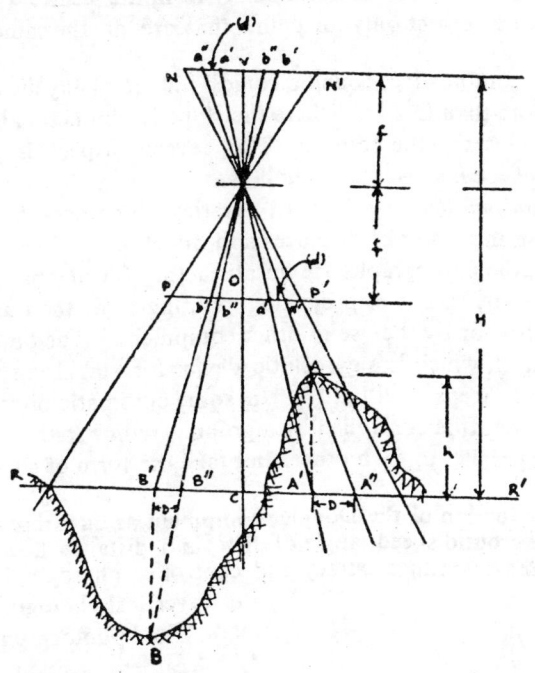

Fig. 2.2

N N′=Negative
P P′ =equivalent positive          OC=CV=f=focal length of lens
C V′=H=Altitude of flight              h=height or elevation of
                                                    the object.
    A  and  B—points on the ground
    A′ and  B′—image of A and B respectively in plan, on the horizontal
                                                    plane.
    a′ and  b′—actual position of image on the photo.
    a″ and  b″—displaced position of image on photo due to parallax.
    (On the negative).  The parallactic displacement 'd' is given by the equation ;

$$d = \frac{h}{H-h} \times O a'$$

1.   Hence it will be noticed that the amount of displacement, is directly proportional to the relief and also to the distances of the objects from the plumb point, and inversely proportional to the altitude of flight.

2.   With reference to the plumb point, all points above the datum are displaced outwards and all points, below the datum, inwards.

3.   The amount of distortion also increases with increasing distance of the image from the centre of the photo.

4.   As a result of parallactic displacement, resulting from relief, directions are only true when measured radially from the centre (plumb point), and distances are correct only for points that are at the same elevation as the datum.

5.   The contour lines round a conical hill are highly distorted when h is anything more than O.2 H. Where the slope is symmetrical the contours appears to be closer at the bottom.  The reverse aspect is presented by contour lines of a conical crater or hollow.

(ii) *Distortion due to tilt* :  In vertical photographs tilt causes distortion, both in respect of distances and directions. Where the angle of tilt exceeds 4°, the photographs are unsatisfactory for mapping and require to be "rectified" by means of projection printing or by the Canadian High Oblique rectifier or by the use of other equipment.  The Smith three axis automatic control, which is a gyrostatic device for stabilising the aircraft, reduces tilt and keeps it within $\frac{1}{4}°$.  In some automatic pilot systems, the course of the aircraft is controlled by a ground radio station so that the flight lines are parallel to each other, and take the form of concentric circle of large radii.

The introduction of the Doppler equipment as an airborne navigation aid, giving the ground speed, angle of drift, and distance flown, is a major advance towards ensuring accuracy and controlled photography.

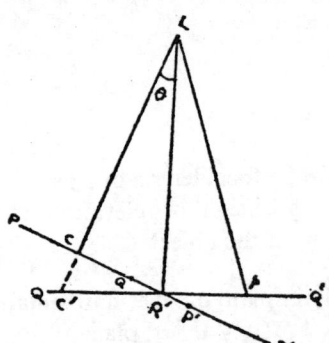

Fig. 2.3

PP′ = vertical photograph.

QQ′ = tilted photograph.

L   = camera position.

C is centre of the vertical photograph.

P is centre point of the tilted photograph.

R is intersection of the tilted and vertical photos is the *"axis of tilt"* (mid point of the axis of tilt the isocentre).

Since, the scale of the tilted photograph is equal to the scale of the vertical photograph only along the axis of tilt, it is possible to find out, by geometrical construction, the corresponding scale on the tilted photograph along any line, parallel to the axis of tilt.  It will be found that the scale of the uptilted side, when compared with that of tne vertical photograph decreases gradually from the isocentre, and vice versa is the case on the downtilted side. It will be seen that angles, measured from the isocentre of the oblique, are equal to the respective angles measured on the ground from the ground isocentre.

It will also be noticed that distortions due to tilt are radial from the isocentre, and that the distortions due to height are also radial, but from the plumb point.

The amount of tilt can be roughly estimated, by a simple device known as Stevensons template or by means of the more complicated Thompson's Comparator etc. The contour tracer and other instruments can also be used.

(*iii*) *Distortion due to the inclination of the air base* : These are caused by variations in flight altitude, between consecutive photographs. This again is either due to faulty navigation or necessitated by atmospheric conditions. Such variations, in the altitude of flight, will cause variations in scale of the respective photographs. Rectification, either by enlargement or reduction is thus necessary, with a view to make the scale uniform. The height of the aircraft is recorded by the altimeter, which is an aneroid barometer, calibrated to read heights directly. In these instruments, with appropriate correction for temperature, a high degree of accuracy is possible. The most common instrument, employed for reading heights in air surveys is the statoscope, which works on the principle of the manometer, filled with amyl alcohol. One make of statoscope is reported to read to within ± .5 metres. These instruments are designed, either for reading absolute heights or for differential or relative readings.

## PRINCIPLES OF STEREOSCOPIC PERCEPTION

Stereoscopic perception is a characteristic of binocular vision. Stereoscopic vision is an invaluable asset in the making of contour maps from aeial photos. Stereoscopic vision depends on, (1) accommodation of the lens of the eye in relation to distances of objects, (2) amount of convergence of the lines of sight from each eye towards a point on the object, (Fig. 2.4) and fusion or apparent merging of the images seen by each eye. This aspect is, however, more psychological. The three dimensional stereo model results from the fusion of the two separate images seen by the left and right eyes. The minimum amount of

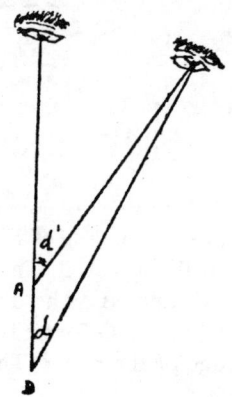

Fig. 2.4(*a*)

convergence detectable by the human eye in about 30 seconds.

In consequence of this limitation, stereoscopic vision is lost gradually with increasing distance or height, and when the height above ground level is about 10,000 ft. variations in normal ground level becomes un-detectable, unless the eye-base is artificially increased by the use of a ste-reoscope.

If in Fig. 2.4 *b* the left eye is fixed on the dot A and the right eye on B, then after some time, it will be seen as if the two dots have merged or fused into one, and a third dot appears in between A and B.

This principle of fusion is employed in stereoscopic photographs, and 3-dimensional images or stereomodels can be obtained. How this perception

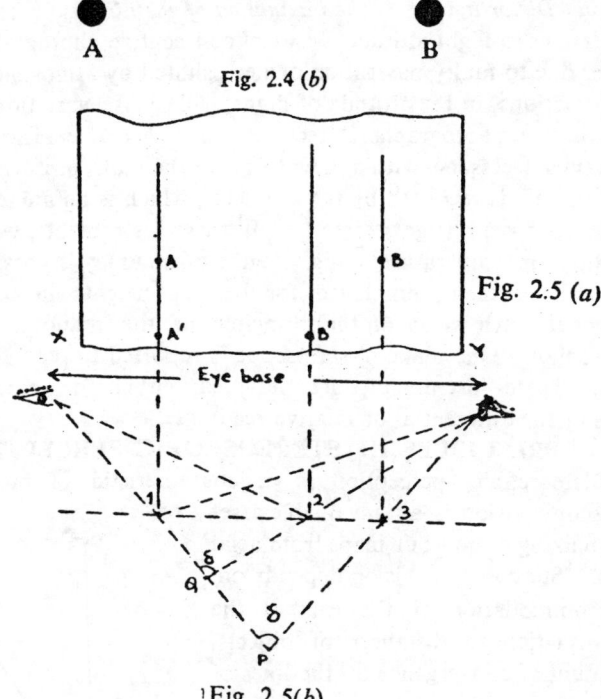

A            B

Fig. 2.4 (*b*)

Fig. 2.5 (*a*)

!Fig. 2.5(*b*)

Here, as in the case of Fig. 2.4, dots A and B "fuse" at P, similarly dots A' and B' fuse at Q. Thus two additional dots appear between A and B and A' and B'. The dot P formed by the fusion of A' and B', appears to "float" above Q formed by the fusion of A and B. In figure 2.5a, XY is the edge of the paper. The location of points A, A', B, B' are plotted as 1, 2, 3.

The line of sight, of the left eye through A, meets the line of sight through B (from the right eye) at P. Thus P apparently has a location in space. Again the line of sight, from the left eye through A, and that through B' from the right eye, converge at Q, which also appears to be located in space. The points of convergence of the lines of sight, may be considered to represent the points of fusion. Since the angle of convergence of the line of sight d has increased to d' this induces the apparent decrease in distance (or depth as in this case).

## PARALLAX IN STEREOSCOPY

As explained earlier d and d' are called angles of convergence for A and B, Fig. 2.5b. The difference between the angles (d−d'), called the angle of differential parallax, is a factor which affects the estimation of depths or distances. The difference is parallax, varies directly as length of the eye base, i. e., distances between the eyes and inversely as the distances or depths of A and B from the observer (Fig. 2.5b). The angles of convergence, generally become less than the detectable minimum for objects located more than 0.4 miles away, and thus binocular perception of depth vanishes gradually with increasing distances. In order to artificially increase the perception of depth or distance, the stereoscope is employed. This instrument helps, either to increase the length of the eye base or magnifies the distant object. This magnification, in effect, amounts to bringing the objects nearer, and thus induces a detectable amount of differential parallax for stereoscopic vision. However, it is the "tolerence" of the brain, which helps to overlook, within certain limits, any distortion or imperfections in a stereopair.

Further, the difference in spacing of the dots A, B, A' and B' Fig. 2.5a known as parallax difference, is not only a measure of the differential parallax, but also of the convergence, as well as the relative depth. Thus for the same set of conditions, the parallax difference is constant, for points having the same difference in depth. This principle is taken advantage of in the design of certain stereoscopic apparatus for tracing contours (see under instruments and apparatus).

## CONDITIONS FOR STEREOSCOPIC VISION

1. The observer should use both eyes.

2. The photographs should have sufficient, overlap i.e., about 66%.

3. The two photographs, of the stereoscopic pair, should be of equal legibility and illuminated equally, The eyes of the observer should be of equal power and properly corrected.

4.  Shadows on the photographs should fall away from the source of light.

5.  The photographs should be set in the proper position. *i. e.* the left hand photo on the left side, and the right hand photo on the right side. The camera base and the eye base should be parallel.

### Factors Affecting Stereoscopic Vision

(*a*)  Wrong orientation of stereopairs. This may even produce pseudo-stereo fusion, or reverse stereoscopic effects.

(*b*)  Injury or markings on the photos.

(*c*)  Uniformity in texture and tone of the photos.

(*d*)  Difference of scale between photos.

(*e*)  Excessive and unequal tilt in one or both photos.

(*f*)  Cliffs and very steep slopes are likely to interfere with fusion.

(*g*)  Sudden changes in topography of large magnitude hampers fusion of the high and low levels simultaneously.

(*h*)  Moving objects, like cars, trains and waves which shift their position between exposures, disturb stereoscopic vision.

### Means of Inducing Stereoscopic Vision

The common apparatus for aiding stereoscopic perception are of three types — the first type makes use of stereopairs or stereograms (*i. e.* photo positive) in correct orientation, *e. g.* stereoscopic. The second type is the anaglyph. In this method, the two pictures are super-posed and printed in complementary colours with the necessary parallactic displacement. Red (primary) and blue-green (secondary) are the two colours used in making the anaglyph. To the unaided eye the imprint looks hazy. Special spectacles with blue-green glass on one side and red glass on the other are worn. The red glass is placed on the eye towards which the red print is displaced and the blue-green glass on the other eye. The image stands out in relief, in shades of grey and white. This method is, however, costly and troublesome. The principle is, however, utilised in one type of plotting machine, known as the multiplex machine. For this purpose transparencies are employed instead of photo prints. A third method for inducing stereoscopic vision is by the use of polarised light, *i. e.* light waves vibrating in one plane. The stereoscopic vision is obtained by causing light to pass through one picture to be polarised in one plane, and the light to pass through the other picture to be polarised in a plane at right angles. Polarisation is affected either by using "polaroids" or by employing special polaroid base films as for making the transparent positives. This method is also not in common use.

### Some Simple Stereoscopic Equipment

(*a*) **Stereoscopes** : Since consecutive photographs, in a strip, are taken at long intervals, it will not be possible to make out the relief, unless one eye of an observer is in the position of the first exposure and the other eye in the position of the next. The same effect is achieved, by artificially extending eye-base and this is the purpose of the stereoscope (Fig. 2.6) below. Hence the stereoscope also exaggerates the relief. Stereoscopes are of two types, (1) Refraction type, (2) Reflection type.

(1) Refraction type consists of two lenses of same power mounted suitably on a stand for comfortable viewing. The normal magnification is 1.25 to 4 times; and the separation between the lens may be adjusted to suit individual requirements.

(2) Reflection type consists of 2 pairs of mirrors mounted at 45° to the plane of the photographs, and sometimes prisms are utilised instead of mirrors ; to achieve the same end.

In certain instruments lenses are also included to achieve better results by magnification (Fig. 2.6).

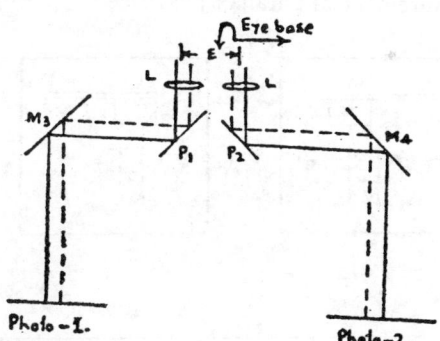

Fig. 2.6

Precision stereoscopes, such as these made by Messrs. Barr and Stroud, the Delft scanning Stereoscopes—have several refinements, besides the parallax grid which enables contouring.

(*b*) **Other Apparatus** : Floating mark apparatus, which enable the measurement of relative depth and height by employing the principle of parallax difference, have been referred to earlier (see parallax in stereoscopy). Simple instruments of this type are the parallax bar, or stereometer,

Abrams Contour Finder, Fairchilds stereocomparagraph or the Tracing Stereometer made by Messrs. G. de. Koningh. A few of these equipment are relatively simple in construction and operation. These have been described below.

## (1) Parallax Bar or Stereometer

The parallax bar or stereometer consists of a rigid frame or bar to which are attached, two glass plates, each having a characteristically shaped mark at the centre (Fig. 2.7). One of the marks Q can be moved in a direction parallel to the axis of the bar by means of a micrometer screw (S), so that the spacing between the two marks can be varied. Thus after orienting the photos for stereo-fusion, the marks are set so as to obtain fusion at ground level, on an identical pair of points (A and B); one being located on each photo. Once again the marks are set to fusion at the ground level at another pair of identical points E and F as before.

The difference between the two settings of the micrometer, is a measure of the differential parallax and this is usually read off in 100ths of a millimetre, One of the mark is also capable of movement along the arm Y in·a direction perpendicular to the axis of the bar (R), and this helps to compensate for small amounts of tilt, The Hilger and Watts SB 180 mirror stereoscope incorporates a built-in floating light dot, which makes direct measurement of parallax possible.

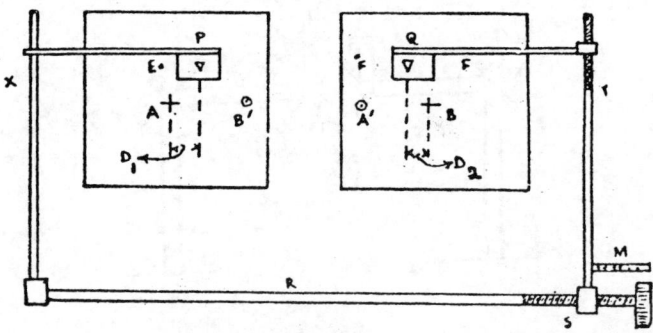

Fig. 2.7

The contour tracing mechanism, in the multiplex plotting machines also employs the principle of stereoscopic fusion.

**Procedure for use :** The instrument is placed on a pair of vertical photos and the spacing between the marks P and Q is so adjusted as to obtain fusion. The marks then appear to rest at a definite height, in relation to the topography (as seen on the photos). Fusion of floating marks is again obtained for a pair of other identical points, on the photographs, and the two readings of the micrometer are noted.

$D_1 - D_2$ = Parallax difference, Fig. 2.7 and this can also be read off from the micrometer scale M.

Parallax differences, for more points, can be determined in a similar manner. The values of differential parallax, so obtained, can be correlated with the variation in ground levels.

### (2) Abrams Contour Finder and other Apparatus

Abrams Contour Finder, and the Fairchilds Stereocomparagraph or the G. de Koningh's contour tracer, consist of the following essential components—(1) Stereoscope, either of the lens mirror type. (2) Parallax bar, with micrometer for the measurement of differential parallax. (3) Drawing attachment. The drawing attachment consists of pantograph with a pencil fixed to the alignment mechanism, which permits drawing of contours on a separate sheet. (4) Parallel guide or alignment mechanism— the alignment mechanism is a contrivance to keep the stereometer parallel to any one fixed straight line, e. g., the eye base.

(1) The centre point and transposed centre points are marked on the respective photographs. (2) The instrument is placed over a sheet of white paper and then by adjusting the micrometer the floating marks are brought to fusion. The distance between the marks is measured accurately to find out the stereo base of the instrument. (3) Next, the photos are aligned by means of centre points and the transposed centre points. Either the instrument or photos are so adjusted that the two index marks are in alignment with the stereobase line.

Now, if the floating marks are "fused" and appear to rest at ground level (stereoscopically), it is possible to trace out all the other points (with the same parallax difference) where the fused marks appear to rest at ground level. The drawing attachment, which also moves along with the stereometer simulataneously, traces out the course of two "fused" marks continuously thus reproducing forms lines or contours, corresponding to the topography.

**Absolute Parallax :** If H is the altitude of flight, the average length of air base along the strip is B (distance between consecutive exposure stations), focal length of camera is f and the height of any one point on the ground is h, then the absolute parallax P can be determined.

$$P = \frac{f. B}{(H-h)} m. m.$$

P the parallax is expressed in m. m., f is expressed in m. m.

If the absolute parallax of the point A is $P_1$ and the difference of parallax $(P_1-P_2)$ between two Points A and B is also known, then the absolute parallax $(P_2)$ of B can be obtained.

### Apparent Displacement due to Parallax : (Fig. 2.8)

PP′=Negative.

X and Y are points on an undulating ground.

X′ and Y′ are the respective projections on a horizontal plane.

P′ and p are the images of X and Y respectively on the photo (negative). CA=f=focal length of lens. X″ and Y″ are the positions on the negative of the images of X and Y, only if distortion, due to elevation, is eliminated. AB=H=Altitude of flight B is the plumb point.

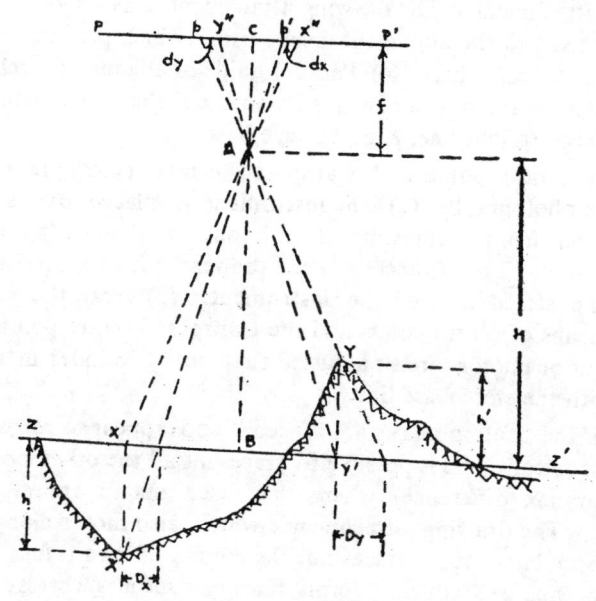

Fig. 2.8

Y Y′=elevation, X X′=depression.

    h=depression of X.

    h′=elevation of Y.

Dx=Apparent displacement of X.

Dy=Apparent displacement of Y.

p′ X″=displacement of the image of X on photo=dx.

p′ Y″=displacement of the image of Y on photo=dy.

### Absolute Parallax—Its Measurement (Fig. 2.9)

Suppose LL' and RR' are the left and right side photos. $X_1$ and $X_2$ are points on the ground at different elevations, in respect to datum $D_1$, D.

X' Y' are the images of XY on photo R R'.

X" Y" are the images of XY on photo L L'.

Through A', draw A' X''', A' Y''' parallel to A X', and A Y' respectively.

X" X', and Y" Y' bear a relation to the absolute parallax of points X and Y respectively.

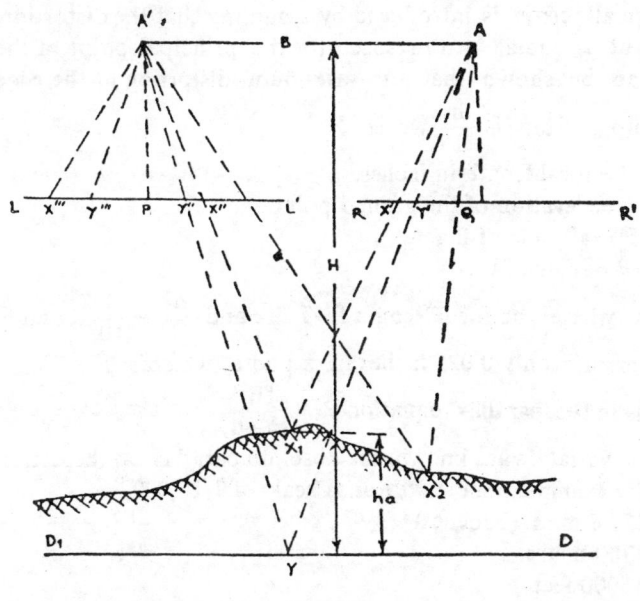

Fig. 2.9

But the parallax equation is more commonly used in determining absolute parallax as shown earlier.

Differential parallax between the points $X_1$ and $X_2$ on the ground can be determined, either visually or by using a stereoscope with a parallax measuring device, e. g., the stereometer or contour finder etc.

X''' X"−Y''' Y" is a measure of the differential parallax (dp). The differential parallax is more often used in contouring than absolute parallax (dp).

Thus, if the absolute parallax of a few points, within the overlap are known, the differential parallax, between any of these as well as of other points, can be determined with considerable facility by the use of a stereo-meter, and the absolute parallax of the unknown point can also be determined.

If Pa is absolute parallax of a point whose elevation is known, then Pb the absolute parallax of any other point can be determined.

### Determining Heights by Absolute and Differential Parallax

1.  In making maps from aerial photo, it is well to remember that the effect of tilt alone, is much less than distortions due to variations in ground elevation. Thus a tilt of 2° on a 7″×7″ photo is negligible.

2.  A small error is introduced by assuming that the distortion due to height is radial, with respect to the principal point of the photo. It can be shown that the maximum distortion at the edges of the

    photo $= f \tan \theta \dfrac{h}{H}$, where,

    f = focal length in inches.
    h = elevation of the ground point.
    H = altitude of flight.
    $\theta$ = angle of tilt.

Thus where the focal length f=7 in. and $\dfrac{h}{H} = \dfrac{1}{10}$, and $\theta$=2°

the error is only 0.025 inches in a photo which is 7″×7″.

Thus in the parallax equation $P = \dfrac{fB}{(H-h)}$

If the variables are known, the absolute parallax can be determined.
If, for example, in a survey on a scale of 1/25,000,
f = 177.8 m.m. (7 inches)
B = 1700 metres
H = 15000 feet
h = 500 feet

$$P = \frac{177.8 \times 1700 \times 3.28}{(15000-500)} = 68.37$$

Thus from the parallax equation the value of the absolute parallax can be computed.

For photographs in which the topography is not excessively rugged and the scale is 1/25,000, the minimum requirements for fairly accurate contouring are that,

    (1) the tilt should be less than 2°, and
    (2) the height of atleast 9 ground points should be known, i.e., 6 should be within the overlap or adjacent photos.

With these as initial points, the levels of other points are determined by measuring the differential parallax.

Then Pa—Pb=P=differential parallax

Pa is known, and P is also obtained from the stereometer measurement, thus Pb, the absolute parallax of b, can be obtained.

By adopting a similar procedure, the absolute parallax of points in the overlap area of a stereopair can be determined.

Once again in using the parallax equation, the unknown factor will be (H—h) of which H, the altitude of flight is known and so 'h' can be determined.

Instead of calculation, tables known as parallax tables have also been published, thus simplifying computations. These take into account the 3 variables of the parallax equation p, fB, and (H—h).

Instruments, employing the floating mark or parallax bar are used in photogeology for evaluating profiles, slopes and dips. Thus enabling the preparation of structure contours, isopachytes and geological sections.

The stereoslope comparator developed by the USGS, directly measures slope angles as seen in the stereomodel; and by reference to a conversion graph the exaggeration in the vertical scale can be eliminated.

The multiplex (projective—plotter) can be adapted for use as an exaggerated profile plotter (upto 5 times) for direct measurement of distances and thickness of beds, etc.

## Apparatus for Detecting and Evaluating Tilts

(a) Stevenson's-Template consists of a rectangular transparent template with a graduated circular scale. The centre of the scale is indicated by a circle, and the quadrant radii parallel to the edges of the template, are also marked.

Procedure for use : The template is placed over the central photo of a triad (set of 3 photos), with the centre of the template over the centre point of the photo and the edges of the template parallel to the line of flight. Next, the position of certain points of details, lying on the circular scale are noted. Observing the same procedure, the left and the right hand photos are now placed under the template successively and the position of the points, marked earlier, are noted on left and right halves of the template respectively. If there is no tilt the plots of identical points from the three photos, on the circular scale, will be coincident. In case of a tilt, however, the plots (made on the template) lie on a circle, which will cut the circular scale of the template at two points. The position of intersection determines the amount of tilt.

(b) Stereometer with movement along an axis perpendicular to the direction of the stereobase.

When using a stereometer, tilt is indicated by a "want of correspondence", unless one of the plates, with the floating marks, is moved in a direction at right angle to the "base line". If the amount of this "want of correspondence" is measured, the tilt can be calculated.

(c) Other instruments such as Thompson's Comparator etc. are more complicated and hence not advocated.

## SIMPLE METHODS OF MAKING PLANIMETRIC MAPS FROM VERTICALS

A.   For approximate and preliminary work :   When it is felt that for a particular work there is no advantage in preparing accurate maps, a rough map may be prepared from single photos, either by (a) direct tracing, or (b) by using the camera lucida arrangement such as the sketchmaster, or (c) by inking details on the photo with waterproof ink and then bleaching with any photographic reducer, so that the ink marking stands out against a white background (Thus the positive or print is bleached in the operation).

The sketchmaster and the Air Survey Model Grant Projector are designed for transferring planimetric detail from a single vertical photograph, on to a base map.   These instruments do not make any allowance for displacements due to relief, but some compensation for variations in scale, between the photo and the base map, is possible.   The Watts Hilgers stereosketch permits the transfer of details from the stereomodel to a base map, employing stereopairs.

B.   For more accurate work : (a) Arundel method (b) Radial method are used for verticals (c) Obliques are also used.

(a) Arundel method is a particular modification of the radial line method.   The principles utilised are dependent on the assumption that (1) all the displacements due to parallax and tilt are radial with reference to the plumb point, (2) that the plumb point, isocentre and centre point (Fig. 2.3 & 3a) are coincident, provided the tilt is not excessive, and not more than 3°, (3) that the horizontal angles as measured from the photo centre are true, irrespective of the distortion in distances.

The procedure consists of four stages, (a) base lining, (b) centreing, (c) marking of photo control point, marking of minor control or wing points in the common overlap of the photos in the strip and (d) preparation of overlaps (Fig. 2.10).

Base lining is done by taking photo (I) of a strip and locating either by inspection or with the stereoscope the photo centre of photo (I) and then locating in the overlap area on photo (I) the photo centre of (II). The line joining the two points gives the base line for photo (I). Similarly in photo (II) the photo centres of both (I) and (III) are located. Now the centre of (II) is joined with (I) and its continuation forwards in marked

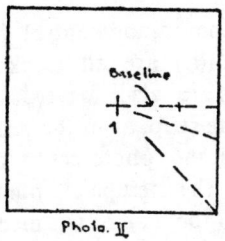 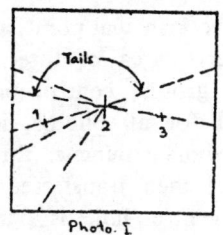 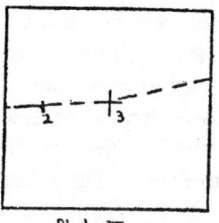

Photo. II          Photo. I          Photo. III

Fig. 2.10

as a "tail" near the edge of the photo. So also the base line III-II is produced backwards and indicated by a "tail" in the rear edge of the photo. The photo centres are pricked with a pin.

In making the minor control plot, at least 3 points are carefully selected in the common overlap of three photos. They are pricked with a fine pointed pin; and radial lines about ½″ long, are drawn from photo centre. The radial lines are not drawn through the points but broken.

In preparing overlays for minor control plots, the strip is placed in the correct stereoscopic position, so that the principal point traverses are continuous. A piece of transparent material or celluloid or "Kodatrace" is placed over it. The base line (I)—(II) is set in alignment and traced out on the Kodatrace with red pencil, and the radial lines or rays are drawn to the minor control points. Photo (I) is removed and photo (II) is placed in position under the Kodatrace. The base line is set in alignment. Base line (II)—(III) is traced and the rays to the minor control points drawn. The photo (II) is removed and photo (III) is placed under the Kodatrace, and aligned with base line (II)—(III). The base line (III)—(IV) is drawn and the minor control points are marked. In will be found, on drawing rays to minor control points from photo (III) that each set of three rays will intersect at a point, and these will give the location of the minor control points. If there is a "triangle" of error caused by imperfect intersections, then suitable adjustments are to be made.

If, in the strip, there are at least two ground control points (*i.e.* points on the ground), which can be identified on the photographs then the plotting can be done to any required scale. For rectification of the principal point traverse, of the overlays, the ground control points are plotted to the desired scale on a grid and the points of the traverse are located by resection (see section on plane table survey).

2. The slotted template method is also a modification of the radial line principle. In this method, templates of the size of photos made of

celluloid are required. The Principal point, and the minor control points or wing points, (in the overlap area of three photos) are all marked by pricking with pin. The ground control points are also located. This procedure is carried out for all photos, in the strip, as in the Arundel method. Next, a circular hole is punched through the photo centre of the template. The template is then transferred to the template punching machine. The punched centre hole is fixed to the axis of the machine. The template is then rotated, so that the pin hole either on one of the radial minor control points, or wing points, engages (with a faint click) with the pin in the middle of the sloth cutting blade. The slot connecting the photo centre to the radial control point is punched by operating a lever. The template is again rotated about the central pivot, so as to bring another pin hole on the template to engage with the pin on the blade of the lever. By lowering the lever another radial slot is made. This process is continued and radial slots are made for all the minor control points in all the templates.

The ground control points are located on a grid, whose scale very nearly equals that of the photos. The slots, corresponding to a particular minor control point, are brought together and the templates fastened with specially designed pins. The templates are adjusted so that the ground control points on the template fall on the corresponding points plotted on the grid. During this operation, the template pins move in the slots and adjust themselves. After this, the position of the minor control points are pricked through on the base sheet below, and their positions obtained. If any particular template prepared from a tilted photo buckles up; that corresponding photo will have to be rectified and the template to be prepared anew.

3. Map making from obliques : The oblique is a perspective and hence, in order to reduce it to a planimetric map the effects of (a) the curvature of the earth, the atmospheric refraction and (b) the prespective will have to be eliminated. Definition of various kinds of obliques and some of the properties of such photos have been given earlier.

In order to eliminate the curvature effect, the following procedure is adopted.

If H is altitude of flight (Fig. 2.11) and d is dip of the apparent horizon, due to curvature effect, as measured from the true horizon, then $d = 58.82\sqrt{H}$ or $60\sqrt{H}$. Thus the position of the true horizon in relation to the apparent horizon can be located; the one being parallel to the other.

The second step will be to calculate, the position of F and the "fore ground line". The procedure is illustrated in Fig. 2.11.

C is the position of the Camera

P is principal point (of the photo)

PC=f=focal lenth of lens

  d=curvature effect

  $\theta$=tilt of camera axis from vertical

  $\alpha$=apparent inclination of the photo to the true horizon

CV=H=altitude of flight

  V=plumb point, V' is the image of V

  G=fore ground point on the ground

  F=the image of G

XY is the edge of an oblique.

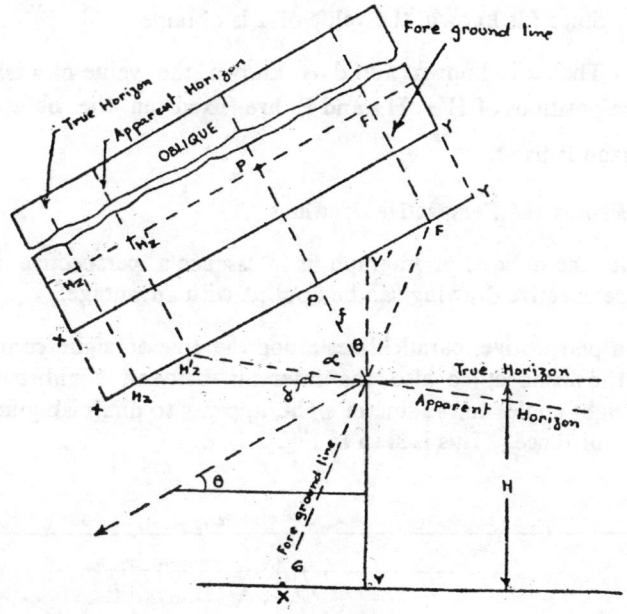

Fig. 2.11

  The projection of XY is drawn, and on it the photo centre P, as obtained from the collimating marks, is plotted.

  Since the focal length, f=PC is known, and since angle d between the apparent horizon and true horizon at C is known, the position of Hz true horizon can be fixed in relation to H'z apparent horizon. Thus $<\alpha$=HzCP, which is the apparent inclination of the photo, is determined and $<\theta$ = PCV' = 90° — $\alpha$; the inclination of the axis of the Camera. is also obtained.

The third step is to form a prespective grid on the photograph, to a definite scale. As indicated earlier, the scale on the oblique decreases from fore ground towards the horizon. A convenient fore ground line is fixed, on the photo and projected to F. In case it is decided to plot on a a scale of 1 inch=660 ft., the position of F can be located from the following :

$$H_z F \text{ (in inches)} = \frac{H \text{ Cosec } \theta}{660}$$

H=altitude of flight in feet.

$\theta$=inclination of the axis of the camera.

$$PH_z = CP \tan \alpha = f \tan \alpha.$$

Since f is known, the value of $\alpha$ is obtained.

Thus $\alpha$ is known and d is known the value of $\theta$ is obtained.

The positions of $H'_z$, $H_z$ and P are fixed on the oblique and the true horizon is fixed.

### Oblique Photos and Perspective Drawings

Since the oblique photograph, is in essence a perspective image, the rules of perspective drawing can be applied with advantage.

In a perspective, parallel lines along the line of sight converge to a point at the distance; so also the intervals between equidistant parallel lines, at right angles to the line of sight, appears to diminish gradually with increasing distance. This is seen in Fig. 2.12 a.

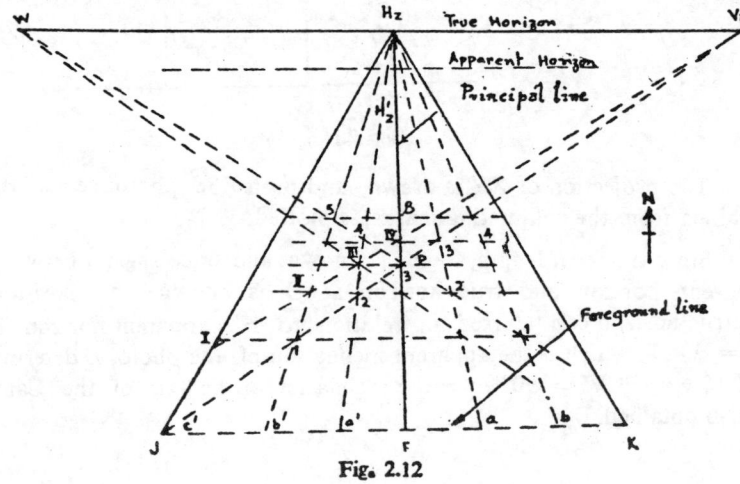

Fig. 2.12

Parallel lines AO, BO, CO, DO, EO converge towards O. So also parallel and equidistant lines $P_{1p_1}$, $P_{2p_2}$, $P_{3p_3}$, etc. appear to get nearer to each other with increasing distance or in other words the scale decreases.

Thus applying this principle to the oblique (Fig. 2.12), it will be seen that the scale decreases from the principal point P towards the true Horizon point $H_Z$. In the reverse direction the scale increases from P to F the fore ground point.

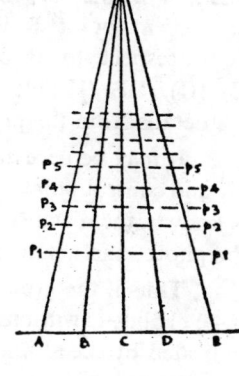

Fig. 2.12 (a)

### Preparation of the Perspective Grid

In (Fig. 2.12) it is seen that while the N-S parallels apparently converge towards $H_Z$, the diagonals between the N-S, and E-W parallels converge towards $V_E$, and $V_W$ on the east and west. These are called the vanishing points.

Thus if the angle B, *i.e.* angle $H_Z P V_E$ is known, then $H_Z$ $V_E$ can be calculated, and the perspective grid can be prepared.

Since for the usual square grid $H_Z V_E = f$ Cosec $\theta$, the points $V_E$ and $V_W$ can be located.

Where,

$f$ = focal length of camera lens in inches,

$\theta$ = inclination of the axis of the camera.

In order to form the grid for a particular oblique photo, on a particular scale, points $H_Z$ $H'_Z$, and P and F are plotted as in Figure 2.11.

The vanishing points $V_E$ and $V_W$ are located.

The methods of calculating the positions of $H_Z$ $H'_Z$, and F has already been indicated $H_Z$ F (in inches) $= \dfrac{H \text{ Cosec } \theta}{660}$ (for a scale of 1 inch = 660 ft.)

From F, *abc*, *a'b'c'* are plotted on JK, with intervals of 1 inch, and joined to $H_Z$, and these constitute the N-S parallels of the grid.

$PV_E$ and $PV_W$ are drawn and produced to intersect the N-S, parallels at 1, 2, 3, 4, 5 etc. and (I), (II), (III), (IV), (V) respectively. The corresponding points 1(I), 2(II), 3(III), 4(IV) etc. are joined and these constitute the E-W parallels of the grid (Fig. 2.12).

The perspective grid when laid on the oblique, in effect constitutes a system of rectilinear coordinates, in respect to which all points on the oblique can be located. It is thus also possible to transfer points from the prespective grid to the square grid and obtain to location of plots in plan. For this purpose, a grid with one inch intervals is perpared. Any convenient point P is located to correspond to P of the perspective grid. With respect to P on the square grid, intersections, 1, 2, 3, 4, 5 etc. (I), (II), (III), (IV) etc. are marked. With respect to these intersections, the details from the photograph are transferred to the square grid.

It may be necessary, to check the scale with actual measurements of ground control points. If the distance between two points, in the photo, is known, the distance can also be computed from the measurement on the photo, and a suitable correction can be applied.

Thus if the ground measurement is 75 chains and the photo distance is 70 chains, it will mean that the actual altitude H is in excess of that indicated by the altimeter reading $H_1$

$$H = H_1 \times \frac{75}{70}.$$

**For Fairly Accurate Work :** A system of triangles is laid out on middle and lower portions of the photographs, where the details are clear. Prominent station points are carefully selected, for this purpose, so that these can be readily identified in the field. Each station should be visible from at least 3 other stations.

A plane table triangulation is carried out, starting with the base line marked out on the photo, so as to connect up the other points in the net. Thus only the triangulation net is prepared in the field.

Later by transposing of details, from the photograph, the interior filling in of the plane table triangulation survey map, is completed.

## CONTOURING

(1) Contour sketching : From the existing ground control points, it is possible to obtain other control points by means of parallax measurements, carried out either with precision stereoscopes or with a stereometer or other device. Looking through the stereoscope, contours can be interpolated by inspection.

(2) Stereosketching is a similar process to that mentioned above, but instead of contours, only the shading is done in a suitable manner so as to depict the topographical relief.

(3) Simple instruments—*e. g.*, Stereometer, Abrams contour finder, Fairchilds stereoplangraph, G. de Koningh contour tracer, Precision topographical stereoscope etc. have been described earlier and these can be used for contouring maps, obtained from verticals or from rectified obliques.

Contouring can also be done from obliques by Cronés method if the tilt is <25° and if elevations of 3 ground control points are known. The focal length of the Camera lens is also required.

**Contouring from Obliques**

For scales of mapping not exceeding 1/50,000, Cronés method has been found useful.

**Principle 1 :** The principle of this method is very similar to that of levelling or contouring with the clinometer. In other words the position of a point P in space can be fixed, with reference to a datum or horizontal plane AB, if both the horizontal distance OQ from a fixed point 0 and the angle of elevation θ are known (Fig. 2.13).

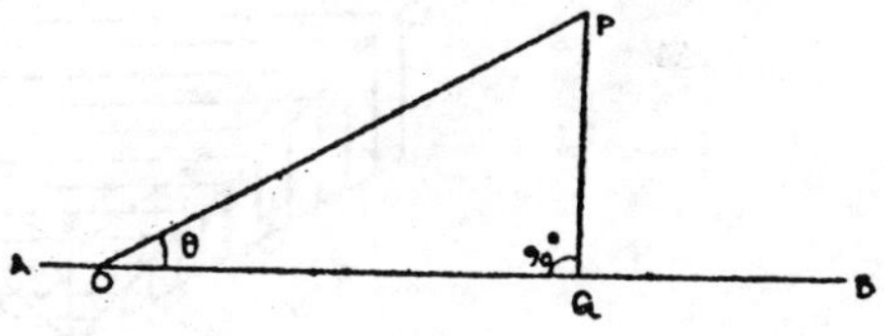

**Fig 2.13**

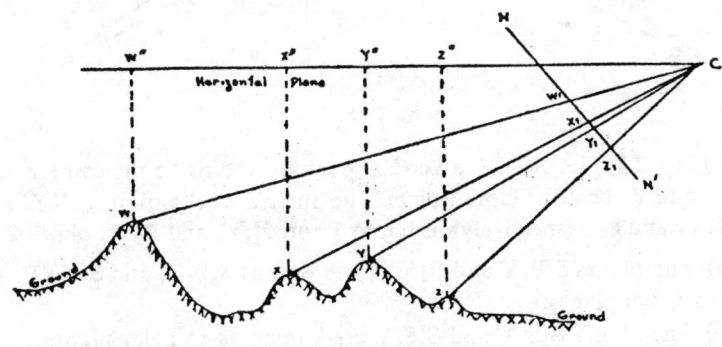

Fig. 2.14

Suppose in Fig. 2.14, C is the camera position, NN′ is the position (of the negative) of the oblique, and θ is the angle it makes with the horizontal.

W, X, Y, and Z are points on the ground.

If W″ X″ Y″ and Z″ are the images of W, X, Y and Z then WW₁, YY₁, XX₁ and ZZ₁ if produced will all meet in C.

If on the other hand, (1) the position of C is known in respect to the horizontal plane, the position of the negative NN' and the value of $\theta$ are known, (2) the positions of the images of the ground points W, X, Y and Z on the inclined negative W, X, $Y_1$ and $Z_1$ are known, and (3) the horizontal distances CW", CX", CY", CZ" are known then the rays drawn from C, through the image points W', X', Y', and Z' will meet the normals to the horizontal plane CW" at the ground points X, Y, Z, W respectively. Thus the ground levels at these points can be obtained.

## Map making with Stereoscopic Instruments

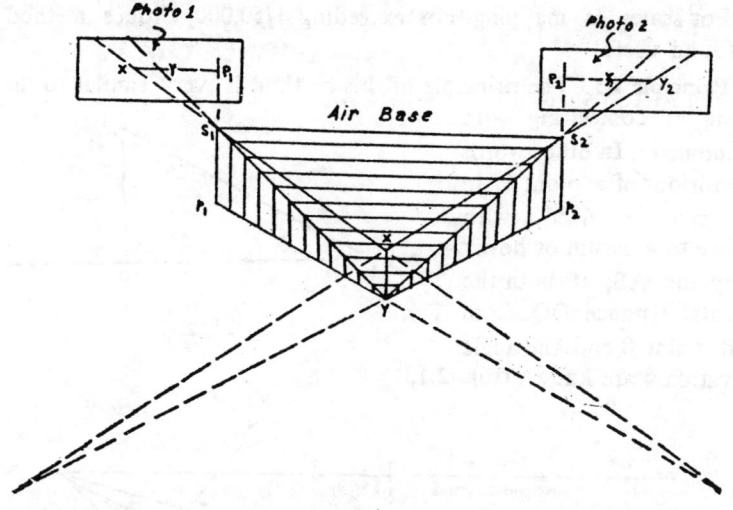

Fig. 2.15

In (Fig. 2.15) let XY be a vertical post, $S_1$ and $S_2$ be the camera stations, $P_1$ and $P_2$ be the plumb points. The images of the points XY and P are $X_1Y_1$ and $P_1$ respectively on photo 1 and $X_2Y_2$ and $P_2$ on photo 2.

(a) The planes $S_1P_1Y$ and $S_1XYP_1$ as well as $S_2P_2Y$, and $S_2XYP_2$ are known as radial planes.

(b) The planes $S_1S_2X$ and $S_1S_2Y$ are known as epipolar planes.

Machines for plotting may be considered as aids for either locating (a) the intersection of the radial planes or (b) for tracing the epipolar planes.

The operating principle of the machines of the first group, can be linkened to two vertical planes rotating about the plumb lines, e.g. $S_1P_1$ and $S_2P_2$ etc. as the axes, after the photographs are set in stereographic relation, in space. Thus it will appear that the point is located by rotating these planes and making them to intersect, so as to obtain the orthogonal projection line of the point required e.g. X is orthogonal projection of Y on the ground plane $P_1P_2Y$.

The operations of the equipment for the location of epipolar planes, may be said to consist in fixing the stereopairs in exact mutual relation in space (Fig. 2.14). In this case the plane bounded by the air-base, $(S_1S_2)$ and the point in question(X), rotated about the air-base as axis and thus the elevation of (Y) and location (X) are obtained simultaneously. Some of the equipment (both radial plotters as well as epipolar plotters) employ the

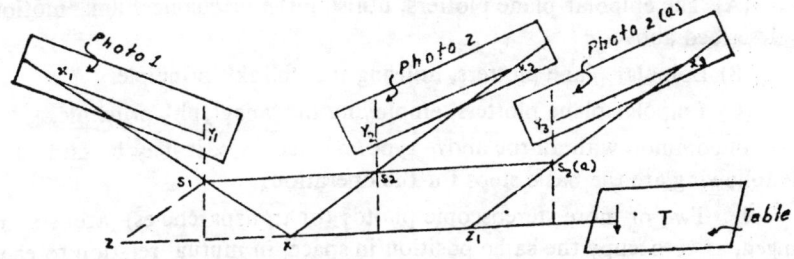

Fig. 2.19

parallelogram device for locating, and also for tracing contours etc. The following are the basic principles of the procedure adopted :

1. The photo 1 and photo 2 are set in the exact stereoscopic alignment, in respect of the photo centres, plumb points etc.

Let photo 1 and photo 2 (Fig. 2.16) be in stereoscopic alignment, so that $S_1S_2$ the air base, is fixed. $S_1X$ and $S_2X$ are rods that are hinged at $S_1, S_2$ and X. Further, the assembly can be rotated about $S_1S_2$ as axis. Thus $S_1X$ and $S_2X$ form the boundaries of the epipolar plane with $S_1S_2$ as air-base. Consequently, the position of X is fixed in space.

If, however, the air base, $S_1S_2$ is extended to $S_2$ (a) and the photo 2 is shifted to a new position photo 2(a), then the new rod $S_3Z$ forms the counterpart of the rod $S_2X$, while $S_2S_2(a)$ and XZ form the link rods. Thus the arrangement is akin to a pantograph, and if the length of one of the arms is altered, e.g., either $S_1S_2$ or $S_1X_1$ then the scale is also altered accordingly.

Once the position of $S_1S_2$ and X, and the position of the distance link $S_2S_3$ are fixed and if a pencil point T is attached to the link XZ, then every movement of X will be reproduced to scale by T, and recorded on the table.

Since the value of absolute parallax is same, for all points, at the same reduced level, the scale remains constant for that contour. For plotting a contour, the floating mark is set at ground level, by stereoscopic observation, and it is now moved in such a manner that it remains on the ground constantly and continuously. The tracing point T will automatically reproduce the contour, on the set scale.

(ii) Some of the most accurate plotting instruments are based on the principle of locating the epipolar plane.

Such equipments are widely employed in the preparation of planimetric and contour maps. Here again, these plotters may be classified under three broad groups.

(A) The epipolar plane plotters, utilising the mechanical link motion as described above.

(B) Epipolar plane plotters, utilising the "blink" principle.

(C) Epipolar plane plotters, employing the 'anaglyph' principle.

In common with all the above types of machines, it may be said that the following are the basic steps for the operation.

1. Two or more stereoscopic photos (or transparencies) are so arranged, as to occupy the same position in space, in mutual relation to each other, as at the time of exposure, with reference to the air base or horizontal reference plane.

2. The stereoscopic viewing device enables the operator to obtain a 3-dimentional model of the field.

3. A floating mark, which may consist of an arrow head or cross or even a light point, is made to rest at "ground level", on the stereomodel. This is linked to a pencil mechanism, which traces out the contour outline to a particular scale, on the drawing board.

The examples of mechanical epipolar instruments are

(1) the Fourcade stereogoniometer and Thompson plotter ;
(2) the Wild Autograph $A_5$ and $A_6$ ;
(3) Zeiss Stereoplanigraph ;
(4) Gigas-Zeiss Orthoprojector.

In the "blink" type of plotters, the stereomodel is obtained by projecting the adjacent stereopictures (transparencies positives) alternating in a rapid manner. The flickering, of the images, induces stereoscopic fusion of the images. When the projection table is moved up or down, the flickering is no longer perceptible, while the stereo-image persists. When the stereomodel is formed, the topographical and other details can be readily demarcated and reproduced, by a link device, on a separate plotting table, e.g., the Gallus-Ferber-Stereocartograph, the Nistri Photocartograph and the Ziess Comparator.

In the anaglyph plotters, advantage is taken of the accomodation of the human eye, to what are known as "complementary" colours so as to induce stereoscopic vision. An anaglyph is obtained by the fusion of overlapping photographs, made in complementary colours. The most common complementary colours employed are red and blue. In fact, the anaglyph comprises overlap of two adjacent pictures, one giving the view as seen by the left eye which may be in red, while the right eye view is taken through a blue filter. The displacement, between various points on the complementary Aero pictures, when projected over each other, is so adjusted that it is equal to the differential parallax.

When this is achieved and coloured spectacles (with red glass on the left and blue glass on the right) are worn, the rays from the left side appear to intersect the rays from the right at definite points in space, and thus induce the formation of the stereoscopic model. When the stereoscopic model is obtained, a tracer point, together with floating mark, is employed for drawing the contours, as in the case of the earlier described equipments. The most important apparatus using the anaglyph effect is the Multiplex machine. A miniature version of the "multiplex" is the Kelsh plotter which is a very popular "mapping" instrument with the U.S.G.S.

# PART III

# PHOTOGEOLOGY

# PHOTOGEOLOGY

When undertaking studies in photogeology one should be fully aware of factors that tend to vitiate the observations made, in respect of the image of objects seen in the photo, and also try to eliminate the discrepancies to the extent possible or desirable. In this context reviewing what has been stated earlier in Part II—Chapter on Photogrammetry, will show that two factors play a significant role in the examination of stereopairs (verticals) and interpretation of topographic details. These are (a) vertical exaggeration and (b) distortion. The extent to which these two factors affect photogeologic analysis is briefly given below :

A  **Vertical exaggeration**

   (a) In a particular set of photographs, since the focal length of the camera lens is constant, it does not affect vertical exaggeration.

   (b) In general, the flight height, of a particular set, of photographs, is uniform and since the height is often large as compared to the relief, the differential relief is low, except where short focal length cameras are employed.

   (c) In any set of photographs, a change in the extent of the overlap, introduces a corresponding inverse change in the air base. It is safer to check on these variations during the course of examination of the photographs.

   (d) If several sets of photographs are used the vertical exaggeration can vary from set to set. When such a contingency is envisaged, it will be necessary to "tie up" one set with another by sufficient mutal coverage, so that the ratios of the various factors determining the vertical exaggeration can be established. In the absence of such coverage the vertical exaggeration is to be determined separately for every set.

   (e) Since in practice the examinations are normal routine and since the same stereoscope is being used by the same person ; stereoscopic variables are, for all purposes eliminated.

Thus vertical exaggeration (Ve) may be taken to depend :

1.  Directly as the air base B.
2.  Inversely as the focal length of camera f.

3.  Inversely as the height of the camera H.
4.  Directly as the viewing distance d.
5.  Directly as the photo separation S.
6.  Inversely as the eye base E.

$$Ve \propto \frac{BdS}{fHE}$$

For ordinary purposes, a very good approximation of the vertical exaggeration may be obtained by determining the ratio B/H. Graphical methods have also been evolved for the purpose, but the value of the vertical exaggeration can also be obtained from the following formula.

$$Ve = \frac{d(E+S)}{f} \frac{(b+d)}{(E)} \frac{(b+d)}{(E+md)}$$

       b = Photo base (ground distance travelled between exposures)
       E = eye base
       d = image displacement (differential parallax)
       md = magnification of the stereoscope.

Where much of the relevant information is wanting, such as camera focal length, height of camera etc., alternative methods will have to be employed, such as (a) map method, (b) field method, (c) temporary exaggeration methods, as outlived below ;

(a) In the map method, fairly large scale contoured topo sheets or maps are used, in conjuction with air photos covering the same area. A few slopes are selected in the middle of the photos, and the angle estimated stereoscopically. These very slopes are measured on the corresponding map sheet. The ratio of the two values, is determined and serves as a measure of the exaggeration. If the slope as measured from the photo is 25° and the slope as determined from the map is 15° then the scale of exaggeration is 1.66.

(b) The principle underlying the field method is much the same as the map method ; the difference being that instead of using a map for obtaining the slope or dip, the measurement of the slope or dip is carried out in the field. Thus the exaggeration ratio is obtained as in the case of the map method.

(c) Temporary approximation method, can only be used for preliminary or rough work. This is based on field observation as well as field experience. To cite an example, if the photo includes a chain of *peidmont* deposits, comprising alluvial fans, and the slope as obtained by stereomeasurement is 60° ; since it is known, that for such loose material with moisture, the angle of repose under normal conditions is about 30° ; the approximate exaggeration factor is about 2. Since the exaggeration factor affects all quantities, there is no structural distortion.

## DISTORTION

Distortion can occur, when the eyes are moved to a side when look-

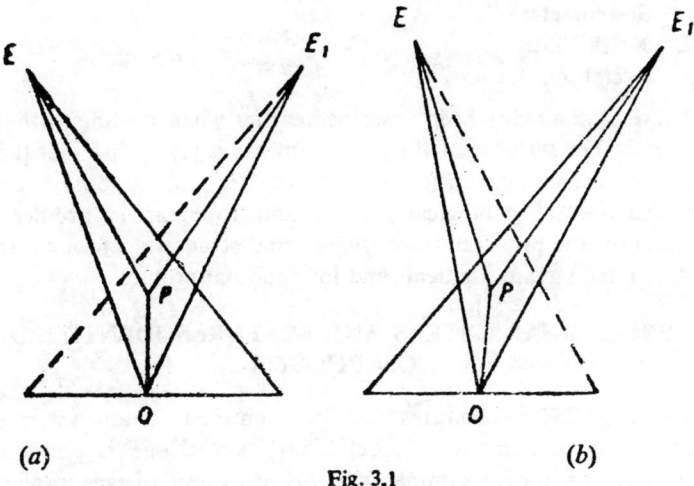

(a)                                                    (b)

**Fig. 3.1**

ing through the stereoscope, (Fig. 3.1, a & b) even though the ground appears to be horizontal, and, the height of the object P is unaltered.

No distortion is obtained, if the eye is directly above the mid point of the line joining the photo centres—known as the perspective centre. Ground points in the vicinity of the perspective centre are slightly distorted, but the distortion increases with distance from the photo centre.

In general, as pointed out earlier, it will be seen that displacement due to distortion of any point on the photo is radial, in respect of the line joining that point to the "perspective centre" (Y) Fig. 3.2. Since this is the case, a large amount of the distortion can be obviated, if the stereoscope can be set up either over the mid point of the radial vectors (M), or over any convenient point on the radial vector QY.

Thus it may be said that the slope or dip depends on (a) the distance from the perspective centre, and (b) the orientation in respect of the radial vector. All slopes, irrespective of the amount of direction, can be converted to equivalent slopes in a direction, at right angles to the radial vectors.

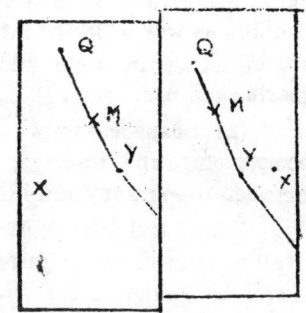

**Fig. 3.2**

**Instruments used in photogeologic work**

The instruments required for carrying out the measurements in connection with photogeologic work are :

1. Stereometer
2. Kelsh Plotter
3. Sketch master

These have already been described earlier when dealing with photogrammetry. The purpose of the measurements is (1) to find out the strike and dip, (2) to calculate stratigraphic thicknesses, (3) to obtain data for the construction of geological section and topographic profiles, (4) to obtain data for the preparation of *isopach* and structure contour maps, and (5) to determine stream gradients and longitudinal profiles.

## PRELIMINARY STEPS AND FEATURES IDENTIFIED ON PHOTOS

Before geological features can be identified, or any interpretation undertaken, a certain amount of preliminary knowledge is essential with regard to the area under examination. (*a*) In respect of geographic location and geological setting, at least the broad features should be understood; whether the area comprises crystallines or the sedimentries; which particular region it represents, *i.e.*, the Assam Tertiary belt or the Gondwana basin etc. Such knowledge will help in understanding and identifying any topographic features. (*b*) Familarity with published literature is essential, as this will assist in deciphering the geomorphology and up warps or down warps etc. can be identified. (*c*) A knowledge of the climates, both past and present will also help in topographic and geomorphic analysis ; and (*d*) Some data in regard to the vegetation if available will also be of additional value. In air photos, the main analysis comprises in the identification, as well as in the interpretation of (*a*) shape and form of features, (*b*) variation in tone and texture, (*c*) also the drainage pattern, (*d*) marine features.

(*a*) **Shape and form.** It is possible to draw relationships (a) between the structure and drainage, (b) between structure and topography, and (c) between topography and lithology.

Faults and folds can be identified very often by the topographic expression or drainage pattern. The following criteria may however, be useful in identifying faults when present on photographs :

(*a*) presence of a linear marking (often straight),

(*b*) offsetting of beds,

(*c*) presence of linear scarp,

(d) linear texture parallel to known major faults,

(e) abrupt change in lithological units about a line.

It may also be pointed out that—

(a) the contacts of horizontal beds are parallel to the surface contours,

(b) the contacts of vertical beds are not affected by the topography. Various bodies of igneous rocks can be identified by their form, texture and tone on the photos.

In a folded plunging sequence with beds of varying resistance, the more resistant beds stand out prominently as zig zag ridges. Basalt capping often presents a flat mesa like structure. Faulting, comparatively recent in origin, may even cause the softer sedimentary formations to stand out prominently against the harder, and more resistant crystallines, as in the case of the Mahadevas lying between the Eastern and the Western portions of the Bokaro Coalfield ; whereas the Iron ore ranges, in south Singhbhum stand out against a background of softer Kolhan and Iron ore series phyllites. Vertical photographs can be considered to be planimetric maps, when used singly. As indicated earlier, the accuracy of such mapping, is not very great due to increasing distortion away from the perspective centre. Yet, these photos are very useful for establishing one's location, in relation to the terrain, especially in heavily forested areas where features such as stream bends, large boulders etc. are extremely useful land marks.

It is advantageous to use stereopairs so that a 3-dimentional picture of the area can be obtained. Several features, which may go unobserved in using single photographs, now become prominent such as scarps, dips, and geologic contacts etc.

Even though a large amount of information can be obtained from the study of air photos, there is no need to emphasise the fact that the geologist cannot forego ground studies for determining details regarding lithology, mineralisation etc. and that details as obtained from the photos are only a means to an end.

Dykes of dolerite, often form prominent rigdes in sandstone as also in granitic country. The Salma dyke in the Raniganj coal field can be traced for several miles. Dykes of Newer Dolerite in the Singhbhum Granite in the copper belt form long ridges extending for miles. Quartz veins are also very often represented by discontinuous ridges, as in the Bundlekhand gneissic complex. Fault rock, with silicification also forms prominent ridges, as seen in the area to the west of Pareshnath hill in the vicinity of

Nurungo, Baraganda etc., in the Hazaribagh district of Bihar. Dyke swarms in the Chittor district of Andhra Pradesh, are also characterised by a regular pattern.

Intrusives generally form prominent masses, and this is the case of the ultra basic intrusions which have been emplaced in the Iron series rocks, to west of Chaibasa, in the vicinity of Jojohatu. The chromite bearing peridotites, and pyroxenites stand out as prominent hills due to differential erosion.

Contact metamorphism, is often noticed by a bolder relief of the aureole rocks. In a way, the prominent relief of the Dhanjoris in Singhbhum district, Bihar, near the contact of the Singhbhum granite, may be attributed to silicification consequents to metamorphism.

The relation between topography and structure is very well displayed, especially where folded or dipping sendimentary formations, comprising beds of variable resistence, are faulted. Displacement of ridges and valleys and the lack of correspondence on either side of the fault is prominently displayed on the photograph. Abrupt termination of such ridges is also noticed against a line of fault. When beds of variable resistance are brought together, the sculpture and erosional characteristics are different in respect to each, especially when sedimentary beds of comparatively recent age are faulted against the gneisses. Wonderful examples are metwith in the Southern part of the Garo Hills district of Assam. Certain features of smaller dimensions are also important. In a glaciated region, it may be possible to identify moraines, eskers or kames or cirques, and in an area characterised by Kaarst topography, swallow holes or sink holes, with subterranean drainage can be located, as seen in the areas coverd by the Cherra limestone in Khasi and Jantia Hills district of Meghalaya.

An asymmetrical slope can be due to various factors. The formation of the scarp may be due to a resistant bed overlying a sequence of soft rocks (Fig. 3.3). The famous Kaimur Scarps in the vicinity of the ancient Rohtas fort in the Shababad district of Bihar are classical examples. Here the hard quartzitic Kaimur sandstones overlie the softer formations such as the

Fig. 3.3

Bijaigarh Shales and the Rohtas limestones etc. The asymmetry is very pronounsed in low dipping formations, and even though there is a cover of scree or talus covering the foot hills on the scarp, the variation in slopes is noticeable. In the case of strata with steep dips of 45° or more the slopes are symmetrical (Fig. 3.4). Examples of such slopes are common, and can be seen in the Rajgir Hills of Patna, Bihar.

In analysing slopes, it is useful to know that there are exceptions. It may

Fig. 3.4

so happen, that recent movements have caused the tilting of peneplained blocks of granite (Fig. 3.5) which may sometimes be capped by sedimentaries. Example of this type of block movement, can be seen in the Garo Hills, where cappings of the Tura sandstones, containing coal seams, are seen resting on uplifted and tilted granite blocks, especially in the east Daranggiri coal field.

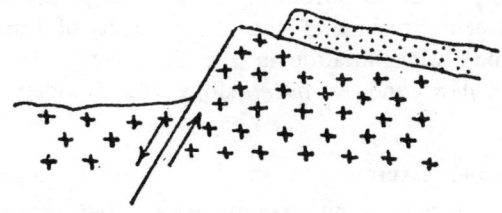

Fig. 3.5

Another case, where asymmetric slopes can develop in practically homogenous rock, is when a stream system on one side is more active than that on the other side of a divide. Slopes of this type are well illustrated on the Kundahs and in the Sisipara section before descending from the Nilgiris (Tamil Nadu) into the Wyanaad (Kerala) area to the west. The streams on the western scarp are generally more active, due to the abundance of rainfall. In addition, the western side is also exposed to the full blast of the high winds during the monsoon. In contrast, the rainfall on the eastern slopes is lower, and these slopes are better protected from the action of the wind. In consequence the western side is rugged and steep while the eastern flank is relatively a gentle slope.

It may also be noticed, in certain areas, where the amount of dips varies in the same group of rocks, the height of the ridges varies inversely, as the amount of dip, e.g., if the dip is high, the height of the ridge is lower and vise versa. This criterion may prove useful, especially where outcrops are few.

Where a gently dipping sedimentary sequence comprises several resistant members, (Fig. 3.6) the dip slope is as usual, gentle and is usually overlain by eroded remnants of the upper strata ; while the scarp slope is steep and the outcrops of the resistant beds stand out as prominent ledges.

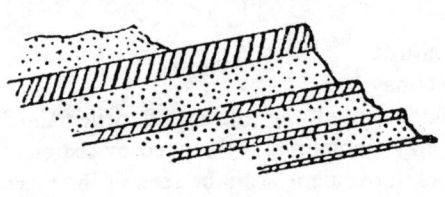

Fig. 3.6

In other words, under these conditions, the scarp side is marked by elongate ledges, while the dip side may show eroded remnants. On the other hand where the dips are steep, then the dip as well as the scarp slope are symmetrical but the ridges have a less prominent scarp side.

The variation in the drainage type, provides valuable criteria in identifying the dip (resequent) from the scarp (obsequent) slope. On the *resequent*, the streams are smaller, and the intensity of drainage is greater. There can also be a wide variation in the drainage pattern, if the rock types on the *obsequent* and *resequent* slopes have widely variable joint patterns etc.

### (b)  Tone and texture

Tone and texture of a photograph, very often show a strong relationship to the rock type. So also the vegetation or the drainage may be related to the rock type. Yet, colour photography is not commonly practiced, mainly because of the costs. The tone of a photograph is the relative intensity or the depth of colour, comprising generally greys and shades of deep brown. The texture of a photograph is also easily percieved by observing the grain in the prints. This again, can be of various kinds, *e.g.*, smooth, rippled, mottled, irregular, or lineated. Texture alone, may not be characteristic of the lithology. It is generally a combination of tone and texture, in conjunction with other criteria, that aid in defining the lithological units.

Weathering, can introduce a change in the tone of the same rock type, hence care in this regard is essential. An outdrop can be recognised by textural and tonal characters which may be caused by lineations due to foliation or rectilinear pattern due to jointing or fracture etc. The tone of the photograph, is rendered darker by irregularities while smooth surfaces present a lighter tone. Another factor which, affects the tone is the orientation of the slope in respect of the sun, at the time of photographing. The surfaces lit by the sun are bright, while the opposite side is dark. This is invariably the case though, during the photography, every effort is made to eliminate such differences. The effects of light and shade are accentuated by the presence of mist which settles rapidly on the shady side. As a consequence the type of vegetation can vary and this also has a corresponding effect.

Soil often varies in colour and composition, and this change is generally reflected in the photographs by a marked difference in tone. This contrast is further accentuated by the differences in the pattern of erosion characteristic of each soil type. Since, there is often a relation between the soil type and bed rock the characteristics indicated above, together with other criteria, aid in identifying the possible bed rock, in areas which are under cover of soil.

Vegetation is an important factor in aiding interpretation and mapping. It is influenced by the soil which in its turn is related to the bed rock geology, *viz.* rock type and structural features. Variation in the moisture content of rocks produces a change in the tone of the photograph. This variation is sometimes direct, but may also be indirect, as variation in moisture content induces a difference in the vegetation, either in species or in the rate of growth etc.

Vegetation may differ in type, from dry sparce type or luscious types, depending on the temperature and precipitation. The colour of foliage is also registered by a change in tone.

Variations in moisture contents may also be due to seepage along a fault line, which may be indicated by the presence of a linear distribution of a particular type of vegetation. An increase in density of trees and plants is often noticeable, (a) in valleys and slopes, (b) where seepage occurs along formation contacts, (c) along river courses, and (d) along traces of joint planes. Density of vegetation may occur as patches, showing the presence of (a) sink holes, which generally contain silty materials at the bottom, and (b) slopes (only apparently).

With reference to vegetation, its nature and type, some factors of interest and consequence in aiding identification may be mentioned.

1. It should be remembered that the foliage is variable with the season. Thus in the major part of the country, comprising the Peninsular part, monsoon plays a very important role. The forests are thin during the winter months in North India, and trees, at least most of them, are bare. With the advent of spring, in mid February, short and young leaves appear. When the early monsoon rains arrive, the forests are covered by a mass of foliage. Thus the variation in the density and kind of foliage introduces a corresponding tonal contrast on the photo, and the same type of forests may appear different in two photographs of the same area, taken during different seasons.

**2.** As indicated earlier, vegetation should be studied in conjuction with topography, slope (since the same density of population, may appear greater with increasing slope), moisture, besides geology and soil.

**3.** What has been said of trees, has its counterpart in examining photos containing grass and pasture lands. Grass grows luxuriantly in most parts of the Peninsula during the monsoon season. In summer, however the meadows are parched and acquire a yellow or brown colour ; but soon after the first few showers, they burst into verdant green.

**4.** Another factor, which requires to be considered is the variation in texture, which results due to cultivation of the areas, which during the dry season lie fallow, but are covered with standing crops during the wet season.

**5.** In early summer most lakes, ponds, swamps are dry, while soon after the break of the monsoon, the character of the country changes enormously. Swamps may extend over several square kilometres. A striking example of this, is the extensive swamp covered areas of North Bihar or the Mymensingh and Sylhet districts of Bangla Desh.

**6.** In the Peninsula, which comes under the influence of both the summer (S.W.) and winter (N.E.) monsoons, the colour changes in vegetation will be complementary in respect of the seasons.

**7.** Even though there is a marked influence of the rock type on the soil, and finally on the vegetation, this may not be very obvious, since there is no way of deciphering and separating arborous trees from each other. Yet, occasionally it may be possible, where each type of trees blossoms at different periods, or sheds its leaves etc. This variation may be helpful in identifying the geological influences, such as rock type or structure etc,

(c) **Drainage.** The study of drainage is an important aspect of geomorphology. If erosional or depositional features, connected with fluvial action, can be recognised on the photograph, then the photogeologist has an additional means of unravelling the geological history. For this purpose a thorough understanding of the fluvial cycle is essential. A thorough comprehension of the erosional and depositional features, associated with each of the stages in the life of a stream, is essential. Taking the youthful stage, the type of valley (generally V-shaped) with rapids and cascades etc., is to be anticipated. Meandering is uncommon. Similarly features associated with the mature and senile stages should be understood. In addition, topographic expression caused by uplift and subsidence, either in the upper reaches or in the lower reaches, near the estuaries and deltas, should be understood. In other words, the development of river cut terraces, river built terraces, entrenched meanders, and antecedent drainage, should be known, so also the presence of wave cut terraces, wave built terraces, submerged valleys etc.

Another feature of importance in photo interpretation is the influence of structure on drainage. The origin of the following drainage patterns are significant, (a) dendritic, (b) rectangular and its modifications, (c) trellis, (d) radial, (e) centripetal, (f) annular with certain amount of radial. Lastly, the erosional and depositional features due to *eolian* action cannot be ignored. The study of *dunes* is of great importance in desert control, and photogeology can play an important part, if interpretations are rightly made.

(d) **Miscellaneous.** The application of photogeology to marine features, is also well known. Good photograph, taken is calm waters, often shows the depth of water near the beach. This method was used by the Allied armies in planning the landing on the beaches of Normandy. A similar technique was used for landings on the Mediterranean beaches.

Glacial features are also amenable to identification from photography. A familarity of the general forms of such features is, however, essential, before undertaking identification. Forms of deposits, such as Kames or Kettles or Eskers, or Moraines, both lateral and median should be known. Erosional features such as U-shaped and hanging valleys, cirques, *roche moutonnees*, and evidences of solifluction are also of importance.

Familarity with features of volcanic origin, is also necessary in order to identify flows of various types, craters, cones and calderas, fumaroles or solfataras.

### Photogeology—its advantages

From what has been said in the earlier paragraphs, it is clear that photogeology in combination with photogrammetry, has a wide application to geological problems, whether of petroleum geology, mining geology or engineering geology or hydrogeology. Leaving aside obliques, both high and low angle, and confining the present discussion to the use of verticals and stereo-pairs, it may be pointed out, that a single photo can be used as a map for general purposes, but the geologist has to be fully conscious of its limitations. This aspect has however, been dealt with under distortions (see photogrammetry).

It is a distinct advantage if the geologist adapts himself to the use of stereopairs, and if possible, without the aid of a stereoscope. The lens or combination stereoscopes no doubt aid stereoscopic vision and provide additional magnification. To carry a stereoscope and use it in the field is rather inconvenient. Even pocket lens stereoscope is not convenient to handle. Orientation of the photos and placing the stereoscope is an effort. When direct stereoscopic fusion, without a stereoscope, is obtained, the fused strip may be small but it has decided advantages—especially in saving time and effort which are of primary importance in the field.

Vertical photos, whether single or a pair or a strip, is a birds-eye view of a large area. The area covered is much larger (depending on the scale of the photo) than what an observer, standing at ground level can see. It may be considered to be, in a very general sense, a plan of the area, as the engineer may call it (save the effects of distortion). Where a stereopair is used, the stereo model so obtained by fusion, may be likened to orthogonal projection. Similarly an oblique is the counterpart of prespective projection wherein greater emphasis is on the elevation rather than plan.

Some of the advantages of photogeological studies in the field are :

1. Information collected from the ground work can be extended to other areas covered by the air photographs, by interpretation.

2. Interpretation and identification can be checked up readily without much effort by examining a few spots in the field.

3. Work can be planned and priorities assigned, since by judicious interpretation, the geology can be unravelled to a large extent.

Some of the important factors which obscure details on the air photographs are mainly due to the presence of :

1. Alluvium and mantle comprising, fluviatile, *lacustrine, eolian,* deposits.

2. Glacial deposits.

3. Forests but occasionally the vegetation is related to lithology.

4. Shadows and glare.

5. Clouds.

6. Snow and ice.

7. Lakes.

8. Cultivation.

9. Cliffs and scarps, and lastly

10. Scratches and scars on badly handled negatives.

## USES OF AIR PHOTOS

Air survey methods are of great advantage, not only to the civil engineer, in charge of a project, but also to the project geologist, and the same photograph can be used by both. To cite a few examples :

(a) Alignments for roads, railways, canals, or transmission lines.

(b) Site survey for towns, defence work, factories, dams.

(c) Construction materials survery for fine aggregate, coarse aggregate, rip rap, limestone etc., timber.

(d) Hydrological survey for surface water studies and ground water assessment.

(e) Irrigation and flood control projects, including water supply, hydel, drainage schemes.

(f) Marine coastal protection, harbour engineering projects and a host of other civil engineering projects.

(a) With regard to the selection of alignments, air photos, besides providing information in respect of the shortest route etc., taking into account both the distance and the gradients, also afford valuable data on the rock outcrops, soil, drainage topography and availabilty of constructional materials. Details of vegetation, found on photograph, can again be interpreted in terms of the geology and the structure. This aspect of the air photos has been discussed earlier and hence it is not repeated. It may, however, be indicated that in the case of such surveys the scale adopted is large e. g., 50 ft. = 1 inch or 100 ft. = 1 inch, with a contour interval of 2 ft.

A point of importance in regard to irrigation canals, is the depth of cutting. If the canal is at a depth, then irrigation is also difficult, since pumping will be required. Again where tunnels are necessary these should be located as far as possible mainly in hard competent rocks. The alignment of tunnels in respect of jointing, foliation, bedding, shearing, or thrusting should be carefully selected so that the over-break is not excessive.

(b) Photographs aid in urban planning with particular reference to the development of infra-structure, vis a vis the topographic features etc. Photographs are very valuable in terrain evalution for defense purposes, so also photographs are of great utility in locating dams, factory sites, etc.

(c) The utility of air photos is not only in helping to locate deposits of constructional materials, but also in estimating the reserves of such material including the amount of timber available. Further the best methods of working and transporting these materials to the work site can also be planned ahead.

(d) Considerable saving in time and effort can be affected in hydrological work by the application of air photo studies. Taking into account the seven main aspects of the hydrological examinations *viz.*, (1) precipitation, (2) transpiration, (3) evaporation and interception, (4) percolation, (5) run off, (6) surface water, and (7) ground water, it can easily be said, that air photos are of prime importance.

The value of air photos, with regard to studies in precipitation, is limited and the main utility of such studies will be in locating, either new sites for main guage stations, stations for snow measurement, river guaging points etc., or for reviewing the suitability of the existing locations of such stations. Air photos also help in assessing the value of such measurements and records of precipitation. If photographs are taken during various periods of the year ; then a better picture of the precipitation, in an area, can be obtained.

Transpiration is connected with plant life and hence air photographs can afford indirect measurements when supplemented with ground studies. During transpiration, water is lost to the atmosphere by the plant mainly through the leaves. Losses from transpiration can be so serious as to affect the surface water supplies and lower the water level perceptibly. Certain plants growing in water can cause loss (like lotus, water lilies, hyacinth etc.), besides moisture loving plants with deep and extensive root systems, like the tamarisk, are a source of danger.

With regard to forests, the question of transpiration is not of major consequence, and the problem is not simple. Forests serve to preserve the soil, provide a large amount of humus and organic manure as well as help to conserve run off etc. Hence, an assessment of the overall benefits is called for, and the question of transpiration losses cannot be isolated in this case.

Evaluation of losses, due to evaporation, can also be made from the study of air photos. Evaporation takes place on exposed surfaces of water as well as from the soil. The rates of evaporation for various surfaces will have to be assessed taking into account the temperature and relative humidity. When this is known, the areas pertaining to each type of surface can be computed, from the photos, and evaporation losses can be worked out. It is safer to work out losses for each type of area separately, so that any error will be localised, *e.g.*, ponds (with and without vegetation), cultivated lands, fallow lands, stream surfaces, and forest land etc. Interception losses are due to evaporation of water standing on the leaves and branches of trees. This water may be assigned to rain or snow, or dew, respectively.

Percolation or seepage is, what in common terms, described as soaking. This depends on the gradient, soil characters, such as texture, grain size, mineral composition, permeability, etc. It is also indirectly related to the vegetation, which directly controls the run off, and to the moisture content, which depends on the depth to the water table, the amount of snow fall, or rate of intake etc. Other factors include temperature. This study can be made from photographs and a fair approximation obtained for the percolation.

Regarding the evaluation of run off, air photos can be used for correlating drainage area, with guage readings, drainage density etc. From this information the run off may be computed. The scale of photo used is 1:10,000 or 1:12,000 and when low gradients are encountered the scale is 1:6000.

(e) An important aspect of hydrological work is (i) water-utilisation and (ii) water regulation or control studies. Here air photos can play a primary role if interpretation can be made.

(i) Water utilisation comprises, water supply projects, hydel projects, navigation, irrigation, pisciculture etc.

(ii) Water regulation includes, flood control projects, drainage, sewerage, land erosion, land reclamation and land conservation.

The importance of air photos in such studies, is the facility and ease with which measurements of certain parameters required for computation of data, for the assessment of water utilisation or regulation, can be carried out with a precision, which is satisfactory for this type of work. Such measurements include (1) the pattern and type of drainage system, (2) lengths, order, and stage of stream development, including profiles etc., (3) area of catchments, nature of divides, basin profiles, shape, hypsometric measurements, etc. Measurements made from air photos are often more accurate than ground measurements, in certain cases, especially in swampy and inaccessible areas. The data may be obtained either from mosaics or from over lays. Where the time permits, and a higher order of accuracy is necessary, then the stereoexamination may be carried out. Where the work is of a reconnaissance type vertical control can be effectively obtained from aneroid readings or by the use of *Shoran* or air borne profile recorder where the areas are large enough, i. e., about 100 sq. miles or more.

Somewhat similar, to the above, is the technique of linear measurements for obtaining the length of streams, gradients etc. This is difficult if there are no ground control points. With sufficient ground control, the profiles can be worked out by the use of simple instruments, such as the parallax bar or with the multiplex. Alternatively the air borne profile recorder (APR) is to be employed during the photography.

Air photographs can also be used for estimating widths, and even depths of major water courses, and this data combined with arial measurements, extent of drainage, gradients etc. will afford a very close approximation to run off calculations.

In addition, air photos can provide information on (*i*) Geology, (*ii*) Soil, (*iii*) Vegetation, which are required in connection with the construction of dams, as well as for planning measures for checking silting up of reservoir etc.

(*f*) Harbour engineering and coastal protection is another sphere where photo geology (air photo study) is of considerable importance. These studies include marine erosion, transportation, sedimentation of beach materials by wave action, current action, tidal influence etc. Studies of off-shore profiles (bars, terraces channels etc.) is also essential for the construction of piers, break waters, lighthouses, etc. as well as groins, and setting up buoys etc.

Air photos often bring out several features of interest and utility in respect of :

(*a*) shore profile—showing width of slopes, crops ;

(*b*) nature of waves—length, frequency, height, refraction, wind effect etc. ;

(*c*) nature of beach deposits, depths of water, turbidity, currents etc. ;

(*d*) other features such as bars, estuaries, deltas ;

(*e*) tidal features such as mud flats, swamps, marshes ;

(*f*) underwater features as well as topography.

Air photos can be used for investigating the cause of land-slides and in planning safety and preventive measures. While large slides can be easily spotted out in air photos sometimes as prominent scars the smaller ones are not easily recognisable. The common scale of photography is 1:20,000 but 1:5000 is occasionally required.

In order to detect slides the most likely spots on the photo should be searched with great care, and these areas are invariably marked by scars and furrows of earlier slides. Deep soil covered valleys, as in the Darjeeling Himalayas, or in the Western Ghats, are potential spots, so also areas of seepage, where the dip is steep, and in the same direction as the slope.

Studies regarding fluviatile cycle, comprising erosion, transportation and sedimentation, can also be carried out on air photographs. These studies, in order to have the requisite precision, should take into consideration the period, or time and place.

From the photos the alluvial deposits can be identified. It is even possible to classify them as pebbles or shingle, sand coarse or fine, and clays etc. ; mainly by the tone and texture. If these studies are combined with the time and period studies, *i.e.*, during high water, normal flow and low water, then the behaviour of the stream can be analysed. Even the amount of load, carried by the stream can be assessed from the tonal variation, in the stream, during various periods, as well as variation in the width of the stream. This data is more accurate than mere guaging. Location of stream erosion, in the past and as well as present can be made out ; and the stability of the banks, effect of meanders can also be assessed. Such studies are useful in siting and locating weirs, bridges etc. on streams.

In relation to ground water studies, air photos have an indirect application. Studies of tonal contrasts and texture, with interpretation regarding the same, will yield results. This in effect will mean the identification of soil and vegetation. The possibilities of ground water can thus be assessed by a geologist of experience. Then again structure can very often be identified by foliation trends, dips, faults, dykes, joints and the like. These are either found as lineaments directly, or as liner trends emphasised by vegetation and drainage. Seepage invariably shows up by tonal contrast (by a darker shade). The contrast is very often accentuated by vegetation. Such seepages may be associated with faults or with the intersection of the water table surface and topography.

The application of air photo studies to engineering projects for flood control etc. as mentioned earlier is obvious. In the case of multipurpose projects data on (1) Dam site and spillway as well as (2) on the reservoir conditions, extent of submergence etc. (3) availability of construction materials can be easily obtained.

Often air surveys also include air borne geophysical investigations. The methods in exploration geophysics, for which air borne equipment is available, are as follows :

1. Radiometric employing scientillation counters,
2. Magnetic,
3. Electro-magnetic,
4. Gravity which is used to a limited extent.

For details of these methods of geophysical exploration a reference may be made to the chapter on Geophysical methods Part V.

**Advantages of air borne survey are that :**

(*a*) Profiles are continuous and do not comprise spot values.

(*b*) Inaccessible areas such as swamps, lakes, forested regions etc. do not hamper measurements.

(c) These are distinctly faster than ground methods.

(d) More sophisticated equipments can be carried as weight and space are no serious limitations and this partly compensates for other shortcomings of the method.

(e) These are best suited for large areas. Due to this the regional picture is clearer and aids interpretation vastly.

**Limitations and disadvantages of air borne survey are that :**

(a) It is not final but the anomalies obtained require confirmation by other methods of exploration.

(b) Interpretations can vary, due to the number of variables which have to be taken into account.

(c) Support by geological ground work is essential.

(d) The resolution and accuracy vary with the speed and height of aircraft.

(e) Locations are also subject to lack of high precision but this can be greatly circumvented by the use of helicopters.

**Surveying from high altitude aircraft and satellite**

Surveying from high altitude aircraft and the use of earth resource satellites for the same purpose is of recent origin. The photos taken by the satellites Gemini, and Apollo have invoked great interest among earth scientists. Photos taken by satellites of our country clearly depict certain broad structural lineaments which can be utilised in deciphering the tectonic evolution in addition to providing clues to the controls in mineralisation. It is hence of interest to know the basic principles involved in this method of surveying.

**Principles involved**

All materials emit E.M. (Electromagnetic) energy of varying intensity and covering a particular segment of the E.M. spectrum. Radio waves and visible light are emitted by the sun and stars, whereas the planets emit reflected light from the surface. In addition energy is also radiated by the surface and this is a function of the surface temperature. Since this energy is dependent on the physical and chemical properties of the materials, this induces a variation in the nature and intensity of the E.M. energy radiated or reflected.

The temperature of the earth's surface is $300°K$. Hence the energy emitted falls beyond that of visible red light, and most of this energy is confined to the Range 2 and 15 microns. Solar energy is also reflected by the earth's surface and the amount of energy reflected and transmitted depends on the nature of the surface, colour, moisture, etc.

The earlier technique in remote sensing, employed infra red cameras with sensitive films, but of slow speed. Later the use of radar, with a narrow E.M. spectral band, yielded images which could reveal topographic details, drainage patterns ET. Alternatively infra red thermic sensors were used, and a high degree of precision has seen attained by the use of solid state type which is comparatively inexpensive and yet extremely sensitive and accurate. These sensors can separate and detect very small differences in wave length of the emitted or reflected E.M. energy. The receiving instrument has filters each of which covers a definite portion of the E.M. spectrum ranging from the I.R. to the U.V. from the filters 5 or 6 channels of E.M. radiation emerge and they are recorded either on films or on magnetic tapes. The magnetic records can be deciphered with the aid of a computer in a very short time to yield the images for map production.

Some of the advantages of remote sensing are the speed of interpretation due to the high degree of automaticity and the consequent rapidity of map production. Further a fund of details in regard to topography, drainage, variations in soil, geological structure, ground water potentiality etc. can also be obtained.

# PART IV

# GEOCHEMICAL PROSPECTING

# (A) GEOCHEMICAL PROSPECTING FOR METALLIC MINERAL DEPOSITS

## Fundamental concepts

Geochemical prospecting is concerned primarily with the examination of the earth's crust, comprising not only the rocks, but also the waters and the gases, with a view to locate mineral deposits. Hence, it is essential to understand the processes involved in the formation of the materials of the crust etc. Taking this aspect into consideration, it is possible to sub-divide the geochemical processes as (a) Primary and (b) Secondary.

The primary process are connected with magmatism as well as with the processes of metamorphic evolution. Secondary processes are mainly those associated with *supergene* agents of rock degradation as well as magmatic waters etc. derived from the primary processes. In general, however, both processes are cyclic in nature. Starting at any one point, on this geochemical cycle and proceeding in a certain direction the same point is reached after going through the evolutionary circuit, e.g , starting with the igneous phase, next is the sedimentary followed by the metamor-phic phase and then deep geosynclinal and its associated magmatism. Thus the cycle is complete.

## Dispersion (Primary)

In geochemical prospecting, the primary object is to detect diffe-rences, in the distribution of the elements, in the crust. Hence, the factors controlling dispersion are of immediate consequence. This aspect again takes us to the question of mobility of the elements. Dispersion is influenced by both mechanical as well as chemical processes and this can happen under the primary as well as the secondary petrogenetic conditions. Primary concentration of ore minerals is mainly according to the P and T conditions etc., when the elements under normal magmatic differentiation separate out in groups such as oxyphyllic or chalcophyllic, lithophyllic etc. In addition various controls like structural, lithologic and stratigraphic also play a significant part as loci for emplacement and concentration giving rise to economic mineral deposits. Superficial structures are mainly the controls for secondary dispersion. Factors influencing chemical dispersion are those which affect the equilibrium of the petrogenetic system, in the case of primary dispersion. Whereas mechanical dispersion is also possible when a sufficiently differentiated magma is emplaced in the earlier formed solid phase. The same will be true of the pegmatitic and hydrothermal phases. In the case of secondary chemical dispersion the chief factor is the solubility, while the mechanical aspect of secondary dispersion depends on the resistance of the rock, to weathering, decomposition and disintegration.

Thus for the rock components, to become mobile a change of phase is required, so that migration can take place. Besides, there is also the concept, which is fast gaining ground, that ionic migration in the solid phase is also possible, in the case of igneous and metamorphic rocks. As in the case of sedimentry rocks, however, the liquid phase will essentially account for the mobility of the constituents. The process is not as simple as stated above. It is not a mere question of P and T condition but far more complex, as the amount and nature of the volatiles and gases available in this system, form an unknown variable and the effects of these are not usually found in the earlier phases. The formation of *eutectics* and other connected thermodynamic conditions are to be accounted. Again, in the case of trace elements, i.e,, those other than the major rock forming elements, in general, it is seen that there is a certain pattern of distribution, depending on their affinities, valence, ionic radii, coordination number, and electron configuration in the cation. Taking the above properties into account it is possible to say why U and Th enter into the structure of ziron. In general cations possessing comparable ionic indices show the strongest tendency for mutul replacement.

Reverting to the problem of dispersion of ions in relation to ore deposits e.g. sulphide deposits, it will be noticed, that under conditions of oxidation a high concentraion of sulphates is likely to result and in consequence leaching and migration take place. Mobility, it may be said is only relative, but in general, it may be taken that in the series Pb, Au, Cu, Zr, and Ag, Pb is least affected while Ag is most mobile. This, however, is too generalised an inference as, the influence of the environment is an important factor. Thus, copper is immobile in a calcareous environment, while lead and iron are immobile in both siliceous and calcreous environments.

As indicated earlier, differentiation whether igneous or metamorphic in character, involves the segregation of elements into groups and such associations are so distinctive that they have been identified in ore deposits—e.g., oxyphile, chalcophile or lithophile. Thus the presence of one member of a group should lead to the suspicion that the associated elements are also possibly there.

**Path finder elements—background values etc.**

In the case of certain minerals, especially those containing the elements which do not respond to chemical weathering and are poor in concentration, e.g., Au, the practice has been to find out the concentration of an associated element known as the "pathfinder" element. The pathfinder employed in the case of Au is As, but Hg has also been used with success taking advantage of newer improved spectrographic techniques for the detection of the element in traces.

Some of the more important pathfinder elements are given in the table below :

| Name of element | Nature/position of sample | Nature/occurrence of ore |
|---|---|---|
| 1. As | Wall rock, soil, stream sediment | Auriferous vein |
| 2. Hg | „ „ | Multimineral deposit comprising Pb. Zn—Ag |
| 3. Se | Gossan „ | Sulphides of epigenetic origin |
| 4. Ag | Soil | Auriferous lode |
| 5. Mo | Water, stream sediment, soil | Copper associated with porphyry |
| 6. So₄ | Water | Sulphide deposits in general |

A helpful guide, in assessing the relative values of the various elements present, is the 'background' value. The background value is the amount of a particular element present in the parent rock not affected by dispersion or migration. In igneous rocks, a fair idea of the average expected background value can be had from data available in published literature e.g., expressed as parts million $Sb=0.3$, $As=2$, $Ce=40$, $Cr=117$, $Co=8$, $Au=0.00$, $Hg=0.06$, $Mo=1.7$, $Ni=100$, $Pb=900$, $Pt=0.005$, $Se=0.01$, $Th=13$, $W=2$, $U=2.6$, $Zn=89$, $Zr=170$. Thus in a rock if the concentration of any element is in excess of the background value obtained, it is to be assumed that this is due to the effect of either primary or secondary dispersion and concentration.

In the case of soils, background values can be very unreliable as there are too many uncertain factors. In the earlier paragraphs, the various factors causing the anomalies and also those which serve to help in locating and assessing the "anomaly" have been briefly indicated. It is to be emphasised that in geochemical exploration, it is the anomaly which is of utmost importance and that it is the magnitude of the anomalous distribution of the elements that requires interpretation, in the light of the geological conditions. Apart from the normal background values, the threshold value or the upper limit of the fluctuations in the background values, should be identified. Here again the threshold, for the region, can be identified as also the local threshold for the area. Taking the case of an ore body, the values of a particular element are determined in certain critical cross sections as shown (Fig 4.1) and graphs plotted. The threshold value for the area, (2) within its area of influence (b), caused possibly by dissemination associated with the dispersion due to ore body (a), should be established and then the threshold for the area as a whole (1), also determined from the graph.

In order to arrive at the figures for the background and threshold values, and eliminate the possibility of the influence of local erratic values, the application of statistical analysis provides a valuable tool. The relevant portions on statistical methods in PART IX—Examination of mineral properties—may be usefully consulted. When the cumulative frequency distributions are graphically represented, then it becomes clear whether the population sampled is of the normal type, or long-normal type, or whether the sampling included more than one population. This latter aspect is revealed by the inflection in the cumulative frequency curve. Standard deviation is used for identifying the anomalies indicating the threshold etc. The value for the threshold is taken as $X + 2S$, where $X$ is background and $S$ is the standard deviation.

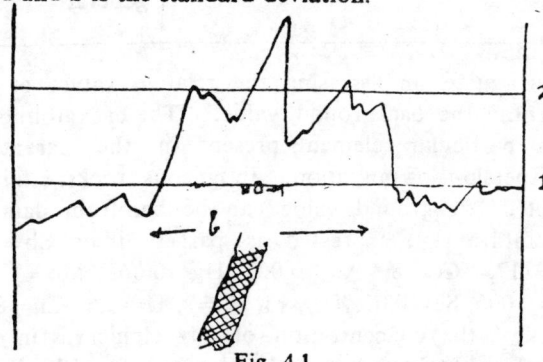

Fig. 4.1

**Primary dispersion** can be classified by means of certain distinguishing criteria such as (1) genesis, (2) causes, (3) type of mobilisation, (4) composition, (5) pattern and extent.

(1) Genetically primary dispersion may be grouped under (*a*) syngenetic and (*b*) epigenetic.

Since syngenetic dispersion is associated either with igneous intrusions or with processes of granitisation or sedimentation, it can spread over large areas including a wider range of rock types irrespective of the stratigraphy. When the dispersion is the result of an emplacement of small dimensions then its aerial extent may be limited. So is the case when sedimentary formations are restricted in area.

An aerially large dispersion is termed a geochemical province.

(*i*) Geochemical provinces are the counterparts of metallogenic provinces such as Aravalli-Mysore (Dharwar), Singhbhum—Keonjhar etc. which constitute extensive areas of iron ore mineralisation.

(*ii*) Localised syngenetic types are those which are not as extensive as the geochemical provinces, but are of the order of a few miles. Such occurrences can be attributed either to petrogenesis or magmatism or metamorphism. Asbestos mineralisation of Andhra Pradesh can be considered as an example.

Epigenetic dispersion may again be identified as (*a*) hydrothermal and (*b*) reconstitution due to dispersion.

(*a*) Hydrothermal and pneumatolytic phases of a magma/migma are characterised by the action of water, at high temperature, steam as well as solutions and gases. The types recognised are (*a*) wall rock anomaly, (*b*) leakage anomaly, and (*c*) mineral zoning. The loci of such dispersion is in the vicinity of ore deposits in the form of halos, aureoles and diffusion patterns depending on the deposition from the various phases.

The concept of wall rock alteration is pretty well established even since the classical work of Lovering.

Wall rock alteration has also been described in the gold bearing lodes of the Kolar Gold Field of Mysore. The same has been suggested for the formation of biotite and chlorite in the copper belt of Singhbhum districts of Bihar. No information is, however, available on the geochemical anomalies associated with wall rock alteration. This is an important feature as the metal content, in the altered zone, decreases from the vein outwards towards the walls.

(*b*) Reconstitution is also indicative of dispersion, and it differs from those described above, where deposition is the main characteristic. In this a new set of minerals may be formed from those existing due to reaction with either the solution/gases, which is in thermodynamic equilibrium. Thus the dispersion is indicated either by alteration haloes or by the presence of secondary minerals. The dispersion is characterised by a fall in

isotherms away from the source which may be linear or otherwise. Associated minerals such as quartz, clay minerals, garnet etc. may
be used as geothermometers for estimating the temperature of formation.

### Anomalies

Leakage anomalies are associated with hydro-thermal deposits under
certain definite influence of structure. As the name implies, the relation,
of the ore deposit to the anomalous zones, as revealed by the structure,
is such that it appears that a leak has taken place from the main ore body

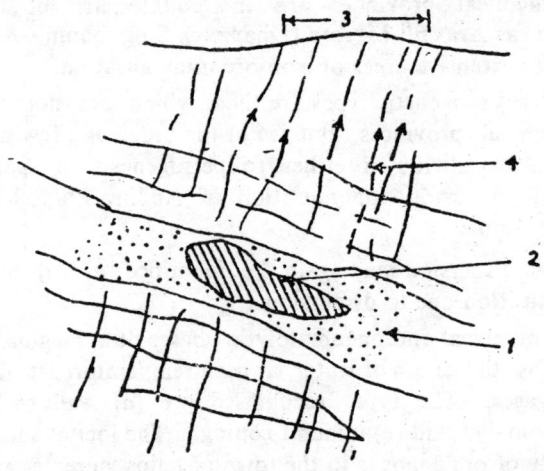

Fig. 4.2

lying at depth to produce the halo Fig. 4.2. The ore body (2) is apparently formed by stratigraphic or lithologic control and lies in the bed (1).
The halo, at (3) on the surface, is evidently due to leakage along the
secondary fracture planes (4), solutions being the main media of transport
dispersion giving rise to leakage anomaly.

Another kind of leakage is also possible as shown in Fig. 4.3, when
the movement of the solutions is confined to very narrow channels.

The first type of halo has possibly been encountered in certain areas,
where the geochemical anomalies are very strong but no ore body is
encountered by drilling to depths of 300-350 ft. This is possibly true in
the case of Rajdah, (Singhbhum district, Bihar) where strong anomalies
for copper were encountered in geochemical traverses. Drilling done
subsequently, showed the presence of only minor disseminations or pods
of chalcopyrite. Instead of secondary planes of cleavage forming the
leakage channels, the leakage may be caused by faults and the associated

shear planes intersecting the ore body. In another case secondary migration or leakage can take place through regular joints, or foliation planes.

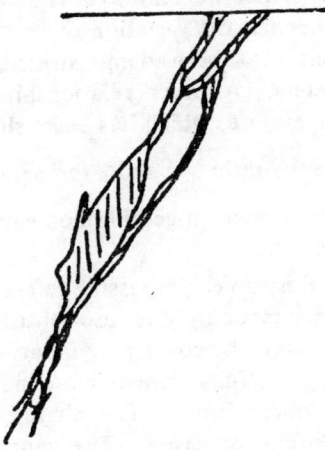

Fig. 4.3

Mineral zoning has also been established in several ore deposits. Theories supporting the accretion principle as well as the secretion principle have been put forth with evidences in support of both. Whatever be the causative, or the actual factor, zoning has been established. Geochemical variations, as indicated by the presence of characteristic minor elements in each of the zones, however, cannot, as a rule be detected by the visual methods. Careful systematic sampling, at selected intervals, and analysis by highly sensitive techniques such as spectrometric are essential for establishing such variations. Not much work has been done in this direction, in India, so far, and zoning is yet to be established definitely in relation to deposits.

The effects of hydrothermal activity may be direct and primary minerals corresponding to the equilibrium conditions may be formed as a result. On the other hand, the possibility of the indirect effect of hydrothermal activity on the invaded rocks cannot be minimised, and secondary minerals, resulting from the metamorphic or metasomatic changes, could be expected. In India, no extensive work has been done in relation to this aspect of mineralisation, except for brief references made with regard to the possibility of such influences in respect of mineralisation in Kolar Gold Fields and Singhbhum Copper belt, where chloritisation and biotitisation are considered as associated metomorphic processes. The classical paper of Sales and Mayer is, however, a standard. A second aspect of this type of alteration relates to the temperature, as indicated by the composition of some ore minerals.  Sphalerite and pyrrhotite have been cited as examples of such minerals.  The composition, in respect of Fe and Zn content in sphalerite, varies with the temperature.  With an increase in temperature sphalerite shows a relative increase in the Fe content.  In a similar manner, the variation of Fe to S content in pyrrho-tite with temperature has also been demonstrated.  Such minerals are known as geothermometers.  Another relationship, between the thermal composition and the temperature, which has been shown to exist, is the inverse variation between the ratio $\dfrac{O^{18}}{O_{16}}$ and temperatures of formation. This relationship has been used in geothermomentry and has been called isotopic geothermometry.

While dealing with this subject, a passing reference may be made to temperature determinations made by the use of the decrepitation tech-nique.  In this method minerals, containing liquid and gas inclusions, are collected and heated under carefully controlled conditions, on a heating stage of a petrological microscope.  The changes in the mineral are observed as the temperature is increased.  The vanishing temperature for the liquid inclusion is determined as well as the decrepitation temperature, at which the bubbles or gas inclusions burst.  These two temperatures can be employed in establishing the temperature of formation of the ore body.

### Sampling

Having obtained a knowledge of the dispersion patterns and the types, the next step is to undertake geochemical sampling.  No definite rules can however, be laid down for sampling of the bed rock.  As a principle, the material sampled should be such that it can be used in the tests covering both syngenetic and epigenetic possibilities.  Where an epigenetic origin is suspected chip sampling, of the cap rock, above the ore body is undertaken.  Samples pertaining to each rock type, zone of fracture or shearing, are collected separately.  The sampling should also include certain amount of the areas beyond the zone of mineralisation.  If there are seepages in the area, samples of water drawn from these sources may also be included.  Sampling may be so designed and executed that, an attempt can surely be made to unravel the type of mineralisation. Working out the ratio of $\dfrac{\text{mobile}}{\text{immobile}}$ elements $e.g.$ $\dfrac{Zn}{Pb}$ is a useful criterion in this regard.

Sampling is done across the strike of the contacts, in all cases $e.g.$, wall rock alteration, fault, vein etc.  Examination of samples should include chemical, macro and micro petrographic examination.  It is common practice to compare the results of analysis of soil as well as the bed rock from the area.

Apart from the dispersion of ions, the dispersion of gases, in the soil, is also a matter of interest. Gases escaping from a petroleum reservoir, deeper down in the crust, produce anomalous distribution of the same in both the soil as well as in the atmosphere, immediately above the seepage areas. By careful sampling and analysis of both soil and air, such anomalous areas can be delineated.

It is known that $Rn^{222}$ (Radon) and He (Helium) result from the radio active disintegration of U (Uranium). Similarly $Rn^{220}$ and He are products of disintegration of Th (Thorium), while Ar (Argon) is derived from isotope $K^{40}$ (Potassium). Uranium also gives rise to minor amounts of $Kr^{85}$ and $Xe^{133}$ (Kryton and Xenon). The gases are inert and under pressure enter into the circulating waters and are released when the pressure falls as they rise to the surface. Thus these gases can be found associated with ground water, as well as in the soil and air, at the places where they are escaping. The methods of prospecting that employ a quantitative assessment of Rn have not yet been extensively used for want of simple field technique for collecting samples.

Mercury is vapourised readily and when found in association with a hydrothermal deposit, it shows up. Thus liquid mercury is found in sphalerite and galena. In fractured areas, the mercury moves along fractures and pore spaces. Mercury anomalies are very helpful in locating Pb-Zn ore bodies as well as auriferous deposits.

## Secondary cycle and secondary dispersion

In the secondary cycle, the major factors are (a) weathering with the formation of *regolith* or *eluvium*, (b) transportation and deposition by various agencies e.g., water, air etc. As the term implies, weathering is the effect of weather on the rocks. To the geologist the zone of weathering extends upto the level of the water table.

The general treatment of the subject is covered in many standard works on physical geology and as such no more than a passing reference is called for here. It is, however, necessary that the broad principles be enunciated for a better understanding of the geochemical methods.

The major types of weathering can be classified under (a) physical, (b) mechanical, (c) chemical and, (d) biological. The last may not be purely a factor in weathering even though biological agencies are important from the point of view of rock disintegration.

Weathering affects minerals differently. Under the same conditions the relative stability, of common rock forming minerals, is similar in arrangement to Bowen's reaction series. After arranging the *mafics*, in the order of crystallisation with the co-existing plagioclases, followed by alkali felspar, muscovite and quartz, it will be seen that weathering starts at the basic end of the series, quartz surviving to the last. During the course of weathering all primary minerals decompose and give rise to secondary minerals. As weathering proceeds further, the chemical changes, which are induced by hydration or hydrolysis, base exchange, pH and redox potentials, become predominant. All the major rock forming minerals (except quartz) degenerate into minerals of the clay group, including the chlorites and vermiculite, in addition to the kaolinite, montmorillonite and *illite* groups. Under certain conditions, the mafics as well as the calcic plagioclases may give rise to the hydrated aluminous or ferrugenous oxides, commonly found in laterites and bauxites such as gibbsite, boehemite, goethite, turgite, etc.

The effects of weathering on sedimentary formations is rather complex. It is just possible that a conglomerate is made of diverse types of pebbles or may be that a graywacke is composite in character. Fortunately, however, sediments are more often bedded, hence rather uniform in hardness and the variation in mineral composition limited. An arkose or an ortho-quartzite consists of a quartzo-felspathic matrix with a cement which may be more often clayey, but occasionally siliceous or ferruginous. In the case of carbonate rocks the chances are that these rocks, whether dolomitic or calcitic, tend to be either siliceous or psammatic. Depending on the overall composition, the rock varies in its resistance to weathering. Taking into consideration the common minerals found in sediments, the minerals of the evaporite suite, and sulphides are most soluble ; these being followed by the carbonates, in the order of decreasing solubility. In this sequence, next are the mafic minerals, followed by felspars and quartz. Of the clay group which comes next, illite is least stable and kaolinite the most stable. Where the normal reactions are interfered with by environmental conditions, the solubilities are bound to vary. Other conditions which affect solubilities are (1) the presence of fissures or fractures, in the rock, which increase the area of attack, and (2) temperature of the solvent, an increase of which, generally causes an increase in chemical activity.

Sometimes, it is a problem to decide whether the topography and drainage are consequent to weathering and erosion, or whether the reverse is true. It is however true that weathering is controlled by movement of ground water and it may so happen under certain conditions, in a granitic country, that the weathering may start from a sheet joint below and work upwards to the surface. This is possibly the case in certain areas covered by the Singhbhum granite of South Bihar. Where kaolin is associated with the bogs and lignite deposits, the action of acids, produced by the decomposing organic matter, is suspected.

Having discussed the residual products of weathering, it is necessary to know whether any dissolved material is also involved. Very often, this is the case. Chemical attack, on carbonate rocks by water charged with $CO_2$ carries away bicarbonates in solution. Fe and Mn are also transported, and on release of $CO_2$ the form precipitates, which abstract Cu, Zn and Pb giving rise to false anomalies. From the mafic rocks Ca and Mg are removed, while from the more acid rocks Na and K. Sulphides yield $H_2SO_4$ and sulphates of metals are produced in consequence.

The texture resulting from weathering is sometimes characteristic especially when related to ore deposits. The well known occurrence of the iron hat or *gossan*, characteristic of sulphide lodes, is a very useful guide, in locating such ore bodies. So also box work, associated with weathered carbonate and sulphide minerals, is of diagnostic value.

While Na and K are stable, with varying conditions of pH, normally obtaining under geological environments, Ca is less stable and other metallic ions are even more sensitive to such changes and are precipitated as hydroxides and basic salts. This happens when the pH of Cu solutions exceed 5.5, and in the case of Zn, the critical pH is 7.0 and above. An increase in pH influences the solubility of the amphoteric oxides, thus aluminates and chromates may be formed. Ideal conditions are, more often than not, absent in nature, and the pH is affected by the local rock types. Thus it is susceptible to rapid variations, e. g., the presence of carbonates may affect the pH of acid solutions emanating from a sulphide ore body etc.

A factor of importance in the formation of anomalies is absorption of cations by clays and organic matter, which behave in a manner similar to Permutites (zeolites) in this respect. It is known that peat can obscure anomalies caused by ore bodies occurring below. The presence of such clays may even change the bases and the Ca, in the clay, may be substituted by Na present in the solution.

Redox is the abbreviation for reduction-oxidation potential and its value is a measure of the oxidising power of a chemical system. It is measured in volts. Eh is capable of experimental determination. It is the difference in potential between the platinum electrode and standard hydrogen electrodes, placed in the same solution. An increase in the positive valence of an ion or conversely a decrease in the negative valence is known as oxidation. Thus redox measurements can help in the reconstruction of geochemical equilibrium obtaining at the time of formation of the mineral deposit as well as during secondary dispersion.

Where there is free circulation of water, oxidation is dominant. Decreases in oxidising environment are either due to restricted movement or due to the absoption of oxygen by organisms. Such reducing conditions are possible in sheltered lagoons and deep narrow channels, inland seas etc. where $H_2S$ formed can cause the reduction of iron bearing solution giving rise to pyrite. Changes in valence, due to redox reactions, can change the equilibrium, and in consequence affect the solubility of elements.

Sorptive capacity depends on the unsatisfied electric charges on the surface of particles ; and this property is characteristic of clay size fraction e. g., clays, colloidal manganese, iron oxides as well as organic gels. Such gels can give rise to organic complexes, with metallic elements e.g. Cu, Pb, Zn, as well as U. Thus these play an important role in dispersion.

Examples of colloidal dispersion are those of silica, alumina and hydrous oxides of iron and manganese. The nature and modes of formation of colloidal suspensions have been studied in great detail. It is known that such molecular suspensions are maintained by the uniformity of electric charges on the particles, which cause mutual repulsion and which prevents coagulation. The type of charge, whether positive or negative, depends on the colloid. Silica, hydrous manganese, and humic colloids, are negatively charged. Flocculation sets in either when the charge in the colloidal matter is removed or when a colloid with an opposite charge is mixed.

Vegetation also has a role in dispersion. Elements are reduced to the soluble forms, in soils, and in this state are taken up by the plants. The soluble constituents, taken up by the plants, are either absorbed or reprecipitated in the soil. The less soluble constituents, obtained by decay, tend to accumulate ; being retained by the organic matter.

The part played by microorganisms, especially bacteria, algae, fungi, is also of primary consequence. Fixation of nitrogen in the soil is one of the earliest known bacterial actions. Precipitation of manganese, iron, copper are some of the examples of bacterial action. Sulphur reducing bacteria are also known, which reduce sulphates to sulphur.

Methods used in geochemical prospecting are as follows:
1. Detection of wall rock alteration by various techniques, including X-ray analysis for the recognition of silicitification, felspathisation, sericitisation, associated with Cu-Mo mineralisation.

2. Separation and analysis of magnetites found in the parent rock for detecting anomalous amounts of TI and Zn.

3. Assessing the amount of Mn which can be correlated with Cu content.

4. Assaying niobium in moscovite for locating columbite bearing pegmattes.

5. Examination of the A and B horizons in soil and basal glacial till for the presence of Zn.

6. Examination of surface and subsurface waters for the presence of Cu, Pb, and Zn, subsurface waters may also contain Mo, a path finder for Cu, in addition to uranium.

7. Gases are used in the petroleum field, so also Radon 222 is used in prospecting for uranium.

8. Hg is a pathfinder for Au and sulphides. Because of its high volatility, it is found in soil gases as well as in the atmosphere immediately above. The origin of Hg is considered to be ionic if the ratio of Hg organic to Hg in the clay is less than 2. Where the value exceed 2 the source is a vapour.

9. $H_2S$, $CH_4$, $O_2$, $N_2$, RA, A, $SO_2$, F, $Cl_2$, 1, BR. have also been used for detection.

10. Aerosols or fine dust suspended in the air, have also been sampled aircraft.

In addition to the above methods there are also the biographical methods which have been applied in various areas.

1. Dogs can be trained to pick up sulphite trails, since they have very sensitive olfactory organs.

2. Remote sensing, using the near infra red (350 to 1100 nm.) has proved valuable in differentiating vegetation of the same species growing in mineralised and nonmineralised areas. For example pinus rubens or red spruce and abies balsamea or balsam fir.

3. Turmites in the desert areas of S. Africa burrow deep to reach the water table 30 or 40 m. or even to 160 m. and grow fungi with these hills. Turmite hills are used for detecting the presence of Au, Zn, Ni.

4. Fish liver has been investigated for anomalous concentration of Au and Zn and a correlation has been found between the provenance and the habitat of the fish.

## Geochemical field techniques

Of geochemical methods, the analysis of soils is most extensively employed in mineral exploration. The Geological Survey of India started geochemical prospecting in the middle 50's and since then considerable expansion of its activities has taken place and geochemical prospecting has been extended over wide areas of the mineralised belts in Singhbhum, Bihar; Jhunjhunu, Rajasthan; Chitradurg, Kolar, Mysore; Guntur, Nellore, Andhra Pradesh; and the success achieved in regard to base metal exploration has been encouraging.

It is recommended that when undertaking mineral exploration in unknown areas where there is little reason to suspect mineralisation, below the cover of soil, a preliminary survey of the area can be carried out to start with. If possible, the trial survey may be initiated in an adjacent area, where mineralisation has been recorded or noticed, so as to obtain a knowledge of the behaviour and dispersion of ions, in relation to the soil.

Background values are determined as also the threshold values, adopting the standard sampling procedure described later. Frequency distribution applying statistical methods is recommended for achieving a higher order of accuracy.

In order to proceed systematically, it is useful to know the type of soil present, whether it is residual or transported and its genetical aspect. In order to get this information it is necessary to put down pits to touch bedrock which in some cases, may even be about 90-100 ft. below the surface. Where pitting does not provide full information, core drilling or auger holes may be employed. The pits are logged in foot by foot sections, and where a striking change in the grain size, or colour or mineralogical composition is noticed, then samples should be collected for each variety. Sampling of pits, which are to be used for establishing the background values, is also carried out in a similar manner. Since normally, square pits are preferred to circular pits, the four sides of the square pit, if sampled, will be preferable.

The samples collected for analysis are from the seived portions, and they represent the 80-mesh or 100-mesh, fractions. Analysis are directed towards determining ore minerals, as well as any associated path finder. It may also be necessary, in some cases, to carry out several mineral analysis, before firm conclusions are drawn.

On the basis of the analytical data, the optimum depth of pitting is chosen, as well as the mesh size to be used in seiving etc. In order to determine the sampling error by statistical methods, several samples may be taken at a few selected points. The sampling should be extended so as to include not only, the super adjacent anomaly (along the ore body) but also the "base-of-slope" colluvium as well as seepage areas and down drainage, if any. Assessment of other variables, such as degree of oxidation with mineralisation, depth of overburden, associated bed rock etc., is also of importance.

Where the soil is transported, dispersion may be mainly influenced by the roots of trees and ground water. Dispersion, in glacial deposits, is influenced mainly by the direction of transport as well as by the presence of syngenetic fans. The lateral seepage anomalies have also to be examined. Organic soils can also be sampled, as in the case of residual soils. It may just be possible to establish a relationship between the metal contents pH, Eh, and organic matter. Examination should cover the super adjacent as well as seepage anomaly areas. Sampling of lacustrine sediments should be such that the traverse is across the expected trend of the outcrop of the ore body, extending into the lake. Samples should include material representing the deeper layers, nearer the mineralised zone, rather than the upper layers, which, more often than not, consist of organic debris. Deltaic deposits may also be sampled for ascertaining, whether mineralisation is associated with the rocks of the drainage area.

Contamination is a real source of danger in the geochemical studies. The sources of contamination are (1) fertilisers, (2) road metal, (3) industrial plants, (4) mine spoil and (5) drainage. Contamination, from the last source, can penetrate to deep levels and give rise to anomalies of a confusing nature. Detailed study and careful examination will, however, lead to their elimination ultimately. When copper smelters have been working for a number of years in an area, it has been noticed, that there is a superficial enrichment of Cu in the soil ; but this scarcely extends to more than a depth of 6 inches from the surface. Ancient smelting sites, often, are also a serious problem, and the Pb anomalies extending from depths varying from 2 ft. to depths of several feet have been recorded.

The most commonly used layout, of the geochemical sampling grid, is the square type, since it is easy to follow, quick in pegging out in the field, and easy to plot on the map.

When the general strike of the lode or vein is known, then the orientation of the grid is determined by this direction. A control line is laid, more or less parallel to this direction, and near the expected medial part of the deposit. The control line can be extended to any desired length. Cross traverse lines, at equal distances of about 1000 ft. apart, to start with, are run at right angles ; the separation between these depending on the extent of numeralisation and minimum economic strike length not exceeding 1/3 this distance. It is, however, to be accepted that smaller the interval the better ; but this should also not entail undue expenditure in time, man power and money. The distances between the cross sampling

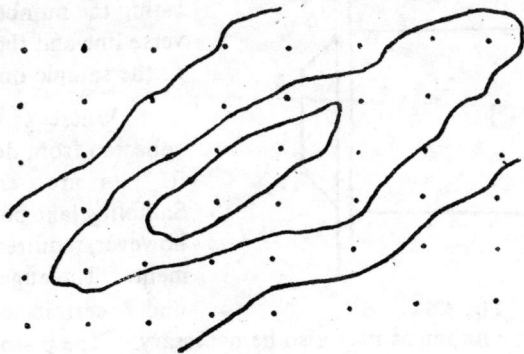

Fig. 4.4

lines and the sample points are decreased gradually in the second and subsequent stages to 500 ft. and even to 100 ft. if necessary, or less ; such that a grid comprising smaller squares results.

Where the dispersion patterns are expected to be linear or fan shaped, a square pattern is indicated, but the sample interval should be such that at least 4 samples lie within the anomaly area (Fig. 4.4). Where the terrain is rugged, it may be difficult to maintain a systematic pattern, in the lay out. In such a case, it is often possible to simplify the procedure by adopting a linear pattern, following the crests of the spurs and foot of the slopes (Fig. 4.5). In a section shown the figures marked (1) to (10) represent the samples points. If considered necessary and if possible; other more or less parallel

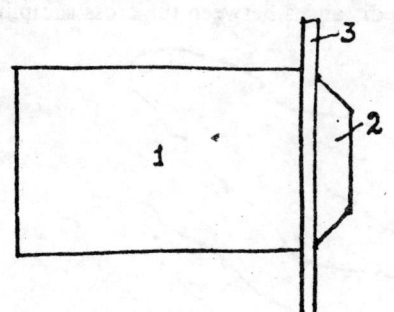

Fig. 4.5

traverses can be run. It has been noticed that in sampling soil, the A horizon gives better results for Pb, Zn, Cu, As, Co, Mu, Hg. Veins covered by 20 ft. or more of fill gave little or no indication of the elements when sampled.

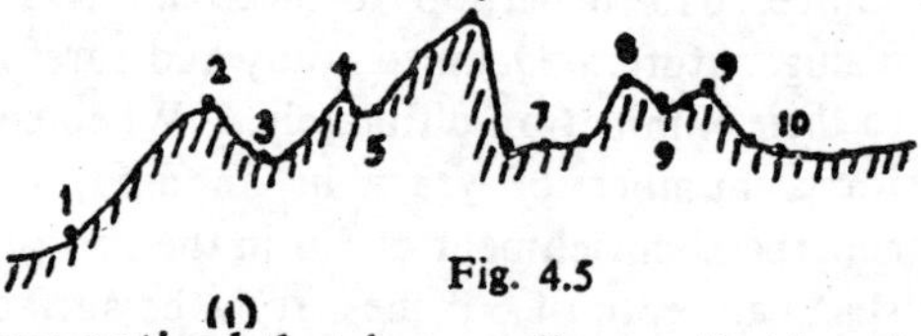

The amount of sample required for geochemical analysis is about 20 to 50 grams. Various organisations have their own methods of packing the samples. Sometimes a thick glazed "packing paper" or strong craft paper envelope (1) (Fig. 4.6) with a metallic strip (3) attached to the flap (2) is employed. After placing the sample inside the envelope, the sealing is simple. The flap is closed, and the top part of the envelope is rolled once or twice over the metal strip. The envelope is sealed by bending the metal strip over the rolled part of the envelope. Alkathene bags, of fair thickness, are often employed. These have to be sealed by the application of heat, say with a lighted match stick. All the samples should be numbered serially; indicating the number of the transverse line and the serial number of the sample on that line.

Where samples are to be collected from depths of 1 or 2 ft. pits are easiest means. Sampling lake bottom deposits, however, requires special equipment. The auger can be used under certain conditions but special drilling equipment may also be necessary. The piston borer used

Fig. 4.6

in undersea sampling can be used, or drilling with vaccum as done in soil sampling may also be employed. The locations of the samples should, however, be correctly plotted by accurate survey methods.

Pace and compass survey is sufficient for most purposes and the contours are determined by the aneroid. Air photos provide sufficient details required for location and can be used for ridge-valley traverses. It is however essential that the stations, on the traverses, and the sample points, be labelled so that they can be identified later. Metal tags or flags can be used for the purpose as these are strong and are capable of reuse. The field book should contain details of the surveys as well as particulars of the samples collected.

Samples, collected from the pits and drill holes, require preliminary treatment. A 2 mm. mesh sieve is employed to eliminate rock chips, pebbles and roots. Finally seiving is done with 20-30 mesh set. Where samples are too wet, preliminary drying is necessary before seiving. This is carried out in the sun. It is safe to use stainless steel, nylon or bolting cloth sieves rather than brass, where metals under investigation include Cu and Zn. Where lead-zinc ores are suspected, it is generally the practice to analyse for Zn because of its broader dispersion pattern. Pb gives a better location because of its relative immobility. Methods adopted in the analysis of the material depend more on the practice. Calorimetric methods for base metals are simple and are often practiced in many organisations e.g., U.S.G.S. as well in the Geological Survey of India.

## Geochemical Analytical Methods

In keeping with the high degree of accuracy required for geochemical analysis, instruments of high precision and sensitivity have been installed in the major laboratories. However, the simpler and less expensive wet methods have their own value, and are still extensively employed in the field, since they provide values with the necessary order of accuracy required for preliminary determinations. Hence the use of the dithiozone for detecting and estimating Cu and Zn has become routine practice.

More sophisticated instruments have however, been installed in several laboratories especially in the U.S.A. and these include spectrophotometer, flame photometer, chromatograph, emission spectrograph, Atomic absorption spectrograph, and the direct ready emission spectrograph. Of all these equipments the atomic absorption spectrograph is the most popular, because of the high order of accuracy and the simplicity of the operations involved in its use. The United States Geological Survey has a mobile laboratory fitted with an atomic absorption unit for field use.

Emission spectrography is used in the preliminary stages for the identification of all major elements in the sample. In order to speed up the work the normal spectrographic technique has been replaced by an electronic system, and this improved version of the instrument is designated direct emission spectrographer. The instrument is yet costly more so for the developing countries.

Laboratories for the identification of hydrocarbons present in soils rocks, drilling mud, and gaseous emanations, require the use of the gas chromatograph, in addition the mass spectrograph may also be used under special conditions.

Neutron activation analysis with its high sensitivity has also been tried in limited areas, with success, and it is likely that this method also will find wider use in the field.

The spectrometric methods are more popular in the USSR. The field data as well as the results of the laboratory analysis are finally plotted on a map in which the topography (drainage included) geology, structure, together with old workings etc. are also depicted. The scale is so chosen as to make the maps neither too big nor too small. In addition, to these maps other maps known as "interpretation" maps are required. These may be in the form of "overlays" and these will show the "geochemical contours" or "iso anomaly" lines. The contours represent anomalies of the same intensity. The contour interval is so chosen as to bring out the maximum in a very prominent manner. Geochemical traverses are also of considerable importance and help in the interpretation of data. Such traverses are often combined with the geological section along the same line (Fig. 4.7).

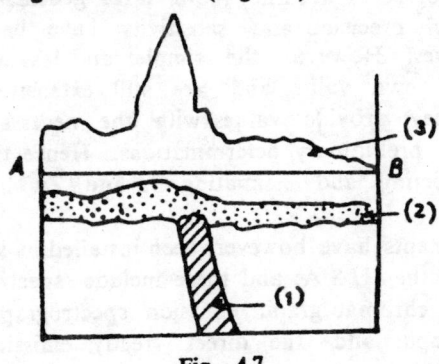

Fig. 4.7

(1) is the ore body, (2) is soil cover and (3) the geochemical values plotted on a traverse AB.

The four main steps in the interpretation of the data are (1) estimation of background and threshold values, (2) distinguishing non-significant from significant anomalies, (3) distinguishing lateral and superadjacent anomalies, and (4) evaluation of the anomalies in terms of the ore body.

The estimation of background and threshold values is a painstaking work as mentioned earlier and it is more so when the general order of the anomalies is low. A comprehensive study is called for, in such cases, and the geochemical relief, as indicated by geology, geomorphology, as well as by the dispersion patterns, has to be considered. In case, geophysical data are available, the same also should be included. It is only when the threshold value is used as the cut off that the anomalies show up.

The non-significant anomalies are those which arise out of (a) the fortuitous occurrence of pieces of rock rich in metal content, (b) contamination due to human agency, which in effect will amount to "salting" even though not intentional, (c) errors in sampling and analysis. i.e. either due to wrong numbering of samples or due to a mistake in computing the results of analysis. Confusion between superadjacent and lateral anomalies can lead also to wrong interpretation. A lateral anomaly, caused by horizontal movement of ground water or soil, should not be taken as a superadjacent anomaly. The two chief types of superadjacent anomalies are (1) residual pattern, and (2) biogenic pattern in transported cover, due to upward migration of ions. In general, if after allowing for some distortion, during the process of dispersion, the anomaly pattern shows certain broad relationship, in trend and shape, with the landforms, but yet conforms to the geological trends then in all probability the anomaly is superadjacent in character. Lateral anomalies on the other hand show a close correlation with the topography. Fossil anomalies, connected with an earlier cycle of soil formation are more difficult to interpret. For example anomalies related to seepages can become isolated due to change in ground water conditions. Similarly downstream extensions of residual anomalies can also be isolated and appear as independent anomalies, as a consequence of changes in drainage pattern resulting from diostrophic or cataclastic earth movements etc.

Appraisal of anomalies has to be undertaken after taking into account all the factors that influence their value. It is not always justified to correlate the intensity of anomalies with that of the associated mineralisation. Areas showing higher values of anomalies will have to be examined again in greater detail for further appraisal.

Subsequent work with regard to lateral anomalies will have to proceed up slope or up drainage depending on the agency responsible. Lateral anomalies have to be drilled for proving the precise location of the ore body, in the final stage. Superadjacent anomalies in their turn will have to be followed up by deeper pitting and detailed sampling. In Fig. 4.8 (1) is the ground surface, and a, b, c, d, e, f are pits, in residual overburden showing the variation of geochemical values with depth, (2) is the ore body, sample pits (b), (c) and (d) show a steady increase in values with depth and show the presence of the ore body in the vicinity. By

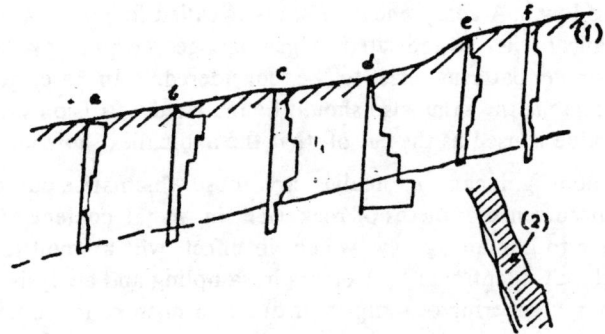

Fig. 4.8.

proceeding in the direction of increasing values it is possible to locate the ore body. In essence this method is akin to "tracing float".

In the case of transported overburden the anomalous values are confined more or less to the contact between the top soil/alluvium and the bed rock. Hence samples will have to be collected right from the surface up to the contact with the bed rock. It will also be noticed that the anomalous values increase gradually in the direction approaching the ore body. This operation is akin to "tracing float" as applied to gold and tin prospecting where the values are confined mainly to the contact of the alluvial material and bed rock.

The use of the geobotanical method, in exploration of mineral deposits has also come to stay. In India the medicinal value of plants has long been recognised and several of the ayurvedic preparations contain either leaves, roots or twigs, either ground or ashed. Plants depend for their growth on the availability of light, $CO_2$, moisture, and minerals. Photosynthesis accounts for the building up of plant tissues and for this purpose light and $CO_2$ are the essential requisities, and the chlorophyl found in the leaf of plants e.g. angioperms (i.e. flowering plants) can absorb $CO_2$ from the atmosphere. Plant growth also is dependent on moisture and salts derived from soils and minerals. The moisture acts as a medium of transport for the minerals found in solution in the soil. Solutions enter the root, by a process akin to osmosis and travel upwards, to the various portions by the fine capillary system found in the plant tissue, impelled by capillary action, as well as by a more positive suction effect produced by the evaporation and release of unwanted atmospheric gases from the leaf known as transpiration.

For plants, to absorb the nutrients and minerals, these materials should be made available in such a condition as to be taken up. Solutions, in soil moisture, and absorbed ions in clays, capable of being released by base exchange, are the most common sources from which the plant acquires its nourishment. As simple as the reactions may appear, the number of variables on which they depend are quite appreciable. Of these (a) the pH of the soil, (b) redox potential, (c) base exchange capacity, (d) the presence of complexing agents both organic and inorganic, are major factors.

Authorities believe, that the simple straight forward osmotic process may not alone explain the means by which the plant derives its mineral requirements, from the soil. It has been suggested, that initially during the process of absorption, an acid environment is created, near the root tips, by release of $CO_2$, by the plant. The carbonic acid which forms (by reaction with $H_2O$) makes available a copious supply of H ions which affect exchange of base and metal ions held in the lattice of clay minerals especially the *montmorilonite* and *illite* groups.

It is known, that the activity of roots may be strong enough to cause the breakdown of primary silicate minerals and induce absorption, even of elements which can be toxic, e.g. Se, Pb, U and V. Toxic elements are evidently deposited in the tissue and thus impede the rate of absorption and circulation. Certain plants show the maximum accumulation of Pb, U, and V in the roots and exceptionally at the tips (top).

From the relative abundance of certain elements in a plant it cannot be concluded that the same order of abundance is true of the soil of that area. Relative abundance of elements may only be a consequence of selective absorption by the plant, of certain elements, that are required for its growth and may be very different to the concentration represented in the soil.

All plants require certain essential elements such as Na, K, P, S, Cu and Mg, besides minor amounts of Cu, Zn, Fe, Mo, Mn and B. A lack of these in the soil affects the growth of plants. On the contrary, excess can also be harmful to the plant. Certain elements provide a short range of toleration by the plant, whereas others are tolerated within a wider range of concentration.

Geobotanical anomalies can be detected by various methods e.g. variation in plant types. It can so happen that particular genus or even the entire class of plants, growing in the same area, show considerable variations in concentration of minerals e.g., a particular kind of grass may show an abundance of one element, in which shrubs in the same area are relatively poor; but the shrubs may show an increase, in contents, in respect of some other element in which the grass is impoverished. This may also be true of other types of vegetation. It is well to guard against this type of anomaly, as correlation made on different genera or classes of plants is liable to gross errors.

The root systems of plants vary with the type of plant and its habitat. Plants growing in an arid country, in general, require to have deep roots, their leaves are also peculiar, generally needle shaped or phyllodes. Plants growing in swamp have "breathing roots" pneu-matophores. The xerophytes growing in desert climates e.g. the cactii or yucca, have a shallow root system, while the phreatophytes have a deep root system which may go down to about 50 ft. or more to reach the water table.

Having located a species of plant for the purpose of geobotanical indicator studies, it is necessary to follow the anomaly in relation to histological association i.e. any particular part of the plant, viz. roots, bark, twig, leaves, flowers, or seed, etc. Since the concentration of elements is not uniform, even in the same plant, and since a variation may appear even within the same part between one side and another one should be prepared for this eventuality. It may so happen that the leaves on any side of the plant may show a greater concentration of a particular element than the leaves on the opposite side of the same plant.

Root penetration is generally greater in well drained soils. The pH of the soil is also affected by the drainage conditions and this in turn affects the solubility of elements. The mineral up take of a plant also depends on the photo-chemical activity which again is dependent on the amount of sunlight it receives. A plant in the shade takes in less mineral substances than one which is exposed to the sunlight. Seasonal variations, in concentration of elements in the plant tissues, are also to be borne in mind when sampling, since concentration of salts can vary from summer to winter.

The accentuation of geobotanical anomalies appears to be dependent on the mobility of the element in soil solutions. It is as high as 100 to 1 for Mo, while for Fe and Ni the contrast is 5 : 1.

The correlation between the geochemical soil anomalies and the geobotanical anomalies is also fortuitous. The soil anomalies are generally homogenous. So also, the form of the anomaly is subject to so many influences and it is difficult to predict or identify which is the main. The geobotanical anomaly may combine with the syngenetic hydromorphic, and biogenic anomalies of the soil, in addition with the ground water anomalies and thus the pattern obtained is a combination. If the roots have entered the ore body or lie within the superadjacent soil, the superadjacent anomaly is reflected in the plant tissue, and the same will be true for the other types of anomalies.

### Techniques used in geobotanical survey

It has been suggested that the preliminary reconnaissance surveys should be carried out with a view to ascertain the depth, extent and dip of the ore body, as well as the orientation and behaviour of ore shoots, grade of ore, nature and extent of secondary dispersion (in rock) and also the influence of the water table and plants. Observations will have to be made on the type of plants as well as rate of growth. It may be determined which type of plant or tree grows well and what type is affected adversely by the soil. Samples should be collected from the mineralised and barren areas. After assembling the geological, botanical and other data (groundwater, drainage etc.) the pattern and interval of sampling can be drawn up.

Sample material should normally be collected from deep rooted plants, and as far as possible, a plant which is wide spread, and occurs over a large area, should be chosen. As an alternative if a relationship between the amount of metal contained, in one plant genus and another can be established, then the scope of the examination can be extended. For most purposes one or two year old twigs will suffice, and taller shrubs are likely to be free from contamination caused by settling of dust as well as splashing of mud (soil) due to rain. Mosses and algae are however free from these dangers.

Sample points may be selected on a grid pattern, wherever possible. This is easy where there is a sufficient density in the type of vegetation. Where the vegetation is sparse then the layout will necessarily have to be irregular, but a correct location, of sample points, will help in analysing anomalies when recorded on the map. Usual type of pruning implements are satisfactory for sample collection, and about 20 grams of material will yield about 1 gram of ash, which is sufficient for analytical work. Where the presence of volatile elements is suspected ashing should be done in carefully controlled conditions. Small plastic tubes with caps are quite satisfactory, but even plastic bags and paper bags have been used for collecting the samples (ash). Maps showing the sampling points may be prepared as an overlay to the geological cum structural maps (showing outcrops).

The analytical method will depend on the equipment available, and the metal content may be expressed in terms of dry plant material or in terms of ash.

Finally when the anomalies have been plotted, the interpretation should be done giving full weightage to the various sources of error. The influence of such factors as local variations in pH of soil, drainage, duration and intensity of sunlight etc., should be identified and eliminated. Determination of the ratio of a pair of elements, in behaviour, under similar conditions, will serve as a means to eliminate such fortuitous values. One example of such a pair of elements is Cu and Zn, whose concentration in the tissue may vary widely, but yet the ratio is consistent within allowable tolerances.

Geobotanical studies are based on plant ecology. Under certain definite conditions of climate and environment only some plants survive. Thus xerophytes thrive where the other plants die. Nature selects. Survival of the fittest is the general rule, but adaptations are possible within limits.

Geobotany entails the study of plants in relation to specific geological environment. There are several factors which influence plant growth and life in addition to those which may be termed geological. These factors include the influence of sunlight, elevation, climate, fires, insect pests. The influence of geological conditions may be indicated by abnormal change in colour, morphology of leaf or flower etc. due to toxicity. Careful examination is essential in order to identify effects of toxicity, and evaluate the same.

Geobotanical criteria may be obtained either, from the abundant occurrence of a species, genus or class, or from the variations in the morphology and habitats of plants of the same species, under the same conditions. Such indications may be employed in the location of ground water sources, saline deposits, hydrocarbons etc. The geobotanic methods have not been applied so far, systematically in India, and only sporadic investigations with disappointing results have been reported from the application to base metal deposits in Eastern India.

It is, however, essential that the method be practiced with greater enthusiasm, undoubtedly it calls for the use of other disciplines, especially systematic botany. Application of plant studies to the North Rhodesian copper belt had yielded encouraging results. The discovery of the so called "Copper plant" Ocimum homblai a cousin of our sacred "tulsi" Ocimum sanctum is of special interest. Further the occurrence of Acrocephalus robertie in the Katanga copper belt, also a Labiatae, of the family Ocimoidae, should afford a clue as to what to look for. Gypsophila patrini is reported from Altai copper; deposits in Central Asia, and gypsophila species is also known in this country but these require yet to be identified as indicator plants.

Air photos often reveal the changes in vegetation, which can be detected either by change in tone or grain in the photos. Verticals show the differences very well and such photos have to be used as base maps for geological as well as for foresty work. Such photos can form the basis for geobotanical work as well.

5. Where plants are used for geochemical prospecting, three criteria may be used in identifying any anomalous concentration of base metals.

1. when an element is toxic plant growth is adversely affected and this is revealed either by the colour of the leaf or by stunted growth etc.

2. When an element is benecial that specifies thrives and spreads.

3. Plants may accumulate a particular element in excess of their requirements. Such plants are called accumulator plants.

Some physiological and morphological defects caused by toxic elements are as follows.

| Element | Defects Induced |
|---|---|
| Al | Short and stubby roots leaf scorch—mottling. |
| B | Dark foliage—marginal scorch of older leaves—shortened inter nodes—deformed—incomplete development—creeping forms—slow maturity—increase in knots and lumps. |
| Cr⁹ | Yellow leaves with green veins. |
| Co | White dead patches on the leaves. |
| Cu | Dead patches on the leaves from tip—purple stems—chlorotic leaves with green veins— poorly developed stunted roots—creeping strile forms in some species. |
| Fe | Stunted tops—thin stubby roots—cell division disturbed in algae. |
| Mn | Chlorotic leaves—reddish colour and lesions on stem and petiole—curling and dead leaf margins—distortion of laminae. |
| Mo | Stunting—yellow orange colouration. |
| Ni | White dead patches on leaves—chlorosis—apetalous stetrile form abnormal forms. |
| Zn | Chlorotic leaves with green veins—white dwarfed forms—dead areas on leaf tips—roots stunted. |
| U | Abnormal number of chromosomes in nuclei—Unusual shape of fruit—stetrile apetalous—stalked leaf rosette. |
| Mo, Cu | Unusual development of black bands on petals of black cross. |
| BITUMEN | Gigantism and deformity—Abnormal and repeated flowering. |

A large variety of plants have been identified either as indicators of various metallic constituents of the soil. A few examples of these are given below:

Copper—Acrocephalus robert 1, Ocimum homble 1, Elshotzia haichowensis, E. Mexicana (Californian Poppy), Polycarpea spirostylus.

Cobalt— Crotalaria cobalticola, silene cobalticola.

Zinc is commonly found in accumulator plants such as equisetum aevense, Lobella Inflata. Plantago Lanceolata (Plantain) Viola Sagittata (violet) Physalis sp., Salvia Pitcheri, Sorghastrum Nutans (Grass).

Molybdenum is also found in accumulator plants such as trifolium repens, Melilotus alba, Lotus Corniculatus.

In general the concentration of metal content in vegetable tissue is as follows:

In trees:      Bark less than root followed by wood, cones, twigs, leaves and needles.

In plants:     Stems less than roots and leaves.

Arsenic is present in twigs and needles and is a sensitive indicator of mineralisation. The Arsenic content increases steadily on crossing the threshold until the mineralised zone is reached. The baramalli tree (Catostemma commune) is a valuable indicator for Cu-Mo.

## (B) GEO-CHEMICAL METHODS FOR PETROLEUM AND NATURAL GASES EXPLORATION

Among the exploration techniques used for petroleum and natural gas, the more important ones are those which employ means for the detection of anomalies relating to gases, waters, rocks, and soils associated with petroliferrous deposits. Analyses are undertaken to determine (a) the presence of petroleum hydro-carbons, including gases and bitumens; (b) the influence of petroleum gases, on waters, rocks, soils and organisms; and (c) substances, conditions associated with oil pools and natural gas reservoirs.

Among the less utilised methods is the geo-micro biological method. The geo-chemical methods may be classed under 'direct' which comprises two groups (a) gas, and (b) bitumens, and 'indirect' which comprises (a) hydrc-chemical, (b) soils-salt. (c) physico-chemical, (d) micro-biological.

Gases used as geo-chemical indicators may be classified according to origin, such as (a) those associated with petroleum deposits, (b) those associated with gas formations, (c) those associated with mud-valcanoes in petroliferrous areas, (d) gases associated with coal formations, (e) gases assqciated with saline formations, (f) marsh gases, (g) peat gases and (h) gases due to photosynthesis etc.

Gases associated with petroleum are mainly methane, ethane, propane, butane, together with minor amounts of nitrogen, carbon dioxide, sulphur dioxide and occasionally helium. In the case of gases, from purely gaseous formations, methane is the main constituent, with smaller amounts of nitrogen and carbon dioxide and a lesser quantities of hydrogen sulphide.

Methane with traces of heavy hydro-carbons, nitrogen, and carbon dioxide is commonly associated with mud-valcanoes.

## Gas survey

Since gas is highly mobile, it moves away from its place of origin, if the surrounding rocks are permeable. Hence the migration of gas is facilitated by faults and fissures which form easy channelways. The movement of gases may also be accelerated/induced by faulting and tectonic movements after gas pool has formed.

The movement of gases is also possible through diffusion of the following types:

(a) Diffusion of gas through liquids;

(b) Diffusion of gas through solid bodies;

(c) Diffusion of gas at liquid-solid inter faces.

During the course of migration, gaseous hydro-carbons may undergo transformation. It is possible that the heavier hydro-carbons are absorbed by the rocks while the lighter fraction, comprising methane alone is transmitted. Hence migration may lead to an enrichment in methane. So is the case, when such gases pass through water wherein the heavier fraction is dissolved. Gaseous hydro-carbons can also be affected by bacterial action, and oxidation, by bacterial action, gives rise to carbon dioxide and water. In the process of migration, hydro-carbons can react on sulphates giving rise to hydrogen sulphide, which may not be present originally in the gas pool.

The most difficult part of gas survey is sampling, since the sub-soil gas, during sampling, can be easily contaminated by gases from the outside atmosphere. With a view to avoid such contamination, special types of equipment have been designed, nonetheless, contamination cannot be completely avoided. The most simplest of techniques, for drawing a sample, utilises the principle of the aspirator. Special water pumps have been designed for drawing gas samples. Equipment suitable for drawing samples from depth going upto 20 meters or more have also been designed.

The gas content, in the rocks, is subject to seasonal variations which affect mainly the uncombustible part of the heavy fractions. In winter, only the combustible gases of the heavy fractions are represented in the sample, while the incombustible components are not. This seasonal variation can be so large as to vitiate gas sampling. Other causes for variation in the composition of gases are (a) seasonal change in moisture content and permeability of rocks occurring above water table; (b) increase in aeration and diffusion/absorption during dry periods, and (c) bacterial action on gases as well as the influence of growing plants.

Since natural gases are complex mixtures, comprising nitrogen, methane, carbon dioxide, and oxygen as well as minor quantities- of other gases, a complete analysis would involve considerable labour and pains. However, apparatus have been designed for total gas analysis. as well as for partial gas analysis, depending on the nature of the investigations. For the purpose of geo-chemical prospecting, it is not considered necessary to go into the technique of gas analysis, since this would be essentially within the purview of the chemist. It is only intended to point out the application of chemical analysis to geo-chemical prospecting. After the results of the gas analysis are obtained, these are plotted on a map with a view to obtain and decipher any anomalous zone. Anomalies are determined by their contrast and the co-efficient of contrast is a measure of the clearness and reliability. The co-efficient of contrast is the ratio, between the average gas concentration within the anomalous zone to the average gas concentration in the area beyond, and it is synonymous with background values. Anomaly contrast upto 1.5 may be considered as weak and those lying between 1.6 and 2 are reliable but values beyond 2 are rather rare.

Anomalies may have various aspects, such as, (1) large circular patch, (2) spoty and patchy, (3) local. A general correlation between the type of anomalies and the associated deposit may be indicated. The anomalies are broadly divided into (a) direct and (b) displaced. The direct anomaly is generally located directly over a structural high and it is generally continuous. A displaced anomaly on the other hand is also associated with structure but displaced to a side, small portion covering the structure. Direct anomalies may again be grouped under continuous, spoty, local and annular or zonal annular. Displaced anomalies can be grouped under continuous, spoty and local.

The origin of direct anomalies is obvious since these are invariably associated with the structure. Displaced gas anomaly is, however, the result of non-uniform migration of gas possibly associated with fractured/fissured rock. Such anomalies can also be caused by movement of ground water and its influence on the gas stream.

### Core gas surveying

Core gas surveying has several advantages over gas surveying since it can be conducted in moist areas where, ordinarily, gas surveying may be a failure. Further, gas extraction, from rocks by the vacuum technique or by heating, always yields a larger percentage of hydro-carbons than the free gas. This can be avoided by core gas surveying. Core gas surveying can be employed, irrespective of season, even in winter. This method is also applicable in fractured rocks and is routine practice in disturbed areas.

Experiments conducted with various types of rocks from various lithological horizons have shown that gas absorption is variable with the rock type. The dark carbonaceous clay, containing traces of organic matter, may absorb as much as three times more gas than ordinary clay, and soil, rich in organic matter, absorbs considerable quantities of gas. The high diffusion co-efficient of gas, especially helium, enables it to pass through glass. Helium can also be associated with various minerals, in varying amounts. Core gas survey takes advantage of these properties with a view to assess the quantity of hydro-carbons, in various positions, of the core, recovered during drilling. The investigation is designed to evaluate the absorpti on properties of the rock as a whole comprising various minerals.

Various methods have been devised for extracting the gas from the cores (a) de-gasing by raising the temperature, (b) de-gasing by vacuum, (c) de-gasing by crushing and, (d) de-gasing by displacing and washing. Generally, a combination of these methods is employed for achieving best results.

## Oil gas logging

In oil prospecting, electric logging is the most popular, but this has its limitations, especially in areas where carbonate rocks are present, when it may be difficult to identify production zones. Again where thin argillaceous sediments are inter-bedded, their identity may not be depicted on the electric log. Even though radio active logging has certain advantages over electric logging in the identification of strata, present in bore-hole filled with mud, yet a further check is often called for.

The methods used, in checking borehole logs, are caliper logs which bring out the variations in the diameter of the bore-hole in relation to the hardness and competence of the particular rock type, and gas oil logging which involves, the direct determination of gas and oil present, in the drilling mud, as it emerges out of the borehole, from various depths.

The logging equipment used in oil and gas surveys is custom built and expensive. It is often truck mounted and provision is made for various kinds of analyses such as (a) examination of gaseous and liquid hydro-carbons, and drilling mud, (b) examination of drill cuttings and cores, (c) determination of porosity and permeability of rocks, (d) determination of the composition etc. of drilling mud, (e) determination of the rate of return of drilling mud in relation to the drill hole bottom. In addition, the equipment is also provided for luminescence testing, caliper logging and electric logging.

Oil gas logs mainly make use of the analysis of (a) drilling mud, (b) drill cuttings, (c) drill cores, but the most popular method is mud logging, since the examination of cuttings and cores present considerable difficulties. Experience has shown that mud logging is indicative of and related to the concentration of hydro-carbons, which pass into solution, from the disintegrated rock, as well as from the productive horizons that are exposed in the drill hole section. It is also possible that the absorption of hydro-carbons is related to the type of clay suspension employed in the mud. Since mud logging cannot be precisely correlated with the distance from the productive horizon, excepting qualitatively, it can only be used for the broad identification of productive and non-productive horizons.

In the application of mud logging, pools of very heavy oil, devoid of gaseous components, are not always detected. Hence mud logging is to be supplemented by bitumen logging. Bitumen logging, in its turn, is not sensitive to purely gas bearing strata, since gaseous hydro-carbons are non-luminescent. However, it is possible to increase the sensitivity of oil gas logging by modifying the method of degasification of mud suspension and analysing the gas so recovered. In this way strata containing heavy oils can also be detected.

The main advantage of mud logging, over other methods, is that mud is necessarily used in drilling, whereas often it may not be able to recover either cuttings or core from the drill hole. The difficulties, in mud logging, are due to (1) contamination caused by the use of oil products as lubricants to improve the quality of drilling fluid. However, it is not necessary that these lubricants affect mud logging to any great extent. The precaution to be taken, in mud logging, is to see that the clay, used in the drilling mud, is free from hydro-carbons or other combustible gases. When the mud flows out of the well being drilled, gases are released due to release of pressure. Further degasing takes place, in the settling tank and other sub-structures. In this process it is only free gas that is liberated, whereas complete removal of dissolved hydro-carbons is not possible. Hence the examination, of inflowing and outflowing drilling muds, will yield a better estimate of the extent of degasification and also indicate the extent of enrichment, caused by any productive horizon. The factors which affect oil gas logging are (1) method of mud degasification and the extent of degasification, as well as the volume of mud degassed per unit time, (2) drilling procedure which includes (a) rate of penetration, (b) rate of mud circulation, and (c) diameter of bit, (3) change in the properties of the drilling mud such as. specific gravity and viscosity, (4) chemical treatment of drilling mud, (5) clays used in the preparation of the drilling mud, (6) nature and extent of lubricants used for facilitating drilling operations, (7) interruption in mud circulation, and (8) influence of higher productive zones on the penetration of the lower productive zones.

In addition, there are certain geological factors which also affect oil gas logging, such as, (a) gas oil ratio, (b) physical properties of the crude whether light or heavy, (c) formation pressures, (d) presence of dissolved hydro-carbons in aquifers and (e) properties of the petroleum reservoirs.

## Bitumen methods

The bitumen methods, used for petroleum prospecting, are based on the determination of the bitumen content in rocks, salts and waters associated with petroliferous formations. Bitumen is the name given to the solid residual product derived from the crude. In a way, this method is a direct method, since bitumens are invariably associated with oil reservoirs. Bitumens are classified as (1) oil bitumens which include (a) oils, including tar-like substances, as well as naphthenic salts etc., (b) ozokerites, (c) asphalts, etc. (2) humus type bitumens comprising (u) plant bitumens, (b) soil bitumens, (c) peat bitumens, (d) coal bitumens. etc. and (3) dispersed syngenetic bitumens, which include bitumens formed contemporiously, with the sediments. The various types of bitumens can be identified by such properties as (a) chemical composition, (b) acid number, (c) saponification number, (d) iodine number, (e) melting point, (f) solubility in alcohol, benzene, chloroform/carbon disulphide/petroleum ether. Bitumen methods are used for building up stratigraphical sections as well as for locating and correlating oil bearing formations. Further, these methods are applicable to surveys, for predicting the possibility of oil pools, within limited areas.

Analytical methods have been designed for the quantitative determination of bitumens. These include determination of (a) organic carbon, (b) nitrogen, (c) bitumens by gravimetric method, and (d) bitumens by colorimetric method.

## Bitumen luminscence method

This method is similar to the bitumen method only that it goes further to testing the luminescence of the bitumen collected. It has the advantage of reproduceability and very high sensitivity, which have led to its widespread use.

Luminescence is a property of certain materials which emit light of a characteristic frequency, when subjected to electro-magnetic radiation, such as X-ray, ultraviolet or within the visible range. Thus we have X-ray luminescence as well as photo-luminescence. The luminescence can be either fluorescence or phosphorescence. In fluorescence, the emission stops immediately the exitation ceases, whereas in the case of phosphorescence the glow continues as an after-glow Fluorescence is characteristic and can be applied in the identification of bitumens.

The luminescence spectrum of a substance depends on the wave length of the excitation used, and in consequence shorter wave-length radiation is used, specially that falling within the ultra-violet region. The following table gives the luminescence spectum of various fractions. It will be seen that the heavier fractions require larger wave-length radiation :

| Name of fraction | Range of wave length in A° | Predominent colour of fluorescence |
|---|---|---|
| Light oils | 4100 to 5100 | Blue |
| Heavy oils | 4800 to 5100 | Azure |
| Light tars | 5100 to 5400 | Green |
| Heavy tars | 5800 to 6200 | Yellow |
| Asphaltenes | | Brown |
| Napthenic Acid | 4000 sharp Max. | Violet |

## Hydro-chemical method

Hydro-chemical oil indicators can be either direct or indirect. This method is designed to determine the nature and occurrence of salts and ions which are characteristic of petroliferous rocks. The direct hydro-chemical indicators of petroleum are materials, which are found in the water, such as salts and ions. These are derived from the water soluble components found in crude oil. They comprise (1) soluble bitumens (napthenic acid), (2) iodine, and (3) ammonia. These, however, cannot be used for detecting gas. The dissolved bitumens, namely, napthenates are by far the most important indicators. These napthenates usually form soaps with cations such as sodium, calcium and magnesium. In addition to these soaps asphaltic soaps are also found but such soaps are not common. It is, however, to be indicated that the hardness of water inhibits the formation of soaps. Bitumens, especially naphthenates, are clearly indicators of the presence of petroleum. This indication may, however, be purely qualitative yet it is positive. But the absence of bitumens, on the contrary, may not signify absence of petroleum, especially when the associated water is very hard. Iodine is present, in water, in the form of iodide, generally sodium iodide. Natural waters are normally poor in iodine, and ground waters may contain exceptionally low concentration, whereas anomalous values are obtained, in waters associated with oil bearing sediments to the extent of $10^4$ or even $10^7$.

Ammonia in water occurs in the form of chloride and is primary in origin. Its behaviour is similar to that of iodine and its origin possibly bio-chemical. *Indirect hydro-chemical indicators*, includes salts and ions, dissolved in waters, and these may be divided into groups (a) comprising hydro-sulphides and other reduced compounds of sulphur, (b) showing the absence of sulphates and (c) containing soda. The second group includes substances which are associated with oil pools but not directly and these include (a) calcium chloride, and (b) bromine.

### Soil salts methods

These methods differ from the hydro-chemical methods, in that the soil is made use of for detecting anomalies, and also in that the indicator used, for example is carbonate which may have no relation to hydro-chemistry. The soil salt method is chiefly used in the preparation of soil maps, delineate the content of (a) chloride, (b) iodine, (c) gypsum, (d) carbonate and (e) sodium.

### Oxidation reduction potential method

Hydro-carbons are known to be reduced compounds. Hence the presence of hydro-carbons in sediments necessarily lowers the redox potential. The effect is more pronounced in the vicinity of an oil pool. Hence, the Eh/pH relationship can be employed as an indicator of the presence of oil. The method is similar to the detection of spontaneous polarisation as indicated in the geo-physical methods.

### Marine Geochemical Exploration

With the advance of technology, there has been an ever growing interest in the mineral resources of the sea. To begin with common salt, being an essential article of food, was extracted in a big way. The extraction of by-products such as the sulphates and the chlorides of magnesium and calcium followed in its wake. Kelp from sea weeds provided the source of iodine, and then the large industrial undertakings, for the manufacture of magnesium carbonate from brine, came into being. With the exhaustion of the land resources for petroleum off-shore exploration has become the routine practice in almost all countries with a fringe of the continental shelf. Scientific exploration of the ocean floor by the various expeditions have brought to light a fund of information in regard to ocean wealth. The discovery of the occurrence of manganese nodules and the provenance has raised hopes of finding a new source of not only manganese but also of metals such as nickel, copper, in addition to precious metals. Researches on the phosphorite deposits have in their turn provided clues for locating similar occurrences, elsewhere.

Under sea geochemical prospecting has been a matter of consideration and many designs of the equipment have been envisaged. One of the designs incorporating the neutron activation technique appears to have every prospect of achieving success. However equipment for use onboard has been in operation, these being largely similar to those used in the land based laboratories.

In the mining field, a firm called The Deep Sea Ventures Inc. have used the R/V prospector for recovering manganese nodules by means of a hydraulic dredge from a depth of 300 to 900 metres. And a bucket type dredge has been tested successfully for recovering manganese nodules from a depth of 3600 metres.

Projects are also under way for the recovery of zinc, copper, as well as precious metals from the 30 ft of the red sea brines and also from the atlantis 11 deep from a depth of more than 2000 m.

PART V


GEOPHYSICAL METHODS IN
MINERAL EXPLORATIONS

# INTRODUCTION

Time flies, but Science advances at its own rate. Thus out of the primitive, all comprehensive group of science termed "Natural Philosophy" emerged 'Physics', which concerned itself solely with the properties of matter. With the growth of knowledge, physics, in its turn, gave rise to various branches—one of which was geophysics. This aspect of the subject, deals with the study of the earth from its centre to the upper limit of the atmosphere. It is divided into various branches viz. geodesy, meterology, isostacy, oceanography, radio activity, seismology, atmospheric electricity forming the pure side while (exploration geophysics forms the applied side of the subject)

This last mentioned subdivision of applied geophysics has made rapid advance in recent times. Petroleum industry was mainly responsible for its phenomenal growth. No stone was left unturned, and every effort was made, in this great gamble, of discovering oil fields.

Despite its glamour, the application of exploration geophysics to spheres, other than oil was slow. The mining man believed in the dictum that "seeing is believing", his trusted friend being the diamond drill, while the engineer, insisting on a comfortable factor of safety, could not place enough trust where "possibly and probably" were involved.

In the wake of the last World War came rapid industrial expansion, in the more advanced countries. Projects involving large sums of money had to be completed expeditiously in the interests of economy. The need for a wizard's wand was great and geophysics seemed to be the answer. Geophysics has since proved its worth and it is now a routine practice to employ geophysical methods in certain types of engineering investigations.

Every science has its limitations, and geophysics is no exception. By geophysical methods a discontinuity in the physical characteristics of the crust can be detected and possibly delineated, but the interpretation of the results requires a strong background of the local geology. It is not thus too much to say that divested of geology, mere geophysical data will appear hypothetical or even fictitious.

The five important geophysical methods relate to the five most common characteristics of the earth, which can be determined from the surface viz. (1) electrical conductivity, (2) density, (3) magnetism, (4) elasticity and (5) radio activity. These are investigated respectively by the electrical (a) self potential, (b) equipotential line, (c) resistivity, (d) potential drop ratio method, (e) electromagnetic, (1) induced polarisation (2) gravity, (3) magnetic, (4) seismic and (5) radioactive methods.

In addition geophysical methods are also used in the logging of bore holes and these are classified as (1) electro-logging (2) radioactive logging.

# ELECTRICAL METHODS

## Principles

All the electrical methods are widely used in the exploration work connected with metalliferous deposits in prospecting for ground water, and in engineering geology investigations. These methods, essentially, involve measurement of (1) electrical conductivity or resistivity of the material of the earth, (2) electro-chemical activity and (3) dielectric constant.

The resistivity of a material is the resistance in ohms of a unit cube and is expressed, either as ohms-meter or ohm-centimeter. Rocks show marked difference in resistivity, the flow of electric current in rocks being mainly electrolytic. Most rocks are porous and contain occluded moisture, e.g., sandstone. Others that appear dry and compact contain included moisture in the component minerals, e.g., granites etc. Thus it is found that the resistivity decreases with the increase in moisture content and for rocks with low moisture content the resistivity values are determined by the component minerals.

Absolute values of resistivity of various types of rocks, igneous, metamorphic, sedimentary (consolidated or unconsolidated) can be determined experimentally. In practice, however, it is the relative values of resistivity that are of greater importance since the same rock can show quite appreciable variations of resistivity, e.g , wet clay has a lower resistivity than dry clay. A weathered granite has a lower value of resistivity than a fresh granite. The values of resistivity as obtained by instrumental measurements is known as apparent resistivity and its value, as obtained by one particular arrangement of electrodes, known as the Wenner configuration is given by the formula : (Fig. 5.8)

$$\gamma = 2\pi \, AE/I$$

where      $\gamma$ = resistivity in ohm-meters

A = the distance between electrodes in meters.

E = Potential difference between electrodes, in volts,

and I = current flowing in amperes.

In general, electrical methods are employed for relatively shallow subsurface exploration, not exceeding depths of 1000 ft.—1500 ft. Each of the methods depends on one of the following electrical properties of the rock formations. (1) The resistivity or its reciprocal known as conductivity defines the amount of current per unit volume, where the voltage is constant. This property is made use of in the resistivity methods. (2) The electro-chemical activity is controlled by the mineralogical composition of the rocks in addition to the composition and concentration of electrolytes in the associated pore ground water. An ore body in contact with electrolytes behaves like a galvanic cell. To start with electro-chemical reactions of varying intensities take place at various parts of the ore body or oil pool. Finally, equilibrium is achieved by a current flowing from the higher to the low potential areas, leading to spontaneous polarisation with a negative centre at the oxidised portions. Similarly, electro-chemical activity or spontaneous polarisation is also employed, in the electrical logging of wells, *i.e.*, bore holes lined with mud..

The dielectric constant is another property utilised in electrical prospecting. Taking a sphere with a flux density of one per sq. cm, the total force will be $4\pi$. The total electrical flux, per unit area, is $(E+4\pi P)=(1+4\pi e)E$, where $(1+4\pi E)=K$, the dielectric constant for the material, and on this depends (*a*) the capacitance of the rock, and (*b*) its static response to an applied field. Mica has a high dielectric capacity and it is used in making capacitors or condensers. The dielectric constant in vacuum = 1. In hard rocks the value ranges between 6 and 16 electrostatic units, while for moist clays it is 40–50 esu. This property is utilised in the inductive methods.

### Description

The electrical methods are subdivided into (1) self potential method (2) equipotential line method (3) resistivity method (4) potential drop ratio (PDR) method (5) electromagnetic method (6) inductive method.

**(1) Self potential or S.P. method :** This method utilizes the natural flow of current and operates on fundamental principle that an ore body, undergoing oxidation, is a source of electric current or telluric current. If a tabular sulphide ore body is present, in the ground, oxidation at the upper levels near P, induces greater chemical activity than at Q. Hence a potential difference is induced and a current flows from P towards Q. The current leaves the ore body and then passes into the soil (charged with electrolytes) and returns to P as shown in Fig. 5.1. In the area, on the surface, immediately above P, say at a

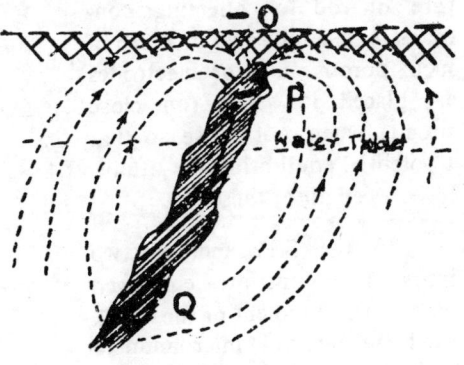

Fig. 5.1

point O, it will be seen that the current enters the ground; and travels towards the positive centre P. Thus at O a negative centre develops.

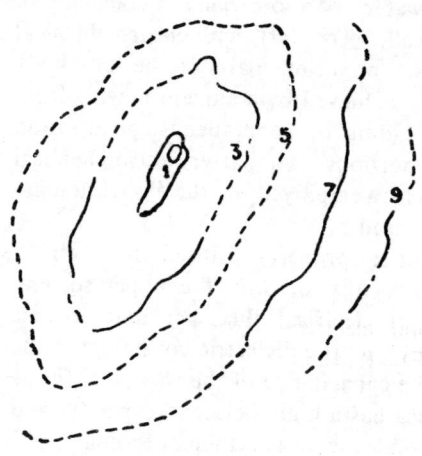

Fig. 5.2

Under such conditions it is possible to obtain the location of O by measuring the drop in potential on the surface and plotting the equipotential contours as shown in Fig. 5.2. The measurement of the relative potentials at various spots over the area is carried out. Since the earth currents are of a low order, the potentials are also weak say of the order of $10^{-3}$ volts. Hence special electrodes are required for the purpose in order to avoid any interference that may be caused by "contact potential" or "polarisation" at the electrode itself. This type of electrode (Fig. 5.3) is constructed in the following manner. A porous pot (1) (Ceramic) is half filled with a saturated solution of $CuSO_4$ (2) and crystals of copper sulphate are also placed inside. A stout copper wire or rod (3) is placed in copper sulphate solution. A clip (4) is attached at the end of the copper wire or rod for affecting contracts. Generally two or more of such porous pots or electrodes are placed in series (on close circuit) when not in use, so that a potential equilibrium is attained between all of these.

In the S.P. method two types of circuits are employed for the measurement of the weak earth currents : (1) microammeter (2) potentiometer.

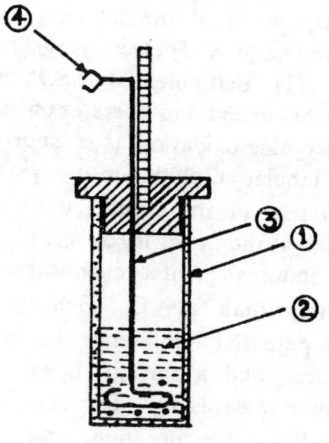

Fig. 5.3

In the 1st method (Fig. 5.4) one of the NP (nonpolarisable) electrodes P, is kept stationary while $P_1$ is moved round and the reading on (A), the microammeter, noted for each station. It is convenient to have an instru-

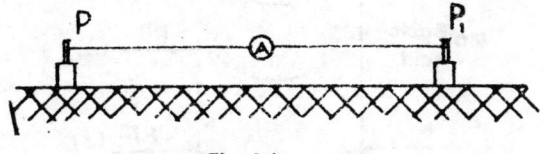

Fig. 5.4

ment which can read on either side of a central zero point, in positive and negative values.

In the 2nd method (Fig. 5.5) the microammeter is replaced by a

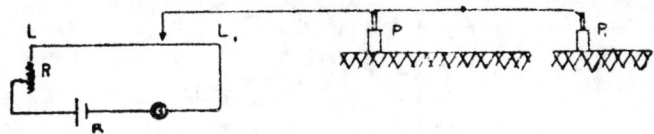

Fig. 5.5

potentiometer circuit. The potentiometer circuit comprises the standard cell B, the variable resistance R and the slide wire $LL_1$ and (G) is galvanometer. As in the former case the NP electrode $P_1$ is moved while P is kept stationary. The instrument may be calibrated so that it can read upto 0.0001 volt to 1 volt. For geophysical work stability is of primary consequence, hence the unpivoted type instrument is preferred.

**Telluric Currents:** Telluric currents are results of the earth potentials due to such causes as infiltration potentials, rotation of the earth, sun spots etc., the electric field produced by these is of the order of 10 mv/km very often these fluctuate in direction and amount.

Telluric currents of this type are, in general, similar to S.P. caused by sulphide ore bodies. They obey Ohm's law. Structural disturbances such as domes, faults, dykes, etc., distort the flow of Telluric currents and as such the presence of such disturbances can be detected by measuring the variations in the electro-Telluric profile.

Taking the example of a gas-oil pool the conditions are normally similar to that shown in the Figure. As is the ground surface, GD is the lower limit of weathering. PQRS is the Geochemical chimney permeated with compaction waters containing hydrocarbons in solution with a reducing environment. The arrows show the direction of the telluric currents caused by the lowering of potentials within the Geochemical chimney as shown by the ET profile 1.

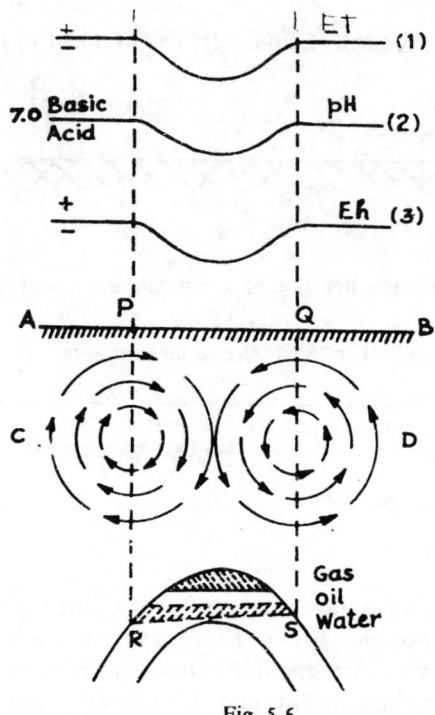

Fig. 5.6

**(3) Resistivity method** : In resistivity method current is passed into the ground by metallic (copper) electrodes and potential difference is measured by using two N.P. electrodes.

Various electrode arrangements or configurations are employed in resistivity surveying. However, common amongst these are as follows :

(1) *Wenner configuration* : This arrangement is most commonly employed, in practice and is theoretically the simplest. In this method electrodes are equally spaced as shown in figure 5.7 the interval being 'a'.

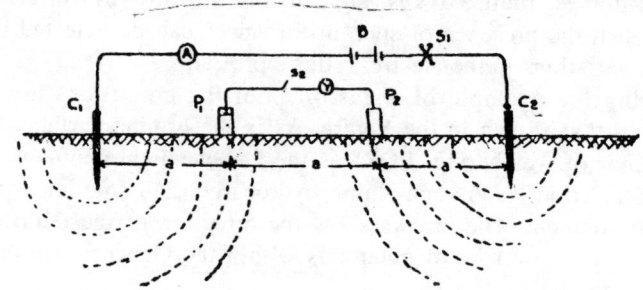

Fig. 5 7

The outer electrodes $C_1$ and $C_2$ are the current electrodes and $P_1$ and $P_2$ are the potential electrodes.

A is a milliammeter, B is a battery, V is a potentiometer, $S_1$ is a commutating switch, and $S_2$ is a switch.

The procedure is as follows :

(a) The potential circuit is closed by $S_2$ switch and the value of the earth current (potential) $(E_1)$ as recorded by the potentiometer (V) is noted. This circuit is broken now.

(b) The switch $S_1$ is closed and the current from the battery now enters the ground through $C_1$ and $C_2$ electrodes. The value of the current (I) is recorded. The current is kept "on".

(c) The potential circuit is closed once again, and the new reading of the potentiometer $(E_2)$ is recorded. Thus the actual potential due to the current (I) is $(E_2 - E_1) = E$ is obtained. The process is repeated reversing the current by means of the commutating switch, $S_1$ and the average value of (I) and E are obtained and substituted in the equation,

$$\gamma = \frac{2\pi a E}{I} .$$

(2) *Lee partitioning method* : This is also essentially the Wenner configuration. In this electrode arrangement, an additional potential electrode $P_2$ is placed half way between P and $P_1$ (Fig.5.7). A constant cur-

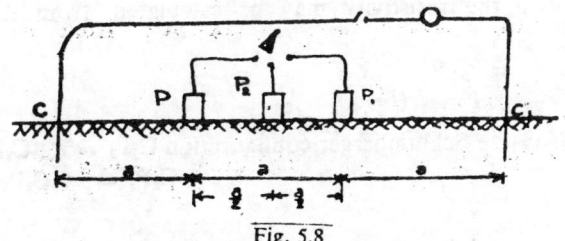

Fig. 5.8

rent I is passed into the ground and the potential difference (V) between P and $P_2$ and also $(V_1)$ between $P_1$ and $P_2$ is measured.

Hence applying the Wenner formula the resistivity of the rocks to the left side is,

$$\gamma_1 = 4\pi a V / I.$$

and the resistivity of the right hand side is,

$$\gamma_2 = 4\pi a V_1 / I.$$

In using Lee partition the central NP electrode, $P_2$, is kept stationary, while the other electrodes (the potential and current) are moved away in the usual manner so as to increase the electrode separation.

Any change in the resistivity can be made out by the use of this method, where the boundary is vertical *e.g.* a faulted contact with two different rock types on either side.

(3) *Unsymmetrical method* : (current electrodes separated by a finite distance) (Fig. 5.9)

RESISTIVITY

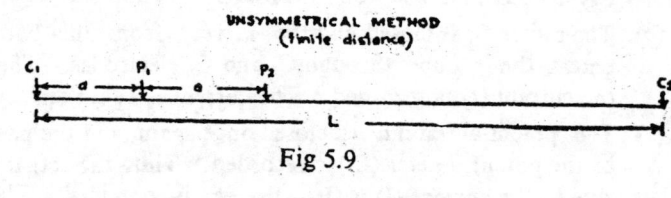

Fig 5.9

$$ r = \frac{2\pi \triangle V}{I} = \frac{1}{\tfrac{1}{2} a + \dfrac{1}{L-2a} - \dfrac{1}{L-a}} $$

Where, $\triangle V$ = the potential difference between $P_1$ and $P_2$, i.e. $(VP_1 - VP_2)$,

$a$ = electrode separation between $C_1P_1$ and $P_1P_2$, and

$L$ = electrode separation between $C_1C_2$,

$I$ = value of current.

In the Schlumberger configuration, (Fig. 5.11) in which $CP = C_1P_1 = a$ and $CP_1 = C_1P = b$, the resistivity may be calculated from the following expression,

$$ r = \frac{\pi \triangle V}{I} \cdot \frac{ab}{(b-a)} $$

(4) If in the Schlumberger configuration $C_1P_1 = a$, $C_1P_2 = b$

$$C_2P_2 = \propto C_2P_1 = \propto, \text{ then}$$

$$ r_a = 4\pi \frac{V}{I} \cdot a $$

This variation is known as the single probe method.

*Single probe method* : In this method the two current electrodes are separated by a large distance (Fig. 5.10). Potential bowls are caused around each electrode and these are picked by the use of NP (non-polarisable) electrodes in the same way as in an S.P. survey. The traverse aligned in the line connecting C and $C_1$ should be at least four to five times the largest electrode separation contemplated i.e. $CC_1 > 5a$. Thus if the electrode separation is 100 ft., the distance $CC_1$ should be larger than 500 ft. The resistivity (Figure 5.11) in this case is given by the formula $\gamma$ :

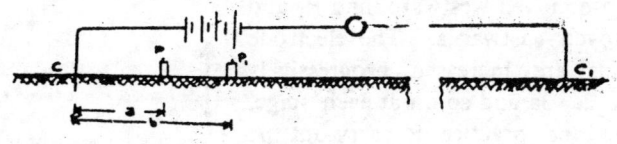

Fig. 5.10

$$\frac{2\pi\, ab}{b-a}\cdot\frac{V}{I} = 2\pi\frac{V}{I}\cdot\frac{ab}{(b-a)}$$ where 'a' is the distance CP and 'b' is the distance $CP_1$,

V=the potential recorded and I is the value of the current used.

In a typical survey where the NP electrodes are used the distance $CC_1$ can be 3500 ft. The voltage of battery can be about 300 V. The current is reversed before making measurements, and the nearest values of V and I are obtained for calculating the resistivity. The value of 'a' can be increased keeping (a—b) = 100 ft.

*Field procedure* : Two types of resistivity surveys using Wenner configuration are normally carried out.

(*a*) *Resistivity traversing* : Keeping constant separation 'a' all four electrodes are moved in a linear fashion. In this way the entire distance of the traverse (however long) can be covered. A second traverse is repeated between the same two points with an increased electrode separation "2a", between electrodes. Similarly, the traverse is repeated a third time with a separation of '3a', and later '4a' etc. This is called resistivity traversing or trenching. Since in the Wenner electrode configuration, the depth of current penetration is proportional to electrode separation, resistivity traversing gives a profile, showing the variation of "average" resistivity with depth. It can be employed for mapping sub-surface topography, say along a dam alignment in a soil covered area.

Transverses with varying (increasing) electrode separations, as described, can be carried out along radiating lines. This method can be used with advantage in engineering geology problems, where the configuration of the subsurface topography is required. It was applied to certain of the investigations in south India and the subsurface topography delineated by this method was confirmed later by close drilling (Ramapada Sagar Scheme W. Godawari district, A. P.)

(*b*) *Resistivity sounding* : The resistivity is measured by increasing the electrode separations about a central fixed point, which is the mid point X, between the two potential electrodes. In this method (Fig. 5.11)

C and P are moved westward, and $P_1$ and
$C_1$ are moved eastwards.   The electrode
separations  are  increased  progressively
from a to 2a, 3a and so on at each stage.
It  is  common  practice  to  carry out the
measurement  on  two  lines  laid  at  the
right angles to each other.   This method
is called electrode probing or 'sondage' or
"well sounding".

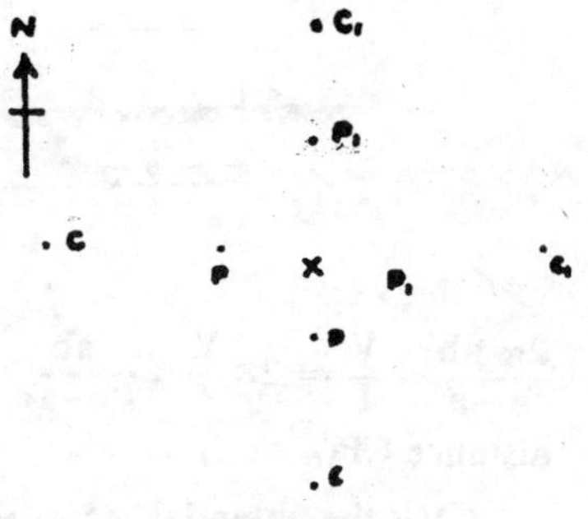

Taking  a  case  where  a  layer of
sandy soil covers a shale bed, the resisti-

Fig. 5.11

vity curve obtained by electrode probing will take the  form  approximating
to that shown in Fig. 5.12. The upper part of  the  curve  shows  that  the
resistivity increases  sharply,  to a depth  of  about  200 ft.  This shows the
variation in clay content, the top  soil,  being  clayey,  is  conductive with
increasing sand content, at  depth, the soil  becomes more  resistive.  This
increase in resistivity continues until the shale  is  reached  at  a  depth of
about 200 ft., which is more conductive.   The  sudden  flexure  at  225 ft.
shows the influence of shale, where the resistivity falls steeply.

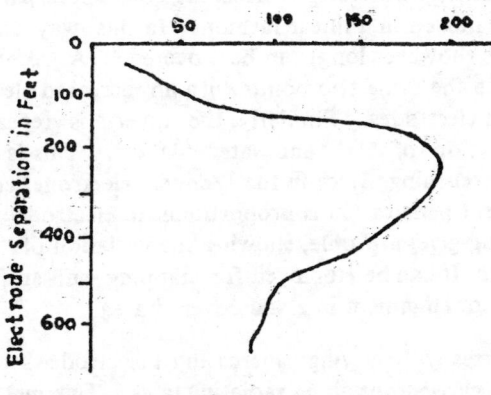

Fig. 5.12

Some practical hints regarding this method are necessary for  using  it
efficiently.

1.   To make good ground  contacts  with  the  NP  electrodes  a  wet
cloth soaked in $CuSO_4$ can be used.

2.   Where the current electrodes (stakes) are driven into sand the
contact resistance may be so high that there can be  very  little  penetration
of current.   Moistening of the ground improves the  contact  and  ensures
passage of current.

3. It is often safe to inspect the contacts as the current electrodes can be placed on high resistance formations and then shifted to highly conductive rocks, which will cause a large current to flow in the potentiometer circuit and blow up the fuses.

4. As the electrode separations are increased, with a view to increase the depth of current penetration, the lengths of wire also increase. The leads to the potential electrodes $P_1 P_2$ should not be coiled as the inductance so produced will cause errors. Coiling of the lead wires for the current electrodes $C_1 C_2$ is permissible.

5. It will normally be seen that the resistivity, for the top layer of soil, in the hot climates, say about one foot or so, is generally high due to dryness. The values of resistivity fall with increase in separation. .

6. In the case of weathered rocks, generally the resistivity is low, say, for example in the case of granite when fresh, the resistivity is high, but on weathering, the value of resistivity falls enormously.

7. With increasing separation or interval of electrodes, the depth of current penetration increases. Thus for all practical purposes a = d, where d = depth of penetration.

8. The average value of $\gamma$ (resistivity) (in ohm-cms) is plotted as ordinate and the value of 'a' the electrode separation is plotted (ft or m) on the abscissa.

**Instruments**

(a) *Resistivity potentiometer* : This consists of a current measuring device milliammeter (Ma) and the corresponding potential measuring device galvanometer (g) placed in a portable box. In this instrument an auxiliary circuit is provided to eliminate self potential present in the ground. Circuit diagram of the instrument is shown in Fig. 5.13.

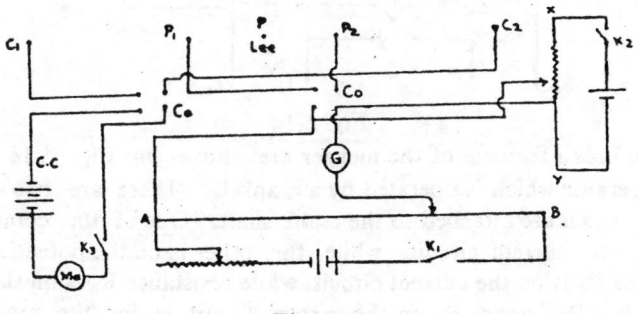

Fig. 5.13

$B_1$ $B_2$ $B_3$ are batteries

| | |
|---|---|
| Ma | = Milliammeter |
| XY | = Auxiliary potentiometer |
| G | = Galvanometer |
| Co | = Commutator |
| AB | = Potentiometer wire |
| $K_1$, $K_2$, $K_3$ | = Contact switches |
| P, $P_1$, $P_2$ | = Potential electrodes |
| $C_1$, $C_2$ | = Current electrodes |

(b)  *Gish   Rooney apparatus* :    In  the  Gish-Rooney  apparatus a
manually operated commutator is introduced  in  the $C_1$ $C_2$ circuit, which
causes the reversals of current at a desired frequency. This procedure
eliminates  (1) the effect of earth current,   (2) the effects of current due
to electrolysis of ore bodies,   (3)  the  effects  of  polarisation or galvanic
action due to metal current electrodes.    In  essence  it is the same as that
illustrated in figure 5.14.

(c)  *Megger earth tester* :   The "Megger" is an instrument used  by
electricians for earth testing in an installation as well as for testing the insu-
lation.    The  equipment  was  originally designed  by  Messrs.  Evershed
Vignoles of London, but has since been modified for earth resistivity
measurements.

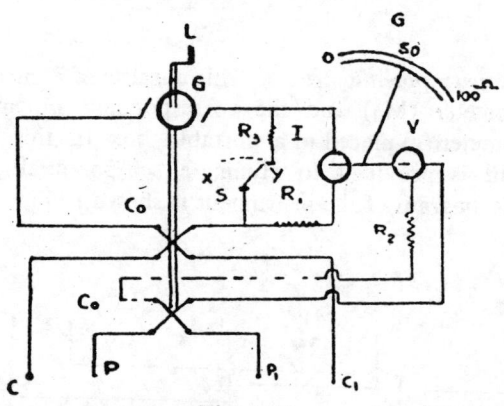

Fig. 5.14

The broad features of the megger are  shown  in  Fig.  5.14 G is a
D.C. generator which is operated by a crank L.  There  are  two  commutator Co which are attached to the crank shaft.  One of  the  commutators
reverses  the  current circuit,  while  the  other is on the potential circuit.
Resistance $R_1$ is on the current circuit, while resistance $R_3$ is on the potential circuit.   Resistance $R_3$ on the current circuit is for the  range switch

(Sensitivity scale XY).

The equipment is generally similar to the Gish-Rooney set up, in that a reversible D.C. caused by commutators is used. Instead of the potentiometer, the megger embodies a galvanometer, which is calibrated in Ohms, since it records the resultant of the reactions between the current coil and potential coil. Thus the value of the resistance E/I is obtained directly, which can be substituted in the formula,

$$\gamma = 2\pi \ a \ \frac{E}{I}$$

As in the case of the Gish-Rooney method the depth of penetration 'd', of the current can be increased by increasing the electrode separation 'a'.

The megger is a very convenient and simple instrument and it is handy for field use. It can be used with facility by a mining engineer or geologist. It can be used only in mineral investigations but also in engineering geology and ground water work.

A point of caution in using the megger is to be noted. It will be seen that generally the values of resistivity are low. The explanation is fairly simple. While operating the megger on various ranges of sensitivity suitable resistances have to be introduced in the voltage circuit, and $R_2$ is varied. When the sensitivity is $1:1$ the value of $R_2$ is 17000 ohms, when the sensitivity is $1:100$ the value of $R_2$ is 1600 ohms, while the resistances of the potential electrodes $(P_1P)$ is about 1000 ohms. Thus the value of $R_2$ increases considerably, affecting the sensitivity also at this range.

The Gish-Rooney method with the Wenner electrode configuration had been modified by Lee and method is known as the 'Lee Partition' as shown in Figure 5.9.

**Interpretation**

Mathematical solution of the resistivity formula is fraught with grave difficulties. Solutions can be worked out based on certain assumptions which may not be true to field conditions. It is possible to subject the field data to mathematical analysis, assuming (a) that the rocks are homogeneous media with considerable lateral extent, (b) that the interfaces between the layers of rock are parallel to one another, and (c) that there are only a definite number of layers—generally not exceeding four.

In solving a problem, under the above conditions, the formula of Hummel is most often utilised. In deriving the formula, the method of images, has been employed. Each interface is taken to be a plane of reflection, and that (1) the current source in one will have its image in the other planes, (2) that each reflection will involve a loss in intensity.

Since the successive reflections, from the more distant sources, fade away rapidly, it is only the first few reflections that are of consequence, in computing the apparent resistivity. If the resistivity of the top most or surface layer is $\gamma$, (which can be determined in the field by using small electrode separations), and if the apparent resistivity, as obtained for an electrode separation 'a' is $\gamma_a$, the value of $\gamma_a$ is obtained by the sum of an infinite series.

Hummel has worked out the theory for two and three layer cases, by using the method of images. In the case, where the surface layer with a resistivity of $r_1$ overlies, an infinitely thick substratum of resistivity $r_2$, the apparent resistivity $r_a$ is

$$r_a = r_1(1+4F).$$

Where F is a function representing the sum of an infinite series and may be written as

$$F = \sum_{n=1}^{\infty} \left| \frac{k^n}{\sqrt{1+(2n\frac{h}{a})^2}} - \frac{k^n}{\sqrt{4+(2n\frac{h}{a})^2}} \right|$$

where, a is electrode separation, for the Wenner arrangement, h is the thickness of the layer and k, the resistivity constant, is

$$(r_2-r_1)/(r_2+r_1) \text{ or } k=\frac{(1-r_1/r_2)}{(1+r_1/r_2)} \qquad ...(1)$$

Tagg has constructed a set of curves for the 2-layer case based on the equation given above, to facilitate computation, of resistivity and layer thickness for the two-layer case. These curves are reproduced in Fig. 5.16. One set is for $r_2/r_1>1$ and other is for $r_2/r_1<1$. These curves are for values of k, varying from +1 to −1.

In applying these curves to actual measurements it will be seen, that the values of 'r' and 'a' are known. $r_1$ can be determined by very careful measurements at small electrode separations. Thus two unknown quantities k and h remain to be determined.

From the data collected by resistivity sounding or probing method, the values of $r_a/r_1$ for each value of 'a' is then drawn as a horizontal line, across the appropriate set of master curves Fig. 5.15. At each intersection of this line, with a curve, corresponding to different k values, the h/a value is read and converted to h, by simple multiplication with a. The k values are then plotted against the h values, on another graph, all points for a given electrode separation being connected by a smooth curve. If the two layer assumption, regarding resistivity variation with depth is actually applicable, then the curves obtained for the various electrode separations, should intersect at a point, i.e., the value of h denoted by the point of

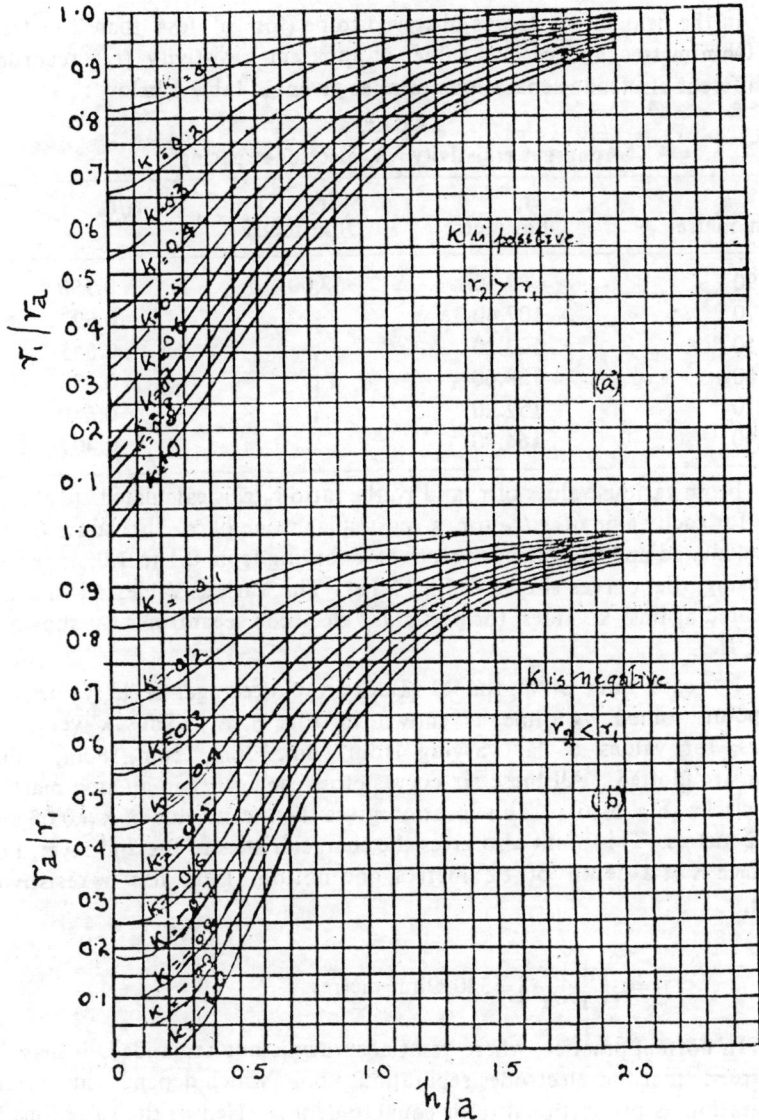

Fig. 5.15

intersection, is the depth to the interface with the values of k and $r_2$ (Fig. 5.16). The resistivity of the deeper layer, $r_2$, is obtained by substituting these values in equation (1).

The method can be illustrated by the following example :

In the field survey $r_1$ was observed to be (for 'a' less than 2 metres) 67.2 Ohm-metres, and various values of apparent resistivity ($r_a$) recorded for different electrode separations were as given in table I below :

## TABLE I
### Apparent resistivity Vs. electrode spacing

| a in metre | $r_a$ Ohm metre | $r_1$ Ohm metre | $r_1/r_a$ |
|---|---|---|---|
| 90 | 89.60 | 67.00 | 0.750 |
| 120 | 107.40 | ,, | 0.623 |
| 150 | 123.20 | ,, | 0.545 |
| 180 | 138.60 | ,, | 0.483 |
| 210 | 152.20 | ,, | 0.440 |
| 240 | 164.80 | ,, | 0.407 |

From various values of $r_a$ and $r_1$, the ratio $r_a/r_1$ is calculated. (Table 1) Now for each value of $r_a/r_1$, for a particular value of 'a', the values of h/a are calculated and the various values of k varying from 0.1 to 1.0, are noted by using the curves shown in Fig. 5.16. The various values of h/a are tabulated against k values, for particular electrode separations as shown in Table II.

For each value of 'a', the 'h' values are plotted against 'k' values and all points joined by a line as shown in Fig. 5.16. Thus 'k' versus 'h' curves, for values of 'a' varying from 90m, 120m, 150m, 180m, 210m, 240m, are plotted. All these six curves cross each other in an area marked by a circle (Fig. 5.16). This location gives the mean value of k=0.66 and h=82 metres. This point also gives the characteristics of the 2nd layer, i.e., interface is at a depth of 82 metres, and bottom layer has a resistivity value,

$$r_2 = r_1\left(\frac{1+k}{1-k}\right) = 320 \text{ Ohm metres.}$$

In normal practice, where great accuracy is not demanded, it may be construed that the electrodes separation, upon which depends the current penetration, is proportional to or equal to depth. Hence, the variations in the resistivity curves may be correlated with electrode separation/depth.

There may be cases where it is possible to identify a resistivity curve from its shape and say whether it is a two layer, three layer or even four layer curve. Where a two layer curve is indicated, the depth to the lower layer may be determined either by the use of Tagg's curves (Fig. 5.16) or by the use of standard curves prepared by Mooney and Wetzel. These latter

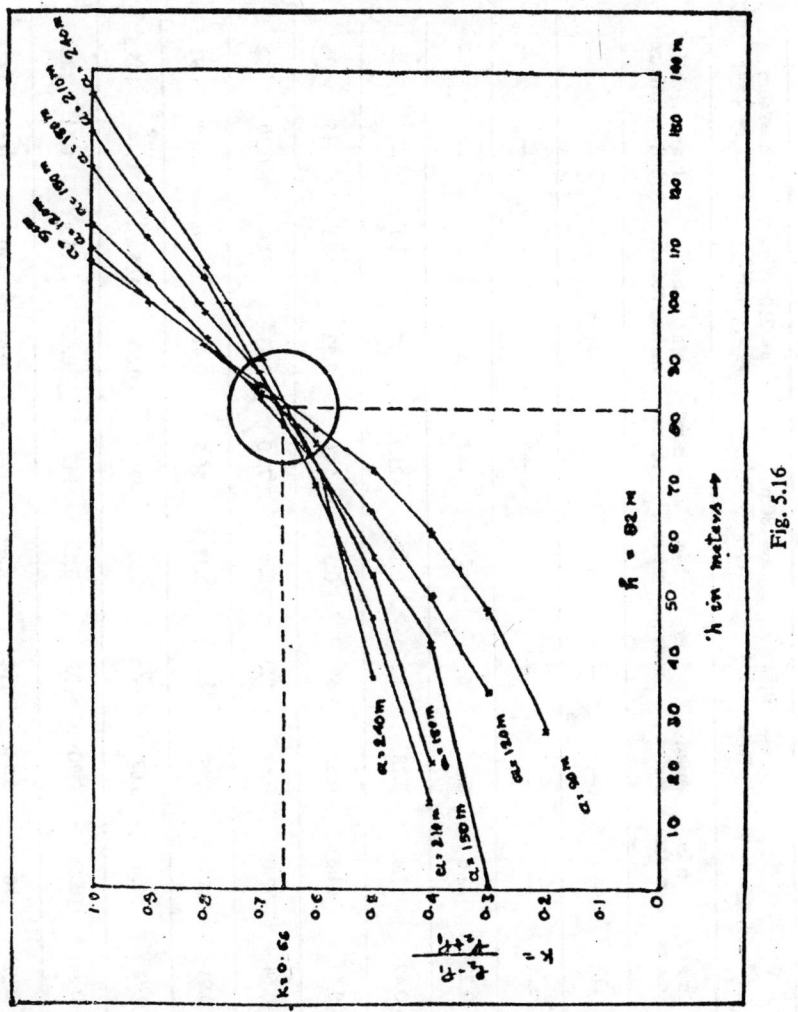

Fig. 5.16

authors have, in fact, prepared curves for various conditionscovering three
and four layers, the total number of curves being about 2400. Such curves
are indeed very useful. When the field data, in respect of resistivity and
electrodes separation, are plotted on transparent double logarithmic paper,
the curve drawn can be conveniently superimposed on the standard curves,
so as to find out which of the standard curves gives the best fit. Once
the fit is obtained then the values of H and $r_1$, $r_2$ and k can be obtained.

In certain cases, curve fitting may become difficult, especially when
the field resistivity curve pertains to 3 or more layers. In this case a

## TABLE II

| K | For a=90m $r_a/r_1$=0.750 | | a=120m $\frac{r_a}{r_1}$=0.623 | | a=150m $r_a/r_1$=0.545 | | a=180m $r_a/r_2$=0.483 | | a=210m $r_a/r_1$=0.440 | | a=240m $r_a/r_1$=0.407 | |
|---|---|---|---|---|---|---|---|---|---|---|---|---|
| | h/a metre | h in metre | h/a metre | h in metre | h/a metre | h in metre | h/a metre | h in metre | h/a metre | h in metre | h/a metre | h in metre |
| 0.1 | — | — | — | — | — | — | — | — | — | — | — | — |
| 0.2 | 0.3 | 27 | — | — | — | — | — | — | — | — | — | — |
| 0.3 | 0.52 | 47.8 | 0.28 | 33.6 | 0 | 0 | — | — | — | — | — | — |
| 0.4 | 0.68 | 61.2 | 0.42 | 50.4 | 0.28 | .42 | 0.18 | 21.2 | 0.07 | 14.7 | — | — |
| 0.5 | 0.80 | 72·0 | 0.54 | 64.8 | 0.38 | 57 | 0.3 | 54.0 | 0.22 | 46.3 | .15 | 36 |
| 0.6 | 0.88 | 79.2 | 0.64 | 77.0 | 0.48 | 72 | 0.4 | 72.0 | 0.33 | 69.4 | 0.30 | 72 |
| 0.7 | 0.96 | 86.4 | 0.71 | 85.2 | 0.56 | 84 | 0.48 | 86.5 | 0.42 | 88.4 | 0.38 | 91 |
| 0.8 | 1.04 | 93.6 | 0.78 | 93.6 | 0.63 | 94.5 | 0.55 | 99 | 0.50 | 105 | 0.44 | 105.5 |
| 0.9 | 1.12 | 101 | 0.84 | 100.3 | 0.70 | 105 | 0.62 | 112 | 0.55 | 116 | 0.51 | 122 |
| 1.0 | 1.2 | 108 | 0.92 | 110 | 0.76 | 114 | 0.69 | 124 | 0.62 | 130 | 0.57 | 137 |

modification has been suggested, where the three layer curve is imagined as comprising two parts, an upper and lower, each part is taken to be a two layer curve. Thus two separate fittings, on two standard curves, may be obtained. In any case, the standard curves, prepared by Mooney and Wetzel, serve as a rough guide as to the number of layers present. It is usually most expeditious to plot the curves on log-log paper, corresponding to a dimensionless form for both coordinates. Roman has used this type of plot, constructing a log-log version of Tagg's curves for the two layer case.

Even after the thicknesses of the various formations have been identified by means of the standard curves, it is very necessary to correlate these with lithological logs, from any bore-hole, in the vicinity. The control is better when the number of bore-holes is larger. One of the shortcomings of the resistivity method, using the Wenner configuration, is the difficulty to differentiate local variations.

### Terrameter

With the improvements in the design of electronic circuits more compact instruments have been made for using low frequency A.C. in resistivity exploration. The "ABEM" terrameter is an example. The essential features of the instrument are shown in Fig. 16A. The instrument comprises three units; A. The Power unit B. The resistivity measuring unit and C. The electrodes.

The power supply unit comprises 1.12 Nos. 1.5 Volts dry cells B, 2. Low frequency transistorised Oscillator 4 C/S, O. 3. Power Transformer, T. with the four terminals, I, II, III, IV. The terminals I and IV are connected to the current electrodes C1 and C2. while the terminals II and III are connected to the Potentiometer P.

The resistivity measuring unit comprises 1. The potentiometer P 2. The switches S and S1. The switch S1 can be connected to either the Potentiometer terminals 1 and 2 or to the potential electrode terminals 3 and 4. The amplifier A is used for measuring the potential difference between the potential electrodes P1 and P2 or the potential drop between the potentiometer terminals 1 and 2. The Galvanometer M, is calibrated in arbitrary units.

The operation of the instrument is simple. The four terminals I, IV, and 5, 6 are connected to the current electrodes C1 and C2, and potential electrodes P1 and P2, as shown in the figure. The wenner configuration is used in the set up. The main switch S2 and PS are closed. The current now passes to the current electrodes C1 and C2. The potential difference between P1 and P2, E is recorded in the meter, C, since S1 and S2 are connected to 3 and 4. Next S and S1 are connected to the terminals 1 and 2 of the potentiometer P. The Potentiometer is now adjusted so that the meter C gives the same reading as before since the potentiometer is graduated in Ohms the resistivity is obtained directly by substituting the value of E in the formula $R = 2\pi$ AE where A is the electrode separation and R is resistivity. The advantages of the terrameter are: (1) Its portability, (2) Easy handling; (3) High accuracy, since values of 0.01 Ohm can be measured. (4) Low frequency permits current penetration upto 600 metres, (5) No special electrodes are required.

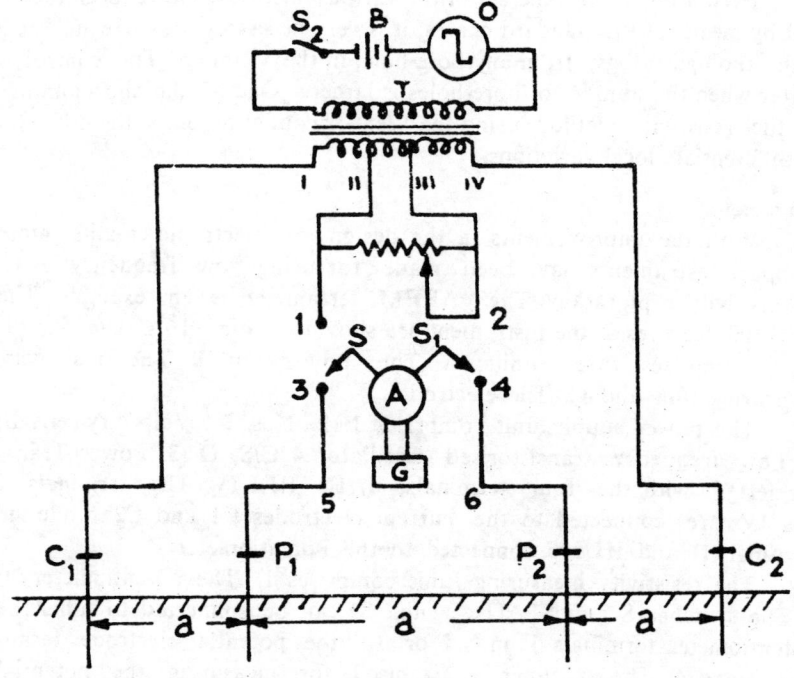

Fig. 5.17

— *Inverse slope method*

Resistivity Survey—Wenner Configuration—Field Data.,

| Depth In m. | $R \dfrac{V}{I}$ | Apparent Resistivity | I/R |
|---|---|---|---|
| 1 | »16.8 | 106 | 0.059 |
| 2 | »6.6 | 83 | 0.15 |
| 3 | | | |
| 4 | »4.08 | 102 | 0.25 |

| 6 | −3.65 | 138 | 0.27 |
| 8 | −3.23 | 162 | 0.31 |
| 10 | −3.10 | 196 | 0.32 |
| 12 | −2.93 | 221 | 0.34 |
| 14 | −2.83 | 249 | 0.35 |
| 16 | −2.70 | 273 | 0.37 |
| 18 | −2.65 | 299 | 0.38 |
| 20 | −2.65 | 334 | 0.38 |
| 26 | 2.71 | 425 | 0.37 |
| 30 | 2.65 | 501 | 0.38 |
| 35 | −2.40 | 528 | 0.42 |
| 40 | −2.28 | 572 | 0.44 |
| 45 | −2.08 | 588 | 0.48 |
| 50 | −1.88 | 590 | 0.53 |

The field data obtained are plotted on a graph paper (semi-log). The apparent resistivity (Ohm-meters) is plotted on the log-scale while the depth in meters i.e. the first column in the above table is plotted on the normal scale. The breaks in the curve so obtained are interpreted in terms of the

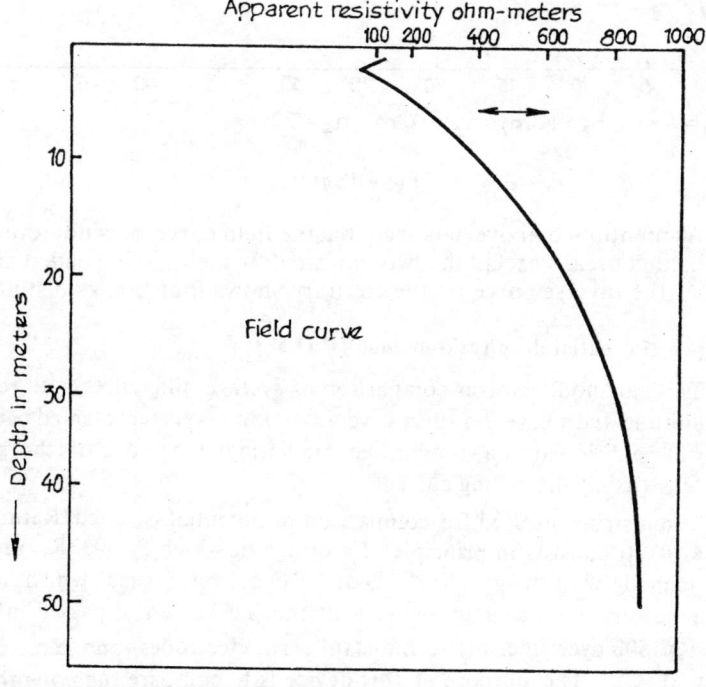

Fig. 5.17(a)

local geology (fig. 5.17 a). As a further check it is a common practice to draw what is termed as the "Inverse Slope Graph" also. In this method, the normal graph paper is used. The electrode spacing/apparent resistivity is plotted on the y-axis, while the depth is plotted on the x-axis. It will be noticed that the inverse slope curve shows very clearly breaks and these can be interpreted in terms of the local geology (fig 5.17 b.)

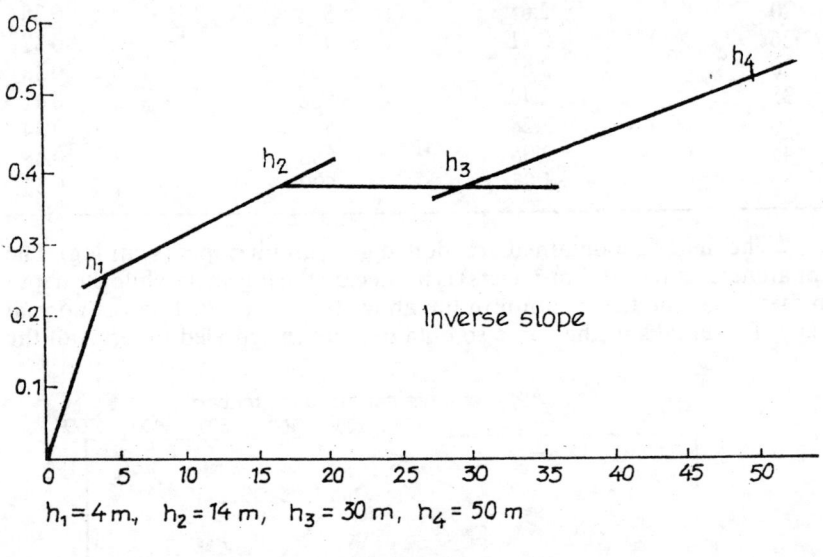

$h_1 = 4\,m, \quad h_2 = 14\,m, \quad h_3 = 30\,m, \quad h_4 = 50\,m$

Fig. 5.17(b)

As mentioned above, it is seen that the field curve does not reveal any very distinct breaks except the two indicated by the double headed arrows, whereas the inverse curve on the contrary shows four breaks distinctly.

### (4) Potential drop ratio method (P.D.R.)

This method involves comparison of voltage differences with respect to magnitude and phase, on successive ground intervals represented between three stakes arranged in a straight line, radiating from one of the power electrodes (using alternating current).

The instrument used for comparison of potential is called Ratiometer (Fig. 5.18). It consists in principle of a bridge in which P, Q, R, are the contacts made with the ground. PO and RO are ratio arms, which have a capacitance and resistance in series. Alternating current is passed into the ground at 500 cycle/sec. between distant earth electrodes (one electrode is shown at C). The purpose of this device is to compare the potentials $V_1$ and $V_2$, between equidistant pair of electrodes, PQ and QR, and also to

POTENTIAL DROP RATIO METHOD

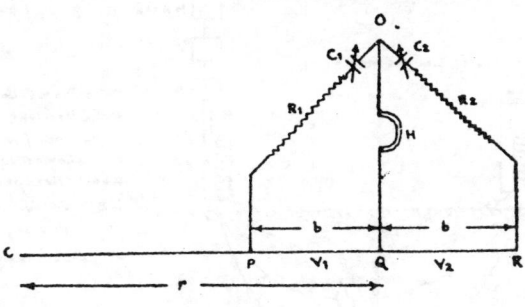

Fig. 5. 18

determine the phase difference (angle) between them. To accomplish this, the capacity $C_1$ and $C_2$ and resistance $R_1$ and $R_2$ are adjusted in each arm, until no sound is heard in the head phone H. In this condition $V_1$ and $V_2$ are directly proportional to the total impedence in the arm PO and RO in homogeneous ground.

$$\frac{V_2}{V_1} = \frac{r-b}{r+b}$$

where r is the distance CQ, and $b = PQ = QR$.

The actual value of the PDR, obtained in the field, is known as the observed PDR. The ratio of observed PDR to normal PDR is obtained, in stratified ground, and is termed as reduced PDR. This reduced PDR is plotted against the distance, and curves are drawn.

This method is mainly used to detect the horizontal discontinuities in subsurface and is comparable with seismic refraction method.

(5) **Electromagnetic Method** (Comparison methods)

(1) The electromagnetic methods measure directly the magnetic field associated with the flow of current in the subsurface. The subsurface current may be generated by creating an alternating field at the surface of the ground. If alternating current is made to flow in a loop or coil of wire suspended either on or above the earth, the current, flowing in a coil, or loop, creates an alternating magnetic field (primary field), spreading out from the coil. This primary magnetic field spreads into the earth, induces varying voltages and also an alternating magnetic field (secondary field), at the surface, which distorts the primary magnetic field Fig. 5.19 below shows the set up.

Superimposition of the secondary magnetic field over the primary field, at the surface, creates a distortion, either in the direction, intensity, or quality of the primary field. The detection of distorted field is done in

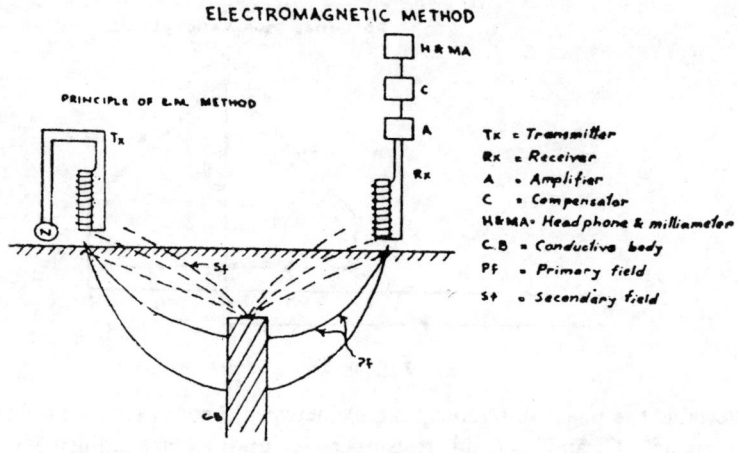

Fig. 5.19

different ways in different instruments. Different instruments measure the following different parameters.

1. Amplitude of the resultant field.
2. Direction of the resultant field (dip angles).
3. Horizontal intensity of resultant field.
4. The in-phase component (or real component).
5. The out-of-phase component (or imaginary component).
6. The in-phase and out-of-phase components.

The instruments measuring the direction of the resultant field (dip angles) are most commonly used for reconnaissance survey. The detailed investigations are carried out with in-phase and out-of-phase measuring equipment. The modern equipment which is available today, can be used 'or both the purposes.

*Field procedure:* — In a virgin area, using the second method, transmitter (Tx) and receiver (Rx) are set up at stations 1-1, 2-2, 3-3, ... etc., in broad side on position, as shown in Fig. 5.19, and the dip angles are measured.

**Comparison Method**

*(a) Horizontal Coil*

In this method two similar coils, with an area of about 6 sq ft. and containing 400 turns of insulated wire are used for the detectors. The coils are connected in opposition with the amplifier-headphone circuit, the amplifier being connected in between the two coils. When the coils are laid one above the other, there is no response in the headphones. If one of the coils is moved 100 ft., along the alignment and placed horizontally on the ground, any difference in the E.M.F. between the two coils caused by the presence of a conducting body, in the ground, is indicated, in the earphones, by a characteristic sound.

When one of the coils is tilted the sound in the earphones is reduced to a minimum. The amount of the tilt as measured from the horizontal is noted. The secant of the angle gives the relative strength of the vertical field at the point, as compared with the field at the point where the other coil (horizontal) is located. The same procedure is repeated along other alignments, all parallel to each other, covering the area. With the data obtained, a relative intensity map may be prepared. If the vertical intensity at any one point is assumed, then the relative vertical intensities of all the other points can be computed, and the iso-intensity lines also drawn.

In the ideal case, where the ore body lies centrally in respect of the energising loop, the induced field is constant and vertical. The horizontal component of the magnetic field, resulting from the flux generated by the secondary E.M.F. within the ore body, increases to a maximum, at the boundary, drops to a minimum at the centre, and rises again on crossing the other boundary of the ore body and falls rapidly beyond that limit. This is depicted in the figure. 5.20

(b) *Vertical loop*

In this method a vertical loop is used and an A.C. with a frequency of 6000 cycles and a wave length of 5000 m. is employed for energising the coil. The examination is carried out on an alignment which is normal to the face of the loop. The detector coil used is of the same type used for the other E.M. methods Fig. 5.22.

The response in the secondary circuit (ore body) varies directly as the current and the frequency, but the higher orders of the frequency are not practicable. Two readings are obtained from the search coil. The first is the 'strike' reading when the plane of the coil is vertical, and the sound in the earphones is minimum, on rotating the coil in azimuth (vertical axis). In this position the plane of the coil is parallel to the horizontal component of the maximum magnetic field. The deflection of the search coil gives the strike. The deflection decreases with increasing depth of the ore body. A second reading known as the dip is obtained when the coil is made horizontal and rotated about the horizontal axis to get the minimum response in the earphones. The strike and the dip values are noted at each of the stations along the traverse. When the ground is free from conductors the values of strike and dip are zero. If a conductive ore body is present the values of strike increase and reach a maximum at a point nearest to the ore body, and show a negative character after a steep fall in the values. The values of dip increase to a maximum away from the centre of the ore body, and then decrease. These features are shown in Fig. 5.22.

The change, in the direction of dip, indicates the presence of the conductor in the subsurface. In order to determine the direction of the conductive body, the set up method, as shown in Fig. 5.23, Fig. 5.24, is used.

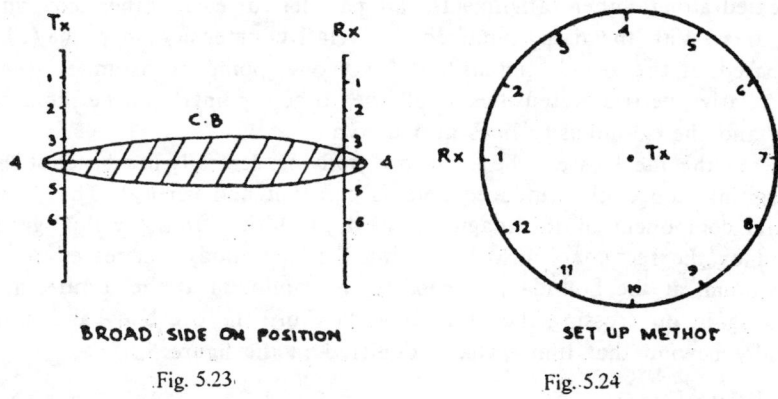

BROAD SIDE ON POSITION　　　　　　SET UP METHOD

Fig. 5.23　　　　　　　　　　Fig. 5.24

The transmitter $T_x$ is kept at one place, $T_x$ on the conductive body, and receiver is taken to stations 1, 2, 3 forming a circle around transmitter. Change, in direction of maximum dip, indicates the presence of a conductor. Once the approximate location, of the conductor, is established, detailed surveys are conducted, by laying traverses orthogonal to the strike, of conductive body and measuring the in-phase and out-of-phase values at stations spaced 10-15 metres apart. The transmitter-receiver separation can be varied between 30 to 90 metres. The middle point between transmitter and receiver is taken as the point of observation. The frequency of transmitting current is generally 400-2500 cycles per second. The depth of penetration is normally 2/3 of transmitter receiver separation, in the case of vertical bodies, and $1\frac{1}{2}$ times Tx-Rx separation, in respect of horizontal bodies. The observed values are plotted in the form of profiles and qualitative-quantitative interpretation is done with the help of model curves.

**Parallel Wire Method** (Fig. 5.25)

In this layout an A.C. generator of more than 100 volts and a fre-

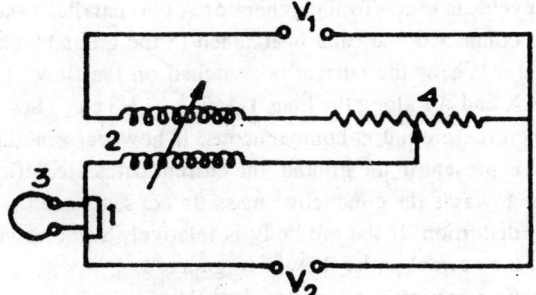

1. Amplifier
2. variable inductance
3. Ear phone
4. variable resistance
$V_1$. Reference voltage
$V_2$. Unknown voltage

Fig. 5.21

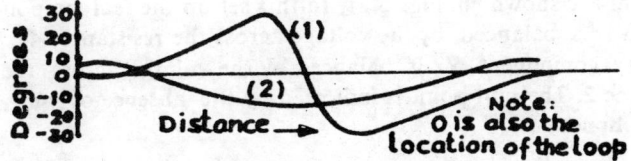

Fig. 5.20

Fig. 5.21(c)

quency of 500 cycles is used. To the generator A two parallel naked conductors AX and AY are connected and each is attached to the ground stakes, at various points 1,2,3,4, etc. When the current is switched on the flow of current takes place between AX and AY along the lines 1-1, 2-2: . . 3-3 etc. These current paths are accurate when the ground is homogeneous. If however a highly conducting mass, such as O is present in the ground, the current paths are deflected and they tend to converge towards the conductive mass. In consequence the equipotential lines also suffer distortion. If the ore body is relatively a poorer conductor than the ground then the current paths show divergence.

The detection equipment, for tracing the E. P. lines comprises two wooden rods (2) fitted with metallic spikes 6 inches long (1) as shown in fig 5.7. The spikes (1) are attached to a radio frequency receiving unit provided with head phones. It will be noticed that when the stake A and B are on the same E. P. line the current is in phase and there is no response in the earphones. If the stakes are on different E.P. lines a buzzing sound is heard on the earphones. Where the response is weak audiofrequency amplification may be needed In practice it is possible to trace the E. P. lines by keeping one of the stakes fixed and moving the other to various positions of 'equilibrium'.

An alternate arrangement of the detection apparatus comprises a 'detector' coil with several hundred turns of wire mounted on a frame 2 ft. by 3 ft. The coil can be rotated both about the vertical and horizontal axes. The terminals of the coil are connected to a headphone. The circuit includes an audio frequency amplifier as well as a variable condenser for tuning in. In order to find the plane of elliptical polarisation, the coil is so oriented as to obtain the minimum response in the earphones. The azimuth of the coil is noted for this position. The coil is now rotated through 90 degrees and then tilted to obtain another position of minimum response. The plane of the coil is now in the plane of polarisation of the magnetic vector. Similar observations are made at various points in the area under examination. This method is suitable for shallow depths.

The audio portion for the comparison of voltages and phase differences includes an A.C. potentiometric circuit. The essential parts of the equipment are shown in Fig. 5.21. In this set up the real component of the voltage V1 is balanced by the voltage across the resistance 4, while the imaginary component V2 is balanced by the voltage across the variable inductance 2. The null point is indicated by the absence of any sound in the ear phones.

**(4)  Horizontal loop method.**

Alternating currents set up electromagnetic oscillations, in the form of electromagnetic waves, which have the same frequency as that of the currents.  The frequency may vary from a few cycles per second (known as audio frequency) to thousands of kilocycles per second (known as radio frequency).  The common frequencies employed, by exploration geologists, are of the order of 50 cycles per second.  Curves such as those shown in Fig. 5.21 may be obtained.

The rate, at which electromagnetic waves penetrate the rocks, fades away with depth and depends on the frequency used and the resistivity of the rock.  The maximum penetration for a given depth and resistivity value $\gamma$ is at an optimum frequency ($f$) and is given by

$$h\sqrt{\frac{f}{\gamma}} = 10$$

h, f and $\gamma$ should be expressed in the same units C.G.S.   If h = 100 meters, $\gamma = 4 \times 10^2$ C.G.S. units (Ohm-cms).

then

$$100 \sqrt{\frac{f}{4 \times 100^2}} = 10$$

$$25 \frac{\sqrt{f}}{400} = 10 \qquad f = 64$$

It is thus seen that low frequency enable greater depth of penetration.

Surveys, using inductive methods, can be conducted either on the ground or from the air.  When ground surveys are undertaken, an area is enclosed by a loop of insulated cable, fed with current from an alternator (A. C. generator) Fig. 5.25.

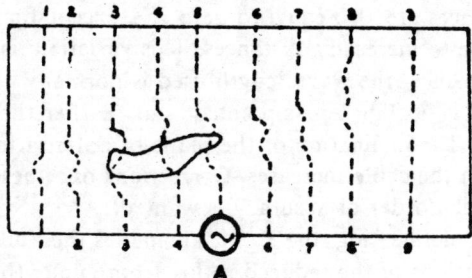

Fig. 5.25

Cross traverses 1-1, 2-2, 3-3, 4-4 etc. are conducted with search coils, comprising several hundred turns of wire.

Where conducting materials, such as an ore body, are present then on approaching a deflection in the trend of the profile is seen. Since graphite present even in graphitic schists, is capable of producing anomalies of this type, the inductive method requires to be confirmed by another method, e.g. gravity or geochemical etc. The detector equipment used is the same.

### (a) Turan method

The equipment known as Turan is made by a Swedish firm, A.E.B.M. and in operation it is very similar to the potential drop ratio [P.D.R.] method. Here again the primary field is produced either by laying parallel cables or by employing a large loop. The detecting equipment comprises two search coils. which are kept at a distance varying from 10 m to 15 m apart. The ratio of the amplitude between the voltage and the phase difference is measured for each position of the pair of coils.

Normally the vertical component of the resultant field is measured by keeping the coils in the horizontal position. The coils can also be kept in the vertical position or one of the coils can be horizontal while the other is vertical. Where the coils are kept horizontal the potential ratio obtained may be shown as $V_1/V_2$, $V_2/V_s$... etc. while the phase differences are $l^2_1 - l^1$, $l^3 - l^2$....etc. at each successive position of the pair of coils, for example stations 1, 2, 3, etc. in order to compensate for the variation in the primary field/P/M with the increasing distance of separation, the measured ratios are divided by the normal amplitude ratios, P1/P2,P2/P3 ... etc. This step is facilitated by the fact that the normal ratio at any point is inversely related to the ratio of the distance from the loop, except for the differences in levels for which corrections are applied. Thus the reduced ratios will be V1.P2/V2.P1, V2.P3/V3.P2,...etc. In the absence of any conductor in the ground, these ratios are all equal to unity, and the phase differences are also equal to zero XX except for the variation in the E.M. field due to increasing distances. This variation is related to the wave length $\lambda$. Since the wave length used is normally several hundred metres the effect is negligible. As pointed out earlier the field is elliptically polarised. The inclination of the plane of polarisation decreases as the distance from the cable increases. Corrections on these accounts are needed when a high order of accuracy is wanted.

Hence for delineating the E. M. anomalies the following data are needed (a) Deviations of the reduced ratio from unity (b) Variations in phase difference from zero. The phase difference is essentially related to the horizontal gradient of the phase, while the reduced ratio bears a relation which can be represented by the function $1 - dH2/1 - H1$, where H1 is the secondary field expressed as a ratio of the primary field, $H_2$ is the horizontal gradient of the primary field, while d represents the constant distance of separation of the search coils.

The Fig. 5.24 shows the general pattern of the voltage ratio and phase difference anomalies.

(b)   *Air surveys.*

The inductive electromagnetic surveys are also conducted from the air. Either a single plane is used, which carries both the transmitting antenna or coil as well as the receiving equipment, or alternatively the transmitting part, is fitted on one plane while the receiving part is trailed in a 'bird' - generally in another plane, about 450 ft. - 900 ft. behind.

The equipment employed (Fig. 5.23), comprises a transmitting unit A and a receiving unit (B), which are generally carried on separate aeroplanes. The transmitting unit has two identical coils of wire mounted at right angles to each other. The combination is supplied with A.C., and on rotation the strength of electromagnetic field generated is uniform about the axis (of rotation).

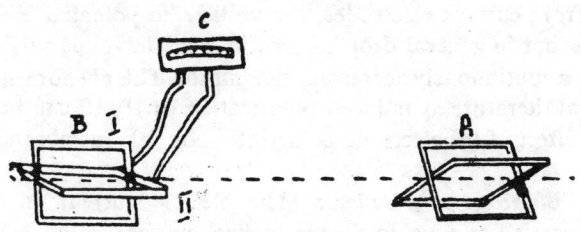

Fig. 5.23

The receiving unit (B) (Fig. 5.23) has two mutually perpendicular coils, one vertical and the other horizontal axis, which is directed towards the transmitting unit. Since the field generated by the transmitting system (A) is uniform in all directions, the field induced in the two receiving coils (B) I and II should be uniform. A recording unit (C), attached to the receiving unit (B), shows the variations in the output of the coils I and II. The recording units picks up both differences in amplitude as well as in the phase of the A.C. impulses.

When no external field, other than that generated by the system (A), is present, the recording unit (C) shows practically no deflections, and the voltage ratio, between the systems (A) and (B), remains constant, even when the distance, between them, is large. Adjustments to the recording unit (C) are made, so that the currents in (A) and (B) are in phase,

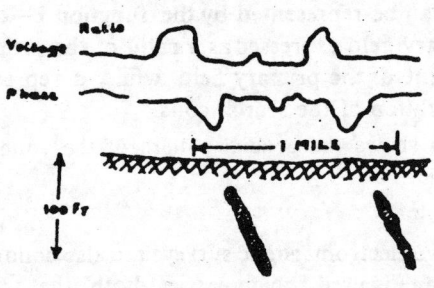

Fig. 5.24

i.e. the phase difference is zero. The presence of a conductor, in the ground below, introduces an imbalance in both the voltage and phase (in coils I & II of (B), and these variations are observed on the recording unit C.

The distance between the aircrafts may be about 600 ft. The height of aircraft is about 250 ft. above ground level (G.L.) while the speed is about 120 m.p.h.

## Induced polarisation methods

It has been observed, in resistivity surveys, that on disconnecting the battery, from current electrodes, the voltage, in potential electrodes, (Fig. 5.25) does not in general drop to zero immediately, but persists for sometime with a continuously decreasing magnitude. This phenomenon is termed in geophysical literature as induced polarization or IP. IP can be explained as being the effect of (a) electrode polarization or (b) membrane polarisation or both.

(a)   *Electrode polarisation :* The electric current, in the ground, is normally carried by ions, in the electrolytes present in the pores of rocks. If the passage of the ions (which as in the case of common metals the transport of current is by electrons) is obstructed, by certain mineral particles, then ionic charges pile up, at particle-electrolyte interface. Positive charges accumulate where the current enters the particle, and negative charges where the current leaves (Fig. 5.25). The piled up charges create a voltage, which tends to oppose the flow of electric current across the

INDUCED POLARISATION

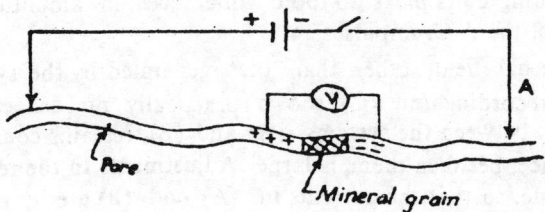

ELECTRODE POLARISATION AT MINERAL-ELECTROLYTE INTERFACE

Fig. 5.25

interface, and the particle is said to be polarised. When the current is interrupted a residual voltage continues to remain across the particle, due to the bound ionic charges, but it decreases continuously as the ions slowly diffuse back into the pore-electrolyte. This process gives the induced polarization effects. Foremost among the minerals having an electronic mode of conduction, and therefore exhibit strong IP characteristics are pyrite, pyrrhotite, chalcopyrite, graphite, galena, bornite, magnetite and pyrolusite.

(a) *Membrane polarization* : If the surface of the clay particle is negatively charged, it attracts positive ions, from the electrolytes, present in

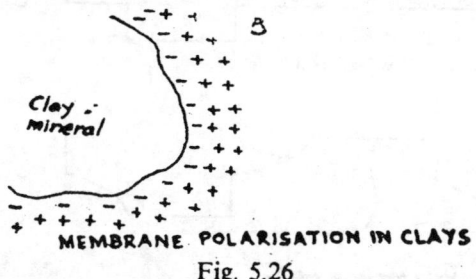

Clay mineral

**MEMBRANE POLARISATION IN CLAYS**
Fig. 5.26

the capillaries of a clay aggregate. An electrical double layer is therefore formed, at the surface of the particle, as shown in Fig. 5.26, the concentration of the positive ions being greatest at the surface of the clay particle. If the positively charged zone penetrates far enough into the capillaries, it effectively repels other positively charged ions, and so acts like an impervious membrane, impeding the ionic movement through the capillaries. When an electric current is forced through the clay, the positive ions are displaced, and on interruption of the current the positive charges redistribute themselves in their former equilibrium pattern. This process of redistribution manifests itself as a decaying voltage. This gives IP effect.

The existence of membrane polarization complicates the interpretation of IP observations, because the IP effects cannot be interpreted as unambiguous evidence, of the presence of minerals possessing the property of electronic conduction (sulphide or magnetite) in contact with the electrolytes in the rock.

Polarization is essentially a surface phenomenon. Hence, greater the surface presented by sulphides or clays stronger are the polarization effects. Thus impregnation type or disseminated or 'porphyry' copper type ore deposits are particularly suitable targets for exploration with IP methods.

*Field procedure* : IP measurements can be made in the different ways.

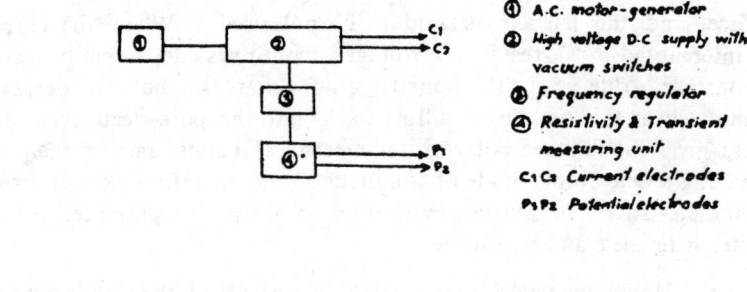

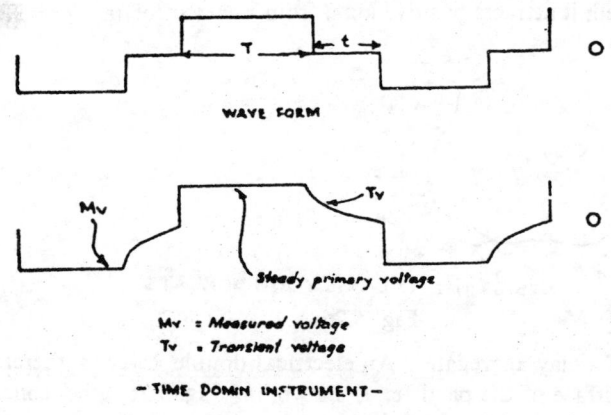

Fig. 5.27

(a) Time domin method : In this method, direct current is sent into the ground and the decay of voltage, between potential electrodes, is measured after the current is turned off.    This is called transient or time domin method. Fig. 5. 27 shows the block diagram of  instrument used  in this method of survey.  Power source upto  on 25 K.V.A., 5000 volt, and 10 amperes have been employed.  The current on time T, ranges from one second to as much as 30 seconds and the current 'off' time may be as much as 3 seconds.  It is not strictly necessary to employ a  cyclic current wave form, but advantages in signal-to-noise ratio are achieved thereby.

(b) Frequency domin method : In this method, variation of apparent resistivity of the ground with frequency, of  the affected  current, is determined. Fig. 5.28 shows the  block  diagram of the  frequency varia-tion field apparatus and the current and  voltage wave forms.

The impedance is measured directly as a ratio of voltage/current on a stable bridge.   In practice the resistivity of ground at two frequencies, one is very low (0.1 c/sec) and the other a higher (10 c/sec), is determined.

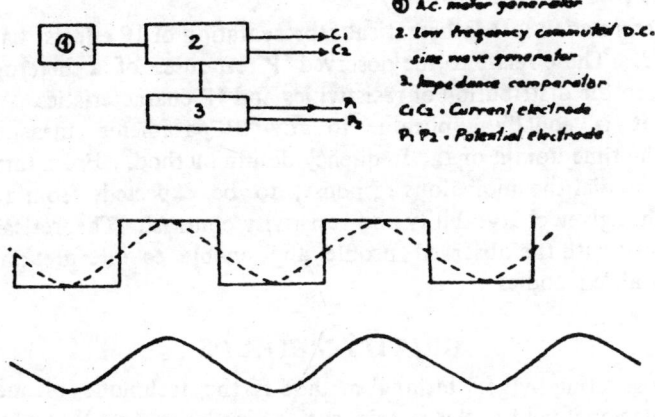

1. A.c. motor generator
2. Low frequency commuted D.C. sine wave generator
3. Impedance comparator
$C_1 C_2$ - Current electrode
$P_1 P_2$ - Potential electrode

FREQUENCY DOMAIN INSTRUMENT DIAGRAM

Fig. 5.28

Common electrode arrays used are shown in Fig. 5.29. CC are current electrodes, while PP are potential electrodes. Eltran method has been in most common use.

IP data may be plotted in the profile form, or contoured, although contouring is not strictly justified. Profile interpretation is superior for shallow confined bodies. Expanding arrays with increasing 'a' keeping the spread centre fixed may also be used. The electrode arrangements commonly used are shown in Figure 5.29.

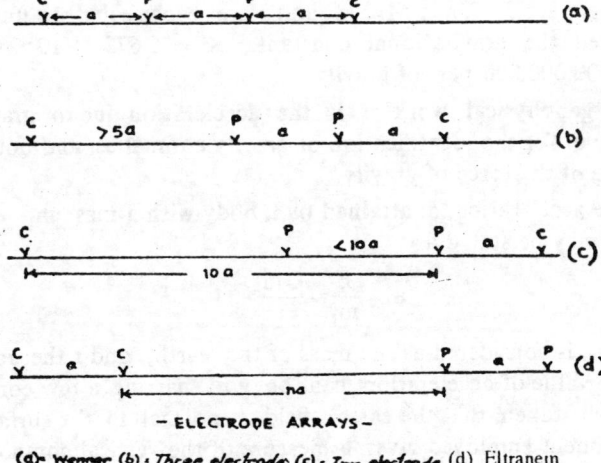

— ELECTRODE ARRAYS —

(a)- Wenner (b) - Three electrode (c) - Two electrode (d) Eltranem

Fig.5.29

*Interpretation* :   Mathematical représentation of IP effects  have been developed.   These relate to the observed IP response, of a  heterogeneous medium, to the distribution of resistivities and IP characteristics.  Approximately,  it  is  equally applicable  to  any  IP  parameter  measured, including the time domin or the frequency domin method.  From this theory one may predict the anomalous response,  to be  expected,  from a specific body with a given chargeability and resistivity contrast.  Theoretical curves are matched with the observed  profile  and  enable  to  interpret anomalies due to localised  bodies.

## GRAVITY METHODS

Prospecting by gravitational method is  the  technique of  measuring the gravitational field at the earth's  surface and the data thus obtained is utilised to predict the nature of subsurface and structure. In  this  method the natural field of earth's gravitation is  used. Gravitation  is  defined as the force  which attracts all  bodies  and  is directly dependent on  the masses involved; and inversely on the distances.

The simple equation impressing this relationship is

$$F = \frac{K \cdot M_1 M_2}{r^2} \quad M_1 \ \& \ M_2 \text{ are expressed in grams}$$

r is the distance in cms, between the two masses $M_1$ and $M_2$

F is the gravitational force expressed in dynes.

Dyne is the force required to move a mass of 1 grm at an acceleration of 1 cm. per sec$^2$.   When $M_1$ , $M_2$ and r are  each  equal to unity,  F = K, K is called the gravitational constant.   $K = 6.673 \times 10^{-8}$ C.G.S. units or 1/15,000,000,000 part of gravity.

In geophysical work it is  the  acceleration due to  gravity, that is made use of for the identification of gravity anomalies, and not  the  absolute value of the force of gravity.

The acceleration, a, attained by a body with a mass $m_1$, caused by a mass $m_2$ may be stated as

$$a = \frac{F}{m_1} = \frac{Gm_2}{r^2}$$

If $m_2$ is considered as the mass of the  earth,  and r the  radius of the earth, the value of acceleration can be  got,  barring a few corrections. It may also be taken, that the earth's field is  normal to the surface.   Hence the instrument employed gives a measure of the vertical component of the acceleration Z, due to gravity as shown in Fig. 5.30.

$$az = G \frac{zdm}{(x^2 + z^2)^{3/2}}$$

*Instruments*

Instruments for measuring gravity are of 3 main types (1) Pendulum, (2) Tortion balance, and (3) Gravimeter.

*Pendulum*

The use of the pendulum, in gravity measurements, is known for quite a long time, and the value of gravity can be obtained from the equation, $g = 4\pi^2 \, n^2 \, L$

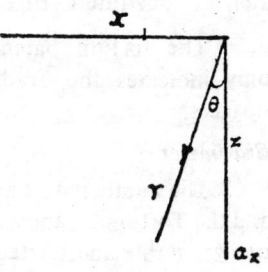

Fig. 5.30.

where g = gravity (acceleration),
L = length of pendulum
n = No. of oscillations of the pendulum in unit time.

The pendulum has been used in Geodetic surveys.

*Tortion balance*

In the Tortion balance, the most popular design for geophysical work is the Eotvos balance. This instrument consists of a beam suspended from

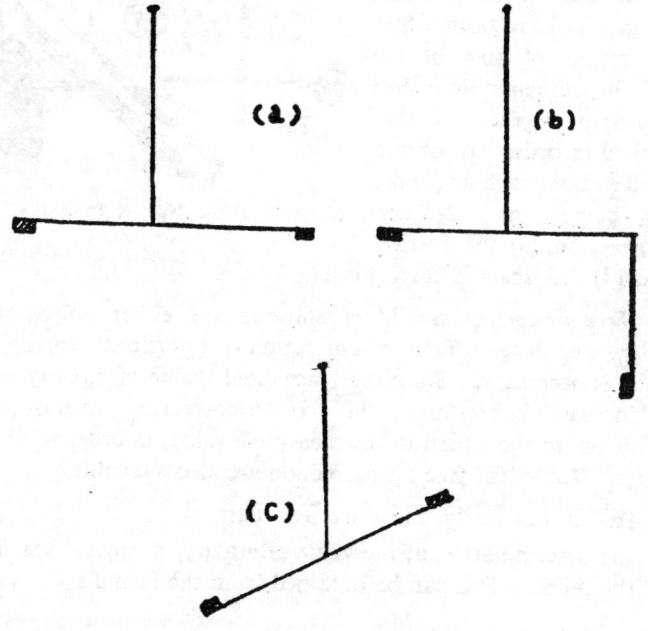

Figs. 5.31 (a, b, c)

a fine wire. The beam is either horizontal or inclined, and two equal weights are attached to the ends of the beam ; either directly or by **suspension** as illustrated (Fig. 5.31 a, b, c)

The tortion balance does not measure the value of gravity but only indicates the gradient, and the curvature of the equipotential surfaces.

### Gravimeter

The cumbersome and time consuming instruments like the pendulum and the Tortion balance were superseded, for normal gravity work, by sensitive yet stable and portable instruments known as gravimeters. Some of these instruments measure variations in gravity by differences in the amount of extension of a sensitive spring, or variation in torque induced in a wire etc. The movements, however slight, are magnified by mechanical leverage as well as by an optical system.

The mechanical principles of one instrument used widely, the Worden Gravimeter, is brought out in Fig. 5.32.

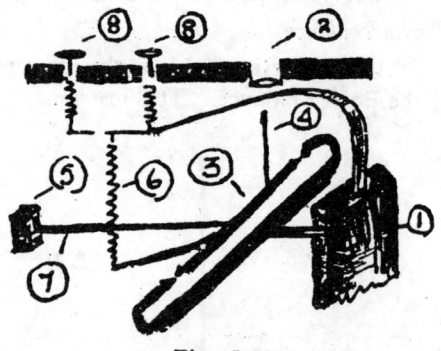

After the instrumental values of gravity have been obtained, by means of any of the types of instruments described earlier, certain corrections have to be applied in order to obtain the absolute value, such as (1) the Free air correction, (2) the Bouger correction, (3) the Terrain correction (4) the Isostatic correction.

Fig. 5.32

(1) *Free air correction* : Most stations are either above or some times below sea level. Thus a correction proportional to the altitude or elevation is necessary. Since the measured value of gravity decreases with the increase in elevation, the free air correction is a positive quantity, and it is to be added to the measured value, in order to obtain the actual value. Tables for free air correction are also available.

The free air correction comprises two parts :

(a) the determination of the value of gravity, at mean sea level, at the point. This can be obtained from the formula

$$g_o = G \frac{M}{R^2},$$

where,     M = mass of the earth,
                 R = radius of the earth.

(b) the determination of "g", the value of gravity, at an elevation of h, above the mean sea level. This can be obtained from the formula

$$g_0 - g = \frac{2g_0 h}{R}.$$

In this formula $\frac{2g_0}{R}$ is a constant for the area.

(2) *Bouger free air correction* : While altitude of a mountain tends to decrease the value of gravity, the presence of a considerable mass of rock, above the sea level, increases the measured value of gravity. A correction is applied to remove the gravity effect of additional mass of material above sea level. Thus gravity is dependent on (1) altitude (2) density of the rock mass. It will, thus, be seen that free air and Bouger reduction can be combined, and the resultant equation will be

$$go'' = g + H(0.0003086 - 0.0000421\ d)\ \text{gals.}$$

where,     $go''$ = gravity at sea level, with free air and Bouger correction.

   $g$ = measured value of gravity.

   $H$ = altitude in meters.

   $d$ = density of rocks found, in the area, between the station and sea level.

(3) *Topographic correction or Terrain correction* : In a rugged country due to large relief, considerable deviations in the values of gravity are caused. Thus a correction has to be applied, so as to reduce value of gravity as if the entire area, is of flat terrain. For this purpose good accurate contoured topographic maps are necessary. In case these are not obtainable, an accurate survey of the area has to be carried out as a preliminary measure.

Charts and tables for easy reference are available.

One common procedure for computing the terrain effect is shown in Fig. 5.33

The area surveyed is divided radically. OR and OS are two such radii. Concentric arcs RS and PQ are drawn with radii $r_2$ and $r_1$ respectively from any point O. The terrain correction for the area is obtained from the following equation.

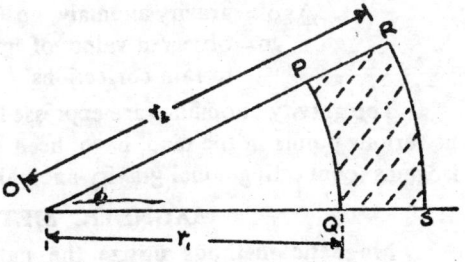

Fig. 5.33

$$\triangle g = kd\phi(r_1 + \sqrt{r_2{}^2 + h^2} - \sqrt{r_1{}^2 + h^2} - r_2)$$

k = gravitational constant
d = density of rocks
$r_1$ = radius of the inner circle in cms.
$r_2$ = radius of the outer circle in cms.
h = average elevation of the area in cms.

(4) *Latitude correction* : Gravity is the force with which the earth attracts all bodies. The unit of gravity is the gal, named after Galileo. It is defined as that force causing an acceleration of 1 cm. per sec². At the equator the value of gravity is 978.049 gals. and at the poles, the value is 983.221 gals. The latitude correction can be computed thus,

$$g = 978.049(1 + 0.0052884 \sin^2 \phi - 0.0000059 \sin^2 2\phi)$$

where, g is the value of gravity, at sea level, and $\phi$ the latitude.

Thus, the theoretical value of gravity can be calculated, for a particular latitude, based on the assumption that the earth is a homogeneous body. In practice, however, the theoretically computed value and the actual measured value may differ, due to local disturbances caused by the presence of rocks with contrasting specific gravity.

This variation, in gravity values, is also partially attributable to isostatic compensation, and will have to be taken into account in regional studies. In smaller problems, isostatic compensation is of negligible nature.

(5) Isostatic compensation is not of great consequence in the study of local problems, but due consideration is required in large scale work, e.g. geodetic surveys etc.

As earlier indicated, the purpose of all the corrections enumerated, is to obtain the true or normal value of gravity, apart from the measured or 'abnormal' value of gravity. This difference between the normal value and measured value of gravity is called the gravity anomaly. This may be expressed as follows.

$$\triangle go'' = go - go'$$

$\triangle go''$ = gravity anomaly, go' = normal gravity
     go = observed value of gravity, with free air, Bouger and terrain corrections.

The gravity anomalies are expressed in milligals. After the anomalies, at various points in the map, have been computed, isogamic contours or isogams (connecting equal gravity-anomalies) are drawn.

## MAGNETIC METHODS

Magnetic methods utilize the natural magnetic field of the earth. The earth's field behaves as if a bar magnet is placed inside the earth with its south and north poles very near to geographical north and south poles respectively. The "magnetic elements" composing the earth's field are d, v, H and Z, where d is the declination or the angle between the magnetic and geographic north, v is the dip or inclination or direction of the resultant field due to the horizontal and vertical components of the magnetic field, H is the value, in Oersteds, of the horizontal component, and Z is the value of the vertical component of the earth's field.

It is possible to find out the value of H by experimental methods, by using ordinary bar magnet, and if the value of v can also be obtained, Z can be computed, by the use of the earth inductor.

### Earth inductor

The earth inductor, an electromagnetic device, can also be employed in the determination of the magnetic elements. In principle, this instrument consists of a coil of wire, of known area, and consisting of a large number of turns. It is rotated at a constant rate, in a uniform magnetic field. The electric potential, that is, induced in this loop is measured by means of a suitable sensitive instrument, say galvanometer, in which the scale division of 1 mm $= 10^{-8}$ amp.

Thus when the axis of the coil is vertical, the current generated in the coil, on rotation, is proportional to the horizontal field, H. When the axis of rotation is horizontal, and in the magnetic meridian, the current generated, in the coil, is proportional to the vertical component, Z.

In practice the axis, of rotation, is placed in the magnetic meridian and inclination of the axis is either increased or decreased so as to obtain various readings of current generated. This enables the determination of the actual angle of dip, with equal deflection in either direction and a median null point. Thus the angle of dip, v is obtained.

An alternative method is employed, in which the axis of the coil is made horizontal and aligned parallel to the magnetic meridian, so as to obtain the value of the current relative to the vertical component, of the earth's field. Next, the axis of the coil is made vertical and rotated, so as to obtain the current corresponding to the horizontal field H. Since $R^2 = H^2 + Z^2$, the value of R and V (the dip) can be calculated, using the principle of the parallelogram of forces (Fig. 5.34).

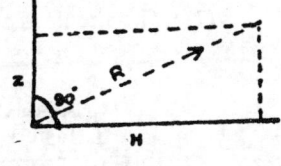

Fig. 5.34

The presence of magnetic minerals in the rocks, especially certain minerals, such as magnetic ($FeO. FeO_3$), Ilmenite $FeO. TiO_2$, and pyrrhotite etc., increases the earth's field locally. This increase is dependent on a factor called the permeability. As u is too small to be measured, in rocks, K or susceptibility is measured

$$u = 1 + 4\pi K.$$

Thus the values of H, Z, v and d can be determined at various points in the area.

The magnetic survey may consist of measuring (a) the dip or (b) the declination.

(a) In measuring the dip, a dip circle or super dip or any similar instrument is used. This instrument consists of a sensitive magnetic needle mounted on a horizontal axis, so that the needle moves freely in the vertical plane. A circle, graduated in degrees and fractions of a degree, enables the reading of the needle. The instrument is oriented at right angles to the magnetic meridian, and the angle of dip is obtained, $Z = K \tan \phi$. K is a constant, depending on weight of the needle, the relative distance between the centre gravity and the point of suspension and the magnetic moment. The value of K is obtained by measuring $\phi_1$ the angle of dip within the meridian ; and $\phi_2$ the value at right angles to the meridian, and substituting these values in the formula

$$K = \frac{H \tan \phi_1}{\tan \phi_2 - \tan \phi_1}.$$

H is the value of the horizontal component, determined experimentally.

Next a magnetometer is employed for obtaining the value of the horizontal component (H).

The Thalen-Tiberg magnetometer is a combination instrument (Fig. 5.35). This instrument consists of a horizontal compass and needle, an auxilliary cylindrical permanent magnet of known moment (M) mounted on horizontal movable graduated arm. By means of the levelling plate, the needle is made horizontal. The instrument is mounted on a tripod.

(1)  The field procedure is to remove the permanent magnet and level the instrument.

(2)  The needle is released, and the reading noted.

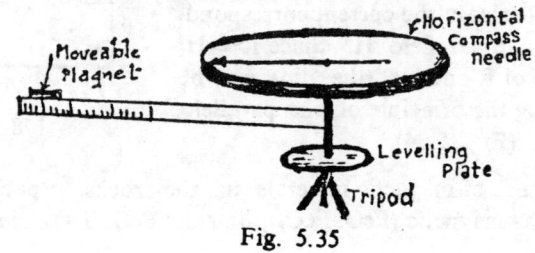

Fig. 5.35

(3) The horizontal arm is turned through 90°. The permanent magnet is replaced and moved back and forth until suitable deflection (a) of the needle is obtained.

$$H = \frac{M \cdot 2d}{(d^2 - l^2)^2 \cdot \tan a}$$

where  H = horizontal component

M = magnetic moment of the magnet

d = distance from the centre of the permanent magnet to the centre of the compass needle.

l = ½ the length between the magnetic poles of the bar magnet. It is normally = 5/16 the length.

a = angular deflection of the needle

Even if it is necessary to vary the position of the cylindrical permanent magnet, at certain stations for minimising deflections, the values so obtained can be mutually correlated.

An alternative method, for obtaining the value of H, is to rotate the horizontal arm, carrying the permanent magnet, to a position, where it lies at right angles to the compass needle. The reading of the needle is noted. The permanent magnet is removed and when the needle comes to rest, the reading is again noted and the angular difference $\theta$ obtained.

In this case, the field due to the permanent is parallel to H or horizontal component, thus $H = K \sin \theta$, where K is a constant, depending on the distance of the permanent magnet. By using the instrument, in a magnetic field of known strength, the value of K can be determined.

Instrumental errors possible are mainly (a) lack of coincidence between the magnetic axis and geometric axis, and (b) frictional losses. Further the measurements should be affected without influence of magnetic objects, e.g., iron articles, and extraneous magnetic fields, e.g., electric mains etc. By utilizing the values of H and Z thus obtained, for every station, contours of equal vertical or horizontal intensity can be drawn.

**Vertical magnetometer**

In practice it has been found that the interpretation of charts showing vertical intensities is more easy, especially in high magnetic latitudes. Hence the dip needle is more popular.

For the majority of cases, however, absolute values of the fields are not necessary, and a measure, of the variation, of field intensity from one station to another, is more than enough. Sensitive instruments with an accuracy of 1 gamma ($10^{-5}$ oersteds = 1 $\gamma$ or gamma) are available, The vertical variometer measures the anomalies of the vertical component Z, only. One of the more popular makes of the vertical variometer is the Askania variometer, and the Watts instrument is identical in principle, and largely similar in construction.

In these instruments the magnetic system (Fig. 5.36) consists of 2 magnets (1), of identical moment etc., separated by a metallic (duralumin) block (2), on which is mounted a mirror (3). There are 2 balancing screws (4), for adjusting the magnetic needle and the entire assembly is mounted on quartz knife edges (5).

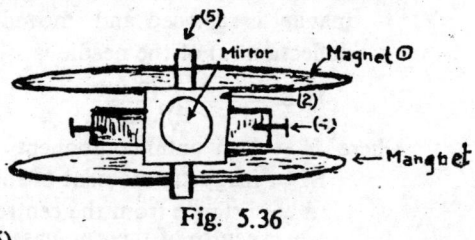

Fig. 5.36

The optical system (Fig. 5.37) is employed for measuring the deflections of the magnetic assembly. Sunlight (1) is reflected by a mirror (2) and then by a transparent mirror (3), placed in the 45° position. The beam is incident on the mirror (4), fixed to the magnetic assembly, passes through a converging lens (5) and falls on the

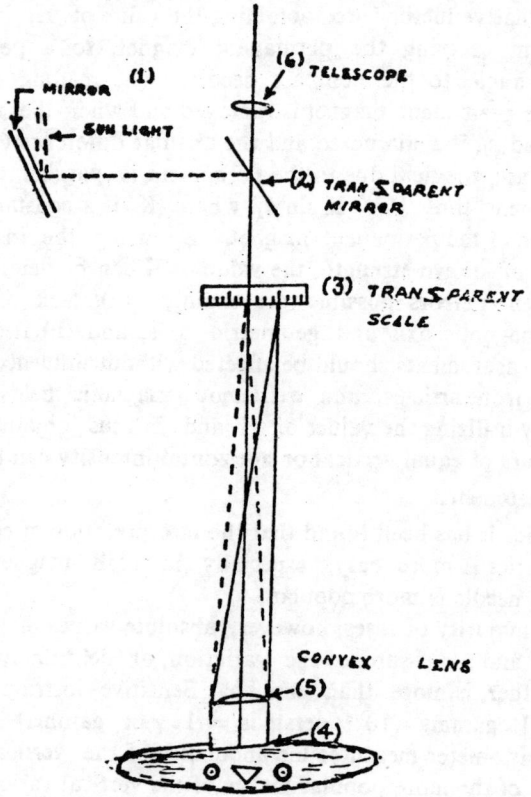

Fig. 5.37

transparent scale (6) graduated to read in gammas. The scale can be brought into focus by a reading telescope system.

There are thermometers, provided in the instrument, for temperature correction.

The deflections, of the magnetic system, can be controlled, when necessary, by means of a cylindrical bar magnet, which is placed vertically below the centre of the magnetic system.

Field adjustments consist of :

(1) Orienting the magnetic system, in the plane of the magnetic meridian.

(2) Carrying out suitable mechanical adjustments so as to make the instrument reading fall within the range of the scale.

The sensitivity of the instrument, i.e., the value of a deflection equal to one small division of the scale, is ascertained. The simplest way for obtaining the value is to apply the electrical method. Here a uniform electromagnetic field, of known strength, is produced by passing a fixed amperage through coil of wire 'a' cms radius with n turns. If the measuring point is x cms., above the centre of the loop, and I is the value of the current, in the coil then the field,

$$F = \frac{2\,\pi\,n\,I\,a^2}{10\,(a^2 + x^2)}\ \text{Oersteds.}$$

**The torsion magnetometer**

This instrument comprises a small highly polished bar magnet, of rectangular cross section, which is suspended horizontally by a strong fibre held taut passing through the smaller axis (of the magnet). Either ends of fibre are fixed to the limbs of a rigid yoke. A circular scale, graduated in gammas, is attached to one of the limbs of the yoke, such that the point of attachment, of the torsion fibre, coincides with the centre of the scale. Since the suspending fibre is actually a composite, comprising fibres of different thermoelastic properties, the influence of temperature on the system is annulled.

For use the instrument is set up on the stand and levelled. The circular scale, attached to the yoke, carrying the magnet, is rotated in a direction, which makes the magnet horizontal. An optical system, similar to that used in the vertical magnetometer (variometer) enables this adjustment, the polished magnet itself acting as the reflecting plane mirror, for establishing horizontality (of the magnet). In this position the torque, induced in the fibre, is equal to the moment of the magnetic system. The instrument is placed over various spots and readings obtained from the circular scale for each. Thus the relative value of the vertical field can be got.

The advantges of the instrument are that :

(a) It can read directly upto 65000 gammas.

(b) It has a very small drift.

(c) Since the suspending fibre, passing through the axis of the magnet, is held horizontal, at the time of use, the effect of the horizontal component of the earth's field is eliminated.

(d) It requires lesser time in setting up the variometers as it need not be aligned in any particular azimuth.

(e) It is lighter than the normal variometers.

Since fatigue is likely to affect the elastic properties of the fibre, after continuous use, calibration of the instrument, by using the normal procedures, is essential to ensure accuracy.

Magnetic measurements are helpful in assessing the nature of the subsurface, in a qualitative as well as quantitative manner. The direction and magnitude of the resultant, at various points on a magnetic traverse line, can often be used for the purpose as illustrated in Fig. 5.38.

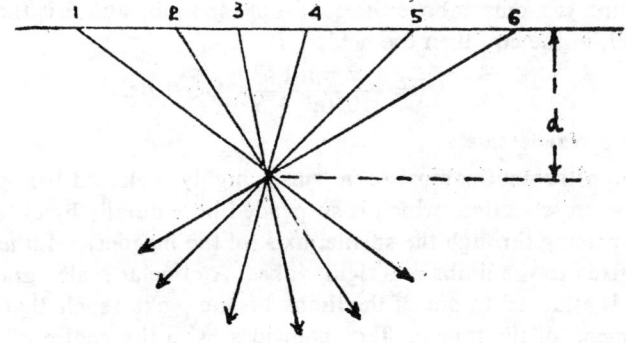

Fig. 5.38

If 1, 2, 3, 4 etc. are stations on a traverse, the direction and magnitude of the resultants are plotted at the respective points.

It will be seen that the directions of the resultants at 1, 2, 3 etc. intersect at P or thereabout, and the indicated depth 'd' is thus obtained with a certain degree of approximation.

### Airborne magnetometers

Two types of magnetometers, which do not use the normal metallic permanent system, are the Fluxgate magnetometer, and Nuclear magnetometer which have been developed for use in mineral and oil exploration.

including ground and air surveys. These are however of special importance in air borne work.

(a) *Fluxgate magnetometer :* Two parallel cores, each with magnetisation curve as shown in Fig. 5.39B, are aligned with their axes in the direction of the earth's field. Identical primary windings in series magnetize the two cores, with the same flux density ; but in opposite direction, since they are wound oppositely, around the respective cores. At any time, the earth's field reinforces the field set up by one of the coils, and opposes the field of the other. Each coil has in addition a secondary winding, the two secondaries being connected to a voltameter, that reads the difference between the two outputs (Fig. 5.39A).

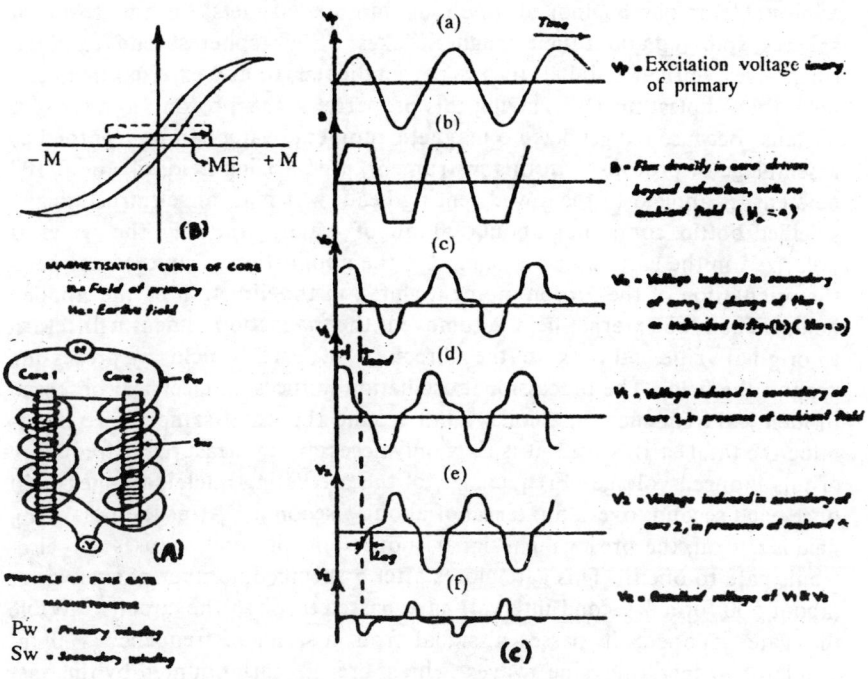

Fig. 5.39

If there is no external field present then there is no net rate of change in the magnetic flux due to the two cores. However, even a small external magnetic field, acting in alignment with the cores, will cause a perceptible change in the sign of magnetization *at different times,* as the internal field while aiding the flux of one of the cores, opposes that induced in the other Fig. 5.39C (d.e.f.). As a consequence an alternating E.M.F. is induced in the secondary, which is proportional to the resultant output from the two primaries, and also proportional to the external field. The output from the secondaries is recorded by a voltmeter after suitable amplification.

For use in aircraft, the system is constantly kept aligned in the direction of total magnetic field by the help of servomotors and it measures the total magnetic field with an accuracy of $\pm$ 1 gamma.

(b) *Nuclear Magnetometer* : The nuclear magnetometer is based on the phenomenon of nuclear magnetic resonance discovered in 1938 by Bloch. About two third of all atomic nuclei have a magnetic moment. (Moment is the turning effect of a force about a point. It is equal to the product of the force and its perpendicular distance from the point to its line of action). Such nuclei might be looked upon as minute magnets, in the form of spheres, spinning about their magnetic axes. Such spheres tend to align themselves, either parallel to or perpendicular to any external magnetic field, the simplest nucleus, having this property, is the proton, a hydrogen nucleus. Because oxygen has no magnetic moment, water can be regarded as assemblage of protons. In this instrument, a polarizing field, of about 100 oersteds, is applied to the towed sensing head, which is an electrostatically shielded bottle, containing about 500 ml. of water. Initially the water is polarized in the earth's field alone. On the application of an external field, the orientation of the proton moment shifts, in the direction of the applied field. When the external field is removed, the magnetic moment will return to original value and back to the direction of earth's field by precessing around the field. The precessional oscillation induces an electrical potential of identical frequency, in a coil wound around the water sample. To determine the total earth's field, it is thus only necessary to measure the frequency of this induced voltage. Frequencies, of the precessing nuclei, are measured once each second, over an interval of about $\frac{1}{2}$ second. After the polarizing field is cut off, the precessional signal, induced in the coil, causes an electronic gate to open. This gate closes after a specified number of sine waves (about 500 for a $\frac{1}{2}$ second interval) have passed through the circuit. While the gate is open, it passes a signal from a standard frequency (100 kc) oscillator, generating sine waves which are in turn counted by the fast counter and converted into values of total earth's magnetic field with an accuracy of $\pm$ 1 gamma.

The development of the high sensitivity alkali magnetometer containing cesium vapour, rubidium vapour, has made it possible to obtain the expression of sedimentary structures, and stratigraphic features, provided these are not obscured by rocks with high magnetic susceptibility, such as basalts etc.

The instrument works on the following principle. A beam of light is passed through a tube containing a small quantity of vapourised rubidium. The manner in which light is absorbed indicates the strength of the magnetic field. The explanation is that the absorption of light depends on the spinning of the electrons in the rubidium molecule, and the spin in turn is controlled by the magnetic field.

Fluxgate and Nuclear magnetometers, in recent years, have also been developed for ground survey operations.

## SEISMIC METHODS

General : This method utilizes the variation in behaviours of waves generated, in the various layers of the earth's crust, by artificial means, commonly by the use of explosive charges, placed at the bottom of a drill hole. In order to understand the fundamentals, it is necessary to know the types of waves generated. Taking a homogeneous elastic medium, a sudden disturbance will cause the generation of waves comprising an expanding series of spherical shells of rarification and compression. The types of waves produced in elastic media may be classed under (1) longitudinal waves, (2) transverse waves, (3) Rayleigh waves, (4) Love waves. Each of these waves, is recognised by the motion of particles under their influence.

(1) In longitudinal waves particles move either in the same direction as the waves or in the opposite direction. The wave form is sinusoidal with particles moving away from the crests (dialation), towards the troughs (compression), alternately. These waves are also known as compressional waves.

(2) With transverse waves, particles in motion traverse the medium in a transverse manner, at right angles to the direction of propagation of the waves. When the particles move in parallel lines, the wave is polarised in that direction. These waves are generally known as shear waves. Transverse waves are not employed in seismic methods of prospecting.

(3) In the case of Rayleigh waves, the particles have a gyratory motion in the vertical plane, as in the case of ocean waves. The amplitude of the gyration decreases rapidly with the increase in depth but exponentially. "Ground roll" noticed in seismic phenomenon is attributed to the effect of Releigh waves.

(4) Love waves are generated in the low velocity layers, at the

surface, due to interaction of elastic waves, travelling in the medium of high velocity below. The motion of a particle under the influence of Love wave is both horizontal and transverse. They are not of much significance in seismic prospecting methods. Of seismic methods, the one employing refraction of elastic waves has been in use for prospecting, especially in the oil fields of Canada. Even though refraction surveys are being largely surperseded by reflection surveys, there is no gainsaying that the method had its advantages.

## (a) Refraction method

In the refraction method greater details, of the smaller subsurface structures, are brought out with the use of explosives at shallow depths. Further, the method is unique in affording the velocity data of the subsurface strata, from which a fairly reliable guess can be made of the rock types as well as the structure involved.) This, however, is to be substantiated by the geology, as far as known, from surface observations and from available drill holes. Refraction is warranted, where a strong marker horizon is present below less competent rocks seen on the surface. Where subsurface data are required for limited areas, refraction is the answer, as for example in exploratory work connected with engineering projects, e.g., dams, bridges etc. In this country, refraction was employed in determining the bed rock contours in the Linganamakki project area of Karnataka, as well as in the Bhramaputra bridge site, near Pandu in Assam, Beasdam project Mandi, and many others.

In practice, however, the refraction method is more difficult than the reflection method as the detector points are spread out over large distances which may be of the order of 12 miles, whereas the distances in the reflection method scarcely exceed the depth to the formations.

*Principles :* The method depends on the fact that elastic seismic waves travel with varying velocities depending on the competence of the rocks. Thus in granite velocity of longitudinal waves may vary between 13,000—18,700 ft /sec., whereas the velocity of transverse waves are of the order of 7000 to 10,800 ft./sec. Longitudinal waves, in limestone, travel with a velocity of 16,000—20,000 ft./sec., while transverse waves have a velocity of only 11,400 ft./sec. In sandstone the velocities of the longitudinal and transverse waves vary from 15,400—17,500 ft/-sec., and 11,000 13,400 ft./sec. respectively. A marked increase in velocities is noticed in salt, where the longitudinal waves travel with velocities varying from 14,400—21,400 ft./sec. Shale has a poor power of transmission of longitudinal waves, and the velocities are 7,500—15,400 ft./sec. and 9,500—10,500 ft./sec. in respect of transverse waves. Alluvium shows a velocity of 10,000—11,500 ft./sec. for longitudinal waves.

It will be noticed, that the velocities of elastic waves vary within appreciable limits for the same rock type, *e.g.* in shale the longitudinal wave velocities range between 7500—15,400 ft./sec. This variation is due to the degree of consolidation. Thus velocities vary with (1) the rock types, (2) degree of consolidation. Since the seismic methods depend on the measurement of the velocities of elastic waves, the lay out is similar. A hole is drilled to a suitable depth and charged with dynamite or other high explosive and tamped ; the quantity of the charge depending on the forma-tions and the depths. Pick ups or geophones or detectors are arranged in a linear pattern. Several lines of detectors can be arranged round the same shot hole or shot point, in a radial fashion.

In order to enable the calculation of the velocities of the elastic waves, it is necessary to know the exact moment of the explosion, and the precise time of arrival of the wave trains at the detector point. The shot moment is recorded electrically by connecting the shot point to the detectors, in the case of reflection surveys, while radio transmission is util-lised, for the same purpose, in the case of refraction work.

The detectors or pick ups or geophones convert the energy of the elastic waves into an e.m.f. (The principles of the various types of geo-phones are given later.) This e.m.f. is magnified by an amplifier and then transmitted to the recording equipment (described later).

### Refraction method

**Examples :** Behaviour of seismic waves, longitudinal (compressional) waves, in homogenous elastic media, disposed horizontally, may be consi-dered for,

    (1) A two-layer case.

    (2) A three-layer case.

*Two-layer case*

    Suppose the upper layer has a velocity of V and the lower has a velo-city of $V_1$. $V_1$ being larger than V.

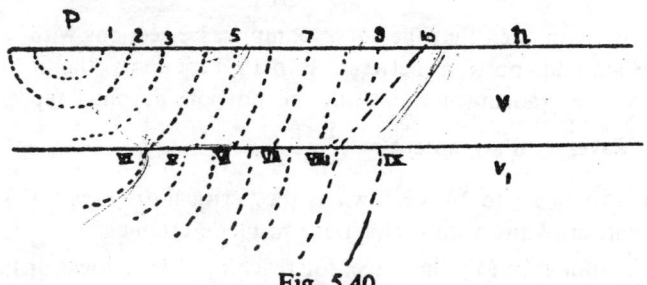

Fig. 5.40

In Fig. 5.40 P is the shot point in layer V. On firing the shot, concentric shells representing successive waves fronts (assuming an interval of 1 sec.) 1, 2, 3......10, are generated. After 4 seconds, wave front 4 touches the layer $V_1$, and another series of wave fronts are generated, in the lower layer ($V_1$), represented by IV, V ......VIII. IX., etc. As this series spreads out, at a certain distance, the wave front becomes normal to the interface, which is the critical angle. Thus refraction occurs at this position e.g., VII, and refracted waves are generated and these also reach the surface. In Fig. 5.37 the refracted wave at VIII of the lower media ($V_1$), reaches simultaneously with the 10th wave front of the upper media (V). The refracted wave after 8 seconds, meets the normal compressional waves, which reaches this point after 10 seconds. At any other position, beyond 10, on the surface, the refracted wave reaches earlier than the normal wave. If the time of arrival of both the waves are noted at successive points, upto n, and if the distances of each successive point from P is measured, e.g., P-1, P-2, P-3.........P-n, then the velocity of the waves can be calculated. When this data is plotted graphically, a curve such as that depicted in Fig. 5.41 is obtained.

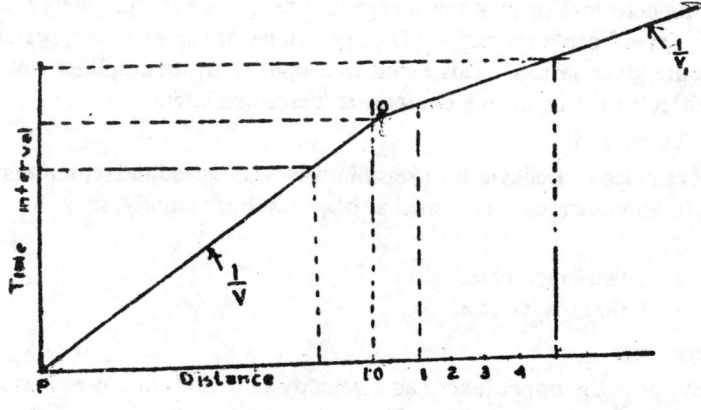

Fig. 5.41

It will be noticed that the curve comprises 2 sections with a kink at 10. The straight portion between P to 10 represents the normal wave velocity, with a gradient of $1/V$, while the portion beyond, represents the refracted wave, with a gradient of $\frac{1}{V_1}$.

The distance P to 10 is known as the "critical distance" (L). Since the gradients are known, the velocities can be determined.

If the time taken by the wave, for reaching the critical distance (L), is T and if the thickness of the upper layer is d, then

$$d = \frac{T}{2} \cdot \frac{V V_1}{\sqrt{V_1^2 - V^2}}$$

In case, the critical distance P—10, (L) is known then,

$$d = \frac{1}{2} \sqrt{\frac{V_1 - V}{V_1 + V}} = \frac{1}{2} L \sqrt{\frac{V_1 - V}{V_1 + V}}$$

*The three-layer case*

Similarly, in the case of a 3-layer or multi-layer set up, the time-distance curve shows as many breaks as the number of layers and the gradient of each the segment gives the reciprocal of the velocity of the normal wave in the first segment, and the gradient of the successive segments represents the reciprocal velocities of each succeeding layer. The velocities, can thus be computed, for each layer as in the case of the 2-layer case.

One of the limitations of the refraction method is, that it is useful only in detecting the presence of rock types, with increasing velocities of transmission (for elastic waves) below each other, in depth. Thus if in the stratigraphical succession, a low-velocity layer occurs immediately below a layer of high velocity the refracted wave is defracted and passes into the low velocity layer below. Thereafter the refracted wave can only reach the surface where it passes from a relatively low velocity layer to a layer of higher velocity beneath as shown in Fig. 5.42.

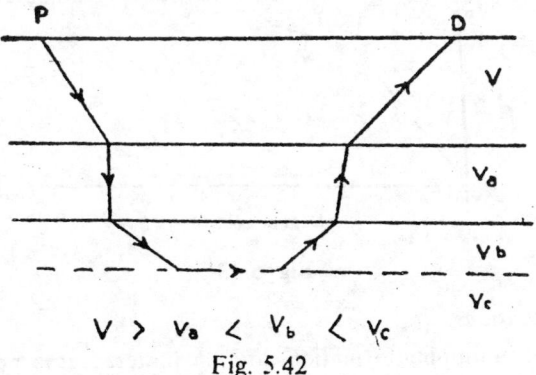

$$V > V_a < V_b < V_c$$

Fig. 5.42

Under certain conditions, it may so happen that the stratigraphical succession comprises rock-types, which show a continuous increase in velocities with depth. In such cases the velocity relation is given by $V_n = V + k d$,

where $V_n$ is the velocity of a layer at depth,

$V$ is the velocity of the top layer,

K is a constant,

d  is depth of the layer from the shot point.

When the ray paths are determined, they appear circular and the locus, of the centres of these circles, is a straight line parallel to the surface. located above the surface at height of V/k.   In such cases, however, the time-distance curve shows no precise kinks, but consists of a smooth gentle arc (hyperbolic).

**Faults :**

It is possible, in certain cases, to determine the throw of a fault from the time-distance curve as shown in Fig. 5.43

The throw of the fault is obtained thus

$$d_t = (T_d - T_u) \ \frac{V \ V_a}{\sqrt{V^2_a - V^2}}$$

where dt = throw of the fault,

$T_d$ = time (refraction) from the down throw side,

$T_u$ = refraction time from the up throw side,

V = velocity of top layer, and

$V_a$ = velocity of lower (faulted) layer.

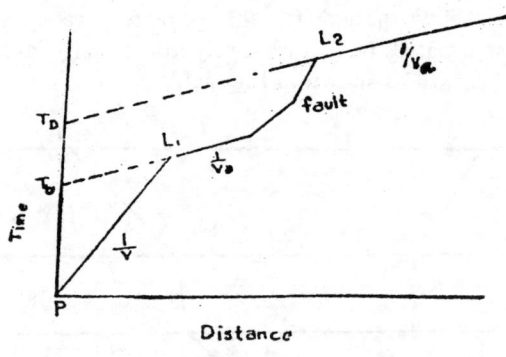

Fig. 5.43

*Dipping Formations:*

Even when dipping formations are encountered, it is possible to employ the refraction seismic method.   The time-distance graph can be used for obtaining the velocities of the formations as well as the dip.   In this case, shooting is done twice.   First, with the detectors arranged in the down dip direction, and next, with detector arranged in the up dip direction.

From the time distance curve obtained with this set up  (Fig. 5.44) we can solve the following equations for C and a.

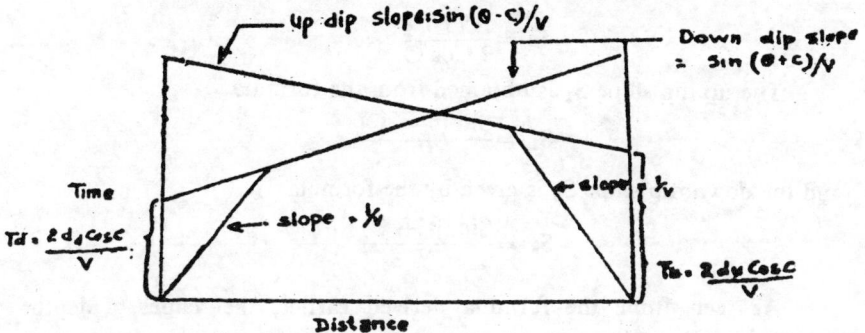

Fig. 5.44

V (down dip slope) = Sin (C + θ)

V (up dip slope)  = Sin (C − θ)

where V = velocity of the upper layer,

C = critical angle, $\dfrac{V}{V_a}$ = Sin C

θ = angle of dip

when C and V are known, then $V_a$ is determined from

$$V_a = \frac{V}{\sin C}$$

where $V_a$=the velocity of the lower layer.

If the depth of cover is du, Fig. 5.45 (with velocity = V), on the up dip side, the time interval Tu, can be obtained from the following equation.

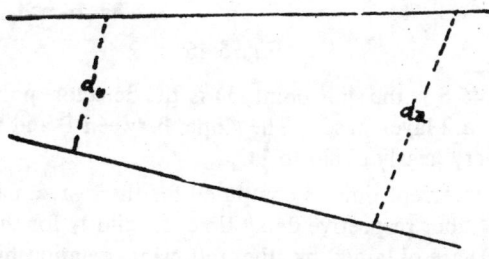

Fig. 5.45

$$Tu = \frac{2\ du\ Cos\ C}{V}$$

The depth, du, can be obtained from

$$du = \frac{VTu}{2 \cos C}$$

The up dip slope $S_1$ is obtained from the formula

$$S_1 = \frac{\sin (\theta - C)}{V}$$

and the down dip slope $S_2$ is given by the formula

$$S_2 = \frac{\sin (\theta + C)}{V}$$

As seen from the formula derived earlier, the values of depths obtained, by using the time-distance curves, are not actual but represent integrated sum of the depths (of the refracting layer) occurring below the shot point and detector. As a consequence, the orientation, of the refracting layer, is of major significance. If the refracting layer is characterised by rolling dips, the computations become more complicated.

In all cases the matter is simplified, if the "delay time" is employed instead of the "actual time". Taking into consideration the path of the waves, the time $(T - L/V)$ comprises two portions viz., that associated with the shot point $(t_1)$, and that associated with the detector $(t_2)$.

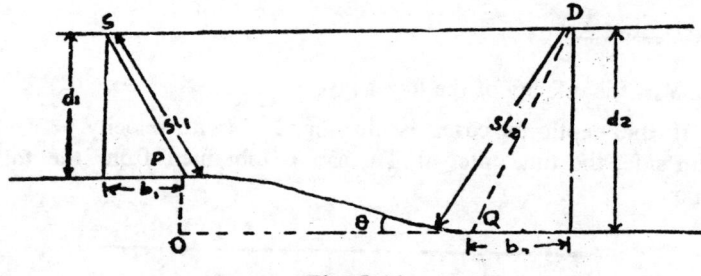

Fig. 5.46

In Fig. 5.46 S is the shot point, D is the detector point and the set up represents a 2-layer case. The slope, between P and Q, is very gentle so that OQ is very nearly equal to PQ.

Since the intercept time is computed for the wave paths represented by $Sl_1$ and $Sl_2$, the respective delay time $t_1$ and $t_2$ for the shot point (S) and detector (D) are obtained by the following relationship, as seen from the Fig. 5.41.

$$t_1 = \frac{Sl_1}{V} - \frac{b_1}{V_a}$$

$$t_2 = \frac{Sl_2}{V} - \frac{b_2}{V_a}$$

When the values of depths and velocities are substituted then

$$t_1 = \frac{d_1 \sqrt{V^2 - V_a^2}}{V V_a}$$

$$t_2 = \frac{d_2 \sqrt{V^2 - V_a^2}}{V V_a}$$

Since the [delay times cannot be isolated, indirect methods are employed, for obtaining a measure of the values, either by fan shooting or by profile shooting as discussed later.

In relation to the shot point the arrangement of the detectors may be in one of the following ways :

(a) Radial (known as fan shooting, Fig. 5.46a

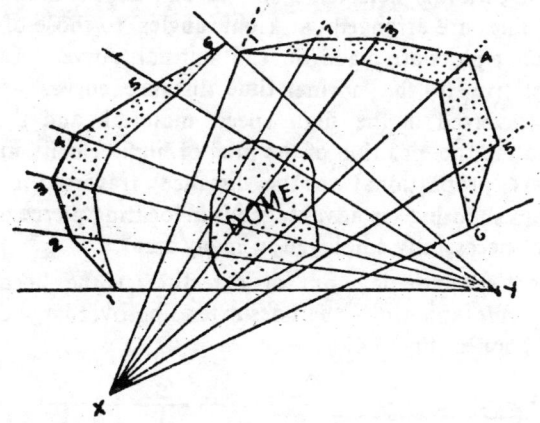

Fig. 5.46a

(b) Linear (known as profile shooting) Fig. 5.47

Fig. 5.47

In fan shooting, the detectors are arranged in radiating lines, along an arc from the shot point, the radius of the arc varying between about 5 to 10 miles. If only horizontal sequences are encountered, then the normal time-distance curve is obtained (Fig. 5.41).

If, however, a piercement dome of salt, (high velocity medium) is present then, along some of the rays, a consistent anomalous increase in

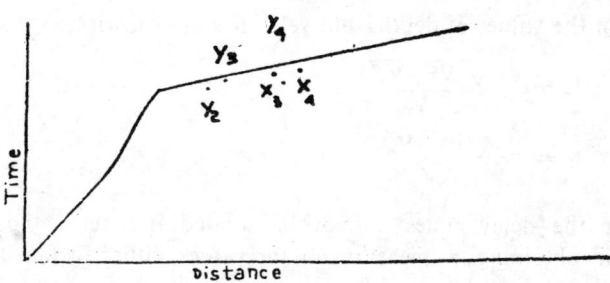

Fig. 5.48

velocity is noticed in these directions. The plots of these points, on the time distance curve, fall away from the normal curve. When such an anomalous condition is detected then a second shot is fired and the second set, of radial lines are arranged, at right angles to those of the first set of rays, and similar plots are made on the normal curve. The location of the plots, relative to the normal time distance curve, is proportional to the distance traversed, in the high speed material, and this affords the means of locating the position of the plug of high velocity material. Since the "leads" are proportional to the distance traversed in the high speed material. "Fan shooting" is advantageous in locating piercement salt domes. and has been successfully employed in many areas.

The most common lay out of detectors, is the linear arrangement known as profile shooting—the lines extending over large distance, i.e., of the order of 7 miles (Fig. 5.47).

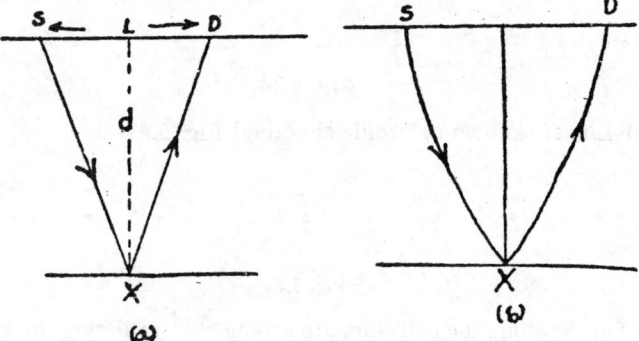

Fig. 5.49

In practice, the intervals, between shot points and detectors, are kept more or less uniform. The "up dip" and "down dip" time distance intervals are recorded as the work progresses. For this purpose (Fig. 5.46) the shot at P is fired first, and detectors are placed at $P_1$, $P_2$ ......etc. and $p_1$ $p_2$ etc. Next, the shot at p is fired and detectors are placed at $p_1$ $p_2$ and P, $P_1$, $P_2$, 1, 2, 3 etc.

Shots are fired successively from $P_1$ then $p_1$, $P_2$ and $p_2$ and the detector points are advanced beyond $p_2$, keeping the interval constant. Thus a linear traverse can be made.

The intervals (between detectors) are so chosen that the first and if so desired the second arrivals of wave trains represent the refraction from a high velocity bed or from the basement. The intervals are selected on the basis of an experimental shot.

In profile shooting, the determination of the delay time, corresponding to the relative depth of the cover, below the shot point and detector, is somewhat laborious. The following rules are useful in this respect.

(1)  When the distance, between the shot point and detector, is not very much larger, than the "critical distance", or when the dips of the marker bed are small, then the two delay times should be about half the intercept time.

2.   Even where the interval is much larger than the critical distance, half the intercept time is a reasonable approximation.

## Reflection method

This method is widely used and has the advantage of great precision, when determining depths to various reflecting beds. It is, however, to be expected that the determinations tend to be less precise, with increasing depth. The method has special advantages in the location of structural traps, in petroleum prospecting, but under certain circumstances 'noise' interference inhibits its use (Fig. 5.49).

## (a) Horizontal formation

Taking the case of a homogenous horizontal medium or bed, the reflected waves travel in straight lines (Fig. 5.49a). Thus a wave from the shot point S is reflected at X located on the interface and is picked up by the detector at D.

If V is the velocity and T is the time interval, then the distance travelled by the wave 2 SX=VT

and
$$2SX = 2\sqrt{d^2 + \left(\frac{L}{2}\right)^2} = VT,$$

where L is the distance between S and D.

$$T = \frac{2}{V}\sqrt{d^2 + \left(\frac{L}{2}\right)^2}$$

$$T^2 = \frac{4}{V^2}\left[ d^2 + \left(\frac{L}{2}\right)^2 \right]$$

$$\frac{V^2 T^2}{4} = d^2 + \left(\frac{L}{2}\right)^2$$

$$\frac{V^2 T^2}{4} - \left(\frac{L^2}{2}\right)^2 = d^2$$

$$d = \tfrac{1}{2}\sqrt{(VT)^2 - L^2}$$         (1)

In case there are a number of interfaces, between several layers, from which reflection takes place, and where the velocity increases with depth, then the path of the wave may be arcuate, in both the incident and reflected portions as shown in Fig. 5.49b.

### (b) Formations with an appreciable amount of dip

When the dip of the reflecting bed is appreciable the computations are a little involved and use is made of mirror images to simplify the procedure.

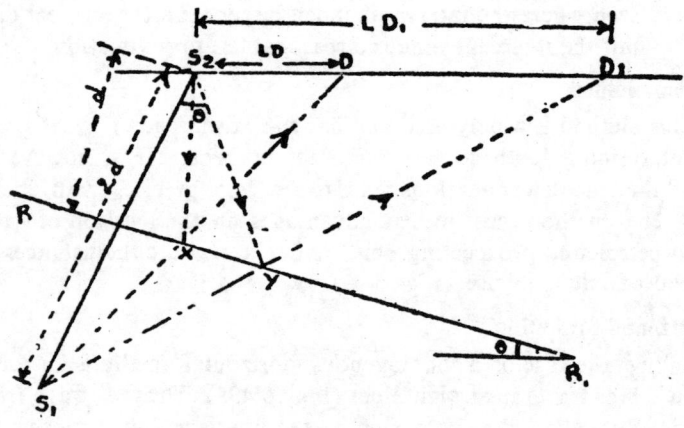

Fig. 5.50

In Fig. 5.50 let S be the shot point and D and $D_1$ the detector points. $S_1$ is the mirror image of S about the reflecting bed, R $R_1$
The rays $S_1P$ and $S_1P_1$ appear to emerge from $S_1$.
If V is the velocity of the upper layer,
then $S_1D = VT$ (where T is the time interval)       (2)
      $S_1D_1 = VT_1$ (where $T_1$ is the time interval).       (3)

If the distances of D and $D_1$, from the shot point S, are respectively LD and $LD_1$ and the dip is $\theta$, then from obtuse angled $\triangle SS_1D$ and $\triangle SS_1D_1$ the following relationship may be obtained, by applying the cosine formula (Fig. 5.51)

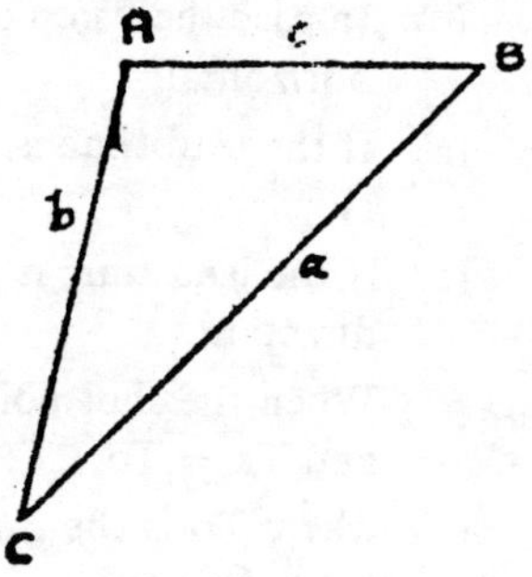

Fig. 5.51

$$a^2 = b^2 + c^2 - 2bc \; \text{Cos } A$$

$$SD = (2d)^2 + (LD)^2 - 4dLD \cos (90 + \theta)$$
$$= (VT)^2. \tag{a}$$
$$SD_1 = (2d)^2 + (LD_1)^2 - 4dLD_1 \text{ Cos } (90 + \theta)$$
$$= (VT_1)^2 \tag{b}$$

but,

$$\cos (90 + \theta) = -\sin \theta$$

Substracting equation (a) from equation (b)

$$(LD_1^2 - LD^2) + 4d(LD_1 - LD) \text{ Sin } \theta = V^2(T_1^2 - T^2) \tag{c}$$

Dividing both sides of equation (c), by $4d(LD_1 - LD)$,

$$\frac{LD_1 + LD}{4d} + \text{Sin } \theta = \frac{V^2(T_1^2 - T^2)}{4d(LD_1 - LD)} \tag{d}$$

$$\text{Sin } \theta = \frac{V^2 (T_1^2 - T^2)}{4d(LD_1 - LD)} - \frac{LD_1 + LD}{4d}. \tag{e}$$

If the average time $\dfrac{T_1 + T}{2}$ is $T_a$; then the "step out" time i.e. $T_1 - T$, is sT.

Then the perpendicular distance from the shot point, S, to the reflecting bed, $d = \dfrac{VT}{2}$. Substituting this value in equation (e), since $L = 0$,

$$\text{Sin } \theta = \frac{VT_a \, sT}{1(LD_1 - LD)} - \frac{LD_1 - LD}{2VT}$$

If $\qquad LD = 0$,

and $\qquad T = 0$.

$$\sin \theta = \frac{VT.sT}{T.LD_1} - \frac{LD_1}{2VT}$$

or $\qquad sT = \dfrac{LD_1^2}{2V^2 T_a}.$

(a) In case the "step out" or lead is normal, the reflecting bed is horizontal.

(b) If the lead time is positive, then the dip is towards the detector side.

(c) If the lead time is negative, then the dip is in the opposite direction.

When the shot point is equidistant from D and $D_1$ then $L_1 = L$ and $Ta = To$,

where To is the time interval for the reflection received at the detector, at the shot point

## Reflection records

After the explosion has occurred, at the shot point, wave trains start arriving, at the detector points, and sometimes it takes several seconds before the motion ceases. The waves comprise partly refracted, partly reflected (after single or multiple reflections), while others are surface waves. It is essential to isolate those due to reflection, from the other types of waves. For this purpose, multiple detectors are employed, and these are connected either in series or in parallel to a single recording galvanometer and a single trace is obtained.

In such an arrangement, the waves form prominent crests or troughs traversing the seismogram. The record of the adjoining galvanometers also yield seismograms with conspicuous crests and troughs, which appear to be the continuation of the former one. A series of matching seismograms may thus be obtained. Any "lead" or step-out noticed, in the arrival times, indicates the effects of dip, of the reflecting bed. The various reflections are recorded as descrete wave trains, along the time axis of the same seismogram. Both reflected and refracted waves act on the detector, in a similar manner, but the reflected waves impinge, on the detectors, making smaller angles with the vertical than their refracted counterparts.

It is often possible to trace the continuity of the reflections, in adjacent seismic records; and when such a programme is adhered to, it is called "continuous profiling". Sometimes, the correlation becomes exceedingly ambiguous due to the branching and splitting of traces.

## Correlation

Often reflections, from known as well as unknown horizons, are traceable and identifiable on the seismograms. Even where such continuity of reflection is wanting, the presence of a distinctive characteristic, in a reflection, may be used for purposes of correlation. In this manner structure contour maps can be prepared.

In preliminary exploration, impersistent reflections are treated as dip segments, but where any peculiarities of dips are encountered, detailed work is carried out.

*Detector grouping*

This is done where reflections tend to be subdued, by interference, due to extraneous wave trains—mainly attributable to surface waves—known as "noise". In order to mitigate the effects of noise, the detectors are so arranged that they have a larger "bias", in favour of "vertical motion" which is characteristic of reflected waves. Thus the horizontal waves caused by the surface and noise waves are eliminated. This is achieved by spreading the geophones, over a distance corresponding to the wave length of the horizontal waves. Depending on the frequency of the "noise" the number of detectors is chosen and this can vary from 36 to over 100.

*Location of geophones in relation to shot point*

It may be indicated, that (a) the sub-surface depth of a bed, obtained from a detector, will be true only for the median position, of the shot point, between the two detectors, when the dip is low, (b) when continuous profiling is required, the detectors are located at constant intervals between adjacent shot points, such that every detector records the reflections on either side of every shot point, as done in the case of "profile shooting".

It is common practice, however, to keep the shot point in the central position, as in the "profile shooting" set up, and move the entire set up along the alignment ; yet keeping the shot point in the middle of the chain of detector groups. This method of profile shooting is known as "split spread", since the grouping of detectors is split up. The "split spread" method is less expensive, since it involves fewer shots for a required length of profiling. Where good reflections are obtained, split spread, with larger intervals, can be effectively employed, and even a grid interval of a mile may be successful.

Where formation dips are high, it is preferable to have two alignments of detectors, intersecting each other at right angles, so that the true dip may be computed from the values of two apparent dips obtained for each of the alignments.

**Average velocity**

In many refraction seismic computations, the average velocity is required. The simplest method is known as "well shooting".

In this method a charge is exploded from shot holes, close to an exploratory well (drill hole) B D. Special detectors, known as shot hole detectors, (Fig. 5. 54.) are placed inside the drill hole, at regular intervals, as shown in Fig. 5.52.

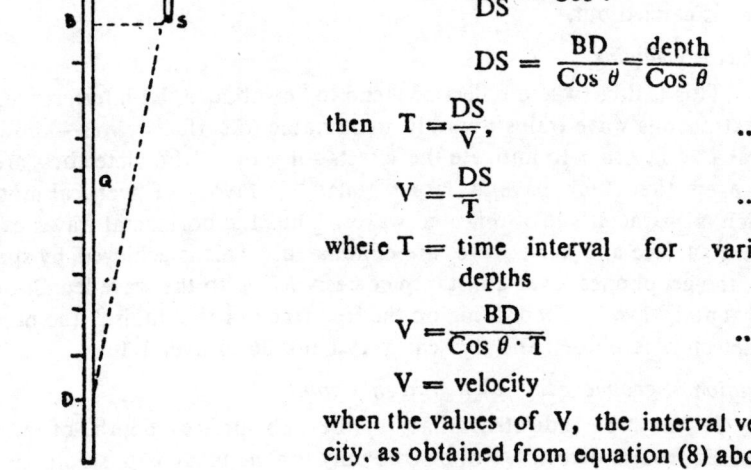

$$\frac{BD}{DS} = \cos\theta$$

$$DS = \frac{BD}{\cos\theta} = \frac{\text{depth}}{\cos\theta}$$

then $\quad T = \dfrac{DS}{V}$,  ...(6)

$$V = \frac{DS}{T}  \qquad ...(7)$$

where $T$ = time interval for various depths

$$V = \frac{BD}{\cos\theta \cdot T}  \qquad ...(8)$$

$V$ = velocity

Fig. 5.52

when the values of V, the interval velocity, as obtained from equation (8) above, are plotted against the respective depths of the detector points, in the bore hole, a stepped curve as seen in Fig. 5.53 is obtained. The curve XY represents the average velocity and this is very

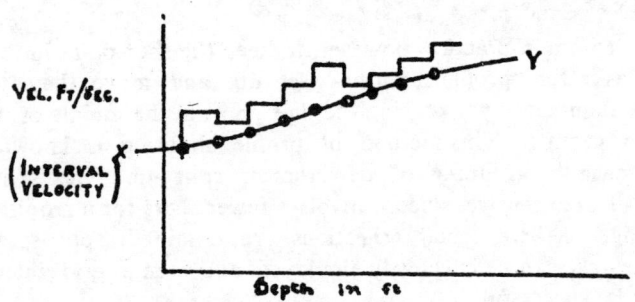

Fig. 5.53

frequently employed in depth computations.

Very often well shooting is done in more than one exploratory hole, covering a large area and then the average velocity, for the area as a whole, is computed and utilised in calculations.

In another method for finding the average velocity, continuous velocity logs are obtained. This requires the use of special equipment (Fig. 5.54) consisting of a vibrator which is attached to one end of a 10 ft. length of rod and a detector attached to the other end. The recording is so arranged that the travel times (interval velocities), of the seismic waves, appear against the respective depths.

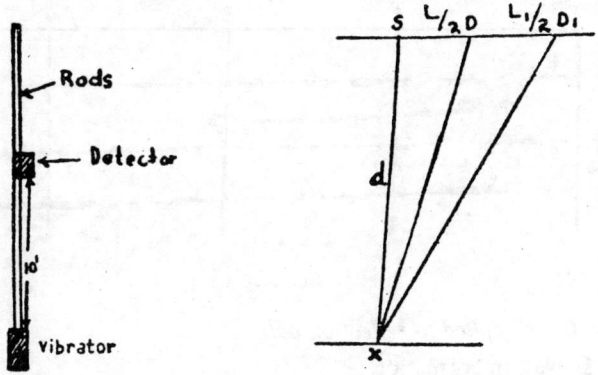

Fig. 5.54·                              Fig. 5.55

A third method for obtaining the average velocity is by computations.

If $T_D$ and $T_{D_1}$ be the respective arrival times of the reflected waves from X at D and $D_1$. (Fig. 5.55) and if d is the depth of the reflecting bed, and V the velocity of the upper layer, then,

$$XD^2 = SX^2 + SD^2$$

$$\left(\frac{VT_D}{2}\right)^2 = d^2 + \left(\frac{L}{2}\right)^2$$

Similarly,

$$\left(\frac{VT_D}{2}\right)^2 = d^2 + \left(\frac{L_1}{2}\right)^2$$

$$V^2 = \frac{L_1^2 - L^2}{T_{D_1}^2 - T_D^2} \qquad (9)$$

*Mapping from reflection data*

As in the case of all surveys, the computed data is required to be plotted and cross-sections, which are known as "time cross sections", are to be prepared. Generally, only the "centre trace" time interval is obtained, from the average of the time intervals recorded by the two groups of detectors on either side of the shot in a linear arrangement *i.e.* profile shooting. The average value is plotted, at the position of the shot point.

The various reflections, for each centre trace, are plotted at every shot point and correlation between the horizons, AB, CD, EF is affected as shown in Fig. 5. 56.

Fig. 5.56

*Correction to be applied to reflection data*

    (1) Elevation correction ;

    (2) soil and weathered rock.

    (1) Where the ground surface is horizontal, no correction need be applied, to the time of the reflected wave. In case, the surface is appreciably sloping, then the time of the reflected wave, requires to be adjusted. Two methods are used for this purpose (a) to assume a horizontal datum plane PD passing through the collar of the shot point (Fig. 5.59a), and (b) to place the datum plane OD below ground level GL as shown in (Fig. 5.57.)

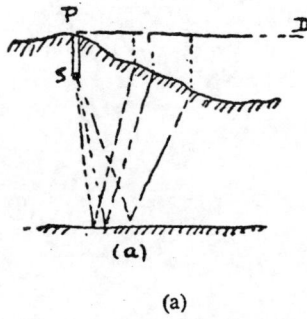
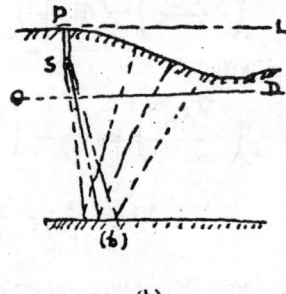

            (a)                                             (b)

Fig. 5.57

    If e is the reduced level of the collar of the shot hole,

        E the reduced level of the detector point,

        h the depth of the shot hole,

        v the velocity (average), and

        d the reduced level of the datum plane, in case (b),

then in case (a), the elevation correction is

$$\frac{h+e-E}{v} \qquad (10)$$

and in case (b), the elevation correction is

$$\frac{E+e-h-2d}{v}. \tag{11}$$

(2) Where a weathered zone is present, the reflected wave is retarded on entering the unconsolidated material.

If the thickness of the soil/weathered cover, Z, is known, and the velocities in the soil ($V_0$), as well as that in the rocks overlying the reflecting layer ($V_1$), are known, then the differential time $T_d$ due to the soil is calculated as follows :

$$T_d = \frac{Z}{V_0} - \frac{Z}{V_1} = Z\left(\frac{1}{V_0} - \frac{1}{V_1}\right) \tag{12}$$

When the average velocity V is known as from equation (9), then charts showing depth can be constructed taking the relation between the two factors, corrected reflection time, and shot detector distance into consideration. The corrected reflection time will be taken into account, along with the elevation factor and the delay element due to soil cover.

If no correction is applied in respect of "step out" time due to horizontal travel, then the times when plotted appear to lie on arcuate curves with the convexity turned upwards, known as umbrella chart, and shot points are located on the crests. Joining of the crests generally, gives the line of the formation concerned.

*Dipping beds*

Due to the influence of the dip, there is an apparent displacement of the points of reflection. In order to eliminate this error, the true dip is determined by using the formula (5)

$$\text{Sin } \theta = \frac{V(T_1 - T)}{2(LD_1 - LD)}.$$

By employing the same horizontal and vertical scales, while plotting of the profiles, a true representation can be obtained.

It is common practice to record the seismic data, on magnetic tape and transform the records into cross sections, after correcting the individual traces, for various factors such as elevation, weathering, and normal moveouts. Truck mounted magnetic tape recording equipment is also available for use in the field.

In preparing cross sections, it is seen that, where dips are steep and the structures are complex, the time data are not only to be converted to depths, but also "migrated" from the apparent to the actual positions. This is seen from Fig. 5.58.

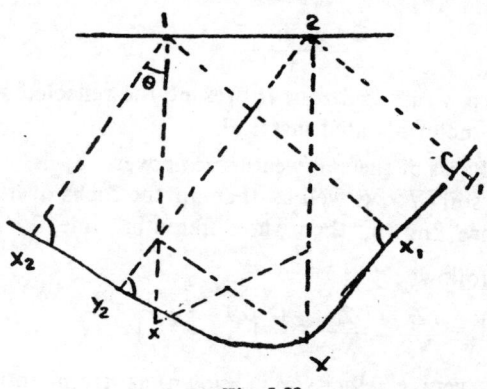

Fig. 5.58

1 and 2 are detector points. The reflections appear to come from points X and Y immediately below, from the reflecting formation, instead of the actual positions, of the reflections, at $X_1Y_1$ and $X_2Y_2$.

Where the velocity has a uniform value V, the angle of dip is $\theta$ and t is the difference, in reflection times of adjacent shot points, at a distance L from each other, then

$$\text{Sin } \theta = \frac{Vt}{2L} \qquad (13)$$

In order to obtain, the location of the reflecting point (1) (Fig. 5.54) a line $1X_2$ is drawn, from the shot point 1, making an angle with the vertical 1X, and the computed value of $\frac{Vt_1}{2}$ is plotted on it, where $t_1$ is the arrival time at 1.

Late reflections often appear on the records. These apparently originate from the top of the basement. It is believed, that such waves are caused by multiple reflections at sedimentary interfaces shown in Fig. 5.59.

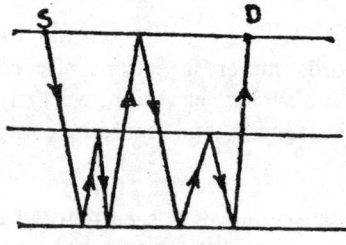

Fig. 5.59

Where V is the velocity in the weathered zone, and $V_a$ the velocity in the rock below the shot point. The return wave reflected at the interface is a "ghost" wave. (Fig. 5.60),

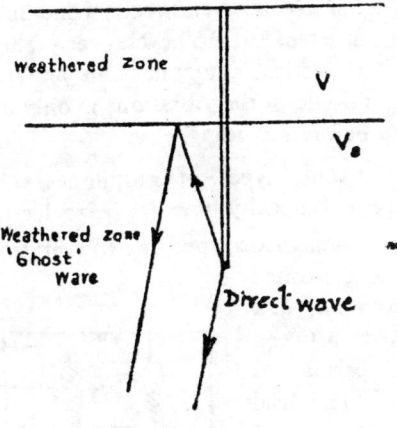

Fig. 5.60

Ghost reflections are produced when the wave travelling upwards from the shot point (Fig. 5.60.), is reflected either at the interface, produced by weathering, or at the surface. The ghost has (1) an amplitude lower than that of primary or direct wave—the value depending on the reflection coefficient, and (2) a phase opposed to that of the primary.

### Conclusion

It will be seen that reflections are complex events and interference should be expected. Reflection, from a single interface, is only an ideal. In practice, it is difficult to obtain such pure and simple conditions. Formations and structures such as unconformities or overlaps may not show up in the reflection records, when the interfaces are separated by less than a wave length ($\lambda$), which is of the order of 200—300 ft i.e., where $\lambda > d$. (Fig. 5.61)

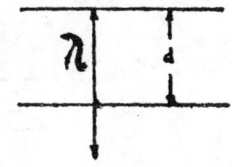

Fig. 5.61

Reflection mapping, when based on the time or depth cross sections, may often end up in certain discontinuous or unconnected patches. Hence, it is necessary to draw a "phantom" or imaginary horizon, parallel to the dip segments, where present or to project the same from the nearest dip segment. The "phantom" horizon behaves in a manner similar to the real horizons and if any discontinuity yet persists then it may be assigned to the incidence of faulting.

### Instrumentation in seismic exploration

Geophones are used for detecting seismic waves. In refraction work, the vibrations vary from 5—100 cycles/sec., while in reflection work, the vibrations are of the order of 10—150 cycles/sec. The instrument is designed with a bias for the vertical component, in preference to the horizontal waves. Since the amplitude of the vibrations is only about $10^{-8}$ inch amplification is necessary before recording.

The principle of some types of geophones is described viz. (a) the electromagnetic type, (b) capacity type, (c) peizoelectric type.

(a) In the electro-magnetic type, due to vibrations caused by seismic waves an e.m.f. is generated by the relative movement, in a fixed coil, located between the yoke of a permanent magnet Fig. 5. 62. The lead wires from the coil are connected to an amplifier. This amplified e.m.f. is transmitted to the recording galvanometer.

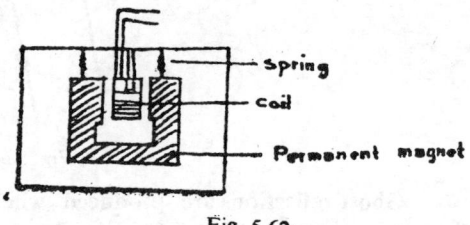

Fig. 5.62.

The records may be made either on ordinary paper by a pen, or by an optical system on photographic paper.

(b) In the capacity type, one plate of a condenser is fixed while the other is capable of motion. Since the separation between the plates varies, the capacitance also varies. Accordingly when the face plate oscillates, due to vibrations, caused by seismic waves, a variation of capacity results and this can be correlated to a variation of the charge, and the same transmitted for recording (Fig. 5.63).

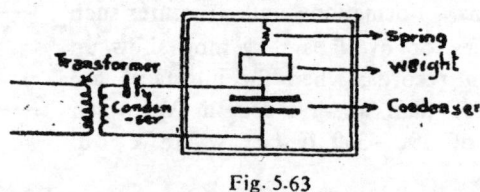

Fig. 5.63

(c) In this type peizoelectric crystals such as quartz, tourmaline, or barium titanate, are employed, in the form of plates cut parallel to the optic axis, and a weight is placed over a pile consisting of such plates. Seismic vibrations produce alternating stresses in the loaded pile and a small variable e.m.f. is generated, depending on the stresses. This is amplified and fed into the recorder (Fig. 5.64).

This type of geophone is commonly employed in bore hole detectors

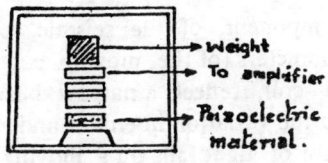

Fig. 5.64

for velocity determination; and invariably barium titanate is used. Since this instrument responds more to pressure changes, it records acceleration rather than displacements and velocities. It is also an accelerometer.

All detectors (geophones) are equipped with damping devices, which utilise the eddy current principle. The earlier models employed the effect of viscous drag, produced by a metal plate suspended in oil.

When an undamped instrument is made to vibrate, at a constant amplitude, and the frequency is continuously varied, it will be found that the e.m.f. output is maximum at one particular frequency. This is the "critical frequency" for the instrument. The maximum damping which can be done, without affecting the oscillatory character of the movements, is the "critical damping". In general, half the value of critical damping is considered as the best value. In cases where undesirable waves (noises) are required to be eliminated, damping is essential.

Seismographs, used in exploratory work, are invariably equipped with electronic amplifiers and recording units. Filters are essential to exclude the disturbing frequencies, both high and low. Sharp cut off can be obtained by feeding the output of one section into the input of another identical unit. Very often, it is required to cut out the low-frequency disturbances, so as to give prominence to the real lower amplitude higher frequency waves, caused by reflection.

It is now common practice to record the data on magnetic tape and then filter out the irrelevant frequencies. This method allows for greater latitude, none the less, it is expensive and time consuming.

Amplification is an essential part of seismic equipment—for both reflection as well as for refraction work. An "expander" is included, in the set up, so that as the signals fade away, with the increase in time interval, the degree of amplification steadily increases. An automatic volume control is also a necessary adjunct. The purpose of the a.v.c. (automatic volume control) is to tone down amplitudes of various signals from the geophones, so that both the weak as well as the strong signals are represented on the seismic records. It is possible to control the action of the "expander" and the a.v.c. automatically.

The recording component, of the seismic equipment, comprises oscillographs and galvanometers (of the moving coil type). The mirror, attached to the moving coil, reflects a narrow beam of light and swings from side to side, due to the variation in e.m.f., and this is registered by the amplifiers. The beam of light falls on a moving photographic paper and thus the movements of the galvanometer are magnified and registered. The photographic unit has about 25 galvanometers which produce 25 records simultaneously, on a paper 8 inches wide. The camera is motivated by the clock work, and the paper moves at a constant rate of 1 ft/sec. Timing is indicated, on the record, by small lines—the time interval being 0.01 or 0.005 sec. The timing is controlled by a vibrator, generally a tuning fork type.

## RADIOACTIVE METHOD

Radioactivity is caused by the disintegration of the atomic nucleus leading to the production of energy (electromagnetic), known as γ-rays, as well as to the release of certain particles of matter (the particles), of the nature of the helium nucleus α-rays and β-particles of the nature of electrons.

There is in the atom the dense positively charged nucleus, which accounts for the greater part of the mass of the atom, and the negatively charged electrons arranged in various orbits round the nucleus. The electrical charge, on the nucleus, may be one or more, and this is equal to the number of electrons.

The nucleus consists of two types of particles—the protons and neutrons. The proton is the smallest single particle, carrying a positive electrical charge, which determines the atomic number of the particular element. When the same element has a different atomic number, caused by a variation, in the number of neutrons, it is known as an isotope. Most isotopes are stable but others break up or disintegrate, emitting α-particles, β-particles and γ-particles. The process of transformation of the radioactive atoms, is a violent one, and it is accompanied by the ejection of an electron from the nucleus of the atom—the β-ray. The resulting helium atom, α-ray, stripped of its 2 outer electrons, forms the γ-ray. Electromagnetic radiations are generated, in the process, and these are the γ-rays.

Of these particles, the α-particles are of little use in prospecting, because of their relatively large size. They have very little penetration power and a large degree of ionisation. The β-particles have comparatively larger velocities, being about 10 times that of the α-particles and have greater penetrating power, about 100 times, but even these are stopped by 5 mm. of alluvium, or a few m.m. of rock. The extent of ionisation is, however, less than that of the α-particles. The γ-particles have larger energy together with higher frequencies. They have 10 to 100 times the power of penetration of the β-particies and can pass through several centimeters of lead, and about 1 foot of rock. The ionisation produced is comparatively less than that of even the β-particles. These are the most effective, in prospecting for radioactive minerals.

Uranium is highly radioactive and so also is thorium. Thus the occurrence of uranium and thorium in rocks and minerals can be detected if radioactivity, as indicated by the presence of disintegration products (α, β, and γ particles), can be detected. Minerals, containing zirconium beryllium, columbium, tantalum, (niobium) as well as the rare earth groups *e.g.* cerium, lanthanum, samarium, gadolinium etc. will fall under this group which requires special detectors. The other geophysical methods. described earlier, are not of much use as these radioactive minerals are present in such minor quantities as not to permit detection.

The equipments, used in the prospecting of radioactive mineral deposits, and indicating the presence of radioactivity, are of two types (1) the Geiger-Muller counter and (2) scintillation counter. In addition, the gold leaf electroscope, used in physics laboratories, can also be employed under certain conditions for qualitative assessment of radioactivity.

The principle of the Geiger counter, is based on the fact that when atoms collide with the high velocity radioactive particles, ionisation takes place and the presence of electrons and protons so produced is detected.

The Geiger counter or G-M counter (Fig.5.65) consists of a partially evacuated thin walled cylindrical tube of either aluminium or glass, containing a co-axial cylindrical metal cathode. A wire passing through the axis of the cylindrical cathode constitutes the anode.

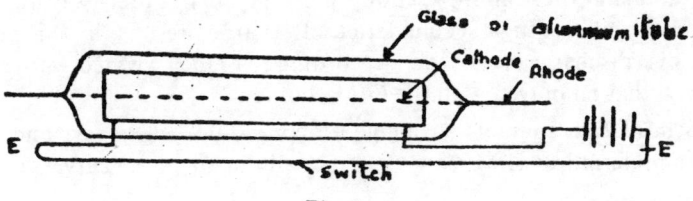

Fig. 5.65

The glass tube is filled with an enert gas such as argon or with halogen such as bromine. The pressure, of the gas, within the tube, is sufficiently low as to be in the atomic state and a "quenching agent", such as ethyl alcohol, is added, when argon is employed.

The cathode is maintained at a fairly high (negative) potential, of the order of about 1000 volt. When the tube is placed in an area where radioactivity is present, radioactive emission (especially $\beta$ and $\gamma$ particles) enter the tube and collide with the atoms of the inert gas within. Electrons and protons are thus produced. The electrons acquire high acceleration, by the impressed voltage, and are attracted by the anode. The protons move towards the cathode. The high velocity electrons collide, with other atoms, and further ionisation occurs. This process becomes continuous. The charge thus accumulates at the electrodes, and if it is not for the presence of the quenching agent, the discharge will be continuous. When the charge reaches an optimum value, the discharge takes place, and the resultant voltage is amplified and transmitted to the recording equipment. In other types of counters the discharge may be indicated either by a flash of light, or by the familiar "clucking", similar to that of a hen. The values so obtained are recorded as counts/minute. Where at deflection on a meter is recorded, it is in milli roentgens/hour (MR/hr). Counters are so designed that they can be made sensitive to $\alpha$, $\beta$ and $\gamma$-rays, or to $\beta$ and $\gamma$ rays only, and so on.

**Scintillation counters** are so called, since during their operation scintillations or minute flashes of visible light are produced, the intensity of which depends on the rate of bombardment by radioactive emanations. The scintillation counters employ certain substances like zinc sulphide (sphalerite) or thallium activated sodium iodide crystals, which emit visible light when struck by $\alpha$, $\beta$, $\gamma$ radiations. Such substances are known as "phosphors", and are employed in making luminous paints. Phosphors trigger off the action and the scintillation produced is made to fall on an electronic light-amplification tube, known as the photo amplifier, similar to that used in sound projection equipment. The light, falling on the cathode of the photomultiplier tube, causes electrons to be ejected. These electrons are accelerated and further electrons are produced at every stage, in the tube, and thus multiplication of a very high order (say $10^9$ times) can be attained. A high accumulation of electrons results, in this manner, and this is discharged through a recording circuit, provided with a meter similar to that employed with the G M counters.

Scintillation counters have high $\gamma$-ray detection efficiency, and hence they are preferred to G.M. counters especially in air borne surveys.

*Field procedure*

It is to be remembered, that a certain amount of radioactivity can be produced by cosmic radiations as well as from rocks containing radioactive isotopes. Such radioactivity which is ubiquitous is known as "background" radiation. This reading is to be deducted from the value of radioactivity as obtained from the meter.

In air borne surveys, the scintillation counter is employed and the speeds are 140—145 m.p.h., while the altitude is about 600 ft. above ground level. Counters are employed in measuring radioactivity in bore holes.

The gold leaf electroscope is employed in detecting ionisation associated with radioactivity. As mentioned earlie α-particles produce the maximum amount of ionisation, while β-particles and γ-rays produce lesser amounts of ionisation. Since most fault zones are invariably indicated by a show of radioactivity, the gold leaf electroscope can be employed, under suitable conditions, for locating fault zones, especially, when mapping is required in connection with engineering investigations.

## ELECTRIC WELL LOGGING

The two properties of sedimentary formations, that have been taken most advantage of in aiding their identification, in bore holes, are resistivity and potential (See Part IV & VI)

Sands, associated with saline waters, are conductive, *i.e.* less resistive than those containing fresh water. The bore hole potentials are. however, related to the electro-chemical reactions as explained under S.P. methods.

The well logging unit enables to record. continuously, the variations in potential, at various depths, in the bore hole. For this purpose, one of the electrodes is lowered by means of a cable and winch, into the bore hole, while the other electrode is earthed by placing it generally, in the mud pit.

Very often, well logging equipments comprise both the potential measuring, and the resistivity measuring units, and are known as two-channel apparatus. The potential and resistivity logs are recorded simultaneously and appear on the same chart side by side. There are two separate circuits, in such logging units (*a*) the (S.P.) potential circuit and (*b*) the resistivity (or resistance) circuit. The potential circuit consists of (1) the alternator, (2) ground electrode, (3) logging electrode and (4) the recording equipment with ancilliaries.

For the purpose of resistivity measurements the single electrode system or the multiple electrode (Slumberger) system can be employed. The single electrode system is simpler, and less expensive but lacks lateral penetration of the multiple electrode system. Hence, it is not possible to differentiate oil sand, from water bearing sands, where mud infiltration has taken place. Except for this disadvantage, the single electrode system is equal to or even superior in operation to multi-electrode system ; especially where correlations are required to be affected between wells, and for determining the depth and thickness of beds.

The single electrode equipment consists of (1) an alternator (with a low frequency 60 cycles), (2) a rectifier circuit with selenium rectifiers, (3) blocking condenser and (4) the recorder with ancillaries.

As the logging electrode is lowered into the drill hole, variations in the formation resistivity, cause relative changes in the electrode resistance and in turn affect the voltage in the logging circuit. The potentials are rectified and recorded as the resistivity curve. The blocking condenser serves to eliminate the natural potential as also the D. C. from the resistivity logging circuit.

There is, however, a limit to the current penetration in the single electrode equipment and this may be taken to be about ten times the electrode diameter. Thus if the diameter of the electrode is $1\frac{1}{4}''$, the penetration may be taken to be 15″ to 18″.

The electric log, which comprises the potential curve and a resistivity curve placed alongside, is extremely useful in geological studies. It can be used for correlation between wells for sub-surface mapping, research in sedimentation for determining the thicknesses of sand beds, shale beds etc.

Figure 5.66 shows the main components of the recorder unit. The main principle in regard to the operation of the recording unit is that the signals received back from the ground are D.C. and these are fed in at the terminals $V_+$. The vibrator (Vi) is an electromagnet with A.C. windings converts the D.C. to A.C. This is amplified in the unit comprising a transformer T, as well as an electronic circuit for voltage (V.A.) and power (P.A.)

The amplified A.C. is fed to a balancing motor (B.M.). The balancing motor is also connected directly to the power main (A.C.) from the generator, the voltage being controlled by a ballast resistance, as well as by a selector resistance. Thus it will be seen that the rotation of the balancing motor (B.M.) is governed by the input (a) from the A.C. mains from the generator, and (b) by the input from the transformer, and amplifier system, depending on the value of the D.C. ground signals. Further, since the D.C. ground signals are balanced against the battery input (b) against a slide wire (W), it is possible to record the movements of the balancing motor by mechanically coupling the slide wire (W) to the balancing motor. When the D.C. ground signals are not balanced, by the slide wire (W), a current passes through the vibrator (Vi) and this is amplified and fed at terminals (a) of the balancing motor (BM). This causes the balancing motor to rotate and move the pen mechanism R, producing a horizontal translation. The graduated paper, wrapped round the drum for recording, moves vertically as it is coupled with the wire

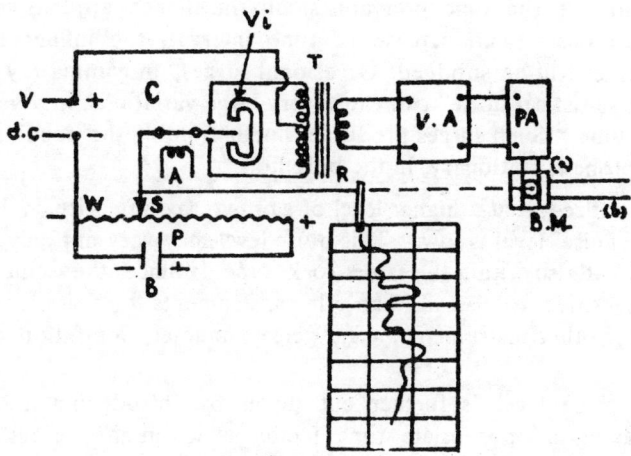

Fig. 5.66

rope and electrode, which are lowered in the bore hole. The vertical scale, of the paper, is thus coupled to the depth of the electrode.

Thus, in the case of the potential recorder, the sidewise movements, of the pen (R), coupled with the vertical movements, of the graduated paper, give rise to the S. P. record. Similarly the resistivity record is also produced simultaneously by the resistivity recording channel.

One of the essential conditions for obtaining logs, capable of easy interpretation, is to have bore holes, in which mud or circulating water is at constant level. Where there is a strong change in the composition of the waters, encountered at various depths, the scales of the various "peaks", seen in the resistivity curve, are different, but this is not a very major factor normally. Where fresh water, of different composition, enters the hole, the potential curve as a whole shows a drift or inflexion and it is flat and does not show up. The resistivity curve is not disturbed. The same characteristics are true of artesian conditions.

### Gamma ray logging

The radio activity of a rock is expressed in terms of equivalent of radium per gram of rock required to produce the same intensity of gamma ray radiation. In general, shales and clays show a higher degree of radio activity than sands, sandstones, limestones and dolomites.

The intensity, of gamma ray emission, is measured by means of a scintillation counter effectively upto a considerable depth, say 5000 ft. or more. At a given spot the gamma emission may not be constant, and it is liable to variation, depending on the amount of radioactive disintegration.

Consequently, if the time interval is small, the chances are that variations will be large, but with an increase in time interval, the influence of the unusual surges will be subdued. Occasional surges, in gamma ray activity, known as "statistical noise" can cause very large variations, in a very short interval of time. Such surges produce a fluctuation, in the readings, even when the probe is stationary, in the bore hole.

It is noticed, that a higher level of gamma ray intensity is indicated when the noise level is low. This noise level may vary not only between rock types but also within the same rock type. Since the scintillometer is more sensitive to γ-rays than other instruments the noise level is less than that for other instruments, e.g., geiger counter, ionisation chamber etc.

The noise level is further cut down by introducing a recording circuit, having a appropriate time factor, which in effect evens out such large fluctuations. By increasing the time interval the effect of the noise can be lowered.

Another feature of importance is that the introduction of time factor, produces a loss in the definition of the formation boundaries, so that the boundaries loose sharpness. In general, if the probe is raised up in the hole, at a constant speed, the formation boundaries, as shown by the gamma ray logs have a tendency for curving upwards.

Taking a hypothetical case of a γ-ray log for a sequence comprising shale with subordinate sandstones in which the statistical noises are excluded it will be seen that the curve obtained at moderate speeds will assume the form given in Fig. 5.67.

This distortion can be reduced, if either the logging speed and/or time constant is reduced. Conversely a lowering of the speed and increase

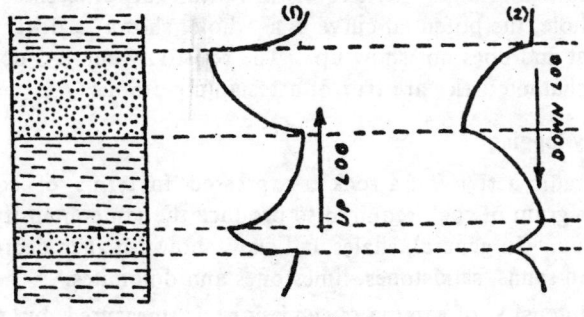

Fig. 5.67.

in time constant, cause greater distortions. There is, however, a limit to the increase in speed, as well as to the decrease in time factor, especially when the thicknesses, to be measured, are small. High speeds and lesser time intervals will impair the accuracy. Thus thin beds may produce little or no impression, in the logs, where high logging speeds and small time constant are employed. In practice, the location of formation boundaries, from γ-ray log, is done by marking the breaks in the curve. These breaks, however, may not coincide with the exact boundaries, unless this has been checked after allowing for the logging speed, the time constant, and the direction of logging (up or down).

The noise level should be determined before the commencement of the logging. This is carried out by lowering the probe and keeping it in a zone of low radioactivity for at least 2 minutes, and recording variations in the γ-ray intensity. This should be done at a sufficiently large depth so as to preclude the effects of cosmic rays—say about 200 ft. If, however, the probe is kept near the bottom of the hole it is likely to get stuck up due to the settlement of mud etc.

In the case of cased bore holes, a good percentage of the gamma rays pass through, even though the intensity is reduced. A reduction of 25% in the intensity is caused by a casing 5/16 inch in thickness. Where mud is employed the intensity is again reduced. The increase of shale and clay contents in the mud, causes an increase in the activity. In a mud of uniform composition this variation is constant, but if the shaly materials have settled at the bottom of the hole then the increase in radioactivity is inevitable.

### Uses of gamma ray logs

1. In cased bore holes.
2. In dry bore holes.
3. In holes where salt water has been found.
4. Where the data obtained by electric logs requires elucidation.
5. In prospecting for radioactive minerals.
6. In prospecting for coal and lignite.

## INTEGRATION OF GEOPHYSICAL METHODS

(a) In an exploration programme the combination of geophysical methods is more effective than a single method used alone. Combination of methods, of course, has to be chosen depending upon nature of problem, and associated physical properties and cost involved. In mineral exploration the combination of electrical (S.P., resistivity, electromagnetic and induced polarisation) and magnetic method is useful, while in oil prospecting reconnaissance gravity-magnetic survey has to be followed by seismic reflection surveys. In the case of ground water and engineering geology problems, a combination of resistivity and seismic refraction methods as described earlier yields good results.

## AERIAL GEOPHYSICAL SURVEY

With the development of sensitive instruments to measure earth's natural and induced fields aerial geophysical surveys are now a prerequisite of any large scale exploration programme. It has the advantage of a tremendously greater speed, and economy in reconnaissance survey. Another advantage is that it is possible to obtain data over swamps, jungles, mountains, lakes and other areas inaccessible to ground surveys.

A combination of different geophysical instruments is made depending upon the nature of problem involved. In oil surveys, magnetometer and gravimeter (only recently) are used while for mineral surveys magnetometer, scintillation counter and electromagnetic system deployed in the same plane.

*Position location*

For all aerial geophysical surveys, it is essential to correlate all instrument readings with the position of the plane, at the instant when the readings are taken. Two methods for correlation, in common use, are (1) Aerial photography and (2) Radio devices.

In areas where suitable air photo maps exist, positions are determined by using continuous strip-cameras, in the survey plane. The path flown is retraced by comparing the continuous film strip with the planned flight lines, on photo mosaic prints, which are used during visual navigation. An electronic intervalometer is used for marking time intervals on all records (including film, visirecorder, magnetic digital or paper record). This device produces fiducial marks at any desired interval ranging from 1 second to 120 seconds. In areas of featureless terrain, such as water or desert or in extremely rugged terrain, where the scale changes in photographs and causes extreme distortion, in mosaic construction or where no photograph coverage is available the electronic techniques are effective.

Shoran, Raydist, Lorac and Decca are different electronic systems involving the use of ground stations. A micro-wave transmitter-receiver unit, operating at two frequencies, (one for each ground station), is installed on the plane and constitutes the 'master' station, of the system, for locating the position of the plane at each moment.

A new system which is self contained, in the plane and most widely used at present is the doppler system. This makes use of Doppler's principle of frequency shift of waves emanating from a moving source. A pulsed or continuous wave is sent diagonally downward fore and aft, and the frequencies are compared, in order to obtain the true ground speed. The reading is obtained from a special magnetic compass and is maintained by directional gyro used as an integrating device.

## Flight pattern

The flight pattern and spacing depends upon the nature of problem. In case of oil exploration, mutually perpendicular sets of parallel lines, separated from one another, by a distance of 5 to 10 miles are flown. All the north-south lines are flown consecutively and then eastwest lines are flown, as shown in Fig. 5.68 Adjustment for diurnal variation is made by least squares, working from the misclosures around each block.

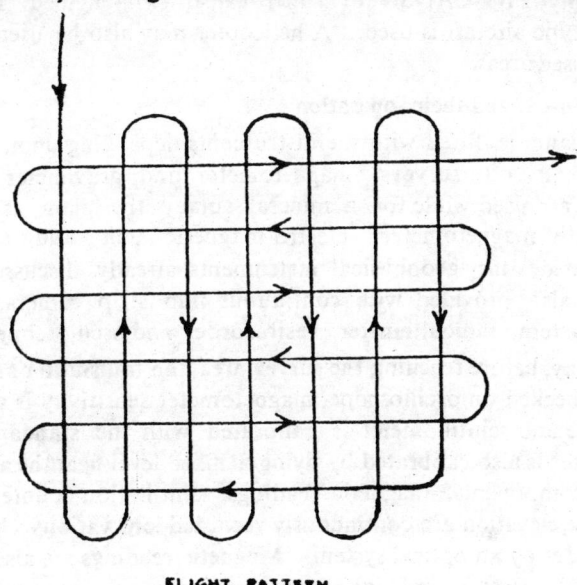

**FLIGHT PATTERN**
**Fig 5.68**

In the case of mineral surveys, lines are flown perpendicular to the general strike of formations or expected ore body, at an interval of ½km to 2 km. The lines are flown connecting all the lines, for applying diurnal and elevation corrections, to magnetic data.

## Height of plane

In case of oil surveys, the hight of plane, above ground, may vary any where from 200 ft to the plane's maximum flying altitude, depending upon the nature of the basement to be explored. However the pilot is instructed to fly at a predetermined constant altitude above sea level.

In case of mineral exploration, flying is done at a minimum constant height, above the ground level. The height, chosen for a given survey, depends upon the nature and maximum depth of penetration of the instruments. The surveys are restricted to heights between 200' to 800' from ground level.

### Type of the aircraft

The aircraft employed for aerial survey must essentially be a stable plane capable of flying at low altitude with crussing speed, as low as 100 to 150 knots with a total flight duration of 8 to 12 hours. The engine must be sufficiently powerful to have a vertical lift of 500 to 1000 ft per minute.

Normally PBY-CANSO or super canso twin-engined high wing amphibious type aircraft is used. A helicopter may also be used for surveys in localised areas.

### Instruments used and their operation

The plane is fitted with the instruments depending upon the nature of problem. In oil surveys, magnetometer and gravimeter (recently introduced) are used while for a mineral survey the plane is normally provided with magnetometer, electro-magnetic unit and scintillation counter. Besides the geophysical instruments already discussed earlier, the plane is also provided with continuous film strip cameras, doppler navigation system, radioaltemeter, vesirecorder and oscilioscope.

Each day, before reaching the survey area, the transmitter and receiver signals are checked on oscilloscope, magnetometer sensitivity is adjusted to suitable value and scintillometer is calibrated with the standard sample. Radioaltimeter is also calibrated by flying at fixed level near the aerodrome. The E.M. components, magnetic readings, scintillation counter readings, and the plane elevation are continuously recorded on various channels of the vesirecorder by an optical system. Magnetic readings are also recorded separately, on continuous moving paper drum by ink pen or in *digital* record form.

A ground magnetometer, measuring total field of the earth, is constantly kept running at the base station, to record if there are any sudden variation of the earth's magnetic field. The flights are repeated, if there are any magnetic storms during the flight.

**Processing of records and interpretation of data :** On the completion of day's flying operation, all the vesirecorder paper, camera film and magnetic record are taken to the base office. The film and recording paper are developed in the dark room, dried and rolled up.

The fiducial marks are numbered and anomalies are picked up and their amplitude is measured. The location of the anomalies, their magnitude and mean plane elevation, at that time, are plotted on a map, showing actual path followed by the plane. The anomaly maps are prepared normally on the same scale as that of airphoto mosaic, showing each anomaly, and the location of important topographical features etc. Geological maps, showing different formations and important structures, are also prepared on the same scale, on transparent paper and these are super imposed on anomaly sheets. Considering the nature, magnitude and direction of anomalies, in relation to the known geological and ground water conditions spurious anomalies are separated, and anomalies, due to probable anticipated causes, are selected for further ground check up and detailed investigation.

Magnetic data and radioactive data, are presented in the form of contour maps and often quantitative interpretation is also done as to the shape, size and depth of the causative body, and checked by further detailed ground follow-up work and drilling.

## Conclusions

The aerial magnetometer method is useful in locating iron ore deposits and non-ferrous ore deposits with which magnetite or pyrrhotite are associated. It is also useful in deciphering geological structures, such as, faults, folds, as well as the presence of intrusive igneous bodies, and metamorphic rocks. This method, in addition, can be used for determining the configuration of the crystaline basement topography, which is of great importance in locating petroleum and gas traps.

Scintillometer survey, is not only useful in locating radio active mineral deposits containing thorium, and uranium, but also useful in locating buried geological contacts. where the formations show variations in radioactivity.

Electro magnetic surveys, on the other hand, can be used with advantage for locating metallic sulphide ore bodies occurring at depths not exceeding 300 ft. or thereabouts, including the height of aircraft.

Often, none of these methods can resolve a problem individually. A combination is called for. The electro magnetic method, for example, can be successfully employed for locating massive sulphide ore bodies, but it will also respond to certain conditions of ground water and hence anomalies will have to be drilled finally.

More recently the infra red sensor has come into use. The infra red sensor is reported to be more sensitive than any of the normal ground geoplysical methods. When used in a high flying aircraft copper ore bodies, not detected by the conventional methods, were brought to light.

## Sonic Survey

Where rocks, such as consolidated sediments or crystalline occur under deep water, under cover of unconsolidated sedimentary low frequency sound waves have been used to determine the thickness of such cover. Low power sound, at a frequency of 14.2 kilo cycles is used.

The maximun depth of penetration is of the order of 140 ft. Sound with a frequency of 11 kilo cycles and a power output of 800 watts was satisfactory. Sound with 6 kilo cycles and a power output of 700 watts gives better results. The interpretation of the results requires a good knowledge of the stratigraphy.

# PART VI

# DRILLING

# DRILLING

Drilling is the process of making holes in the ground or rock. It has been employed in mining and geological work for (1) prospecting, (2) exploration, (3) blasting, and (4) during exploitation for development, shaft sinking, rescue work, etc., Drills may be manually operated or mechanically operated either by compressed air, water under pressure (turbo-drills), petrol engine, diesel engine, electricity, high temperature flames and rarely steam. Based on the principle involved, in the operation, drills may be classified under the following types.

**A. Percussion**

1. Jumper bar or hand drill.
2. Pneumatic drills—Jack hammer, Hammer drill, Wagon drill.
3. Churn drill.
4. Reichdrill or Drillmaster (down-hole type).

**B. Rotary**

1. Auger.
2. Calyx.
3. Rotary drill using rock roller bits, tricone bits etc. and turbo drills using water as well as compressed air.
4. Diamond drills (using, diamond and T.C. bits).

**C. Miscellaneous**

1. Jet Drilling.
2. High temperature flame drill.
3. Banka or Empire drill.
4. Burnside drilling equipment.
5. Soil sampling drills.

Drilling machines can also be classified according to the purpose for which they are used as shown below :

1. Drills for alluvial prospecting.
2. Drills for petroleum drilling.
3. Drills for water well drilling.
4. Drills for Hardrock drilling.

5. Drills for shaft sinking (large diameters and for driving large diameter tunnels.

6. Drills for soil sampling, *e.g.*, ultrasonic drills, vacuum drills.

For the purpose of this discussion, it is proposed to follow the classification involving the principle of operation as given above.

## A. Percussion drills

1. *Manually operated.* Percussion drills are the oldest types of drills and most commonly used. The manually operated drill is used commonly for making blast holes in stone and road metal quarries. It is called a "jumper" bar, and consists of a high carbon manganese steel bar, about 1″ to 1½″ diameter and of varying lengths, between 2 ft. and 4 ft. One end is chisel shaped. It takes two men to operate. One man strikes the jumper with a sledge hammer, while the other holds the jumper on the rock and also rotates the bar. Wet drilling is usual, and the sludge is removed from the hole with a scoop periodically. The rate of drilling per day in 3 ft. or so, in granite.

2. *Pneumatic drills.* In mechanised mining, pneumatic drills, or drills worked by compressed air, are frequently employed. The principle involved is the same as that of the manually operated "jumper" bar. A popular version of the compressed air drill is the "jack hammer". The principal parts a jack hammer are given in figure (6.1). The piston (3) is pushed forward, in one direction, by the action of compressed air under pressure (800-100 lbs/sq. in) which enters the cylinder through the valve (9). The piston in turn delivers a blow on the shank or upper end of the drill (8). Due to the action of the valve, (9) the air supply is reversed and the piston is pushed back.

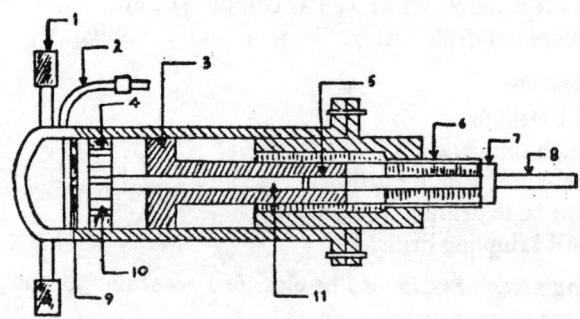

Fig. 6.1

In the forward stroke, the piston (3) does not rotate but the rifle bar (11) in side the piston rotates, and the pawl (10) slips over the rachet

(4) In the backward movement, the chuck, (5) which connects the drill (8) to the front part of the piston is made to rotate, as the hammer slips back on the rifling, on the rifle bar. Thus a rotary motion is communicated to the drill rod, simultaneously with the hammer action.

The exhaust, (compressed air) is also released through the central hole in the piston. The exhaust air passes out through the central small calibre hole, in the rod, and serves to cool the cutting bit, and also to remove the cuttings at the same time.

In the piston pneumatic drills, which were developed earlier than the jack hammer drills, the reciprocating piston and the drill rod were connected, so that the reciprocating motion was directly transferred to the rods. This type of drill is considerably slower in action, and the number of strokes per minute is also less, since the inertia due to the rods together with friction of the drill rods rubbing against the side of the hole is of a high order.

Hammer drills are classified as (a) Hand held drills, (b) Drifters, (c) Stopers and, (d) Wagon drills.

Hand held drills are designed either for wet or dry drilling and weigh 30-85 lbs. but normally 40 lbs. When used for stoping an "air-leg" is added. Drifters are heavier, being 115-125 lbs. in weight. They have an incorporated rotation device, either rachet and pawl or rotation motor. Stopers are specially designed for use in raises or slopes. The machines weigh 80-100 lbs. Wagon drills, are either crawler mounted or mounted on wheels with pneumatic tyres. They are used in open cast mines for drilling blast holes.

Air pressures, normally employed in pneumatic drilling, vary between 70-100 lbs per sq. in. Most effective pressures are, of the order of 90-95 lbs. per sq. in.

The quantity of air in c.ft. at 90 lbs/sq. required for (1) hand drill jack hammer of 30-45 lbs. weight, is 60 to 90 c.ft. (2) drifters—it is about 150-200 c.ft. and (3) stopers is 70-90 c.ft.

A common type of large diameter pneumatic blast hole drill known by various trade names is the "drill master" or Reichdrill "down hole drill" etc. This machine works on the principle of a heavy churn drill. It has a positive rotary motion, which may be as much as 70 r.p.m. A positive reciprocating motion is also communicated to the rods, as well as a powerful thrust (30,000 lbs.). As in the case of pneumatic drills, the exhaust is let into the hole, and the cuttings are also sucked by the positive induced draft (partial vacuum) caused by a "Sirocco" fan. The sediment, which is bought out of the drill hole, is deposited in a box. The drill is provided with a cross chopping bit with T.C. inserts.

One of the most popular types of drilling units, used in the exploitation of ground water, in the hard rock areas, especially in the granitic and the metamorphic terraines, is known as the Down-the-Hole-Hammer. It comprises three units, the prime mover which is a high power diesel engine which runs a compressor, and the compressor in turn motivates the down-the-hole hammer. In principle, this machine may be considered to be a larger version of the very commonly used compressed air drill known as the Jack Hammer. Even here, the hammer which in effect a piston, is actuated by the compressed air and hits the bit which is in contact with the hard rock at the bottom of the drill hole. Because of the rifling within the cylinder, the hammer is given a rotation and so also the bit. A diagramatic illustration of the DTH is given in fig. 6.7a. It will be noticed that the drill consists of 3 units. The upper unit I houses the values 1 and the hollow stem of the hammer 2, which is located in the middle unit II. The lower unit 3 houses the anvil 5 and the button bit 6. Number 7 is the drill hole. (See Fig. 6.7a).

Action of the DTH—When compressed air is let in at P, the units I and II come together thus closing the gap Q. At the same time, the valve 1 closes the hammer stem opening R, and thus the hammer 2 is pushed down and strikes the anvil 5 with great force and this impact is communicated to the button bit 6. When the hammer 2 delivers the blow to the anvil, the valve 1 retracts due to the action of the spring 5, and the vent R is opened and the hammer 2 also retracts due to the action of the spring S2, and the cycle of operations is repeated. The number of blows delivered per minute can vary from 1000 to about 3000 depending on the type of machine.*

Aerial rock drill, for secondary drilling of large boulders produced after blasting a blast hole face, has been recently introduced into the maket by Athey Products Corporation, U.S.A. The unit consists of a self propelled tractor-mounted cabin with controls. The operator sits in mobile cabin which is mounted on a 33 foot boom. The operator can move the cabin along the boom. The cabin is equipped with a high speed drifter, with three times the speed of normal drifter drills. In the. vertical arc, the drifter can operate between 104° to 105°, so that several holes can be drilled without moving the cabin. The drill is operated by a self contained compressor. Thus the rubber hoses connected to the drill are not damaged by wear and tear due to dragging.

The advantages of the equipment are : (1) Cheaper operation which is about 75% less than the normal. This is mainly due to the firm and convenient position of the drill operator. (2) The higher rate of drilling and quick movement, of the operator, from boulder to boulder also makes for efficiency.

The Churn drill (Fig. 6.2), in its simplest form, consists of a lever of the 1st order, known as the walking beam (7). The drilling tools are arranged to one end of the lever, while the other end of the lever is

---

*See addendum.

coupled either to a cranked wheel, known as the band wheel (3), or eccentric through a rod known as the pitman (4).

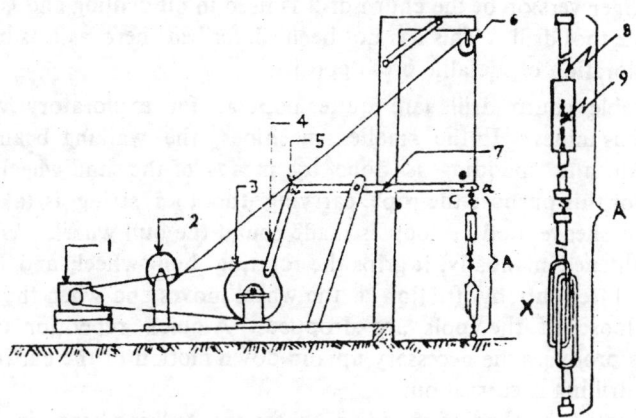

Fig. 6.2

Drilling procedure : When the motor or engine (1) is started, the bandwheel (3) rotates, and actuates the walking beam (7), which moves up and down, through the movements of the pitman (4). Thus the drilling tools (A), which are suspended at the other end of the walking beam, move up and down, derive the percussive action. The major part of the weight, of the drilling tools, is however, taken up by the rope suspension attached to the drum (2) known as the sand wheel. When the point of suspension, of the drill rods, on the walking beam (a c) is varied, the length of stoke also varies.

The temper screw (9) is an arrangement for increasing the length of the string of tools. It also includes the double loop arrangement (x) called the jars which forms part of the string of cutting tools. When the tools are dropped into the drill hole, by the walking beam, hammer action is produced by the jars, due to the impact of one loop against the other. Certain amount of water is used during the spudding or cutting operations. The jars also help in lifting of the rods with a jerk. This action is of special importance, when the rods are jammed by the cuttings or due to caving of the side. The cuttings are removed from the hole by the use of a dart bailer (Fig. 6.3). This consists of a tube, about 4 inches in diameter, (b), a conical valve at the bottom, and a stirrup (c) at the top. For operation,

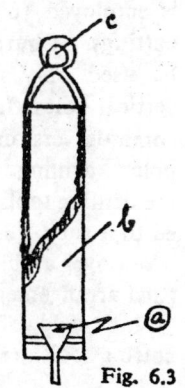

Fig. 6.3

the drat bailer is let into the hole by a rope, attached by a separate drum called the sand wheel (2).

A larger version of the churn drill is used in oil drilling and is known as the cable tool drill.  This has not been described here as it is not used in the exploration of metallic ores deposits.

Portable churn drills are quite popular for exploratory work in metalliferous mines.  In the smaller mechines, the walking beam is dispensed with, and spudding is done by means of the sand wheel or bull wheel.  For this purpose the rope, carrying the tool string, is taken over the pulley sheave and a loop is made round the bull wheel.  When the loop is tightened manually, it grips the rotating bull wheel, and the drill tools are lifted up by friction as the wheel moves and when the rope is suddenly loosened the tools are dropped.  A quick repetition of these operations produces the necessary up and down motion of the cutting tools and thus drilling is carried out.

In another method of spudding (where the walking beam is absent), the sand wheel rope is taken round the crank in the band wheel *viz.*, the crank, where the pitman is attached.  As the crank moves round, the drill tools are raised and dropped automatically and spudding takes place.  The 6 inch bit is used for holes normally up to 600 ft. depth.  For deeper holes up to 1000 ft., the starting bit may be 26″, and the derrick is about 80-100 ft. high.  This method of drilling is often used for shallow water wells, within 100-120 ft.  The wells are cased with 8 inch pipes.

Another modification of this method is "Spring pole" drilling.  Here the tools are suspended from end of a pole, and it works like the rice mortar "Dhinke" used for flattening rice.  A sand pum por bailer. Figure 6.3 is employed to get out the cuttings.  Churn drills can be used only for drilling vertical holes. In the more common version of "spring pole" drilling (Fig. 6.4), the drilling tools are attached to a rope, (3), which is

Fig. 6.4

taken over a pulley (2) and attached to a winch, (4), with a ratchet and a pawl arrangement.  Men standing between the tripod (1), and the winch, (4) take hold of the rope here, pull it down and let it go alternately.  The cutting tools are thus lifted and dropped.  So drilling thus progresses.

## B. Rotary drill

There are various types of machines, included under this class. In these drills the drilling tools are rotated, by a prime mover and at the same time certain amount of pressure is applied. Some of the common rotary machines are described below.

    1. The *auger drill* is simple machine. It is either worked manually or by a prime mover (engine). The auger (Fig. 6.5) consists of a flat steel strip, which is twisted in the form of a screw. In principle, it is similar to a laboratory cork borer, or to the carpenter's auger drill. As the auger cuts into the clayey soil, the material is packed up in the screws, and after the auger has penetrated its full length it is pulled out of the hole and the screw is cleaned for obtaining samples of clay or soil.

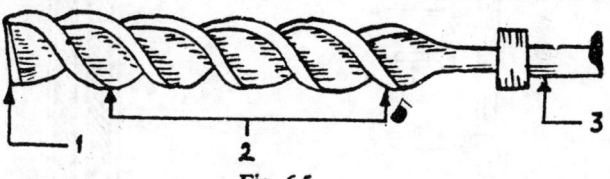

Fig. 6.5

A scoop type (Fig. 6.6) auger is also manufactured. This types of auger is also manually operated, and is capable of making holes up to 50 ft. deep.

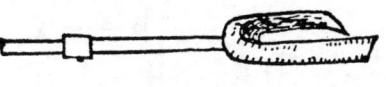

Fig. 6.6

Auger drills can also be operated mechanically. A smaller version of the auger drill is used in coal mines, for making shot holes, for blasting. This makes $1\frac{1}{2}''$ holes about 5 ft. long, and is operated electrically.

    2. *Calyx drills.* This machine is also a form of rotary drill. The calyx drill is either hand operated or machine operated (Fig. 6.7). The drilling tools are actuated by an engine through a bevel gear (7) which meshes with a companion gear (6), hitched to a longitudinal slot (5), on the feed rod (3), by a feather key. The feed rod (8) is driven (rotated) by this gear, and it is capable of moving up and down, by sliding along, the longitudinal slot (5).

    The water swivel (3) has a thrust bearing and rotates freely. Of the two inlet pipes (2), in the water swivel, one carries water into the drill rods for washing, cooling etc. The other carries chilled shot for shot drilling as described later. The water swivel is essential, for it allows the drill tools and feed rod (3) to rotate, while the water pipe and the chilled shot pipe remain stationary.

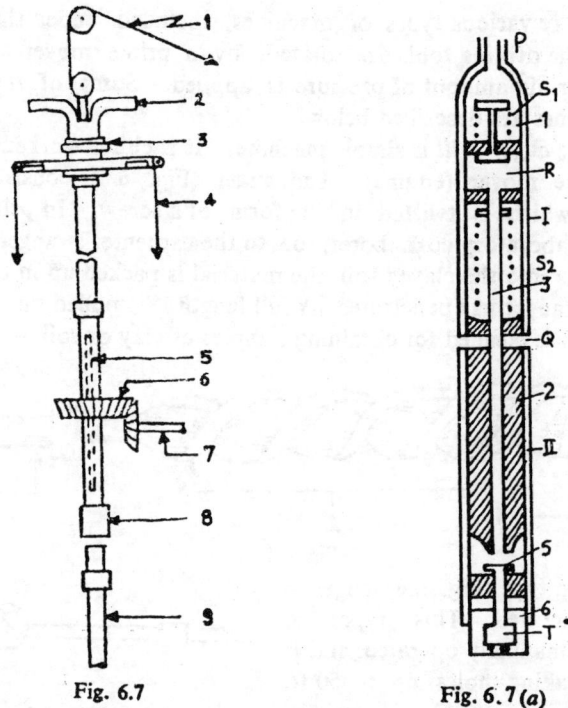

Fig. 6.7                    Fig. 6.7 (a)

The essential parts of water swivel are shown in (Fig. 6.8). An inner tube (2) carrying the T-joint, with the water tube (10), and the shot tube (11), has an eye loop (12), for enabling the swivel to be suspended. The inner tube has a collar (8), and this fits snugly into an outer tube (1). The outer tube has an internal lip (9). The annular space between the outer tube, and the inner tube, lying between the collar of the inner tube (8) and (9), of the outer tube, is tightly packed with greased asbestos rope (7). A thrust bearing with ball raises (6), and (5), encloses the flange (3), of the outer tube. The drill rods are attached to the screw at the end of the outer tube,

When the plates (14) and (4), holding the thrust bearings. are pressed together, by tightening the bolts (15), the inner tube (2) is pushed down by the collar (13), whereas the outer tube (1) is pushed upwards (relatively) by the flange (3). Thus the gland packing consisting of asbestos rope) (7), is made water tight and any leakage is sealed.

The drill rods are attached to the chuck at the end of drive rod. The coring equipment (Fig. 6.9) comprises the cutter, either Davis cutter (2) or

Calyx cutter, with a core barrel (5), which is attached to a sub or connector (7). A sediment tube (3) is also screwed on to the sub. In soft rocks the Davis cutter (2) is used, but in hard rocks the shot bit (6) is employed. The Davis cutter is a steel pipe with a toothed (serrate) margin. When the shot bit is used, chilled steel shot (0·332" in diameter) is fed into the drill rods, along with the water. The shots pass into the drill rods, from the water swivel. At the bottom of the hole the shots are crushed and penetrate the shot bit. During the rotation the shot bit, which gets impregnated with steel splinters cuts into the rock. When the rock is too soft to crush the steel shot, crushed steel is added instead.

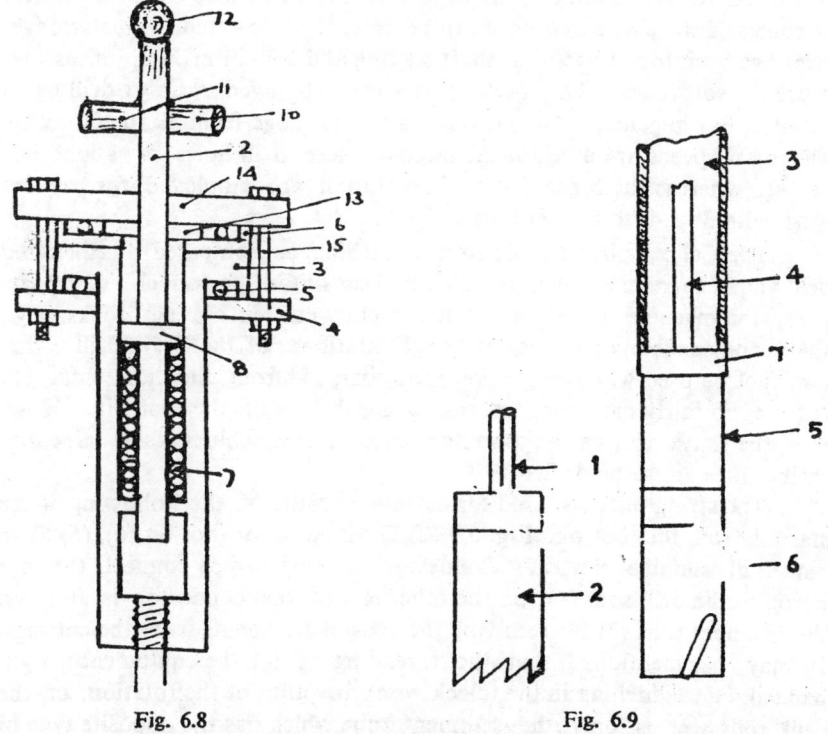

Fig. 6.8                    Fig. 6.9

Tungsten carbide tipped bits also known as T.C. bits can be used in place of the Davis cutter, and separate T.C. inserts are also available for reconditioning the bits. For removing the solid cylinder of core, grout or broken gravel 1/8" diameter is fed into the drill rods, along with the feed water, and gradually the pressure of the feed water is increased, as the grout settles down at the bottom of the core barrel. This causes a wedging action and the core, which is in the core barrel, is jammed. When the drillrods are rotated slightly the core which is firmly held by the core barrel by wedging action of the gravel, breaks off from the rock attachement, and can be hoisted out of the hole together, with the rods.

In the larger diameter drill holes, exceeding 24″, core getting is more cumbersome. After drilling a foot or two, the cutting tools are hoisted, and a small charge of dynamite is placed, in the annular space of the cut, and then blasted, in order to dislodge the core. A hole is drilled in the centre of the core and a lifting bolt is attached. The core is removed from the hole by attaching the lifting bolt to a rope and pulley arrangement.

Some of the larger diameter calyx drills, such as those used in sinking shafts, are mounted on a cage which is suspended from a derrick by rope and the entire assembly can be raised or lowered. This arrangement has been found useful in shaft sinking and also in making raises and winzes in soft rocks. The speed of rotation, employed in Calyx drilling, is variable, but in general, this may be taken to range from 50 to 500 r.m.p. The lower speeds are used for the large diameter drill holes, say about 3 ft. to 4 ft., whereas the higher rates of revolution are employed for routine work with 3 or 4 ″ diameter holes.

The Calyx drill is a more robust machine, as compared to diamond drill and is capable of being handled by less skilful personnel. Comparatively, the machine is cheaper and it is highly efficient in soft rocks, e.g., shales and sandstones. One of the limitations of the Calyx drill is that it cannot be used where angle holes less than 75° from the horizontal are required in fairly hard rocks, necessitating the use of the shot bit. However, the Calyx drill has an advantage over other machines where large diameter holes are to be drilled.

The string of tools, used for cutting, consists of the following items starting from the bottom Fig. 6.9 (2) Davis cutter or shot bit (6), (5) Core barrel of suitable size, (7) Connector or plug which connects the core barrel to the drill rods (4), on the other end of the connector is attached the sediment tube (3) for receiving the coarser fragments from the cuttings. It may be mentioned that the threading in all the equipment is right handed, i.e., tightening in the (clock wise) direction of the rotation, of the drill rods. It is only the sediment tube which has the opposite type of threading i.e., left handed. This is done to prevent the sediment tube from working loose and dropping off when the drill rods are rotated. The drill rods are attached to the connector or plug (7).

Calyx drills have been suitably modified for drilling large diameter holes (upto 1·5 metre diameter). Such machines are used in the Ruhr Coalfields for sinking of staple pits. Advantages of using the machines over conventional methods of shaft sinking are given below :

    1. Easy disposal of debris,

2. Good ventilation is maintained throughout the sinking operation except during blasting.
3. Lesser consumption of explosives.
4. High rate of progress, *i.e.*, 50—100% more.
5. The formations are known well ahead by making the pilot hole.
6. Percolation and flooding is better controlled.

The machine, for shaft sinking, is either driven electrically, or by compressed air. It moves on pair of rack-guides for feeding as cutting progresses. The drill bit is of the rock roller tri-cone type with T.C. tips. The cones are three in number (placed 120° apart). (Fig. 6.10) for diameters up to 610 m.m. For larger diameters, 5 cones are employed.

A bore hole 1220 m.m. in diameter is done in three stages as indicated below :

### Table 6.1

| Stage | Diam. of bit | Manner of boring | Expected progress in coal measure rocks |
|-------|-------------|------------------|------------------------------------------|
| I Stage | 195 m.m. | From bottom to top | 10 m /shift |
| II Stage | 305 & 406 m.m. | From top to bottom | 12 m/shift |
| III Stage | 610, 813, 1050 and 1220 m.m. | From top to bottom | 4·5 m/shift |

Large diameter bore holes have been used not only for sinking staple pits, but also for opening up longwall faces, in steeply inclined seems.

3. *Rotary drills* : These are also known as the rock roller drills. In the case of these drills also, cutting is done by rotating tools. The type of cutting tool used is, however, different. Rotary drills have a characteristic cutter known as the rock roller and tri-cone bit. As the name implies, three bevel gears of high grade steel with T.C. (Tungsten Carbide) tips placed at centres 120° apart are made to mesh, as shown in the Fig. 6.10. When the rod is rotated the gears are rolled, at the bottom of the hole, and cut into the rock. No core is normally recovered, in this method of drilling and the formations are identified by collecting the rock chips, which come up with the circulating water or mud. Special coring bits have to be used, if a solid core is required. For routine work the sludge analysis is carried out. This method of drilling is imminently suited for drilling blast holes in seismic

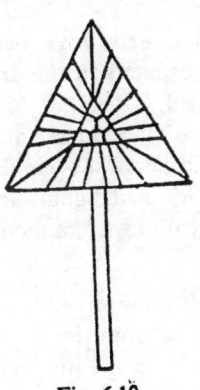

Fig. 6.10

prospecting, for ground water exploration, in soft

rock areas and also for oil well drilling in soft rock areas. Rock roller bits of modified design, with buttons of T.C. are used in harder formations.

Besides the tricone bit described, there are other types of bits used in rotary drilling, e.g., the fish tail bit, the diamond point bit, and the drag bit. The fish tail bit, when rotated, breaks up boulders projecting from the sides of the bore hole, while the diamond point breaks up hard pieces of rock, at the bottom of the bore hole. The drag bit is employed when hard formations have to be penetrated.

Mud flush is often employed with rock roller bits, as it serves to bring up large pieces of the rock cuttings, besides ensuring stability of the bore hole sides, and cooling of cutting bit. The preparation of mud, required for meeting the conditions in a particular area, involves a special technique, and only brief indications can be given. The purpose of the mud employed is, (1) to seal any water bearing formation or oil or gas bearing formation. (2) The mud also prevents the bore hole walls from collapsing by means of the plastering effect. (3) Special muds with barite (known as weighted muds) are used when water, gas or oil under pressure is expected. Bentonite is the most common clay used in forming suspensions. Depending on the salinity of the water encountered in the formations' the pH is altered, and certain neutralising salts are added, so that coagulation and flocculation of the mud is avoided. (4) The mud also helps in bringing up the cuttings, in larger fragments than that obtained by using clean water. Since unlined holes can stand longer when drilling with mud suspensions, this practice is often followed where electric logging is to be done.

The quality of the bentonite, used for drilling purposes is, specified essentially by the 'yield' test. The 'yield' is expressed in terms of 'the number of barrels of mud having an apparent viscosity of 15 centipoises that can be prepared from one ton of clay'. (One barrel is 42 gallons). Sodium bentonites are generally favoured on this account.

Peptised bentonite is also employed since this has an extra high yield, the transportation cost being the same. If the carrying capacity of the drilling water is to be increased, asbestos fibres can be added.

Organic colloids are added to increase the viscosity, reduce infiltration, stabilise the clay, flocculate drilling solids, and serve as emulsifiers and lubricants. Organic polymers are hydrophyllic. Modified gums and starches are also useful, such as sodium carboxymethyl cellulose and modified guar gum.

The rotation of the drill rods, is done by a turn table, which is driven through gearing, by an engine. The feeding is done by means of a hydraulic suspension, which regulates the pressure on the bis, especially, when the depth of the hole is great. The speed of rotation ranges from about 30 to 150 r.p.m., but the rates of drilling can be as much as 200 ft. in a day of three shifts, depending on the nature of the strata encountered. When coring is necessary, special coring bits (also of the tricone type) are employed. But coring is exceptional rather than the rule.

*Reverse Rotary*: This method of drilling has been largely used in India in drilling for ground water, where large diameter holes are usually needed. In this system drill pipes, the counterpart of drill rods in rotary drilling 20 ft. or longer and in O.D. fitted with flange joints are used. However large quantities of circulation water are essential, and water loss is a serious impediment. Caving in shales and clays can be a setback, but the trouble can be minimised by adding sodium hydroxide to raise the pH to about 10.5. Sodium silicate 5 to 10 per cent is more effective.

When drilling through permeable formations 200 to 500 G.P.M of make up water may be needed, the normal quantity being about 500 G.P.M. A centrifugal pump with a large passage way is employed to deal with the cuttings, individual pieces of which, can be 5 in. in diameter. The smallest holes drilled by this method is about 16 in. diameter and wells of 50 in. diameter can be drilled. The drill pipe rotates at about 10 to 40 R.P.M. This method is cheap when drilling is done in soft, unconsolidated formations.

*Turbo drills* utilise a jet of water or compressed air to actuate a turbine, The turbine, in turn, is utilized to rotate a drill. Since the entire assemblage, comprising the turbine and the drill is so compact, that it is easily lowered into the bore hole, there is a large saving in power, as this drill has not to move large lengths of drill rods. Turbo drills are largely employed in oil drilling.

4. *Diamond drills* : These are by far the most popular type of drills employed in mineral exploration. A rotatry motion is transmitted to the drill rods by the engine, as in the case of the rotary drills, but the essential difference is in the type of cutting tools employed. In one of the types of drill bits in use, diamonds (borts) are either set, or sintered in position in a bronze alloy matrix, In another type, sintered diamond strips are brazed on to the blank crown. In the impregnated bit, diamond powder is mixed with metal powder and sintered, under high pressure to form rings which are attached to the cutting bit.

In diamond drilling, a core clip is used in the core barrel, (1) for lifting up the core. This consists of an incomplete circular band, (3) which

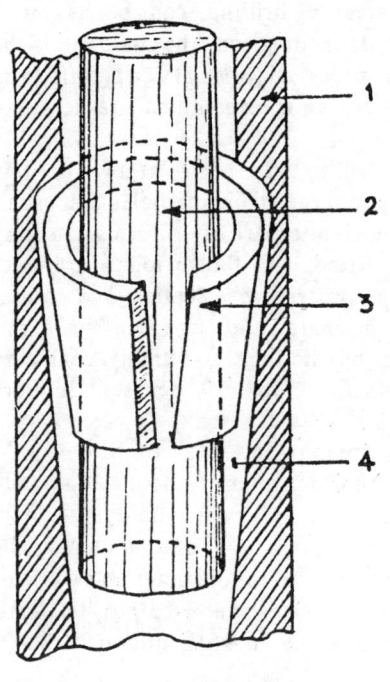

Fig. 6.11

is wedge shaped, in section (Fig. 6.11). The clip allows the core (2) to pass through into the core barrel, when drilling is in progress. After drilling is over the rods are lifted. The core clip slips down the seat (4) when the rods are lifted and thus acts as a wedge between the core and core barrel gripping the core right round. The core is held firmly in position, within the core barrel, from which it is removed by unscrewing the diamond bit.

The diamond drill is one of the tools which, if available, enables to speed up prospecting operations in hard formations. Drilling machines of various capacities are available, and such machines are made in India. The smaller machines, with a capacity of about 100 ft., equipped with AXT tools, i.e., 5/8″ cores, have been made in portable units. The popular make known as X-RAY, made by Boyles Brothers of Canada, is axample. In this range of machines, is the Mindrill made in Australia. This is also a good machine. This type of machine can be used in hard rock areas with considerable success, e.g., in limestone and also in prospecting for base metals. In the medium size group, there are a good number of makes, e.g., Longyear, Joy, Mindrill, Craelius, Acher etc. Some of the machines are made in India, under foreign collaboration, and are quite easily available. This group of machines can be used to drill a maximum depth of about 750 ft. using AX equipment. The larger capacity of the medium class machines are also made in India by Joy, Longyear.

The drill string, i.e., accessories used, consists of the following in the ascending order (Fig. 6.12). (6) the diamond bit, (5) reaming shell for making the hole wider so as to decrease friction caused by rubbing of the tools and drill rods, against the drill hole walls (2 and 9). Core barrel with the core clip (Fig. 6.11) inside it, (Fig. 6.12) (1) sub or connector and (4) the drill rods.

The core clip (Fig. 6.11) has been described. A word is necessary about the core barrel. As in the case of the Calyx drill the core barrel used in diamond drilling is of two types, (a) single tube (b) double tube. Again in this classification there is a sub-class of of thin walled and thick walled core barrel. The single tube type is used in hard formation, which do not dissolve or disintegrate when in contact with flowing water. The double tube core barrel, which can be either of the swivel or ball bearing type, is essential when drilling through either soft friable or soluble formations. The double tube core barrel consists of an inner and outer tube. The diamond bit is fitted to the outer tube, which rotates with the drill rods. The core clip is placed inside the inner tube, which is attached to the sub or connector, either by a swivel or a ball bearing such that when the outer tube rotates, the inner tube is stationary.

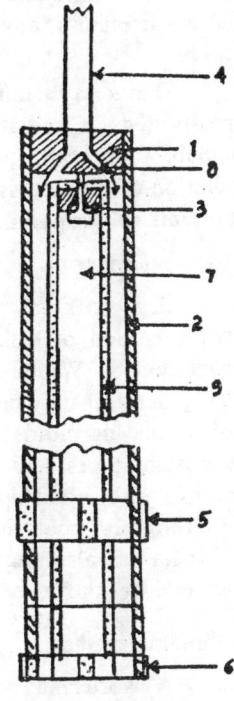

The double tube core barrel employed is shown in (Fig. 6.12). In this type of core barrel, the circulating water from the drill rod (4) does not come into contact with the core but flows through the annular space (9) between the inner (7) and outer tube (2) and comes out at the bottom of the hole. To the outer tube, which rotates, are fitted the coring bit (6) and the reaming shell (5) while the inner tube (7) remains stationary due to a swivel (3) arrangement or by a ball bearing arrangement.

Fig. 6.12

A split inner tube has been recently introduced and the results have been very encouraging. A further modification has been achieved by placing a plastic tube within the core barrel, to take in or enclose the core as it enters. Core recovery has improved by employing this modification.

The prime mover or engine, used in diamond drilling, is either diesel or petrol driven. The speed of revolution is very high, as compared to other methods mentioned, and is of the order of 2000—3000 r.p.m. Even though the diamond bit just removes a thin film from the surface of the rock, in each revolution, the high speed of rotation more than compensates for this, as the overall efficiency is very high. As drilling progresses, the feeding of the rods is done mechanically, either by a screw machanism or by a hydraulic system which lowers the rods at a desired rate, and maintains the desired pressure on the diamond bit. (These mechanisms have been described later). After the entire length of the feed screw is lowered, or when the hydraulic piston has reached the bottom, a short length say 5 ft (in the case of medium sized machines) of rod is added.

One of the other advantages, of the diamond drill, is that, it can be used with confidence where directional drilling is called for. Angle holes, directed at any desired angle, on any azimuth direction, are done easily.

Diamond drilling is also employed when salt deposits are to be prospected. Circulating water is stopped as soon as the saline formation (soluble) say salt, potash or gypsum is encountered, and a saturated solution of the salt is circulated, when drilling through the bed. This prevents solution to a very large extent, and enables core recovery.

### Reverse rotary

Among the diamond drilling methods the *reverse rotary* or Winter-wiess method of drilling may be mentioned. It was introduced in the continent by Winter-wiess and adopted in England and in the U.S.A. In general, it may be stated that, it is a method, in which a high pressure of circulating fluid is used, the circulation is reversed, as the water enters the casing pipes, and comes out through the drill rods. In doing so the core is also lifted up. This method was used, with considerable success, in drilling for coal, in the British Coal measures. The Geological Survey of India had also employed this for the same purpose, with success. This method has, however, not been very popular in India.

### Wire line method

A word may here be included about what is described as the wire line method of recovering core. In this method a special double tube core barrel is employed. After the run is over, say it may be a 5 ft. of drilling, the drill rod is disconnected from the drilling spindle, and a dart, attached to a wire line, is lowered through the drill rods. The V-shaped head of the dart engages, with two claws attached, to the top of the inner tube of the core barrel. As soon as the inner tube grips the dart, it is released from the core barrel assembly. Thus the inner tube, containing the core, is baled out. After emptying the core, the inner tube is replaced and set back in its old position, and drilling is continued. The wire line equipment is useful when deep diamond drilling is resorted to *i.e.*, where the depths are of the order of 2500 ft. or more. A considerable saving of time is affected, as the method avoids the raising and lowering of rods, and also the time taken for screwing and unscrewing.

## Diamond bits

There are various types of diamond bits, available in the market, and each having specific advantages. The variations, that can be affected in diamond bits to accommodate various conditions encountered are, (a) in the hardness of the matrix, (b) in the density of stones, and (c) in the size of stones. The stones are set in holes, made in the matrix, and the metal is squeezed in from the sides of the holes, by gouging to hold the stones. This process is known as caulking. The general rule regarding the choice of a bit, should be the coarser, the grain, the softer and fiable the rock, the best choice of a bit will be one with larger and fewer stones set in a soft matrix, e.g., sandstones, arkoses and rocks of this type. The harder the rock and finer the grain size, then the bit selected should have finer stone size, greater density of stones, and harder matrix. Thus in drilling through hard compact dense quartzite, even impregnated bits will have to be employed, whereas hard granite will be easily cut by a bit set with small stones. The following table is a guide.

Table 6.2

| Quartzite | Granite | Shale | Sandstone | Schist | Limestone |
|---|---|---|---|---|---|
| Impregnated | Impregnated or 80-110 s.p.c. | Surface set 8-10 s.p.c. or T.C. | Surface set 8-10 s.p.c. or T.C. | 30-60 s.p.c. | Impregnated and T.C. |

s.p.c.=stones per carat
T.C.=Tungsten carbide.

Attempts have been made to arrive at the optimum cost of diamond bits. One such basic cost equation for all types of rotary drilling employing any type of bit is given below:

$$C = O/R + (B - to + M)/F$$

where C is the drilling cost/foot in Rs./ft.

O is the true rig operating cost inclusive of direct cost, such as camps, supports etc. in Rs./hour.

R is the rate of penetration in ft. per hour.

F is the bit life in ft.

B is the net bit cost in Rs.

T is the round trip and circulating time at depth in hours.

M is the mud conditioning and added cost incurred on each trip in Rs.

*The general guidelines for the selection of bits is as follows:*

Case 1: .  Conventional penetration rate, 10 to 12 ft. per hour.
Style: Soft formation bit.
Diamond setting: Single row of diamonds per rib.
Diamond size: 1 carat. Diamond quality: Regular.

Case 2:     Conventional bit peneration rate 7 to 10 ft. per hour.
Style: Medium formation bit.
Diamond setting: Two rows per rib.
Diamond size: 1/2 carat. Diamond quality: Regular.

Case 3:     Conventional penetration rate 3 to 7 ft. per hour.
Style: Medium to medium hard bit.
Diamond setting: Two rows per rib.
Diamond quality: Premium.

Case 4:     Conventional bit penetration rate 1 to 3 ft. per hour.
Style: Medium hard formation bit.
Diamond setting: 3 to 4 rows per rib.
Diamond quality: Premium.

### Drill rod feed systems

There are two common systems for feeding or automatically lowering drill rods with the progress of drilling (a) Differential screw feed, (b) Hydraulic feed.

(a) Differential screw feed is one of the common methods. The principle is simple and is based on the fact, that if the nut and bolt are rotated at the same speed, in the same direction, then both stay in the same position. If either, the bolt or nut rotates faster, or if one of them rotates in the opposite direction, then there is a translation or relative movement. In Fig. 6.13 (4) is a square threaded spindle with longitudinal complementary groves for feather key (2). Bevel gear (1) is driven by the engine and the feather keys (2) are attached to it. The feather keys (2) are also attached to the gear wheel (3). Companion gear (5) drives the counter shaft (6) and the interchangeable set of gears $a_1$, $b_1$, and $c_1$ afford a variable gear ratio. The gears, a, b, and c are keyed to a threaded sleeve or nut, (7) which engages with the spindle (4). The variable gears facilitate changing the speed of rotation as explained earlier.

*Operation* : When the spindle (4) and nut (7) are rotating, at the same speed, and in the same direction, the spindle (4) stays in the same position. This happens when the gears 3, 5 and $b_1$ are engaged. The speed of the spindle is the same as that of the counter shaft. If the gear speed, on the counter shaft, is changed, e.g., if $a_1$ and a are engaged then the nut (7) will now rotate faster than the spindle (4) and the spindle (4) will consequently move up or down. The gears (1)

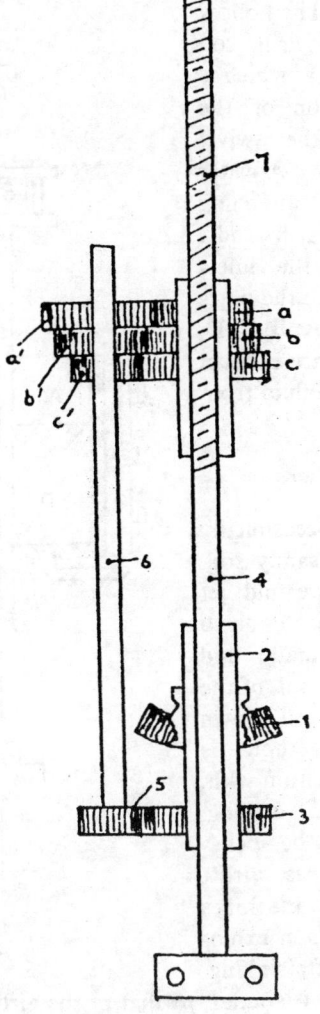

Fig. 6.13

and (3) remain stationary (*i.e.*, do not move up or down) as the feather key (2) slips in the complementary grooves in the spindle (4).

(*b*) The Hydraulic feed (Fig. 6.14) is a much simpler device, than the differential screw feed. The tools are lifted up, or lowered, by means of a hydraulic piston (4). The piston has a hollow stem (7) and the dril rods (6) pass through it. The hollow stem is connected to the drill rod by a collar or chuck (12) which is tightened. Rotary motion of the piston (4) is prevent by the swivel with thrust bearings (9). A water swivel (10) is used for attaching the hose pipe (11) to the drill rods. Cocks a, b, c, and d are the inlets for the hydraulic fluid into the parts of the cylinder marked A, B, C, D, respectively. (d) and (c) are exhaust cocks for letting out the fluid from the cylinder.

## Miscellaneous drilling methods

1. *Jet drilling* : Occasionally where drilling is done in sandy soil, as in a river bed, a powerfuld jet of water is directed vertically on to the sands. A hole is made and as the sand is washed out of the hole, a drive pipe is lowered down simultaneously. This method is often employed in conjunction with, either calyx or diamond drilling, when there is soil cover to be penetrated. In soft rock areas, airlift boring rigs are used. These machines are designed for groundwater exploration and also for shaft sinking. The principle of operation is similar, to that of the airlift pump, described under pumping and drainage.

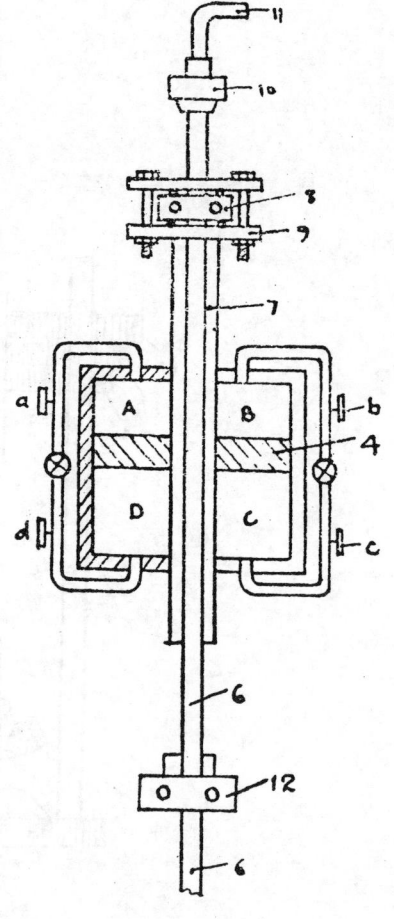

Fig. 6.14

2. *Flame drilling* : This method is largely employed in blast hole drilling, through hard rock *e.g.*, taconite, hard ferruginous quartzite. A high temperature oxyacetylene flame is directed onto the ground through a nozzle and the cinder, so generated, is blown out.

In the Linde process of drilling, liquid oxygen and fuel oil (paraffin) are used as a combustion mixture, to produce a high temperature jet from a nozzle. This method is used to make the blast holes in hard taconite (iron formations). The nozzle requires to be constantly cooled by cold water, which is circulated in a jacket placed round the nozzle.

3. *The Banka or Empire drill* : Figure 6.15 is also a form of rotary drill, where a circular platform (1) is fixed to a string of 4" drive pipes (2) fitted with a cutting shoe (3).

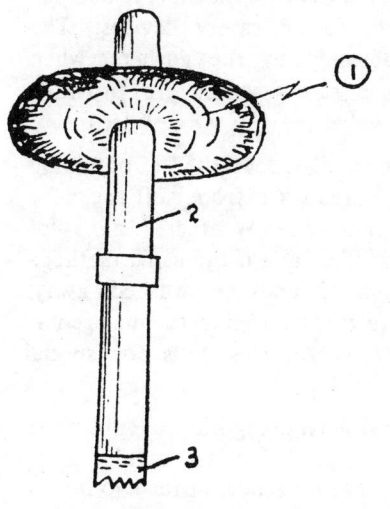

By rotation either manually or by animal power (by means of a Yoke, not shown in the figure) the drive pipe sinks into the ground. After the pipe has penetrated a certain depth men get on the circular platform fitted around the pipe and lower a "dart bailer" or "sand pump" (Fig. 6.3). This consists of a long pipe, (1) about 3" in diameter fitted at the bottom with a flap valve or spring loaded disc valve (a). For efficient functioning of the "bailer" water is poured into the hole, if no water has been encountered. After bailing out the cuttings, the drill pipe is lowered further by rotation as before. The depth of normal

Fig. 6.15

working, of this drill, is about 100-125 ft. and it is popular in alluvial prospecting, especially for tin and gold.

In the case of Empire or Banka drill sampling, the computations involved include the following considerations :

1. Impact of the churn drill which may cause the gold particles to move downwards in the gravel ; due to agitation with water.

2. Some of the particles may not be taken up by the sludge pump or vacuum pump when dry drilling is done.

3. Some values may be lost, in panning, inspite of the care taken.

Thus the gold recovery is likely to be lower. In order to compensate the possible losses, the Radford factor, which has a constant with a value of 0.27, is employed, with reference to the standard pipe area of 0.3068 sq. ft., for a 6" diameter (p.708).

4. *Burnside boring machines* : A special type of drilling machine is occasionally required for use, especially, when driving advance headings, in coal mines, in areas where old workings are anticipated. The machine, known as the Burnside Borer (or boring machine), is, essentially a non-

coring type rotary drill, and is wagon mounted and employs hollow rods. It is so designed that horizontal holes or slightly inclined holes, about 100 ft. long, can be made from galleries or drifts underground. It is generally compressed air operated, but electrically operated machines are also available. In driving advance headings, in areas where large seepage, with danger of flooding is anticipated, the use of such a machine is compulsory by Law. The machines incorporate certain safety devices. The presence of water under pressure can be detected by the gushers, when the reservoir is penetrated by the bit. Since safety valves are provided the water can be sealed off by shutting the valves.

5. *Soil sampling drills/Miscellaneous samplers* : Method of obtaining cores from soft rocks : Cores are often required from soft rocks for conducting various types of tests and analyses. Weathered zones of mineralisation have got to be sampled by drilling, and if the usual methods are adopted the core is likely to disintegrate, and be washed away. Undisturbed soil samples are also often required, by engineers and geologists, for studying the mechanical properties. All this calls for special equipment.

(*a*) In some of the earlier designs for soil sampling kits, a split tube was used. This equipment, in its simplest design comprises a tube 2″-3″ in diameter, cut into two halves lengthwise. The two ends, of the two halves were gripped together by a clamp or a tight fitting ring. When the split tube is driven into clayey soils, it penetrates with ease and a cylindrical core is obtained, when the tube is pulled out of the ground and opened out by removing the binding rings. Soils with higher moisture content (near the liquid limit) are often disturbed, while sampling with this equipment.

(*b*) A plunger type of sampler was introduced for obtaining samples of very loose and incoherent materials. In this equipment, a piston was introduced into the sampling tube. After the sampling tube is introduced into the soil, a suction effect is induced by pulling the plunger up, with the result that the core within the sample tube is consolidated (due to the liberation of excess water) and is removed intact for examination.

(*c*) Further improvements, with the application of new techniques, led to the evolution of the vibro drills, and these have proved to be superior in performance to any other. Samples of soils collected, by this method, were as good as those collected directly from pits. Bouldery soils were found to be unsatisfactory for the operation of this method. By employing this method, a faster rate of drilling than any other method, by 3 to 5 times, is possible. The vibration should not exceed 12 mm. in amplitude, while the frequency is dependent on the nature of the soil. Special rods are used in conjunction with the vibrator, and one of the drills yields 219 m.m. diameter cores, and has a capacity of drilling 40 meters. An immersion type vibrator has also been designed and with this equipment greater depths of drilling can be achieved. In this method of drilling, the entire equipment is lowered into the easing.

## Drill sampling

For purposes of formation sampling, "side wall" samplers, have been designed. These ean be used either with percussion tools or with rotary drilling. These samplers are generally of two types. In one model (Fig. 6.16) the sampler tube (1) is fitted with two blades (3). The blades have two hollow sample tubes (4) inserted into recesses. The sample is taken

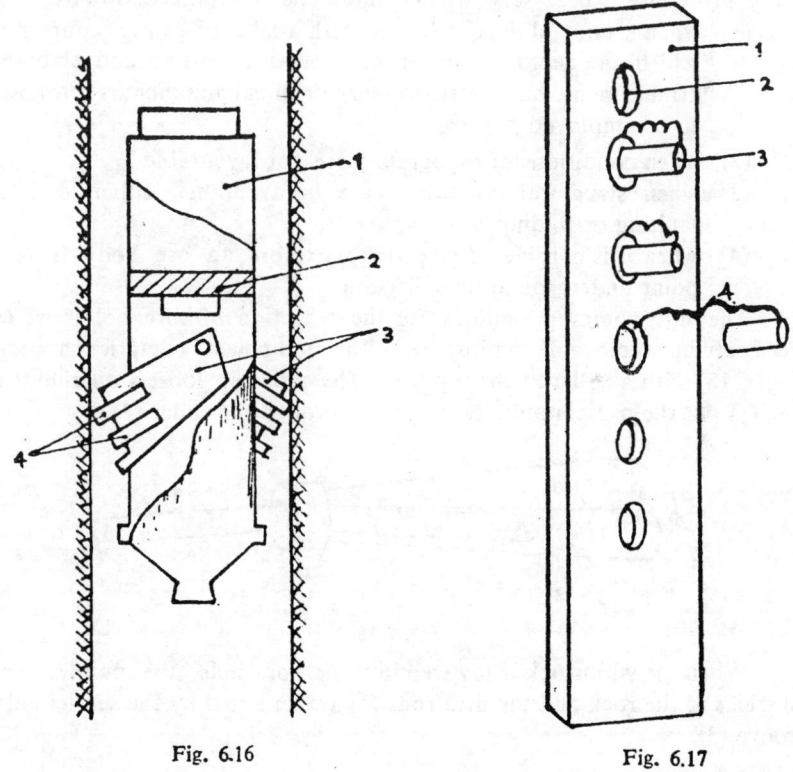

Fig. 6.16          Fig. 6.17

in the following manner. The sampler is lowered to the desired position. Then hydraulic pressure is exerted through the hollow drill rods. The plunger (2) causes the two blades to move outwards and the sampling tubes penetrate into the formation and the hollow tubes are filled with the rock. On release of the hydraulic pressure, the blades retract and the sampler is withdrawn. The sample core from such machines is about 3 cms. in diameter and 4 to 5 cms. in length.

A second type of side wall sampler (Fig. 6.17) is called the sampling "gun". The arrangement is similar to the muzzle loading gun. Hollow shots (3) are fired, from the barrels (2). The entire sampling device comprises 6 or sometimes as many as 36 sampling bullets or shots. The shots are fired electrically, and penetrate deep into the formations. Since the shots are attached to wires, they can be retreived, when the sampler is pulled up. The sample is retreived from the hollows of the 'shots'.

*Directional drilling and planned deflection of bore holes* : A bore hole especially, rotary drill, or diamond drill bore hole, may be deflected from the original direction and angle, according to a preplanned programme. Such a procedure is necessary, when drilling under certain conditions.

(1) when a mineral deposit lies beneath a lake or water course, in which the depth of water may entail expensive and elaborate arrangements but where ordinary drilling equipment is proposed to be employed ;

(2) when conditions of topography are not favourable ;

(3) when structural conditions are unfavourable, either due to faulting or folding ;

(4) when it is considered essential to explore an ore body from a point underground by deflection.

The most common methods, for the deflection of bore holes are (a) the Eastman removable whipstock. The equipment comprises a wedge (Fig. 6.18) with a collar at the top (2). The collar fits loosely on the drill rod, (1) but the rod coupling cannot pass through the collar (2).

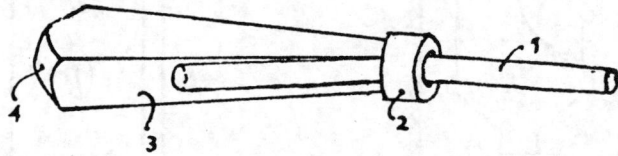

Fig. 6.18

When the whipstock is lowered into the bore hole, the pointed end (4) sticks to the rock and the drill rods (1) are deflected by the semicircular groove (3).

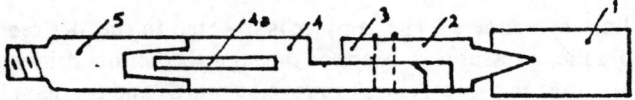

Fig. 6.19

There is another method of deflecting diamond bore holes (in hard rock). In this method (Fig. 6.19) a cylindrical wood plug is first pushed into the bore hole by means of the drill rod, to the desired depth. The plug becomes tightly fixed by soaking and expanding. There is a V-shaped notch on the top of the plug.

In the second stage, a guide wedge (2), to which is rivetted (3) the clinometer holder (4), with the clinometer (4a), is attached to the drill rods (5). The guide wedge (2) fits into the V-notch in the wooden plug (1).

At the third stage the clinometer, after $\frac{1}{2}$ an hour is withdrawn. This is accomplished by sharply driving the rods, so as to shear the copper rivets (3). Thus the guide wedge is left in the bore hole.

In the fourth stage, the deflecting wedge (Fig. 6.20) (7) with the circular median groove (8) and collar (9) is attached by a special stub or

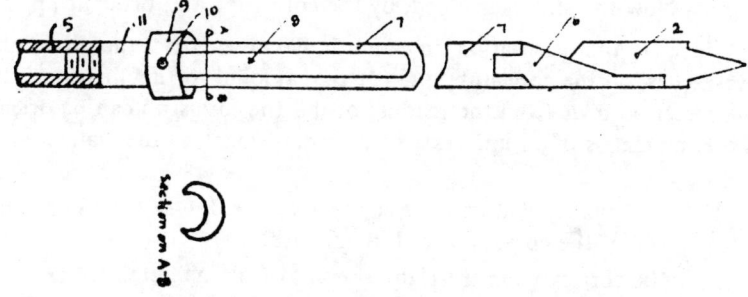

Fig. 6.20

connector by copper rivet (10). The cross section of the deflecting wedge (7) is crescent shaped (X). To the lower end of the deflecting wedge (7) is attached a pilot piece (6).

In the fifth stage of the operation the drill rods with the deflecting wedge and pilot piece are lowered into the hole. The drill rods are rotated so that the flat face of pilot piece (6) comes against the flat face of the guide wedge (2) which is already in the hole. Now the copper rivet (10) is sheared off by a sharp blow to the rods and the rods are withdrawn from the hole.

In the sixth stage the drill rods with a reaming bit are lowered and the opening in the collar (9) of the wedge is enlarged.

In the seventh stage the normal coring bits are attached and drilling is started. The hole is deflected and coring proceeds.

The operation in general is slow and time consuming. It occupies about five shifts, and the deflection is only 1.5° to 2° per wedge.

*Method of orienting cores* :  In diamond drilling, it is often necessary to know the orientation of the cores in space.  The following method can be used with advantage : (also see bore hole surveying)

1.  A short stub or stick of core is drilled, without the use of a core clip.

2.  The rods are removed leaving the stub in place.

3.  Light aluminium rods fitted with core clip, and scratching or marking attachment, together with HF, inclinometer, are lowered.

4.  The scratcher or marking device. makes two shallow notches diametrically opposite to each other, on the core stub. Since the position of the scratches is known, the alignment of these can be fixed, on the surface, when the scratcher is at the bottom of the hole.  The inclination of the hole is recorded, by the H.F. meniscus, in the inclinometer.

5.  Now the stub is gripped, by the core clip, and brought up.

Since (1) the orientation, in azimuth, is known from the direction of the scratches, on the core stub, and (2) the amount of inclination is also inclinometer, known from the reading of the the core stub can be oriented in space, by means of an universal stage arrangement, in the manner shown below :

1.  The plane, containing the scratches **AB, CD** found on either side of the core, is placed in azimuth.

2.  The core is given the right amount of tilt, as obtained from the inclinometer, keeping the plane A B C D in the same azimuth.

3.  Keeping the core in this position, the true dip, as well as the apparent dips, and strike of the bedding planes, or banding etc. can be measured.

Survey of bore hole cavities—*e.g.*, solution cavities, is also done. An accoustical method was developed for this purpose, but it was found that the interpretation, of the result was rather complex, and the caliper logs were more easily obtained.

In the optical method, a narrow beam of light is projected horizontally from the apparatus attached to the drill rods, and this is reflected by a plane mirror, which is rotated about an axis normal to the beam of light. The reflected light from the walls of the cavity is focussed by a lens on to a photo cell or photo tube.  The angle of rotation of the mirror gives the measure of the distance (of the equipment) from the cavity wall.  When a number of such measurements are obtained, the exact configuration, of the cavity, can be made out.

*Data collection and sampling from bore holes* : While drilling, the geologist is expected to visit the site frequently, not only for sampling but also to advise the driller, as to what depth drilling should be continued, as to the nature of the rock, hard or soft etc. The driller on his part records (1) rate of drilling, (2) nature and colour of the sludge as well as that of the sediment coming out of the drill hole (3) quantity of water used *i.e.*. whether water seepage or water loss through fractured zones was observed, (4) length of the run, (5) length of core recovered and the length of core lost with its depth, (6) type of tools used *i.e.*, NX, BX, AX or EX upto what depth, etc. In addition the logs should contain certain general information.

1.  Location of bore hole, *i.e.*, co-ordinates, distance and direction from the nearest village.

2.  Purpose of drilling, *i.e.*, name of project.

3.  Angle at which the hole is drilled, in the case of an inclined hole, as well as the bearing in azimuth.

4.  Elevation of the collar (RL).

5.  The remarks column should contain particulars of any troubles encountered and causes for delays.

In siting or locating bore holes, the geologist should not only take into the consideration the geology and structure of the area, but also certain pre-requisites, for making drilling possible and convenient, *e.g.*, topography, water supply, accessibility, including communication, transport supplies, etc., are important.

Where the core loss is excessive or the recovery is poor, arrangement should be made to recover the sludge. Settling tanks of stone or earth excavations lined with water proof bitumenised fabric or plastic sheet may be used. Lime or sodium silicate is also used for rapid settling of the sediment suspended in the sludge.

In certain cases of doubt (where the core is lost), geophysical methods (electric) may also be employed for logging the bore hole, even though samples are not available for analysis. The methods commonly employed are (1) resistivity logging (2) S.P. logging (self-Potential) and (3) Gamma ray logging. These have been described. (See Geophysics)

### Accuracy of bore hole sampling

Where drilling is employed, contamination of the sample by sludge and extraneous materials, is to be avoided. Hence (*a*) the bore should be cased as drilling proceeds, (*b*) washing out of the sludge should be thorough, at every step. Even though this may not be possible, adherence to these rules is essential.

The sample should represent both the core and the sludge in the correct proportions. Thus the core is weighed and the total amount of sludge is also weighed, and aliquot parts of each are taken as samples. Say, if the core : sludge is 1 : 2 then, the ratio should be the same in the sample. It is also to be noted that core can be obtained easily from hard compact ore, while a fiable ore is represented mainly in the sludge. This condition calls for combined assay of both the core and sludge.

Combined assay value of core and sludge $(C_A)$ from diamond drills can be calculated.

A method for obtaining the average assay value of the core and sludge where loss of core is encountered, is given below :

$V_1 =$ is the volume of core in c.cs.

$V_2 =$　　,,　　,,　sludge in c.cs.

$A_1 =$ Assay value of core.

$A_2 =$　　,,　　,,　of sludge.

$V_1 + V_2 = V$ the volume in c.cs. of the hole drilled.

$A =$ Average assay of core+sludge.

$$A = \frac{A_1 V_1 + A_2 V_2}{V} = A_1 \times \frac{V_1}{V} + A_2 \times \frac{V_2}{V} \qquad \ldots\ldots\ldots(1)$$

If $D/2$ is the radius of the hole in cms.

$D_1/2$　　,,　　,,　　,,　core in cms.

$L$ is the length of core in cms.

$L_1$　,,　　,,　　,,　of run drilled in cms.

Then,

$$\frac{V_1}{V} = L \times \frac{D^2}{4} \times \pi \div L_1 \times \frac{D_1^2}{4} \times \pi = \frac{L}{L_1} \times \frac{D_1^2}{D^2}$$

$$\frac{V_2}{V} = \frac{L_1 \frac{D^2}{4} \times \pi - \frac{LD_1^2 \times \pi}{4}}{L_1 D^2/4\pi} = 1 - \frac{L}{L_1} \times \frac{D_1^2}{D^2}$$

$$A = A_1 \times \frac{L}{L_1} \times \frac{D_1^2}{D^2} + A_2 \times \left( 1 - \frac{L}{L_1} \times \frac{D_1^2}{D^2} \right)$$

$$= A_1 \frac{L}{L_1} \times \frac{D_1^2}{D^2} + A_2 - A_2 \left( \frac{L}{L_1} \times \frac{D_1^2}{D^2} \right)$$

$$= (A_1 - A_2) \left( \frac{L}{L_1} \times \frac{D_1^2}{D^2} \right) + A_2$$

Note :—For general calculation of the diameter of the drill hole 1/16 inch is added to the outer diameter of the shot bit (in case of Calyx drill) and 0.02 inch is added to the outer diameter of diamond drill bits.

## Method of Calculating Assay Values from Diamond Drilling
## Standard Sizes of Core Barrel Bits
### Table 6.3

| | Diameter of bit | | Diameter | |
| --- | --- | --- | --- | --- |
| | Inside | Outside | Hole | Core |
| EX | $^7/_8$ | $1^7/_{16}$ | $1\frac{1}{2}$ | $^7/_8$ |
| AX | $1^7/_{32}$ | $1^{27}/_{32}$ | $1^7/_8$ | $1^1/_8$ |
| BX | $1^{11}/_{16}$ | $2^5/_{16}$ | $2^3/_8$ | $1^5/_8$ |
| NX | $2^3/_{16}$ | $2^{15}/_{16}$ | $3$ | $2^1/_3$ |

Samples obtained from the bore holes will be (1) in the form of solid core, (2) in the form of a slurry or sludge, from which the sediment can be recovered. The sludge represents that part of the rock which has been cut and ground, by the drilling bit, or diamond bit. It also includes that part of the core which has been lost by grinding and abrasion between fractured pieces of core or due to washing out.

Weightage, for combination assay values of core and sludge can be determined either by taking into account (a) the nominal diameters of the core and hole (see table above) or by using (b) either the measured volumes or weights of core and sludge.

If the weight and/or volume of the solid core is known, then the specific gravity can also be determined. The weight of sludge can be determined after drying the sediment. Where some portion of the core is lost, it becomes necessary to assign the appropriate weightage, to the solid core and the sludge, when computing assay values. If 100% of the core is recovered, then no computation is required for obtaining assay values. Where losses occur then (1) % core recovery is calculated, (2) the sludge recovered is also calculated, in respect of the volume of the annular space cut out by the coring bit, taking into account the length of the run.

### Methods

Problem : The data for AX bits as obtained from tables provided by the bit manufactures is as follows :

| | Length | Weight |
| --- | --- | --- |
| Core | 1.7 | 2.1 |
| Sludge | | 10.8 |

### Table 6.4
The following data are known for AX bore holes

| Diameter | Area | Vol./ft. length | Wt./ft. length | % of Vol. of hole |
|---|---|---|---|---|
| Hole 1⅞″=1.875″ | 2.76 sq. in. | 0.0192 C.ft. | 3.15 lb. | 100 |
| Core 1⅛″=1.125″ | 0.99 sq. in. | 0.0069 C.ft. | 1.13 lb. | 36 |
| Sludge OD=1.875″ | 1.77 sq. in. | 0.0123 C.ft. | 2.02 lb. | 64 |
| Annular Ring ID=1.125″ | | | | |

*Method* :—1. Disregarding the incomplete recovery of core and sludge. Weightage may be assigned for volumes, in computing the assay value.

|  | % Vol. of hole | Assay % | Assay product |
|---|---|---|---|
| Core | 36 | 0.50 | 18.0 |
| Sludge | 64 | 1.25 | 80.0 |
|  | 100 | | 98.0 |

Average assay is $\dfrac{98.0}{100}$=0.98 % or 0.98% Cu.

This method is satisfactory where the recovery of both core and sludge is high or complete.

*Method* :—2. If weightage is given for the vol. of core lost; since only 1.7 ft. of core is recovered from 7.5 ft. of drilling, *i.e.*, 10.41% of the vol. of the hole, while the sludge represents $100-10.41=89.6\%$ of the vol. of the hole, drilled, then, the computations will be modified accordingly.

Vol./ft. length of AX drill hole     (1) 0.0192 c.ft. (from table)
Vol./ft. length of core AX     (2) 0.0069 c.ft. ( ,,    ,, )
Vol./ft. length of annular space     (3) 0.0123 c.ft. ( ,,    ,, )
∴ Volume of 1.7 ft. of drill core is $1.7 \times 0.0069 = 0.01173$ c.ft.
   Volume of drill hole is $5.8 \times 0.0192 = 0.11136$ c.ft.
   Volume of 1.7 ft. of annular space is $1.7 \times 0.0123 = 0.02091$ c.ft.

∴ Vol. % of the drill core alone = $\dfrac{0.0117}{0.144} \times 100 = 8.13\%$

∴ Weight of sludge = 91.87%

|       | Vol. of hole | Assay % | Assay product |
|-------|--------------|---------|---------------|
| Core  | 8.13         | 0.50    | 4.065         |
| Sludge| 91.87        | 1.25    | 114.838       |
|       | 100.0        |         | 118.903       |

Average assay is $\dfrac{117.25}{110} = 1.189\%$

*Method* :—3. When weight, of core recovered, is to be assigned weightage, in respect of the weights of the core lost and sludge gained, then the calculations are as follows :

Wt./ft. length of (rock sp. gr.=2.63) AX drill core O.D.=3.15 lb.
Wt./ft. length of     ,,    ,,    ,,    ,,    ,,   I.D.=1.13 lb.
Wt./ft. length of     ,,    ,,    ,,      annular ring=2.02 lb.
Wt. of 1.7 ft. of AX core $1.13 \times 1.7 = 1.921$ lb.
  ,,  (7.5 — 1.7) = 5.8 ft. of AX core lost $5.8 \times 3.15 = 18.270$
  ,,    1.7 ft. of AX annular ring lost as sludge $1.9 \times 2.02 = 3.434$ lb.
∴ total core found as sludge $= 21.71$ lb.
(O.D = outer diameter, and I.D. = inner diameter)
∴ Wt.% of the drill core is $= \dfrac{1.921}{21\cdot7} \times 100 = 8.853$

|                     | Weight | Assay % | Assay product |
|---------------------|--------|---------|---------------|
| Core alone          | 1.92   | 0.50    | 0.96          |
| Sludge + core lost  | $\dfrac{21.70}{26.95}$ | 1.25 | $\dfrac{27.125}{28.09}$ |

Average assay $= \dfrac{28.09}{23.62} = 1.189\%$

*Method* :—4. Where weightage is assigned to volume after allowing for losses in core recovery and as well as sludge :

| Vol. (V) | % Vol. recovered VR | V × VR=C | Assay Value—C | Assay product C × A |
|----------|---------------------|----------|---------------|---------------------|
| Core   | 10.3  | 100 | 10.3 | 0.50 | 5.2  |
| Sludge | 89.7  | 50  | 49.8 | 1.25 | 62.2 |
|        | 100.0 |     | 60.1 |      | 67.4 |

Average assay $= \dfrac{67.4}{60.1} = 1.12\%$ Cu.

Computations involving weightage of drill hole results :

1.   Where the drill holes are placed in the regular square grid system the computation of reserves is relatively simple. The assay foot is computed for each of the 4 bore holes at the corner of a square. The average assay foot value is obtained. The average assay volume is obtained from this.

2.   If the drill holes are irregularly placed then the formation of triangles is resorted to. The procedure adopted in such cases, is as follows :

(a)   where the drill holes are at the corners of an equilateral triangle then the weightage for each is the same (since the angles are equal) ;

(b)   where the drill hole is at the corner of an obtuse angled triangle the weightage is relative to the size of the angle.

Thus if the assay foot value for a drill hole is Y and the angle at one of the corners of the triangle is $\theta$,

then the corrected assay foot value A is

$$Y \times \frac{\theta^\circ}{60^\circ} = A.$$

## BORE HOLE PROBLEMS

1.   When three bore holes and their relative levels are available the resolution of the dip is simple. Where data for only two bore holes, one vertical and the other inclined are available then alternative solutions are possible, by using the general formula given below :

$$\text{Cos } A = \tan B . \tan C + \frac{(\text{Sin } i)}{(\text{Cos } B \text{ Cos } C)} .$$

Where A = the angle between the bearing of the inclined hole and dip direction of the bedding,

B = inclination of the bore hole from the horizontal,

C = dip of bedding as seen in the vertical hole,

i = angle of bedding as seen in the inclined hole.

It will be seen that A can have two possible values ; i.e., clockwise or anticlockwise, from the azimuth of the inclined hole. Further A can be either positive or negative in the equation, but only one of these will satisfy the equation, the other being imaginary.

2.   In the case, where three vertical bore holes intersect the lode or band or vein, the solution given below can be applied (Fig. 6.21).

Let the bore holes 1, 2 and 3 whose reduced levels are known, intersect the sheet like ore body at X, Z and Y respectively. If X be the

deepest intersection, draw X A parallel to the strike, to meet ZY at A.
Let CXA be the horizontal plane passing through XA. The angle of dip
is ZDC=δ

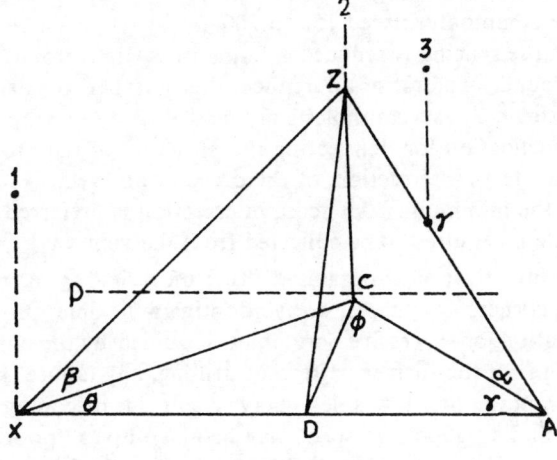

Fig. 6.21

The apparent dip of ZX= $\angle$ ZXC=β

The apparent dip of ZY or ZA= $\angle$ ZAC=α

The true dip is on ZD= $\angle$ ZAC=δ

In figure, the true dip= $\angle$ ZDC

Hence if we can determine ZC and DC this will give the value of tan δ
Thus, the value of the $\angle$ ZDC can be obtained by reference to the tables. In
XCZ, which is a right angled triangle, the value of β the apparent dip in
the direction of ZX is known. Hence, ZC can be determined. So also, in
XCD which is also a right angled. $\angle$ DCX can be calculated, since it is
compliment to the $\angle$ PCX which is the apparent dip on ZX. Hence, DC can be
calculated.

Thus the value of tan δ is obtained from which the value of angle δ is
got.

## II method

Tan True dip=Tan apparent dip×secant of the angle between the
apparent dip direction and the direction of the true dip ( φ )

of Tan $= \dfrac{ZC}{DC}$, Tan apparent dip ( α )$= \dfrac{ZC}{CA}$

secant of the angle between apparent dip direction and true dip direction$= \dfrac{CA}{DC}$

Tan α sec φ $= \dfrac{ZC}{CA} \dfrac{CA}{DC} = \dfrac{ZC}{DC} =$ tan

## Drilling Programme

In deposits of uniform grade, *e.g.*, "prophyry" coppor deposits, churn drill holes are commonly spaced 150 to 200 ft. apart. Where mineralisation is erratic, the spacing is reduced. Sometimes it is found expedient, as in underground exploration to reduce the number of drill hole sites. Thus from a single site several holes may be drilled by varying the azimuth and angle of inclination for delineating the structure of the ore shoot. In any case too acute an intersection, of the ore shoot or vein, is to be avoided, 30° being the minimum. An acute intersection is preferred only when a larger sample is required to be collected from the vein.

Drilling investigation is organised either on a rigid grid pattern, *i.e.*, drilling at the corners of squares, or by adopting a flexible programme, in which the location, of successive bore holes, is guided by the current results of the drilling. In the former pattern of drilling, any failure to intersect the ore shoots, in the first few holes, may result in discouraging further exploration ; and a "good" prospect may be given up as "poor". Even so the rigid pattern affords scope for the deployment, of a larger number of rigs, at a time, and also for utilising rigs according to capacity. In the "progressive development" scheme, relatively larger amount of data can be obtained, with lesser footage drilled. A happy mean, between the two schemes will more often be the most advantageous.

When the controls of mineralisation are clear and enough data is available, then possible locations, of bore holes, can be made as can be seen from the example given in Fig. 6.22. The drill holes A and B are on the plane of ore body or vein.

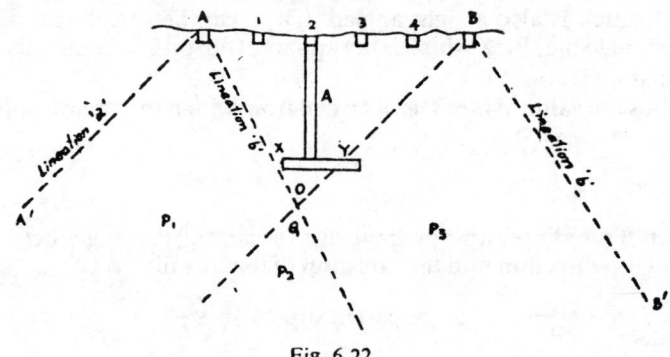

Fig. 6.22

1, 2, 3, 4. etc. are test pits. Points A and B mark the limits of mineralisation, as seen on the surface. The drive, XY, is within the lode. From the amount and direction of plunge of lineations, as seen in the test pits and in the workings etc., the lode may be expected to occur within the area XYQ. The limiting positions AA′ and BB′ can also be obtained by the study of the lineations. Drilling, on the side of Y, may be done for exploring mineralisation, controlled by the other direction of plunge BB′ of the lineations. Drilling may also be done to intersect the ore body at depth, near $P_1$ $P_2$ and $P_3$.

Contouring with reference to an inclined datum is shown in the Fig. 6.23. If oriented drilling is done, *e.g.*, if holes are directed at right angles to the assumed plane of the ore shoots, then the "profiles" may be obtained for several cross sections. XY, X′ Y′, and X″ Y″, are sections of the reference plane, *e.g.*, a bed or marker band. $B_1$, $B_2$, $B_3$ are the points of intersection of the vein, and $R_1$, $R_2$, $R_3$ etc. are points of intersection of the reference plane, in the same bore holes respectively. Thus reduced levels of the top or bottom surface of the vein can be worked out with reference to the marker plane. If the same procedure is applied, to the data obtained from a number of sections, contouring of the surfaces can be done. An iso-pach map of the ore body can also be prepared, if contouring of the upper and lower surfaces of the ore body is carried out.

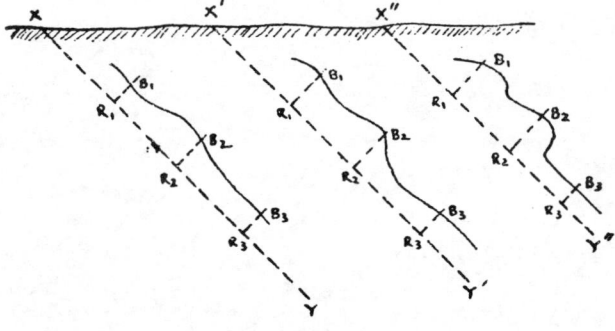

Fig. 6.23

It may be remembered that in a large scale drilling programme, it may not be very necessary to recover cores from each and every drill hole. Some of the holes may be used only for checking the presence of the ore body or coal seam, while others may be required for obtaining cores for sampling purposes. Since core recovery is costly operation, considerable discretion should be exercised in this regard. In prospecting for coal, it is often desirable to evoid coring too frequently. The project can be completed expeditiously, if coring is done in the minimum number of holes.  For the large number of holes, non-coring bits can be employed or rotary methods employed or even a churn drill will do.  During the first phase, a rough strata log is prepared from the nature of the sludge. After this geophysical methods especially, two channel resistivity—S P. logger, is employed to confirm the earlier lithological logging done by means of sludge.  This method of bore hole logging is described under geophysical prospecting methods.

# TABLE 1

## Core to Sludge Ratios in Diamond Drilling 1 to 50% Recovery

Percentage of Volumes of Core and Cuttings Obtained and the Relation to the Percentage of Core Recovered.
(To be Used in Computing Weighted Analyses of Core and Cuttings.)

| Per Cent of Core Recovery | EX | | AX | | BX | | NX | | Per Cent of Core Recovery |
|---|---|---|---|---|---|---|---|---|---|
| | Core | Cuttings | Core | Cuttings | Core | Cuttings | Core | Cuttings | |
| 1 | 0·4 | 99·6 | 0·4 | 99·6 | 0·5 | 99·5 | 0·5 | 99·5 | 1 |
| 2 | 0·7 | 99·3 | 0·7 | 99·3 | 1·0 | 99·0 | 1·0 | 99·0 | 2 |
| 3 | 1·1 | 98·9 | 1·1 | 98·9 | 1·4 | 98·6 | 1·5 | 98·5 | 3 |
| 4 | 1·4 | 98·6 | 1·4 | 98·6 | 1·9 | 98·1 | 2·0 | 98·0 | 4 |
| 5 | 1·8 | 98·2 | 1·8 | 98·2 | 2·4 | 97·6 | 2·6 | 97·4 | 5 |
| 6 | 2·1 | 97·9 | 2·2 | 97·8 | 2·9 | 97·1 | 3·1 | 96·9 | 6 |
| 7 | 2·5 | 97·6 | 2·5 | 97·5 | 3·4 | 96·6 | 3·6 | 96·4 | 7 |
| 8 | 2·8 | 97·2 | 2·9 | 97·1 | 3·8 | 96·2 | 4·1 | 95·9 | 8 |
| 9 | 3·2 | 96·8 | 3·2 | 96·8 | 4·3 | 95·7 | 4·6 | 95·4 | 9 |
| 10 | 3·5 | 96·5 | 3·6 | 96·4 | 4·8 | 95·2 | 5·1 | 94·9 | 10 |
| 11 | 3·9 | 96·1 | 4·0 | 96·0 | 5·3 | 94·7 | 5·6 | 94·4 | 11 |
| 12 | 4·3 | 95·7 | 4·3 | 95·7 | 5·8 | 94·2 | 6·1 | 93·9 | 12 |
| 13 | 4·6 | 95·7 | 4·7 | 95·3 | 6·3 | 93·7 | 6·7 | 93·3 | 13 |
| 14 | 5·0 | 95·0 | 5·0 | 95·0 | 6·7 | 93·3 | 7·2 | 92·8 | 14 |
| 15 | 5·3 | 94·7 | 5·4 | 94·6 | 7·2 | 92·8 | 7·7 | 92·3 | 15 |

| | | | | | | | | | |
|---|---|---|---|---|---|---|---|---|---|
| 16 | 5·7 | 94·3 | 5·8 | 94·2 | 7·7 | 92·3 | 8·2 | 91·8 | 16 |
| 17 | 6·0 | 94·0 | 6·1 | 93·9 | 8·2 | 91·8 | 8·7 | 91·3 | 17 |
| 18 | 6·4 | 93·6 | 6·5 | 93·5 | 8·7 | 91·3 | 9·2 | 90·8 | 18 |
| 19 | 6·7 | 93·3 | 6·8 | 93·2 | 9·1 | 90·9 | 9·7 | 90·3 | 19 |
| 20 | 7·1 | 92·9 | 7·2 | 92·8 | 9·6 | 90·4 | 10·2 | 89·8 | 20 |
| 21 | 7·5 | 92·5 | 7·6 | 92·4 | 10·0 | 89·9 | 10·8 | 89·2 | 21 |
| 22 | 7·8 | 92·2 | 7·9 | 92·1 | 10·6 | 89·4 | 11·3 | 88·7 | 22 |
| 23 | 8·2 | 91·8 | 8·3 | 91·7 | 11·1 | 88·9 | 11·8 | 88·2 | 23 |
| 24 | 8·5 | 91·5 | 8·6 | 91·4 | 11·5 | 88·5 | 12·3 | 87·7 | 24 |
| 25 | 8·9 | 91·1 | 9·0 | 91·0 | 12·0 | 88·0 | 12·8 | 87·2 | 25 |
| 26 | 9·2 | 90·8 | 9·4 | 90·6 | 12·5 | 87·5 | 13·3 | 86·7 | 26 |
| 27 | 9·6 | 90·4 | 9·7 | 90·3 | 13·0 | 87·0 | 13·8 | 86·2 | 27 |
| 28 | 9·9 | 90·1 | 10·1 | 89·9 | 13·5 | 86·5 | 14·3 | 85·7 | 28 |
| 29 | 10·3 | 89·7 | 10·4 | 89·6 | 13·9 | 86·1 | 14·8 | 85·2 | 29 |
| 30 | 10·7 | 89·3 | 10·8 | 89·2 | 14·4 | 85·6 | 15·4 | 84·6 | 30 |
| 31 | 11·0 | 89·0 | 11·2 | 88·8 | 14·9 | 85·1 | 15·9 | 84·1 | 31 |
| 32 | 11·4 | 88·6 | 11·5 | 88·5 | 15·4 | 84·8 | 16·4 | 83·6 | 32 |
| 33 | 11·7 | 88·3 | 11·9 | 88·1 | 15·9 | 84·1 | 16·9 | 83·1 | 33 |
| 34 | 12·1 | 87·9 | 12·2 | 87·8 | 16·4 | 83·6 | 17·4 | 82·6 | 34 |
| 35 | 12·4 | 87·6 | 12·6 | 87·4 | 16·8 | 83·2 | 17·9 | 82·1 | 35 |

(*Continued*)

| | | | | | | | | | |
|---|---|---|---|---|---|---|---|---|---|
| 36 | 12·8 | 87·2 | 13·0 | 87·0 | 17·3 | 82·7 | 18·4 | 81·6 | 36 |
| 37 | 13·1 | 86·9 | 13·3 | 86·7 | 17·8 | 82·2 | 18·9 | 81·1 | 37 |
| 38 | 13·5 | 86·5 | 13·7 | 86·3 | 18·3 | 81·7 | 19·5 | 80·5 | 38 |
| 39 | 13·8 | 86·2 | 14·0 | 86·0 | 18·8 | 81·2 | 20·0 | 80·0 | 39 |
| 40 | 14·2 | 85·8 | 14·4 | 85·6 | 19·2 | 80·8 | 20·5 | 79·5 | 40 |

| | | | | | | | | | |
|---|---|---|---|---|---|---|---|---|---|
| 41 | 14·6 | 85·4 | 14·8 | 85·2 | 19·7 | 80·3 | 21·0 | 79·0 | 41 |
| 42 | 14·9 | 85·1 | 15·1 | 84·9 | 20·2 | 79·8 | 21·5 | 78·5 | 42 |
| 43 | 15·3 | 84·7 | 15·5 | 84·5 | 20·7 | 79·3 | 22·0 | 78·0 | 43 |
| 44 | 15·6 | 84·4 | 15·8 | 84·2 | 21·2 | 78·8 | 22·5 | 77·5 | 44 |
| 45 | 16·0 | 84·0 | 16·2 | 83·8 | 21·6 | 78·4 | 23·0 | 77·0 | 45 |

| | | | | | | | | | |
|---|---|---|---|---|---|---|---|---|---|
| 46 | 16·3 | 83·7 | 16·6 | 83·4 | 22·1 | 77·9 | 23·6 | 76·4 | 46 |
| 47 | 16·7 | 83·3 | 16·9 | 83·1 | 22·6 | 77·4 | 24·1 | 75·9 | 47 |
| 48 | 17·0 | 83·0 | 17·3 | 82·7 | 23·1 | 76·9 | 24·6 | 75·4 | 48 |
| 49 | 17·4 | 82·6 | 17·6 | 82·4 | 23·6 | 76·4 | 25·1 | 74·9 | 49 |
| 50 | 17·7 | 82·3 | 18·0 | 82·0 | 24·1 | 75·9 | 25·6 | 74·4 | 50 |

Example : A X bit used for drilling in iron ore gave a core recovery of 45%. Analysis of core, gave 58% Fc, of cuttings, 60% Fc.

Hence weighted assay $= \dfrac{58.00 \times 16.2 + 60.00 \times 83.8}{100} = 59.68\%$ Fe.

## TABLE 2
### Core to Sludge Ratios in Core Drill Samples 51 to 100% Recovery
Percentage of Volumes of Core and Cuttings in a Core Drill Sample for Each Per Cent of Core Recovered, (To be Used as Multipliers in Combining Analyses of Core and Cuttings.)

| Per Cent of Core Recovery | EX | | AX | | BX | | NX | | Per Cent of Core Recovery |
|---|---|---|---|---|---|---|---|---|---|
| | Core | Cuttings | Core | Cuttings | Core | Cuttings | Core | Cuttings | |
| 51 | 18·1 | 81·9 | 18·4 | 81·6 | 24·5 | 75·5 | 26 | 73·9 | 51 |
| 52 | 18·5 | 81·5 | 18·7 | 81·3 | 25·0 | 75·0 | 26 | 73·4 | 52 |
| 53 | 18·8 | 81·2 | 19·1 | 80·9 | 25·5 | 74·5 | 27 | 72·9 | 53 |
| 54 | 19·2 | 80·8 | 19·4 | 80·6 | 26·0 | 74·0 | 27 | 72·4 | 54 |
| 55 | 19·5 | 80·5 | 19·8 | 80·2 | 26·5 | 73·5 | 28 | 71·8 | 55 |
| 56 | 19·9 | 80·1 | 20·2 | 79·8 | 26·9 | 73·1 | 28 | 71·3 | 56 |
| 57 | 20·2 | 79·8 | 20·5 | 79·5 | 27·4 | 72·6 | 29 | 70·8 | 57 |
| 58 | 20·6 | 79·4 | 20·9 | 79·1 | 27·9 | 72·1 | 29 | 70·3 | 58 |
| 59 | 20·9 | 79·1 | 21·2 | 78·8 | 28·4 | 71·6 | 30 | 69·8 | 59 |
| 60 | 21·3 | 78·7 | 21·6 | 78·4 | 28·9 | 71·1 | 30 | 69·3 | 60 |
| 61 | 21·7 | 78·3 | 22·0 | 78·0 | 29·3 | 70·7 | 31 | 68·8 | 61 |
| 62 | 22·0 | 78·0 | 22·3 | 77·7 | 29·8 | 70·2 | 31 | 68·3 | 62 |
| 63 | 22·4 | 77·6 | 22·7 | 77·3 | 30·3 | 69·7 | 32 | 67·7 | 63 |
| 64 | 22·7 | 77·3 | 23·0 | 77·0 | 30·8 | 69·2 | 32 | 67·2 | 64 |
| 65 | 23·1 | 76·9 | 23·4 | 76·6 | 31·3 | 68·7 | 33 | 66·7 | 65 |

| | | | | | | | | | |
|---|---|---|---|---|---|---|---|---|---|
| 66 | 23·4 | 76·6 | 23·8 | 76·2 | 31·7 | 68·3 | 33 | 66·2 | 66 |
| 67 | 23·8 | 76·2 | 24·1 | 75·9 | 32·2 | 67·8 | 34 | 65·7 | 67 |
| 68 | 24·1 | 75·9 | 24·5 | 75·5 | 32·7 | 67·3 | 34 | 65·2 | 68 |
| 69 | 24·5 | 75·5 | 24·8 | 75·2 | 33·2 | 66·8 | 35 | 64·7 | 69 |
| 70 | 24·9 | 75·1 | 25·2 | 74·8 | 33·7 | 66·3 | 35 | 64·2 | 70 |
| 71 | 25·2 | 74·8 | 25·6 | 74·4 | 34·2 | 65·8 | 36·4 | 63·6 | 71 |
| 72 | 25·6 | 74·4 | 25·9 | 74·1 | 34·6 | 65·4 | 36·9 | 63·1 | 72 |
| 73 | 25·9 | 74·1 | 26·3 | 73·7 | 35·1 | 64·9 | 37·4 | 62·6 | 73 |
| 74 | 26·3 | 73·7 | 26·6 | 73·4 | 35·6 | 64·4 | 37·9 | 62·1 | 74 |
| 75 | 26·6 | 73·4 | 27·0 | 73·0 | 36·1 | 63·9 | 38·4 | 61·6 | 75 |
| 76 | 27·0 | 73·0 | 27·4 | 72·6 | 36·6 | 63·4 | 38·9 | 61·1 | 76 |
| 77 | 27·3 | 72·7 | 27·7 | 72·3 | 37·0 | 63·0 | 39·4 | 60·6 | 77 |
| 78 | 27·7 | 72·3 | 28·1 | 71·9 | 37·5 | 62·5 | 39·9 | 60·1 | 78 |
| 79 | 28·0 | 72·0 | 28·4 | 71·6 | 38·0 | 62·0 | 40·4 | 59·6 | 79 |
| 80 | 28·4 | 71·6 | 28·8 | 71·2 | 38·5 | 61·5 | 41·0 | 59·0 | 80 |
| 81 | 28·8 | 71·2 | 29·2 | 70·8 | 39·0 | 61·0 | 41·5 | 58·5 | 81 |
| 82 | 29·1 | 70·9 | 29·5 | 70·5 | 39·4 | 60·6 | 42·0 | 58·0 | 82 |
| 83 | 29·5 | 70·5 | 29·9 | 70·1 | 39·9 | 60·1 | 42·5 | 57·5 | 83 |
| 84 | 29·8 | 70·2 | 30·2 | 69·8 | 40·4 | 59·6 | 43·0 | 57·0 | 84 |
| 85 | 30·2 | 69·8 | 30·6 | 69·4 | 40·9 | 59·1 | 43·5 | 56·5 | 85 |

(Continued)

| | | | | | | | | | |
|---|---|---|---|---|---|---|---|---|---|
| 86 | 30·5 | 69·5 | 31·0 | 69·0 | 41·4 | 58·6 | 44·0 | 56·0 | 86 |
| 87 | 30·9 | 69·1 | 31·3 | 68·7 | 41·8 | 58·2 | 44·5 | 55·5 | 87 |
| 88 | 31·2 | 68·8 | 31·7 | 68·3 | 42·3 | 57·7 | 45·1 | 54·9 | 88 |
| 89 | 31·6 | 68·4 | 32·0 | 68·0 | 42·8 | 57·2 | 45·6 | 54·4 | 89 |
| 90 | 31·9 | 68·1 | 32·4 | 67·6 | 43·3 | 56·7 | 46·1 | 53·9 | 90 |
| 91 | 32·3 | 67·7 | 32·8 | 67·2 | 43·8 | 56·2 | 46·6 | 53·4 | 91 |
| 92 | 32·7 | 67·3 | 33·1 | 66·9 | 44·3 | 55·7 | 47·1 | 52·9 | 92 |
| 93 | 33·0 | 67·0 | 33·5 | 66·5 | 44·7 | 55·3 | 47·6 | 52·4 | 93 |
| 94 | 33·4 | 66·6 | 33·8 | 66·2 | 45·2 | 54·8 | 48·1 | 51·9 | 94 |
| 95 | 33·7 | 66·3 | 34·2 | 65·8 | 45·7 | 54·3 | 48·6 | 51·4 | 95 |
| 96 | 34·1 | 65·9 | 34·6 | 65·4 | 46·2 | 53·8 | 49·2 | 50·8 | 96 |
| 97 | 34·4 | 65·6 | 34·9 | 65·1 | 46·7 | 53·3 | 49·7 | 50·3 | 97 |
| 98 | 34·8 | 65·2 | 35·3 | 64·7 | 47·1 | 52·9 | 50·2 | 49·8 | 98 |
| 99 | 35·1 | 64·9 | 35·6 | 64·4 | 47·6 | 52·4 | 50·7 | 49·3 | 99 |
| 100 | 35·5 | 64·5 | 36·0 | 64·0 | 48·1 | 51·9 | 51·7 | 48·8 | 100 |

## Bore Hole Logging

Drilling is one of the most costly items in mineral exploration. At the existing rates, the cost per meter, of AX hole, will be in the vicinity of Rs. 350/- upto a depth of 200 m. It is hence very essential that complete record of the entire operations be maintained. Taking a diamond drilling operation, two types of logs will have to be maintained (a) Driller's log (b) Geologists' log or Lithological log. The driller's log will pay particular attention to the technical/cost aspects of drilling the operations, such as equipment/tools used, difficulties and set backs, rate of penetration, etc., including a qualitative description of the rocks encountered and cored. The lithological log (geologist's) will include full particulars of the rocks encountered/cored, colour of sludge, length of core recovered, core lost, total drilled in the run, any special feature such as water loss and the depth of such loss—details of mineralisation, structures etc. It will be advantageous if two types of logs are separated. Typical proformae for the two kinds of logs is shown tables 3 and 4.

Where lithological logs are incomplete, or additional information is sought, or where confirmation is considered essential, other logging methods are employed such as (a) electric logging which has been dealt with under geophysical methods, (b) optical/accoustic methods which have been briefly mentioned under bore hole surveying, (c) mechanical methods — of which the caliper method is the most important.

Caliper logging utilises a pair of calipers, which when lowered centrally into the bore hole, gives the variation in the diameter of the bore hole. The movement of the arm of the caliper is amplified by an electrical circuit, so that the reading of the diameter of the bore hole, at a certain depth, is indicated on the dial of a milliammeter etc. in inches/cms. Caliper logs can be correlated with lithological logs if the rock types occurring in the area are known. Correlation with the other types of logs is also possible.

## TABLE 3

Location : State/Distt./Village/(Coordinates)     Project........Rig. No.........

R.L. of bore hole collar :     Type.......Make.....

Date :     Name of geologist........     Serial No. of bore hole : 

    Year of manufacture........

Angle of hole:

Azimuth/direction of inclination :

In general foot by foot or even 30 cm. should be described in the log column 6.

*Note* : In column 7, inch by inch or cm. by cm. should be described.

| 1 | 2 | 3 | | 4 | 5 | 6 | 7 | 8 |
|---|---|---|---|---|---|---|---|---|
| | | Length of core | | Total depth | Colour of sludge | Rock types present | Coal/Minera-lisation | Remarks Type/size of bi water loss visua estimate of Cu/Au/Zn etc. |
| Date | Depth | Recovered | Lost | | | | | |
| | From | To | | | | | | |

## TABLE 4

Location : State/Distt. / Vill./Coordinate          Project........ Rig. No......Engine H.P/RPM

R.L. of Bore hole collar:                            Type........          Make.........

Serial No. of bore hole ........                     Year of manufacture......

Date :                                               Pump :

Name of driller ........

Angle of hole

No. of helpers :   (1. Asstt.

               (2. Rigmen.

               (3. Labour.

Azimuth/direction of inclination

| 1 | 2 | | 3 | | 4 | 5 | 6 | 7 | 8 | | 9 |
|---|---|---|---|---|---|---|---|---|---|---|---|
| | Depth | | Length of core | | Total depth | Return water colour loss or gain | Nature of rock hard, soft etc. Name | Coal minerals | Details of equipment | | Remarks breakdown/ stoppage Amount of P.O.L issued/shifted |
| Date/ time | From | To | Recove- red | Lost | | | | | Bit/ type | Rods Core barrel | |

Note   Every detail is required in Column 6 & 7.  In Column 5 if mud is employed details may be specified.  Three copies o the form will be made :

      (a) for driller's record ;

      (b) for geologist's record ;

      (c) for Accounting/existing.

### Preservation and Sampling of Cores

Sampling of cores is essential for analysis and examination. Material from the core can be subjected to (1) conventional chemical analysis, (2) non destructive chemical analysis, such as the electron probe, etc., (3) other methods, such as atomic absorption analysis, spectroscopic analyses or DTA, or X-ray etc., (4) petrological examination is also necessary, for identifying rocktypes, ore minerals, etc. In the case of coal proximate/ ultimate analysis is necessary, (5) where sedementological studies are to be made, such as grain/shape/etc. sampling can help in lithofacies mapping.

In sampling cores every care should be exercised to see that no loss occurs, on any account. Core is a precious material and it should be treated as such.

Cores, as soon as they are recovered from the core barrels, are placed in core boxes. These are flat wooden boxes, which are normally 10 ft. or

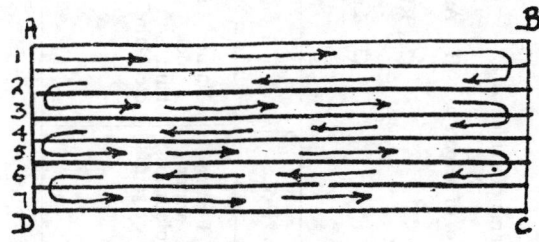

Fig. 6.24

3 m. long and about 3 ft. or a meter wide. The height, of the box, is so adjusted as to accomodate the core and hence it depends on the diameter of the core. The box is divided into a number of logitudinal compartments, by wooden partitions. The width of each partition is also dependent on the diameter of the core. The box is fitted with a lid as well as latch (Fig. 6.24)

In arranging the cores the top, of the core, is placed near A in the box, and the rest of the cores are arranged below in sequence following the direction as indicated by the arrows (Fig. 6.30), in a serpentine fashion, ending at the corner C of the box. As soon as a box is taken, it is numbered and full particulars for its future idendification are painted on the lid as well as one side. In case small pieces of core are placed in the box, the top of the piece is marked by an arrow. The colour of the paint employed for the purpose, will depend on the colour of the rock. Places where the core is missing, wooden blocks indicating the missing length are placed. It is useful to indicate the depth of the cores, in the segments 1, 2, 3, 4......7 along the sides AD and BC of the core box.

For systematic sampling of cores, the rough and ready method of crushing the entire length, of a particular core piece, then coning and quartering the same for getting a sample, is not the most appropriate. It is more scientific to use a core splitter, which is nothing more than a chisel which splits the core longitudinally along the axis (Fig. 6.25).

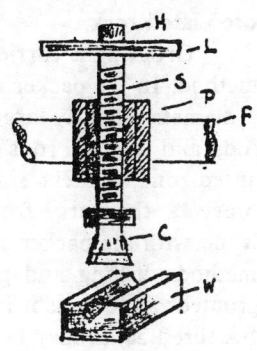

Fig. 6.25

The core splitter comprises the following essential parts.

Hard steel anvil (W) with a V-shaped groove (C) manganese steel chisel, which is attached to a shaft (S) with square threads and it is entailed to the head (H). The shaft (S) is supported by a frame (F) in the vertical position. The shaft can be raised and lowered by the rotation of the wheel (L). The shaft S can be moved by rotating the wheel L, or by pressing the shaft, since it is spring loaded and moves in the sleeve (P).

Operation : The core is placed along the V shaped notch in the anvil (W). The shaft (S) is lowered so that the chisel, C, rests on the core, along the length. A strong sharp blow is dealt, with a hammer, on the shaft head (H). The shaft moves in the sleeve (P) and the chisel strikes the core. The core is thus spilt, along the length.

Once the core is split one half of it is returned to the appropriate place in the core box. The other half is taken up for further examination. Economising, in the amount of sample collected from the spilt core, is necessary. It is a good practice to put by, as much as possible, for further reference.

### Water loss while drilling and its prevention

More often than not whether drilling is carried out in sedimentary or crystalline rocks fractured and pervious zones are often encountered and as a consequence the loss of drilling water may be so excessive as to prevent further work. In the event that the weak zone comprises only of minor fissures or porous material which can stand by itself and require only to be sealed, then ordinary clay or bentonite slurry can be used for preventing water loss. Where larger fissures are suspected then Jute fibres or husk can be added to the clay slurry for plugging the openings. The above remedial measures may not be effective where the rocks are highly fractured or where open joints are present or where drilling is done in cavernous zones. Under these circumstances cement grouting is recommended. Cement is made into a slurry with the consistence of treacle and is pumped into the bore hole. Sufficient time is given for the cement to settle down and set. On the resumption of drilling it will be seen that the first part of the core will consist of hard cement and this will be followed by cemented broken or brecciated rock.

Grouting is carried out either by the packer method or by the stage method. In the packer method tight fitting tough rubber cup washers, with a diameter slightly larger than that of the bore hole are fitted to the drill rods and the drill rods are lowered up to a point just above the fractured zone. Cement slurry is now pumped under pressure and the packer prevents the slurry from leaking. Similarly when another fractured zone is met with the packer is lowered and grouting is carried out. In the stage method drilling and grouting go hand in hand and each weak zone is grouted as soon as it is encountered. In the packer method two or more fractured zones may be grouted at a time if possible.

# PART VII

# ANCILLIARY OPERATION IN PROSPECTING

# ANCILLIARY OPERATIONS IN PROSPECTING

Since during the course of Prospecting operations, especially in the metamorphic and igneous terrain or even in the older sedimentaries, trenching and pitting on a large scale will, at times require the use of explosives, hence, it behoves every mining geologist to be equipped with the knowledge of the more common types of explosives used in blasting, as well as the rules governing their storage and safety. It is not proposed to deal with the latter aspect since this has been published under the explosives rules and regulations by the concerned authority, the Chief Inspector of Explosives, Nagpur and the essentials of the same have been incorporated in the relavant cluases, in the Regulation of the Mines Act, published by the Director General Mines Safety, Government of India. The following paragraphs contain a brief description of the classification and types of explosives.

Explosives are classified under the following types :— (a) Low explosives, (b) High explosives, (c) Sheathed explosives, (d) Permitted explosives, (e) Liquid oxygen, (f) Miscellaneous e.g. AN/FO and slurry types.

An explosive is a substance which has a high rate of combustion. involving an excessive and instantaneous increase in volume of resultant gases. Based on the rate of evolution of gases, explosives are classed as (a) low explosive, (b) high explosive.

Black powder, or gun powder, is the most common low explosive. Its essential constituents are sodium nitrate (Chilean Nitrate), charcoal and sulphur in the following proportions 72 : 16 : 12. Potassium nitrate is used in superior quality gun powder. It is used in quarrying, and also in under-ground mines, where there are no inflammable gases. It is normally used in blasting comparatively weak rocks, or where the extent of shattering is required to be low.

High explosives can be classed into those containing nitroglycerine. Among high explosives, nitroglycerine is well known, and it has been in use for nearly a century. It is formed by mixing sulphuric acid and nitric acid and then adding glycerine—the reaction being strongly exothermic, the vessel containing the mixture has to be cooled, during the process. The excess acid is removed, by washing with water, and neutralising by alkalis. Nitroglycerine is a liquid and as such it is unsafe to handle. Its vapour is poisonous, and is reacts with the skin. It explodes at 392° F and it is sensitive to shock.

Due to these disadvantages, further work led to the discovery of a derivative of nitroglycerine, known as blasting gelatine. This is a jelly like substance obtained by mixing 90-93 parts of nitroglycerine with 10-7 parts of guncotton (nitrocellulose). Blasting gelatine is a stable compound, especially when rendered alkaline, by adding 1% of alkali. It is the most powerful explosive of this group. It is liable to freeze and in this condition it becomes highly sensitive.

Another class of derivatives, from the nitroglycerine base is the dynamite. Dynamite consist of nitroglycerine base, to which is added combustible materials, such as sodium nitrate, wood pulp, or saw dust etc., and these are known as straight dynamites. In straight dynamites, a certain percentage of neutralising agents, such as calcium or magnesium carbonate, is also included in the composition. The so called 50% straight dynamite can contain 50% nitroglycerine, 14% combustibles, 35% sodium nitrate and 1% neutralising agent. Further modification of the series is, the ammonium nitrate class, in which part of the nitroglycerine is replaced by ammonium nitrate.

Since part of the nitroglycerine has been replaced, in these dynamites, by ammonium nitrate, these are comparatively slower in action, but less fumes are produced, and are safer to handle. The deliquescent nature, due to the presence of ammonium nitrate, inhibits the use of this class of explosives, in wet conditions. The cartridges are to be carefully packed in waterproof paper, dipped in paraffin. The range, of these dynamites, is between 15% to 60% strengths.

A family of explosives, known as gelatine dynamites, is derived similarly by adding, sodium nitrate, a combustible substance, and a neutralising agent, to blasting gelatine. Gelatine dynamites, range in strength from 20% to 90%. Some of the permitted explosives, used in coal mines, come under this category. When part of the blasting gelatine, in the gelatine dynamite, is replaced by ammonium nitrate, the explosive is known as ammonium gelatine dynamite. A 50% ammonium gelatine can have 36% blasting gelatine, 20% ammonium nitrate, 33.5% sodium nitrate, 8% carbonaceous material, 0.8% neutralising agent, and 1.4% moisture.

Mining explosives are classified as follows :

1. Gelatines    (a) Blasting Gelatine    92% Nitroglycerine+Guncotton
                (b) Sodium Nitrate
                    Gelignites           30-80%  „        +      „
                                         Sodium Nitrate 28-44%

    (c) Ammonium
        Nitrate Gelignites
        Combustible matter             7-11%
                      Nitroglycerine  25-35%
                      ammonium Nitrate 30-60%

2. Semigela-     Nitroglycerine   12-20%
   tines

               Ammonium Nitrate 65-80%
               Combustible matter 6-9%

3. Nitroglycerine  Nitroglycerine upto 10%
   powders

               Ammonium Nitrate 80-85%

4. Non-nitro-    TNT—15-18%
   glycerine     Ammonium Nitrate 82-85%
   explosives

               Aluminium powder

These explosives are available as powders and also as slurry, i. e., mixed with water.

Chlorate explosives contain about 80% potassium chlorate, and 5% castor oil, with di-nitro-toluene forming the rest. In order to avoid explosions, during manufacture, the chlorate is dipped either in liquid petroleum, nitro-benzene or nitro-napthalene. Nitroglycerine explosives are manufactured by the Imperial Chemical Industries (India) at Gomia in Bihar.

Liquid oxygen explosives, contain liquid oxygen together with charcoal or saw dust. Detonators or electric blasting caps, are required for firing liquid oxygen cartridges. The soaking, of the saw dust or balsa (wood) dust, is done, just before the cartridges are required, as oxygen escapes easily, if the cartridges are exposed to the atmosphere.

For starting the liquid oxygen cartridges, a stick of gelignite or other explosive is used. The liquid oxygen explosives are more popular in open cast workings, where large diameter drill holes are used, for blasting operations. Liquid oxygen evaporates, vigorously and actually boils, when exposed to atmospheric pressures, and ordinary temperatures. It is, hence, stored in specially constructed strong tanks, and conveyed in well insulated containers. It is transported, in thermally insulated tanks or vessels to the work spot. Cartridges of saw dust, packed in jute cloth, are soaked in the liquid oxygen and placed immediately into the bore holes or blast holes.

In India, liquid oxygen is made by two important concerns, Indian

Oxygen and the Asiatic Oxygen, and they have branches in most of the important mining centres. As an explosive, liquid oxygen is used in the iron ore mines, of the Singhbhum district of Bihar, and in large scale open cast coal mines, in Bokaro coalfields of Hazaribagh district of Bihar.

There are in addition special types of explosives, known as sheathed explosives, in which a sheath or casing of magnesium carbonate or lime or of sodium bicarbonate or borax, or calcium fluoride, is placed over the ammonium nitrate, so as to produce a screen of carbon dioxide on exploding. Thus, the chances of igniting, any combustible gases, are reduced. The sheathed explosives also fall under the category of permitted explosives. This type of explosive is more popular, in coal mining, and is largely employed. The ammonium permissibles contain 7% to 15% nitroglycerine, 60-80% ammonia nitrate, the remainder being pith, sawdust, etc. of absorbent materials. "Permissible" or "Permitted" explosives are specially made for use, in gassy coal mines. They include, in their composition, substances containing water of crystallisation, which tends to lower the temperature of the flame, produced by the explosion. Patent names for such explosives are Stonobel and Monobel, Polar Ajax, Viking Drifex etc.

"Cardox" is a safety explosive. This consists of a steel cylinder, filled with liquid carbon dioxide. There is an electric arrangement for igniting. The ignition mixture consists of aluminium powder, charcoal, potassium chlorate. The heat evolved by the combustion, causes the liquid $CO_2$ to be converted to gas, with an enormous increase in pressure. A stopper, on the gas cylinder, is sheared off and the high pressure gas enters the shot hole and causes the bursting. The cardox cylinders are capable of being recharged for re-use. This is a high explosive. "Hydrox" is a patent name for an explosive chemical mixture, mainly comprising ammonium chloride and sodium nitrate contained in a steel shell. An electric ignition arrangement is provided for starting the explosion. The products of combustion are water, nitrogen, and sodium chloride. The hydrox cylinders can be recharged, underground in the mine, and used again.

### ANFO Explosives

Liquid explosives are, now, becoming increasingly popular because of their greater flexibility, in use, and storage. They commonly consist of mixtures of ammonium nitrate, of the fertilizer grade, and some suitable fuel oil, and are popularly known as ANFO explosives.

There are certain features of importance with regard to these explosive mixtures, which help one in understanding their performance. It may be stated at the very outset, that the speed of reaction of an explosive depends to a large extent, on the intimacy of the mixture of the ingredients. This is also the same in the case of a low explosive, such as gun powder. In the case of the ammonium nitrate, and fuel oil explosives, one of the difficulties is to obtain a uniform mixture. Even a reduction of particle size, to—200 mesh, is not sufficient, since the particles of ammonium nitrate are repelled by the fuel oil. Thus the addition of an activating agent is necessary, so that the surface of the ammonium nitrate particles is conditioned, in such a manner, that the molecules of fuel oil can be attached. Agents, such as sodium dinapthyl methane disulphonate, and sodium lauryl sulphate, and sodium dodecyl benzene sulphonate, have been found satisfactory, for this purpose. Only about 0.5%, by weight of the additive, is necessary, for ensuring bonding between the ammonium nitrate and fuel oil. It is also necessary, that the ammonium nitrate and the fuel oil be mixed in the right proportions, so that the explosion is complete, and maximum energy is generated.

One of the main advantages of these explosives is, that they are in the liquid form, and can be piped for distances to the workings, right from the batching plant, to the mining sites, thus saving energy and time lost in transport. Batching and mixing is also carried out at site, where the ammon nitrate (prills), loaded in a truck, are straight away mixed with fuel oil and the additives, and supplied for immediate use. There is, however, a possibility that ammonium nitrate prills may acquire a static electric charge, especially in dry mining conditions.

As apart from the AN/FO (Ammonium Nitrate/Fuel Oil) type of slurry explosives, another type of slurry, containing the following components is also in use. This type of slurry is effective, both as a low as well as high explosive, since it is a matter of only varying the proportions of the components in the mixture. The essential components of the mixture are :

1. Sodium Nitrate 2. TNT. (Trinitro toluene) 3. Aluminium or 4. Hydrocarbon 5. Water 6. Thickener or stabilizer e.g. guar gum/carboxy methyl cellulose.

The percentage of TNT varies from 5% to 40% depending on the strength of the explosive ; but a camposition, corresponding to AN or Sod. Nitrate : TNT : Water as 65.5 : 20 : 14.5, is normal. Where aluminium is used, as the "thermic" agent, instead of fuel oil, it can form upto 40% of the mixture ; but normally, it forms only 1.9 to 20%.

Another variant, of the slurry type explosive, is the NCN (Nitro-carbon-nitrate) type. These are solid hydrocarbon sensitised slurries. They have low strength and low velocities of detonation. The hydrocarbon content is similar to the AN/FO type of explosives.

The choice of an explosive, for mining purposes, is entirely dependent on economic factors, i.e., the total cost involved per cu. yard, or metre of rock broken. In weak rocks, a smaller charge of a high explosive may be used, or a larger charge of a low explosive. The economics will have to decide the choice. Further, the size of material obtained on blasting, or the degree of fragmentation, will depend on the amount of charge, as well as on the nature of the explosive. If it is desired to have a product of small sizes, then, a larger quantity of explosive is used. In hard rocks, like granites, a quick acting explosive is desirable, whereas in tough rocks, like dolerite or diabase, dynamite and blasting gelatine is more effective.

### Equipment and Material Required for Blasting

A. A hole is drilled either by (1) manual method, with jumper bar or (2) by pneumatic drills operated by compressed air or by (3) Rotary auger type electric drills, in coal and soft rocks.

B. Tamping or packing material consists of pellets made with a mixture of clay and sand, with sufficient water. Tamping rod, which is a wooden rod, shod with a copper cap at one end, is employed for pressing the tamping materials into position.

C. Safety fuse or black chord or detonating chord consists, of a thin waterproof jute tubing, which is tightly packed with gun powder. It burns at the rate of 30—40 ft./min. The detonating chord is a similar chord containing nitroglycerine explosives. When ignited these burn fast, so that the length can be adjusted, in order to allow the shot firer sufficient time to get to a place of safety.

The TNT (trinitro toluene) fuse detonates at the rate of 17,500 ft./sec., while the fuse made with (penta erythrite tetranitrate) PETN detonates at 20,035 ft./sec. These fuses are safe to handle, as they do not go off either due to friction, fire, or ordinary shock, yet these are also not in common use

D. Squib is a wick of tough paper, rolled into a narrow cylinder, the core of which is tightly packed with gunpowder. These are not safe, and not used in India, except in the smaller quarries, where sometimes a train of gunpowder is used to light the charge in the hole. This is a highly objectionable and dangerous practice.

E. *Crimping pliers.* When a detonater (described later) is to be attached to a safety fuse, a special type of pliers is required. The safety fuse is inserted into the detonater cap, and then, by means of the pliers, a constriction is made, at the outer rim of the detonator, which helps to attach the detonator to the fuse.

F.  *Detonators.*  There are two types (*a*) lead azide ($PbN_6$). This is usually mixed with styphnate and aluminium powder.  Tetryl, used earlier has been generally replaced by penta-erythrital-tetra-nitrate (PETN). (*b*) Mercury fulminate $Hg(CNO)_2$ is usually mixed with pot chlorate.

The detonators can further be classified as (*a*) ordinary or (*b*) electric detonators.  Under the electric detonators, again there are two types (*i*) those which detonate instantaneously and (*ii*) those which deto-nate with a time lag.  Milli-second detonators have very small time iag, amounting to fraction of a second.

Detonators are high explosives, which can be started of only by violent or sudden impact or by a flame.  Detonators are employed with high explosives to start off the explosion.  There are two kinds of deto-nators.  (1) Lead axide in aluminium caps and (2) mercury fulminate in copper caps.  They are made in various sizes.  These detonators are manufactured with time lag fuses, which enable the detonators to go off after the lapse of a fixed interval once the electric contact is made—lags of O sec., 1 sec., or 2 sec. etc.

Detonators are similar in construction to the "caps" used in shot gun and rifle cartridges, the only difference being their larger size.  The deto-nating substance is placed inside the cap.  As in the case of the rifle

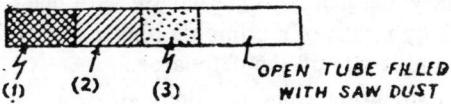

(1)   (2)   (3)   OPEN TUBE FILLED WITH SAW DUST

Fig. 7.1

cartridge, the detonator can be started off by a sharp blow on the cap or by sudden ignition.  The latter principle is employed in blasting.  The ordinary normal detonator is ignited by the safety fuse.  In the case of the electric detonator, the detonator is set off by the heating effect of an electric current.  The ordinary blasting cap or detonator (Fig. 7.1) consists of a metal cap, either of copper or aluminium, 0.22 inch in diameter and closed at one end. A charge, consisting of 3 sections is placed inside (1) the detonating mixture, consisting of either mercury fulminate or lead azide, together with potassium chlorate, trinitrotoluene etc., (2) a priming charge, (3) an ignition compound. Normally No. 6 and No. 8 caps are used for nearly all explosives.

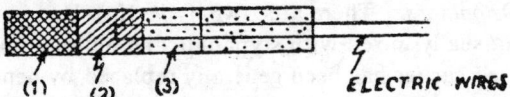

Fig. 7.2

In its simplest form, the electric detonator has 4 units. This also consists of a metallic (copper or aluminium) tube, closed at one end. The detonating mixture (1) and then (2) an ignition mixture over which is a sealing compound (3) (Fig. 7.2) The electric wires are embedded in the ignition compound section of the cap. When current is passed through the wire, it is heated up and the ignition compound is ignited. This in turn sets off the detonating mixture.

In the time lag fuses, a timing element is inserted, between the detonating mixture and the ignition compound, which induces the delay in detonation, which may vary in timing with an interval of 1½ seconds. A series of fuses with 1½ sec. interval delays can be had.

The detonators are connected either in series or in parallel. When there are a number of detonators to be fired at same time, the circuit is so arranged that the total resistance of the circuit is not large enough as to cause an excess drop in the voltage, and cause a misfire. Thus a parallel circuit is also to be employed to keep down the resistance even though the connection is a little more difficult than in the series circuit.

The ordinary detonator is started off with black chord. The detonator is attached by means of a crimping pliers, to the black chord. The chord is ignited and it sets off the detonator.

In electric detonators, time lag detonators are available. The detonators go off at a predetermined time interval after ignition. The interval of time or delay commonly used is one second; and a range of 10 to 15 types of time-lag detonators are available, each "going of" one after the other at an interval of one second.

A further modification, of the delay or time lag detonator, is the millisecond delay. In these the time interval or delay is of the order of 25 to 50 milliseconds, and a series of 14 detonator is now available. The advantages of the millisecond detonator are, better disintegration of rock and greater economy in the long run.

**Exploders**

The electric detonator is set off, either by connecting to the electric power supply main, or by means of a magnetobar exploder or a magneto type generator (rotary). Blasting machines are of two types. They are in fact D. C. generators. In one of the types, the magnetic field is produced by a permanent magnet. This is known as the magneto type. In the second type of machine, the field is produced by the passage of current through insulated wires wound on to a soft iron core forming the stator. This is in essence a small D.C. generator. The second type—generator type, is used where only a few detonators are to be fired. The generator is rotated by means of a crank and lever, and when it has developed its full voltage a switch is used to close the circuit. Where a large number of detonators are to be ignited, then a magneto or "bar machine" as it is called, is employed. In this machine a rachet bar is employed to move the rotor. When rachet bar is pressed down, with full force, the generator rotates and the full voltage is generated, by the time, the bar reaches the end of its travel. On touching the lowest point of its travel the rachet bar automatically closes the circuit and the detonators are fired.

Another type of exploder, which is quite popular, is the 'Condenser' type. This compries an AC generator, worked by a crank handle. The generator is used for charging a condenser. When the condenser is fully charged the circuit including the detonators is closed through a press button switch. The discharging of the condenser causes a large current to flow through the detonators, which are thus set off.

G. *Explosives*. These have been discussed earlier and the choice is made depending on the conditions of mining, i.e., the type of mine, either it is open cast or undrground, whether coal or metal mine, whether the mine is gassy etc.

**Method of Charging Explosives**

(1) Gun powder is generally carried to the work spot in a metal container, in the form of cylindrical packets $1\frac{1}{4}''$ diameter and 6" to 4" long. Loose gun powder is not convenient for use.

(2) Gelignite and dynamite as well as the ammonium derivatives, are packed in water proof paper sheaths. The cartridges, are $1\frac{1}{4}'' \times 4''$.

(3) Liquid oxygen cartridges are made by soaking cylindrical jute canvas bags, packets of saw dust or balsa dust, in liquid oxygen, just before use.

(4) The ANFO and slurry are piped down, into the mines and run into the blast holes or drill holes.

For charging holes with gun powder, a convenient length of safety fuse is taken and fixed to one of the cylindrical packets and the mouth of the packet is tied with a string, so as to hold the safety fuse in position. This is known as the primed cartridge. The cartridge is lowered and one or more cartridges are also placed into the hole, depending on the requirements. Tamping material (consisting of pellets of wet clay and sand mixture) is placed above the gun powder cartridges, and the hole is packed tight by mean's of the tamping rod. The hole is now ready for firing.

In the case of the explosive classed in categories (2) and (3), the safety fuse of desired length is taken, and to one end of it, the detonator is fixed, by means of crimping pliers, as described earlier. The fuse, containing the detonator, is placed in one of the cartridges, by making a small hollow with a pointed piece of wood shaped like a lead pencil. The safety fuse is fastened to the cartridge cover with a thread. Sometimes, it is desirable to fasten the fuse to the primed cartridge, with a string so that it lends further strength and support. The primed cartridge can be placed in any position, either at the bottom, or in the middle, where 2 or 3 cartridges are employed.

In the case of the ANFO explosives, the primed fuse is placed in contact with the explosive. Where electric blasting is adopted, the equipment used, has been described earlier. The detonator may be connected. either in series or in parallel, or by using a combination of these.

The current strength recommended for the different firing of circuits is as follows :

1. Series connections not less than 1.5 amps.
2. Parallel connections not less than 0.8 amps. per cap.
3. Series in parallel not less than 2.0 amps. per series.
4. Parallel in series not less than 1.5 amps. per cap in series group.
5. Graded series in parallel, not less than 2.0 amps. per group.

Normally direct current (DC) is preferred to alternating current (AC). When AC is used 60 cycles is the minimum desirable frequency.

Even though the series connection is simple, it has certain undesirable points. When a large current passes through every cap, and as soon as the first cap in the series explodes the current may either leak, or the connection may be broken off, so that the other caps may not explode.

Where electric power is available, the parallel circuit is desirable, as each cap explodes independently. Where the voltage is not high enough, series cum parallel arrangement may be adopted.

For weaker rocks, either a smaller charge or a weaker explosive is used. Irrespective of the toughness of the rock a smaller or weaker explosive is used when only fracturing of the rock is required. When the rock is to be shattered or broken to small pieces, a large charge or a very high explosive or both, is used (see also driving tunnels, galleries etc.).
There are two common types of arrangements of the blast holes.

(a) The pyramid cut, (b) the wedge cut or V-cut.

In both these patterns, the drill holes are made in three sets and they are fired or blasted also in three rounds. The first set of holes may be called "easers" and these make a central cavity. The second set is called "sumpers". These are also slightly inclined inwards and serve to deepen the cavity. The third set known as "dressers" is driven vertically and these holes make the sides of the excavation more even. In the pyramid cut, the first or inner round of holes (4 nos), point towards the centre, so that the hollow produced on explosion will resemble a pyramid. In the V cut or wedge cut the "easer" holes comprise parallel sets pointing towards the centre line, so that the excavation will resemble a V.

The same arrangement of blast holes is also practised in shaft sinking operation, irrespective of whether the shaft is in hard or shoft rock.

In order to take maximum advantage of the explosive charge a template is sometimes provided to the drilling crew using the jack hammer or stoper. This device is called the "hole director" (H.D.). It consists of a light steel frame. The use of the hole director enables the drill man to automatically align the drill.

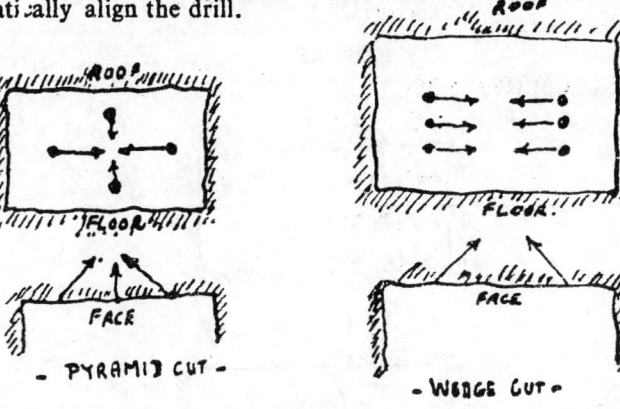

- PYRAMID CUT -

Fig. 7.3

- WEDGE CUT -

Fig. 7.4

Besides the pyramid (Fig. 7.3) and wedge, cut, (Fig. 7.4) , there are other patterns of drilling employed so as to obtain efficiency and economy in explosives, *e.g.*, the burn cut, (both the draw cut and toe cut). In the draw cut the central holes are drilled but these are directed upwards; whereas in the toe cut, the central holes are drilled pointing downwards, i.e., towards the floor of the tunnel. Burn cut may be arranged in the box pattern or line pattern (Fig. 7.5).

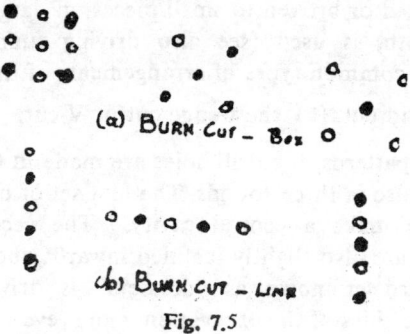

(a) BURN CUT — Box

(b) BURN CUT — LINE

Fig. 7.5

The other common types of "cuts" are :

1. Drag Cut (Fig. 7.24)

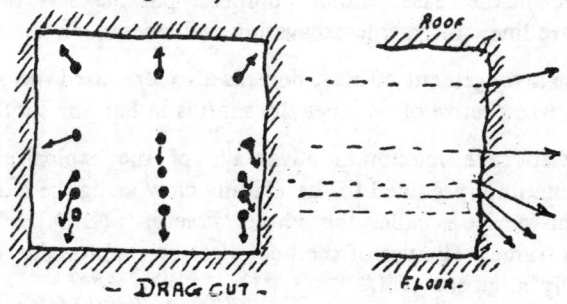

— DRAG CUT —

Fig. 7.6

2. Fan Cut (Fig. 7.25).

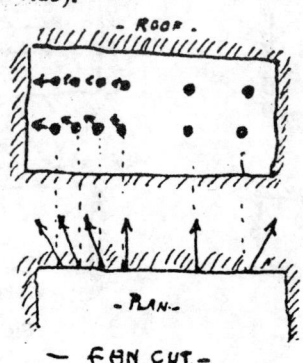

— ROOF —

— PLAN —

— FAN CUT —

Fig. 7.7

In general, when using the various cut patterns for the middle holes, it may be remembered : (1) that the extent of break is dependent on the depth of the cut ; (2) since the holes, for the cut, are drilled at an angle to the axis of the drive, the limits or set by the size of the drift, as well as by the length of the drill rods used; (3) the depth of excavation cannot exceed the least dimension of the drift, whatever be the pattern of holes.

In the burn cut, sometimes, the central or cut holes are drilled parallel to each other ; but one or more holes are blanks, while the rest are charged with explosives.

Pneumatic drills or Jack hammer drills are commonly used for drilling holes underground. In this machine, there is a reciprocating piston type hammer, which is moved backwards and forwards by compressed air. The hammer hits a drill rod which makes the hole. The drill rod is hollow and compressed air goes through and blows out the dust and cuttings from the drill hole (see drilling). Larger machines, known as stopers, are used in the stopes in conjunction with air legs. In open cast mines larger drilling machines known as blast hole drills are employed for making 6″—8″ diameter holes which may be 40 ft. to 60ft. deep.

*Misfire* is a term employed to describe an explosive charge which does not "go off".

*Hangfire* is a term used to signify an explosive charge which goes off after some delay. This is particularly dangerous, if a shot firer has missed count of the number of charges, he had placed originally.

*Magazine* is a temporary storage place for explosives. In some metal mines underground magazines are found. The general practice is to have the magazine on the surface. Some important rules of the Explosives Act with regard to the magazine state, that

1. Magazine should be at a safe distance from habitation.
2. Detonators should not be stored in the same magazine as the explosives.

It is preferable that double brick walls be used, in the construction, and the intervening space be filled with sand. The floor and foundation should be sufficiently raised so as to be water proof. Good ventiliation is essential. Certain amount of *humidity* is desirable. The person entering a magazine should be either bare footed or may use rubber soled shoes.

Note :  Anyone handling explosives should be thorough with the Act, as well as with the relevant portions of the Regulations under the Mines Act.

*Efficiency of blasting* : The most effective method of blasting is obtained by experience in the area. Certain general guiding principles may be formulated.

(*a*) The amount of charge used depends on :

    1. The nature of the rock, i.e., whether hard, soft or jointed.

    2. The kind of explosive, i.e., high, low etc.

    3. The required degree of fracturing or breaking up of the rock, i.e., whether large pieces are required or small fragments, etc.

(*b*) The line of least resistance should be observed before placing the drill hole.

(*c*) Smaller charges are required, when the free faces are larger in number,

(*d*) Economy, in the consumption of explosives, can be affected by keeping a line of benches or faces.

(*e*) Charges can be reduced by employing simultaneous shot firing.

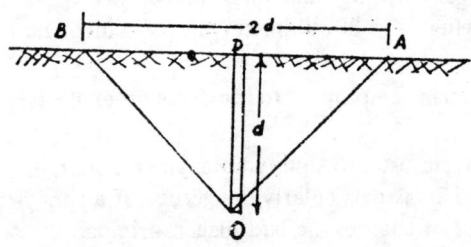

Fig. 7.8

From Fig. 7.8. it will be seen, that the volume, of the cone of rock excavated $= 1/3 d \times d^2 = \pi d^3$ (approximately). In general, however, if the hole is placed along AO, there is the advantage of a free face. In Fig. 7.9. since there are two free faces, a hole drilled along DA and charged with explosive will yield the maximum volume of rock as is the case when $\angle A = 45°$.

Thus the volume blasted out depends on the following variables viz., free face available, the depth of the hole, and the nature of the rock, and the inclination of the hole (max. inclination$=45°$)

$$V = nmd^3$$

where    $V =$ Vol. of rock blasted.

        $m =$ rock factor or rock coefficient which can be got by trials.

        $n =$ the no. of free faces and

        $d =$ depth of hole.

The above relationship holds good where the charge is equidistant from A and D, if the charge is placed at $O_1$, then the break will possibly follow the line $AO_1D_1$ (Fig. 7.9).

In quarries, blast holes are used in breaking large quantities of the materials. Large diameter holes, about (9″ in diam) are drilled to a depth of 30—40 ft. at 15 ft. centres along the quarry face. The holes are charged in a line, and blasted simultaneously. Some thumb rules are adopted in blast hole firing, e.g.,

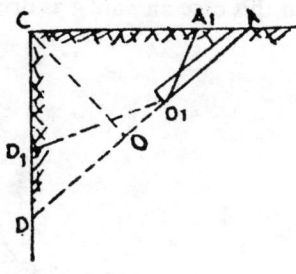

Fig. 7.9

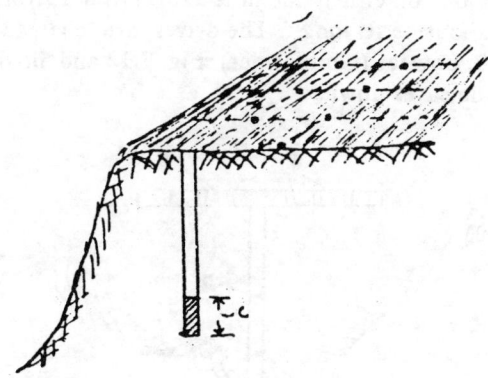

Fig. 7.10

if (a) The length of the charge L should be 12 times the diameter of the hole. (b) The rock coefficint is determined by drilling 3 ft. holes in uniform (homogenous) rock and by varying the charge. (c) If $d$ is the length of the line of least resistance then

$$W = Cd^3.$$

C = rock coefficient.

*Snake hole* are holes drilled more or less horizontally at the foot of a high bench, in a large quarry and charged with explosives. Under certain conditions snake holes can be more effective than blast holes (Fig. 7.11).

---

*See addendum.

Gopher hole or Tunnel or Coyote blasting is also large scale blasting In this case an adit A is driven to about 45 ft. in the hill face, and two

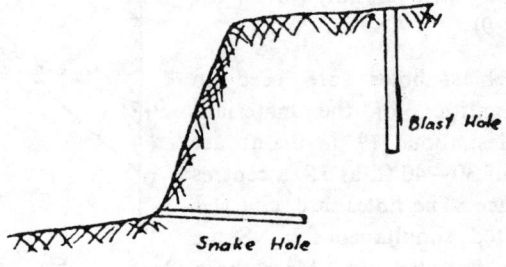

Fig. 7.11

drives D are made, one on either side in the form of a T, from the end of the adit, say each about 45 ft. long. The drives are charged with explosives, generally B.B. powder (Gunpowder) Fig. 7.12 and fired, so that the entire mountain side caves down.

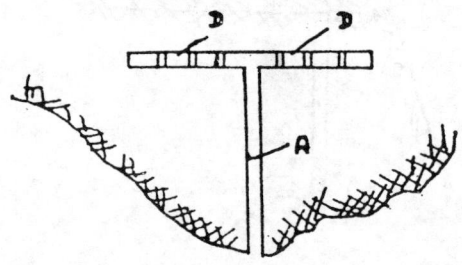

Fig. 7.12

Springing or chambering is the process by which the bottom portion of vertical blast holes are widended, so that additional charge of explosives can be placed. For this purpose powerful explosives are employed, e.g., 50% dynamite or 60% ammon. dynamite etc. Water is normally employed for stemming spring holes.

Further notes–See annexure.

## DRAINAGE

It is often necessary, during large scale prospecting operations, to deal with water that percolates into the pit or trench. Thus, the mining geologist will have acquired sufficient knowledge as to how to cope up with such a situation.

In the case of a prospecting pit located in hard ground, where it is needed to deal with surface water, it is best to make an embankment right round the pit. The height of the embankment will depend on the local conditions. Otherwise, a trench can be made around the pit—the dimentions depending on the local conditions. If the seepage is from within the pit, bailing out with buckets may be sufficient, in some cases, if this method does not succeed then pumping is the answer. Occasionally, it will be possible to use a syphon for drainage if the topography is favourable.

If it is a question of dealing with storm water, it will be necessary to examine the surrounding areas to find out if it will be easier to divert the drainage, even at the upper reaches, by contour channels into an adjacent one. The question of safety should not be neglected or minimised in all cases.

In case shallow pits are to be made for prospecting either alluvial deposits, or beach placers, stiff bamboo matting generally serves the purposes upto a depth of 1—2 m., beyond this depth systematic shaft sinking techniques will be needed.

# PART VIII


# MINING METHODS

# MINING METHODS

**Mining terminology :** employed in exploration and exploitation activities.

**Exploitation or winning :** is the process of extracting the ore or economic mineral from the earth.

**Development :** is the first stage in winning the ore, when further drilling may be undertaken and the various tunnels or roadways etc. are prepared according to a definite plan. In the case of open cast workings, additional drilling is done, and benches made. A certain amount of reserves of the mineral can also be blocked out, and proved as a result of these activities.

**Shaft :** is a vertical or inclined excavation, which is either square or rectangular in section, lined or unlined. Shafts are openings which serve as a means of entry etc. Inclined shafts are called "inclines" in coal mining. Shafts which do not reach the surface are called "blind-shafts". In the case of vertical shafts, these are termed sub-vertical or staple pits. If a shaft is partially vertical and partialy inclined it is termed as a compound shaft. Shafts can be away from the ore body or a mineral deposit but connected by tunnels known as cross cuts. (Fig. 8.1)

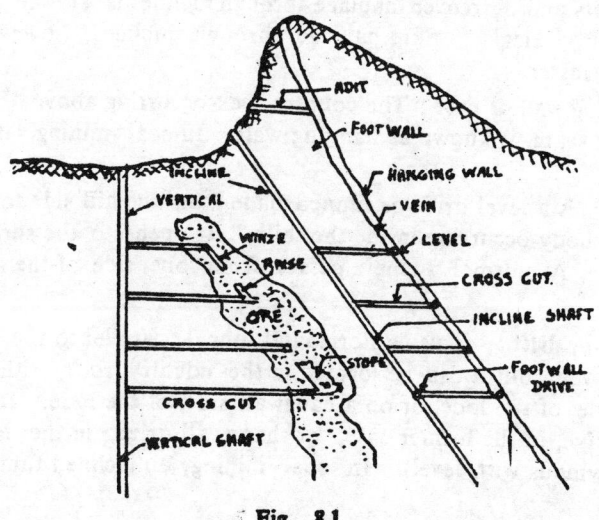

**Fig. 8.1**

Vertical shafts are used for hoisting or winding men and materials and also for drainage and ventilation (either up cast or down cast, *i.e.*, intake or exhaust). Shafts are also employed to carry cables, for (electricity) lighting, power, and also compressed air mains. Auxiliary shafts are used for stowing, carrying cables, materials etc. They are also known as service shafts, whereas shaft, used for bringing out the mineral, are called production shafts. Blind or sub-vertical shafts serve to connect lower levels, in a mine. Vertical shafts are necessary, (i) where the deposits does not outcrop within the lease area or property, (ii) where haulage distances are long and when it is essential to cut down time taken in raising the mineral won from the underground workings.

Inclines or inclined shafts are openings made in the deposit but follow the vein or seam, down the dip. Such shafts may either start from the outcrop, or they can be located parallel to the dip. Normally inclines or inclined shafts are located in the ore body, or within the deposit. It is sometimes desirable, for various reasons, that the incline be placed away from the deposit, either on foot wall or on the hanging wall side. Some of the reasons for keeping the inclined shafts preferably on the foot wall are,

(1) to prevent oxidation of a massive sulphide ore body,

(2) to minimise the effects of stoping and subsidence,

(3) and also to maintain the permanent roadways, where the ore body is liable to crushing or is weak, etc. A hanging wall shaft may achieve the first objective but not the latter two.

The inclines suffer more from the effects of lateral pressure than vertical shafts and moreover, haulage through inclines is slower than winding through verticals. Skip haulage through inclines, however, saves time as it is faster.

**Hanging wall or roof :** The country rock occurring above the vein or ore body or seam, is known as hanging wall. In coal mining, it is called roof.

**Adit :** is a level drive or tunned made from the hill side to meet the lode or ore body occurring inside the hill. It opens to the surface, only at one end. A normal tunnel passes from one side of the hill to the other.

**Drive or drift :** is also a horizontal tunnel parallel to the strike, of the lode or vein, but it can be located in the country rock, either on the foot wall side, of the lode, or on a hangwall side, of the lode. It is called foot wall drive, in the former case, or hangwall drive, in the latter case (also synonymous with level). In coal mining, an inclined tunnel driven through barren rock to connect the workings, in an upper seam, with those in the lower seam, or vice versa is called a stone drift.

**Level :** is a tunnel, in the ore body or vein, which follows the strike. It is also level. In coal mining this is termed a level gallery.

**Cross cut :** In mining unstratified deposits, the term cross cut is used to describe a tunnel or drive, which leads from the shaft or level and passes through the country rock, in order to intersect the lode. The "stone drift" defined earlier, is in fact a cross cut also. In the pillar and stall method of coal mining, when a roadway is made to cut across the general direction of the level galleries, then it is called a cross cut. The cross cut lies within the seam, and runs oblique to the strike. The main purposes of a cross cut are : (1) to reduce hauling and walking distances, to the workings, (2) to reduce the gradient, when the dips are high, and (3) for arranging for ventilation and drainage etc.

**Tunnel :** is an opening, which is nearly horizontal. It may be circular or semicircular in cross section. It is driven from one side of a hill and opens out, on the other side. Any underground excavation, either level or drive, is also called a tunnel.

**Raise :** when from a level, in the lode or ore body, an excavation is made (upwards) in the lode, to meet the level above, the excavation is called a raise. In coal mining, the term employed for describing such a working is rise gallery or working.

**Winze :** when from a level, in the lode, a tunnel is made downwards, to meet the lower level, the excavation is known as a winze. The relative coal mining term, used for describing a similar working, is dip gallery.

**Ore bin :** is an excavation (lined with masonry or concrete) underground or a masonry construction of special design for storing ore mined from a part of a mine. Ore bins may be located, either at the main shaft bottom, or at the top.

**Chute :** is a lip or small channel shaped projection at the bottom of an ore storage bin, or from a stope, where ore is stored. It enables the broken ore to be directly loaded into tubs, or mine cars or skips. The term is also used to describe lined (wooden or concrete) portions, of raises or winzes, through which ore waste filling, is dropped. It is then called an ore pass.

**Stope :** is an underground excavation made for removing or winning the ore.

In Figure (8.1 a) the width of the ore body is W, *i.e.*, true thickness in the geological terminology.

The stoping width is the convenient height of excavation that will afford miners enough height to work with efficiency. Thus, in the case of

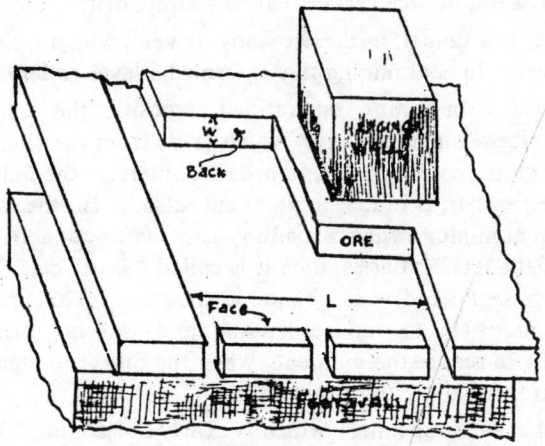

Fig. 8.1 a

narrow ore bodies, the stoping width is very much in excess of the width of the ore body. Normal stoping width for satisfying the above conditions should be about 3 ft or 1 m. In the case of minerals with high unit value stoping width may not be such a major consideration as in the case of minerals of low unit value, where dilution factor may render the deposit uneconomic.

Width of stope is indicated in (Fig. 8-1a) by W. It is in this case equal to the width of the ore body. The minimum width is about 4½ ft or 1½ m. The length of the stope is measured in the direction at right angles to the raise (R). It is indicated as L in the Figure.

**Over hand stoping :** (See mining methods). When an excavation is started upwards, *i.e.* up dip, from a level, the miner raises the pick over his shoulder, to break the ore. The method of extraction is known as overhand stoping. This is a mining operation and not developmental, *e.g.*, making a raise mentioned above.

**Under hand stoping :** (See mining methods) When an excavation is started downwards *i.e.*, down dip, from a level, such a method is known as underhand stoping. This is a mining operation, as apart from winze described earlier.

**Air crossing :** is the arrangement by which the intake or (fresh air current) is made to cross the return (or used air current) in the underground workings. A large diameter circular concrete conduit, is employed for conveying the return air current.

**Ventilation stopping :** is a wall built, across an underground excavation, or tunnel, to prevent the air current from flowing into unwanted places, and for coursing it to working places, according to the ventilation plan.

**Fire stopping :** is a solid masonry wall, used for sealing off workings and tunnels, leading to any area of a mine under fire. It has pipes passing through it, for obtaining samples of air, from within the sealed off workings.

Excavations made either during the exploratory or development stages can be classified as follows :

(*A*) Excavations lying outside the mineral deposit :

    (1) Parallel to the strike.
    (2) Parallel to the dip.
    (3) Across the dip strike.

(*B*) Excavations made in the ore body :

    (1) Parallel to the strike.
    (2) Parallel to the dip.
    (3) Across the strike.

(*A*) **Excavations :** lying outside the ore body can be grouped as under :

| | |
|---|---|
| 1. Parallel to the strike | (*a*) Footwall and hanging wall levels or drives. |
| 2. Parallel to the dip | (*a*) Shafts—in foot wall or hanging wall. |
| 3. Across the dip | (*a*) Cross cut. |
| | (*b*) Vertical shaft. |
| | (*c*) Adits. |
| | (*d*) Drifts (coal mining). |

(*B*) Excavations made in the ore body can be grouped as under :

| | |
|---|---|
| 1. Parallel to the strike | (*a*) Levels, drives (in the lode). |
| | (*b*) Sub levels. |
| | (*c*) Stopes and chutes. |
| | (*d*) Level galleries (coal mining). |
| 2. Parallel to the dip | (*a*) Raise and winze. |
| | (*b*) Inclines (shafts). |
| | (*c*) Dip or rise galleries (coal mining). |

3. Across the strike and away          (a) Cross cut (coal mining)
   from the dip direction                   within the coal seam.

                                             (b) Cross cut winze/raise (metal
                                                   mining).

**Assay width** : is the width of the entire mineralised zone whether strongly mineralised or not. In the case of a vein it can so happen that the core may yield rich values whereas the assay values at the walls may be low.

**Stoping width** : is that width of the ore body which can be economically mined. Hence, it is generally less than the assay width. There can be cases where the assay width and stoping width are equal.

**Cut off grade** : is that grade of ore below which it is considered that economic mining is not feasible. Hence it is only ore with values above the cut off grade that is mined,

**Average grade** : is that grade which is obtained by working out the arithmetic mean, giving weightage to the quantities available in each grade taking either the stoping width or the assay width as it is decided. Hence if the assay width is taken then the average grade may be lower than the cut off grade.

**Mill grade** : is that grade of ore which will ultimately be fed into the beneficiation plant, very often the mill grade will be higher than the average grade and necessarily higher than the cut off. This is so because only ore of a higher grade than the cut off grade is mined.*

**Factors influencing the choice of the method of stoping for achieving the maximum amount of economy, efficiency and safety.**

1. The method selected should provide maximum efficiency in drilling and blasting. This in effect means ability to employ deeper holes and larger charges with maximum output.

If slightly inclined holes are drilled upwards, the workmen can stand in a comfortable position, and the hole is also kept clean by gravity. The face is also free after the blast.

2. It is also clear that longer the working face, the larger is the output. So also larger the number of faces, the larger is the output. Efficiency of clearing up the face, of the smoke, dust and muck, after the blast, also affects the output. Conveyance is costly, and must be generally avoided. Gravity should be taken advantage of, wherever possible.

3. Rate of progress affects the economy. A higher rate of working saves time, and possibly also expenditure, involved on support, drainage, ventillation, lighting. When working at higher rates, fewer faces will be required, for the same output.

4. The method should aim at the recovery of the mineral deposit with the minimum amount of losses, due to blasting, loading, etc.

5. As regards the effect of contamination of ore by gangue, overhand methods are safer, but sorting when resorted to increases cost.

6. Facility of access to workings, and the transportation of ore, and other materials, is also an important aspect.

7. In the method selected, the requirements for support, especially timber, which increases cost of materials and labour, and increases dangers of fires etc. should be minimum.

8. Facility for stowing or filling is an important consideration.

9. Facility for ventilation is also of paramount importance.

## CLASSIFICATION OF MINING METHODS

1. **Alluvial mining**

   1. Pan and batea
   2. Rocker
   3. Longtom
   4. Sluicing, in general
      Ground sluicing
   5. Derrick and Cableway
   6. Hydraulicking with Hydraulic giants or monitors
   7. Drift mining
   8. Dredging

II. **Open cast mining or Quarrying**

   1. Loading by hand (manual)—general layout using

      (a) Trucks/Dumpers
      (b) Direct haulage
      (c) Aerial rope way

   2. Loading by machines—general lay out using

      (a) Drag line    (b) Power shovels
      (c) Scrapers    (d) Land dredges
      (e) Over burden bridges
      (f) Bucket wheel excavators
      (g) Conveyors (flight/belt)
      (h) Spreaders

   3. Glory hole
   4. **Kaolin mining**

**III. Underground mining**

**Open Stopes**

    1. Gophering
    2. Breast stoping
    3. Open underhand stoping
    4. Open overhand stoping
    5. Underground glory hole
    6. Pillar and chamber
    7. Sub level method

**Overhand Stoping—with supports**

  A. Timbered stopes (square set method)
    (i) Flat backed
    (ii) Domed
    (iii) Rill
    (iv) Vertical face
    (v) Underhand

  B. Filled stopes
    (*i*) Filled flat back (vein)
    (*ii*) Filled flat back (wide ore bodies)
    (*iii*) Baltic dry wall (wide ore bodies)
    (*iv*) Resuing
    (*v*) Cross cut method
    (*vi*) Inclined cut and fill—Filled rill-stopes

  C. Shrinkage stopes
  D. Mitchell slicing system

**E. Caving methods**

    (*a*) Top slicing
    (*b*) Sub level caving
    (*c*) Block caving

## ALLUVIAL MINING

**1. Pan and batea**

    This is manual mining method, where picks and shovels are used in excavating sands; sandy clay etc. Pans of various shapes and sizes are employed in recovering the values. For normal work the pan used is about 10″ wide at the top and 3″ deep, with sloping sides (Fig. 8.2). The soft alluvial material dug up, is placed in the pan or batea, and washed. The process is called panning. In first stage of operation, a gentle

Fig. 8.2

rocking of the pan enables the heavies to settle at the bottom, near the base of the sloping sides due to the tilt. After this stage, a gentle gyratory motion of the water, in the pan, helps to float up the lighter material, which is carried away as the water spills out. Finally, only the heavies remain from which, the gold is separated from the magnetite etc. by magnetic methods. About 1 cu. yd. can be handled in one hour/man.

### 2. Rocker (Fig. 8.3)

This is also employed in mannual mining when the pick and shovel are the main equipments used. The rocker consists of a metal screen mounted at the bottom of a strong wooden box. The box is made to stand on two semi circular iron hoops. A handle is fitted to one side of the box. The alluvial material is poured on the metal screen ($\frac{1}{4}''$ to $\frac{1}{2}''$) together with water, and with the help of the handle the rocker is rocked from side to side. During this process, as the rocker rolls on the semicircular iron hoops, the finer portion of the alluvium, containing the values, passes through the screen, along with the water, and gets into a riffle box. The lighter fraction flows over the riffles, while the heavier fraction is retain

Fig. 8.3

ed, by the riffles, from where it is collected later on. The gold is separated from magnetic materials contained in the heavies by magnetic separation.

### 3. Long tom (Fig. 8.4)

This consists of an open box 10 ft. to 12 ft. long, from 15" to 20" wide, at the upper end, and nearly double this width, at the lower end,

Fig. 8.4

where there is an inclined screen. The slope given is 1 in 12. The finer material, passing through the screen, falls on to a riffle box. Here the separation of the heavies is affected. This equipment is also used where manual excavation of the soil/alluvial materials is carried out.

### 4. Sluicing—Ground sluicing (Fig. 8·5)

Water from flowing stream (R) is diverted into the area occupied by the alluvial deposit (D) through a channel (C) and men standing on the banks, of the channel, shovel the placer material, into the water. The material disintegrates and the muddy water now flows into a sluice box (S),

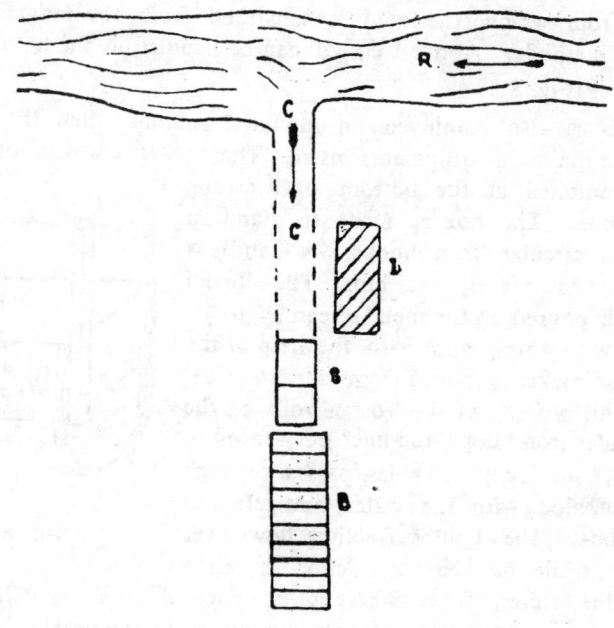

Fig. 8.5

and riffle box (B), from where the concentrate is recovered. Thus, this is also a manual method of mining. Sluicing equipment in general comprises,

(a) sluice boxes, (b) ground sluices, and (c) riffle boxes.

A sluice box as used in placer, or in hydraulic mining, (described later), is a wooden box often rectangular in section. It is generally made in

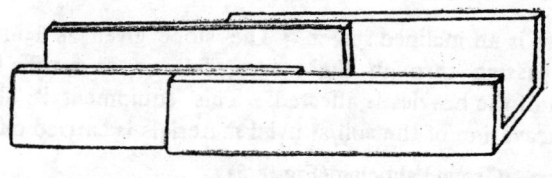

Fig. 8.6

lengths of 12ft. One end of the box is made slightly narrow, so that a chain of boxes is formed (Fig. 8.6) by fitting the narrower end of one to the wider end of the adjoining box. Within the sluice boxes, riffles are placed across, which help to retain the values. It is found that as the sluice boxes become older they retain appreciable amounts of values (especially fine gold). This is recovered by burning the boxes.

A riffle box : This is also a box (B) with a ractangular cross-section. At the bottom of the box, narrow wooden strips or riffles (R) are fixed

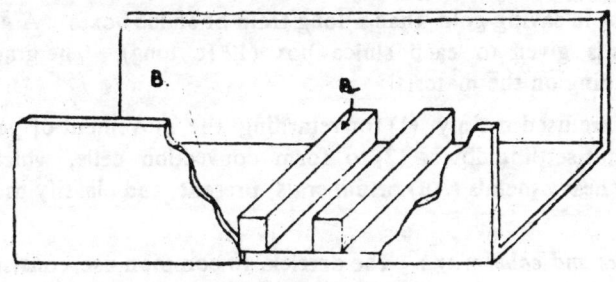

Fig. 8.7

transverse to the length of the box (Fig. 8·7). Instead of wooden strips, cobbles or rock riffles, are used and sometimes wooden blocks or steel rails are also employed for the purpose.

Alluvial materials can be excavated and supplied for the sluicing operations by (a) using spades, picks etc , (b) wheel barrow, (c) horse scrapers, (d) in larger quantities by means of aerial rope, (e) derrick and bucket, (f) power showels or drag line exacavators, which are described later on. For successful working, the gravel should not be more than 9-10 ft. deep and plentiful supply of water is essential.

Favourable conditions for ground sluicing are :

1. Shallow depths of the deposit, which can vary between 6 ft. to 8 ft.

2. A gradient of sufficient magnitude to facilitate the transport of the loosened gravel and clay, e.g., of the order of 17 to 25 ft. per mile.

3. Copious supply of water, since any defficiency, in this regard, is to be supplemented by pumping, from the settling tanks.

4. Facilities for the disposal of tailings which normally accumulate at the sluice end, due to fall in gradient. Disposal of tailings by scrapers may be necessary to improve efficiency.

*Booming or hushing* : Where the supply of water is poor or insufficient, arrangements are made for impounding and releasing the water at

intervals by an automatic arrangement. Ground sluices are also employed in this instance, for saving the values comprising, e.g., gold, or
tin, etc.

In common practice, the sluice boxes are made in sections, 12 ft. in
length, and the length of a section is normally 36-72 ft. At times, drops
or "falls" are also introduced, and such arrangements are considered to be
more efficient in saving gold than a long train of sluice boxes. A gradient
of 6-7 inches is given to each sluice box (12 ft. long). The gradient is
varied depending on the material.

Riffles are used mainly (1) for retarding the movement of particles,
so as to permit settling down, (2) to form convection cells, which help
to retain the heavy metals (Au) or minerals present and classify the same,
in sizes.

5. *Derricks and cable way* : The derrick, in common use consists of a
stout wooden mast, which can be used to lift loads, by means of a rope and
pulley arrangement. At times, this can be improved by attaching an
inclined member, which can move round the vertical mast, as well as move
up and down. By this arrangement loads can be moved outwards, from
the mast as centre, as well as concentrically. The cable way comprises,
two masts, the tops of which are connected by a tight cable. A pulley
arrangement is mounted on this rope, and this can run on the cable, from
one mast to the other. The pulley can be used for lifting the loads, while
the wire rope (or cable) is used for moving the loads outwards. The
derrick and cable way are used in open cast methods, for winning clays
etc., or for removing clays overburden, from the excavation to the surface,
for storage/dumping.

6. **Hydraulicking** : is a method of mining, in which hydraulic giants or
monitors, are employed. The head of water should not be less than 200 ft.
or about 8.7 lbs/sq in.—and the giant must be securely anchored, to prevent
any accident to the operator, if the monitor loses stability.

The Hydraulic Giant comprises an assembly of heavy steel pipes,
which terminates in a tapered nozzle, for throwing out a jet of water, as
done by a fire engine (1). The giants or monitors are made in various
sizes, with nozzles ranging from 2-10 inches in diameter. The quantity
of water required, varies from 100 cubic ft./min. to about 6000 cubic ft./
min. The rotation of the nozzle, is about a joint, with ball bearings, which
is held together by flanges (2) and king bolt. Since the water is under
heavy pressure, it is essential to provide the nozzle, of the giants, with a
moveable tip or deflector (3). The reaction of the jet as well as a long
lever attached to it enables the operator to move the deflector with ease.

For most effective work, there should be no rotating motion. Hence giants are placed, as close as possible to the bank.

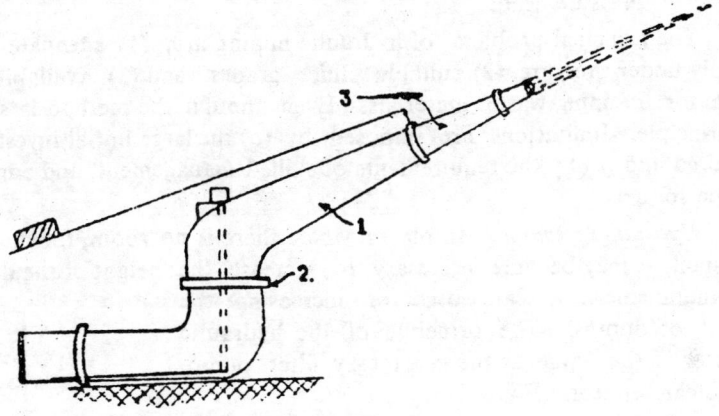

Fig. 8.3

The water jet, from the monitor, makes scallop shaped cuts, in the alluvial bank. The area, under exploitation, is called a paddock. (Fig. 8.9).

Fig. 8.9

Hydraulicking requires certain preliminary arrangements :

1.  Construction of reservoir, which may entail putting up a dam, which may be either a masonry, concrete, or earthen structure of any standard design, for providing a continuous and copious supply of water, is essential.

2.     Since high pressure is required, for producing the jet from the monitors, water has to be conveyed, to the paddocks, or work spot, by means of open channels or flumes, or pipes, or penstocks, from higher elevation, so as to obtain the necessary pressure head.

The principal problems of hydraulic mining are, (1) adequate water supply under pressure, (2) suitable sluice grades, and (3) availability of room for dumping waste materials. Even though the method is simple, in principle, limitations are imposed by (a) the large initial investments required and (b) by the requirements of skilled management, and engineering personnel.

*Hydraulic elevator*: In places, where there is no room, for dumping the spoil, it may become necessary to increase the height of the dumps Hydraulic elevators are used for increasing the height of dumps. The principle of the hydraulic elevator is the same as the laboratory filter pump, or a steam ejector. (Fig. 8.10).

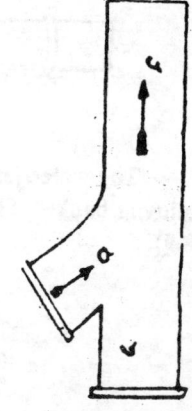

,   A powerful jet of water is directed, by the nozzle (a), into the pipe (b). The feed consisting of debris, is carried by pipe (c) upwards, by the jet, into the elevator pipe (b), and delivered into a sluice, from where, after the recovery of the values, the spoil is dumped. The arrows show the movement of water.

The largest size of elevator is about 9 inches in diameter. The hieght to which the gravel can be raised economically, varies from 1/10 to 1/6th the head of water employed. The velocity is about 5 ft./sec.

Fig. 8.10

Since gravel forms only 5% of the total weight lifted, and if the total quantity of water used is known, then the capacity of the hydraulic elevator can be worked out :

$$M = \frac{H \times N}{C},$$

where,            N = cu. yds. lifted/hr.
                   H = available head in ft.
                   C = efficient working head in feet × 15.

**Drift mining :**

Drift mining is more expensive than either sluicing, or hydraulicking. It is employed in the exploitation of placers occurring underground at depths, such as buried beach placers, and ancient river channels. Forepoling is employed in advancing the workings in the deposit. The main entries, in most cases, are vertical shafts, (for description of forepoling see under 'Mine Supports'). (Fig. 8.11).

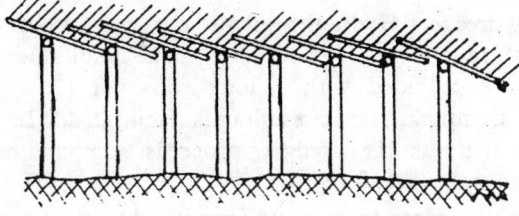

Fig. 8.11

### Dredging

This method is essentially employed for mining placers particularly tin, as in Malaysia. The dredge is large flat bottomed barge or pontoon (4). These dredges are provided with a chain of large shallow buckets (1), which is lowered down to the bottom of the pond, from a boom The buckets bring up the soil from the bottom of the pond, just as water is carried up by a Persian wheel. The buckets can be raised or lowered by a rope and pulley arrangement (6), attached to a mast (5). This form of dredge is advantageous, in clayey soil, as digging is not diffi- cult. Head lines, or stay wires are used, instead of spuds, to keep the dredge anchored. Sometimes mechanical harrows, delivering 60 strokes per minute, are employed for loosening the mud. Some dredges are fitted with classifiers, and jigs (see under ore dressing) for recovering values (2). Trommel or revolving screen, feeds into a classifier, where final concen- tration is affected. In the new types, there is no classifier, and the ore goes from the trommel to a jig (2) (Fig. 8.12).

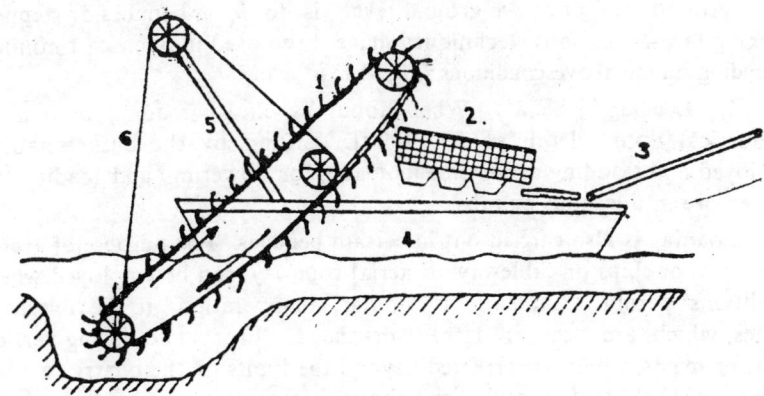

Fig. 8.12

For effective dredging, the depth of gravel should not usually be greater than 50 ft. to 75 ft. Ground, containing large boulder etc., should be cleaned up, before dredging is undertaken, with a view to prevent damage to the buckets.

**Resoiling dredge :** (Hydraulic giants are now used, in the rear of some dredges, to level the gravel and make it suitable for cultivation. Some dredges are provided with a long conveyor (3) which carries the waste rock for dumping\ Hence resoiling is facilitated.) In such cases, it is so arranged that, as the dredging proceeds, the pond behind is simultaneously fiilled up. (Fig. 8.12).

**Dredging in frozen ground :** Thawing by steam is a slow, and costly process. ,Thawing by a cold water has been found to be more economical under favourable conditions. For the best results, the temperature, of water must be from $50° - 60°F$.

**Centrifugal pump dredge :** In this type of dredge, a centrifugal pump is used, to suck up alluvials. This is useful for working sandy materials, or for materials, which have been rendered loose, by means of mechanical harrows. The pumps used in these dredges are rubber lined, to prevent or minimise damage due to abrasion.

## II. OPEN CAST MINING OR QUARRYING

Open cast mining by manual methods is adopted, (1) where large investment is not warranted, (2) where mechanization is likely to prove inefficient, either due to the large size of pits, or to the shape of the pits, and their location, (3) where mineralization is erratic, and where selective. mining, with sorting is called for, thus inhibiting the use of mechanical mining and glory holes, (4) where large scale work is possible with cheap labour. A general pattern for the lay out for open casting, whether it be below ground level or above ground level, is to form benches or stepped working faces. Various techniques have been used in open cast mining, depending on the above conditions (Fig. 8.13).*

**1. Loading by hand :** When labour is cheap, loading by hand is economical, upto a depth of 50—100 ft. Lifting by the buckets can be employed and skidding arrangements made, for lowering and loading the broken ore into wagons or tubs.

Loading is also carried out in certain benches. For purpose of transport, cars, or skips or cableway, or aerial ropeway, can be employed where conditions permit. *Cableway* is similar, in operation, to the overhead cranes, which are used in large workshops. From two strong vertical posts or masts, which are errected beyond the limits of the quarry, a stout wire rope is held taut. A pulley and sheave arragement, mounted on wheels

---

*See addendum.*

is placed on the wire rope. This pulley block can not only transport loads, along the wire rope, but it is also capable of lowering and lifting

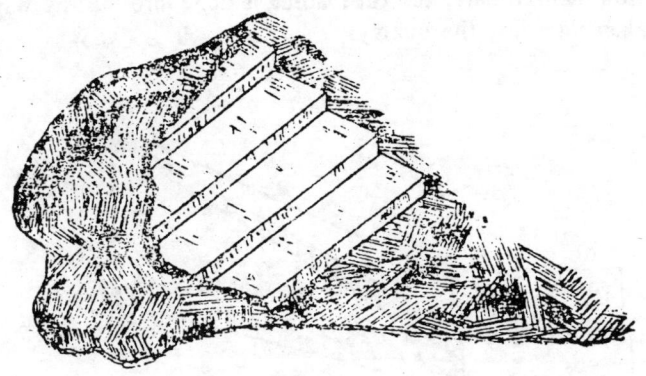

Fig. 8.13

loads. Thus, when the loads, from the quarry, are lifted up, the pulley block can be moved out of the quarry, for discharging the same.

Transport by aerial rope has been discussed later.

The holes may be drilled either by hand or by jack hammers for purposes of blasting. (Please see Chapter on Drilling).

Open cast mining is practised, in the Manganese Mines at Tirodi, M.P., and the overburden benches are 30 ft. high. Blast hole drilling is employed, and power shovels are used for loading into the dumpers/trucks. Bull-dozers do the levelling of the waste. Jack hammers are used in drilling the ore, which is blasted, handsorted and then loaded into tubs or mine cars.

Waste disposal is a serious problem, as the ratio of overburden to ore is very high, especially at Shivrajpur, Kandri, Goamgaon, Mansar, etc. where diesel locos are used. At Shivrajpur, tractors are employed, for the purpose, while at Kandri, steam locos are employed. Self acting inclines, or jigs, are used for the transport of ore, from higher levels (from hills) to the foot of the hills. Electric haulage is used at Goamgoan, Kandri, Mansar for transporting ore from the quarries below ground level to the surface.

In the iron ore mines of Singhbhum, e.g., Noamundi, blast hole drilling is done to break vast quantities of ore, by using large charges of liquid oxygen explosive. The broken ore is loaded into dumpers by electric power shovels (for details see later).

Working, in the Kargali seam, one of the thickest coal seems in the country (Bokaro coalfield), are mainly of the open cast type. Blast hole drilling is also practised here, and the loading is done into railway wagons, which are taken right into the quarry.

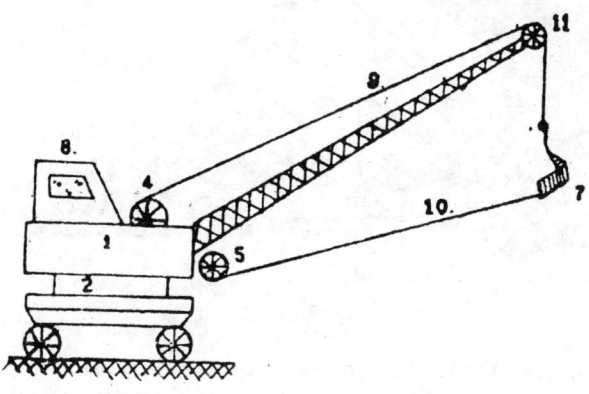

Fig. 8.14

## 2. Open cast mining with loading by machinery :

(a) *Dragline* : A dragline (Fig. 8.14) consists of a prime mover (electric motor, diesel or petrol engine) housed in a body (1). The engine activates the drums 4 and 5, as well as the ropes (9) and (10) respectively. The rope (9) passes over pulley (11), at the end of a long arm or boom (6). The two ropes, (9) and (10), are attached to either end of a heavy scraper or bucket (7). The operator, sitting in the cabin (8), moves the pulley (5), so as to make the scraper or bucket, (7), to swing to and fro from pulley (11), while the rope (9), is held tight. When the swing, of the scraper, has attained a large amplitude, the ropes (9) and (10), are suddenly released. The scraper (7), is flung far away and lands on the ground, with a great impact—digging deep. Now the rope (10) is tightened, and the scraper (7), is drawn along the ground gathering the debris. The operations are repeated, so that the desired excavation, is made.

The dragline is capable of turning round, on a turn table (2). It is either mounted on wheels or skids, or caterpillar etc. There are also walking machines. The dragline has a longer reach, than the power shovel but it is not so convenient, for loading into cars and wagons as it cannot lift up the broken rocks and excavated materials. It can dig below the level of its base, on which it stands, to a limited extent only.

(*b*)  *Power shovel* : The power shovel or mechanical shovel (Fig. 8.15) comprises a prime mover (1), which actuates all the movements. It has a boom (3), and a dipper arm (4), to which is attached the shovel (5). The dipper arm (4), has a rack and pinion arrangement at position near (7), by which it can be thrust forward or pulled backward. The dipper can also

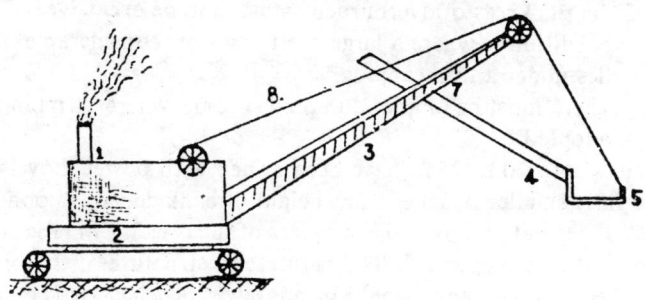

Fig. 8.15

be moved up or down along the boom (3), so as to increase or decrease the distance or depth of operation. The machine turns round on the turn-table (2). In operation, the shovel is placed at a convenient place, close to the work spot. The rope (8) is kept loose. The dipper arm thrusts forward into the debris. The boom (3) is lifted and so also the dipper. The rope (8) is tightened by the drum (6) and so the debris is collected by the shovel, and loaded into trucks or dumpers.

The shovel (5) is provided with strong steel teeth, at the front, for digging. It has a hinged bottom plate which opens downwards and facilitates emptying. Thus the debris falls into the truck or dumper with considerable force and the vehicle should be able to take the impact.

Power shovel is more positive in action than the dragline. By the action of the dipper and boom, the shovel can be made to dig into the broken rock, and lift the same to a considerable height, so as to facilitate loading into truck, dumpers, or wagons. It can operate on fairly hard materials, e.g., consolidated clays, boulders, sandstone etc. Thus power shovels are often employed for stripping overburden, before opening up quarries (Fig. 8.13). This is used in the lignite mines of Victoria, Australia, and also in Neyveli, Tamilnadu on the surface bench and at the Iron Ore Mines at Noamundi, Bihar, Pale, Goa.

Some of the conditions required for the efficient and economic deployment of power shovels are :

1.  There must be sufficient quantity of ore to warrant the use of the power shovel. Further its deployment entails large investments of

capital equipment, as well as preliminary stripping. If the shovel is properly used, the returns are enormous. There should be proper arrangements for servicing and maintenance of the machines or else the time lost, in breakdowns, can be very large and make working uneconomic.

2. The thickness of overburden must not be excessive. The ideal condition is where a large ore body (of considerable thickness) lies under a thin capping.

3. There must be ample dumping room, where stripping is to be adopted.

Banks upto 20 to 25 ft. have been mined with power shovels, but it is better to have smaller benches. The height of bank depends upon : 1. Total thickness of deposit; 2. Physical characters of the ore; 3. Climatic conditions; 4. Method of blasting; and 5. The nature and structure of the ore body.

In theory, high banks would mean fewer benches, lesser movement of the shovel, and greater percentage of working hours, for the shovel. The common heights of banks are 12 ft.-35 ft. The benches should be wide enough for accommodating the loading track, and for allowing shovel movement and 30 ft. is probably the minimum width. The height of the bank and width of the bench should be such that the overall slope, taking into consideration all the benches of the mine, is not too steep. It is customary to estimate the general slope, by the inclination of a line drawn through the outer edge of the top, and bottom benches.

*Casting over* is the term applied, for establishing benches that have either been covered or have caved in. The term is also used to imply the cutting of a high bank into one or more smaller banks. Since there is not enough room on the bench, the material is simply cast or dropped over to the bench below. Where power shovels are used, power shovels can dig 3 ft. below rails, and also load 12 ft.-17 ft. above. After the cut has been made for the full length, the loading track is laid in the first cut. The process is repeated.

The cut is then carried down until sufficient length and depth are reached to start bench work.

*Course hacking* is a term meaning shovel operation, in which no ground is handed away. The depth of the quarry increases greatly, as compared to the area. This sometimes limits working space, in benches, which may in consequence become so steep, that with increasing depth, the haulage costs may become prohibitive, and underground methods may have to be adopted.

Types of shovels :

1. Small revolving machines have a capacity ranging between ½ to 1¾ cu. yd.

2. Rail mounted machines have a capacity ranging between ¼ to 6 cu. yd.

3. Large revolving machines have a capacity ranging between 3 to 16 cu. yd. with booms 85 ft.-425 ft. long or more.

Shovels, mounted on rail, are good for general purposes. Most power shovels are equipped with caterpillar, which eliminates the extra labour on rail laying and shifting and makes the machine mobile. Walking machines are also made, but their efficiency on soft ground is adversely affected. Power shovels can be operated by electricity (A.C/D.C., diesel, or steam. Of all the types, the electrical machines are most advantageous, when suitable source of power is available.

Chief advantages of electric shovels : 1. Cleanliness of the face and pit, as there is no need for supplying water and coal, and for disposing of the ash (where steam shovels are used) ; 2. There is no boiler trouble due to the cold and to the freezing of water lines in bad weather ; 3. There is no delay due to lack of fuel supply; and 4. Cost of repairs are generally low.

Action of power shovels :

1. Digging : the powerful engine, connected to the arm of the dipper by a cable which passes over the end of the boom, when the rope is pulled by the engine, at the same time, dipper digs, into the material while moving up. The dipper is urged forward by the rack and pinion.
2. Crowding motion holds the dipper against the bank and the shovel fills up.
3. The swinging motion is imparted by the turn-table, and the loaded shovel turns away from the face, for emptying the material into dumpers etc.
4. Emptying of the bucket is done by opening a flap or door at the bottom of the bucket.*

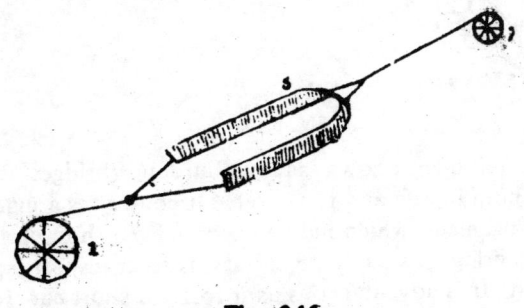

**Fig. 8.16**

---

*See addendum.

(c) *Scrapers* are also used for surface mining. A scraper (Fig. 8.16) comprises essentially a U-shaped steel plate (3), which when dragged along the ground collects broken material to a desired spot, Chains are attached to the two limbs of the plate as well as to the back. Wire ropes are attached to the chains, and passed over pulleys (1 and 2) so that by adjusting their lengths through pulleys, the scraper can be made to move back, and forth. Scrapers are used essentially for collecting loose material and not for excavation (Fig. 8.16). Scrapers are also employed in underground mining in stopes, for pushing ore into ore passes where the dips are low, e.g., breast stoping. These are called slushers.

(d) *Land dredges* (Fig. 8.16a) are similar, in operation, to pontoon dredges employed in alluvial mining. The dredge, in this case, is mounted on caterpillars (1) and is stationed at the edge of a quarry bench, so that the buckets touch the face. As the chain of buckets (2) move upwards, the material, from the quarry face, is excavated and carried along. The material is dropped from the buckets into a revolving screen or tromel (3) and then into a conveyor belt, (4) for loading into wagons for transport by rail. This type of dredge was used in the lignite mines of Neyveli, in Tamil Nadu, and is used, in the lignite mines of Australia and Germany.

(e) In some of the larger open cast workings, as in the extensive lignite mines of Germany, where a large amount of overburden stripping

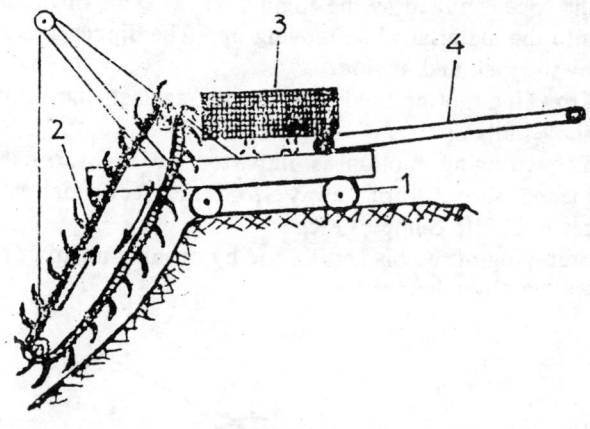

Fig. 8.16a

is necessary, equipment known as overburden "bridges" are employed, for removing both lignite and top cover. It comprises a gigantic T-shaped cantilever arrangement, which moves back and forth on rails. The main purpose of the bridge, as its name implies, is to transport spoil/ore across the mine or pit. It is advantageous as it forms a short cut for back filling the pit, after excavating the coal/lignite, simultaneously with mining operations moving forward. In one type, grabs and buckets are moved on the rails by means of wire ropes, which are operated from the "bridge". The "bridge" itself can be moved forward as it is mounted on rails (Fig. 8.17). The land dredges located at points (1) and (2) on the right are employed to strip the overburden O, and expose the underlying coal or lignite (B). The debris now falls on to a belt conveyor (C), and then to conveyor C, and finally dumped at S, where the lignite has already been excavated.

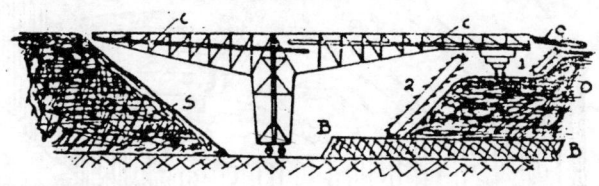

Fig. 8.17

3. *Glory Hole.* In the glory hole method, the pit or quarry is opened up, and developed in such a way, that the working faces are arranged in the form of concentric steps, descending to the deepest or central portion. Sometimes, at the centre of the pit, a shaft (S) is made to connect the glory hole workings with an adit L below (Fig. 8.18). Glory hole is a cheap method of mining and loading ore. It is suitable for massive and thick ore bodies. The nature of the ore should be such that it does not jam the raises. It should run freely when broken. The ore should preferably slide down, in the broken condition, on moderate slopes. Large glory holes become unsafe, due to the danger of pieces of rock falling from the sides, even though the walls are carefully dressed down.

Adverse weather conditions seriously affect the glory hole mining, as wetting of the ore, makes it to clog. Freezing is also a disadvantage. This method of mining is used in some lignite mines in North Bohemia, and also in Mount Lyell mine, Australia.

4. *Kaolin Mining in Cornwall* : This is special modification of open casting and hydraulicking. The operation consists in sinking two shafts, one in the kaolinised ground, and the other in the hard granite, in the vicinity. These two shafts are connected by a cross cut tunnel. A ram pump is placed in the pit, located in the granite country. In the kaolinised patch, a quarry is started by enlarging the shaft. The kaolinised rock is broken up by picks and shovels, and washed out with jets of water directed from monitors. The water carrying the sediments is passed through a sluice box, known as the "sand box" where the coarse sand is separated. The clay slurry is passed through a fine wire screen. It then drops, through

the shaft, and flows into the cross cut, and from there on to the pump shaft, from where it is lifted to the surface. As the quarry expands, circular benches are formed and hydraulicking is continued, on a large scale.

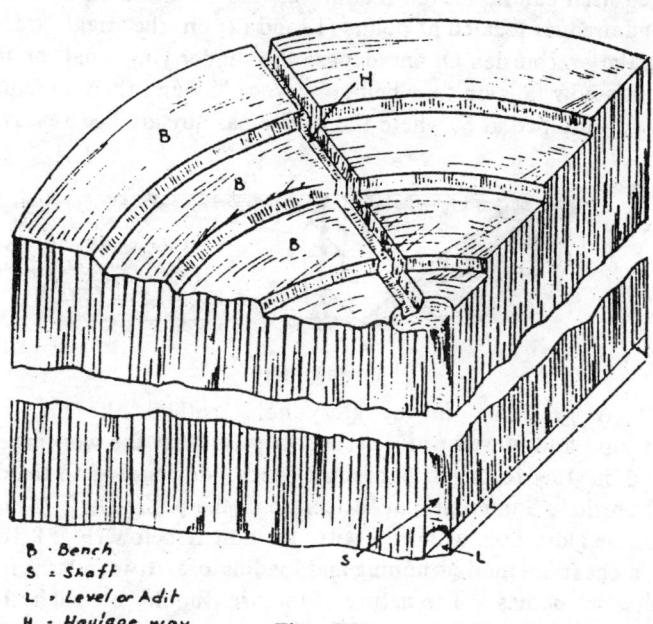

B · Bench
S = Shaft
L · Level or Adit
H : Haulage way

Fig. 8.18

The clay water is conveyed, by shallow open channels, to the shallow tray like tanks—the drags. Here the fine sand is separated. From the drags, the clay water passes into the 'micas', which are also similarly constructed shallow tanks. Here the micaceous minerals separate out, and the water, containing the kaolin minerals, is taken to the settling tanks. After the clay has settled down, the super natent water is drained away, and the clay slurry pumped into filter presses. The "filter cake", from the presses, is transferred to drying floor, heated by steam or hot air. When the clay is dry, it is stocked or bagged for shipment.

*Strip mining* has been mainly used for winning coal. It is also a specialised method of open cast mining. It has been applied with advantage to extract coal seams, which occur under relatively shallow cover, to yield a large output. The method of layout has been described under "coal mining methods". *(See Annexure)*

### III. UNDERGROUND MINING
**A. Open stopes**

1. *Gophering*: The term is applied to small sized, irregular and unsystematic underground workings. It may just comprise drifts, or other

openings, which follow the ore shoot or vein. Such a method may be justified, if applied to work only small rich spots, or ore bodies. However

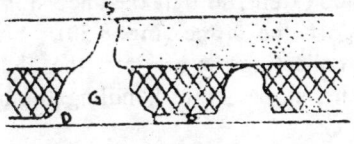

Fig. 8.19

if the very rich streaks, in a large ore body, are removed, it may leave the mine in such an impoverished condition, that the remaining ore cannot be mined profitably afterwards. This method has been described as "plucking the eyes". The method has been adopted by the ancient miners of our country with great skill, e.g., Rajasthan and Singbhum copper areas, diamond mines of Banganapalli, Kurnool, and also in the gold bearing areas in Chota Nagpur (old Jashpur State) (Fig. 8.19).

2. *Breast stoping:* Breast stoping is a mining system, in which the working face is vertical, with a maximum height of about 10-12 ft. The stope advances horizontally. It is employed in horizontal, or low dipping ore bodies, upto 15-18 ft. thick, when mined by open stoping method, i.e., no supports are used in the stopes. (Fig. 8.20).

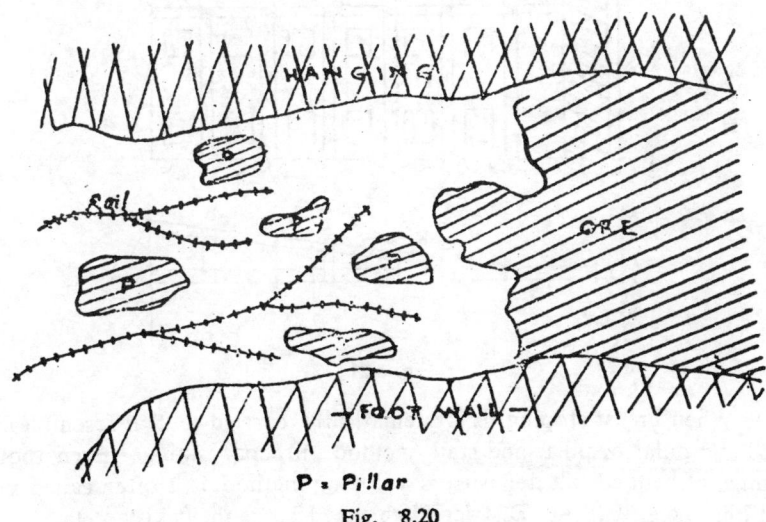

P = Pillar

Fig. 8.20

Breast stoping can be employed when the ore body is not too thick (not exceeding 12 ft.) and when the hanging and foot walls are strong. Thicker deposits are also amenable to breast stoping, if benching is resorted to. Thus a deposit over 100 ft. in thickness can also be mined in this way.

The common method of support employed is by leaving pillars. It is a cheap method, especially suited for the extraction of low grade ores, requiring "selective mining" to a limited extent, so that the uneconomic ore is left untouched as pillars, to support the stope (Fig. 8.20). Scrapers are often employed, in the stopes, for collecting the ore into the chutes or ore passes. Tracks are also laid, in the stopes, for handling the broken ore.

This method is extensively employed in the Rand Gold mines. Here the stopes are developed systematically and symmetrically, on either side of a main raise R. In plan the layout has a herring bone pattern. Hence it is known as the Herring bone system (Fig. 8.21.) In order to facilitate transport of broken ore from the stopes, chutes C are made to connect with an adit drive, on the footwall A. The ore is loaded in the adit A, and taken to the main haulage level L.

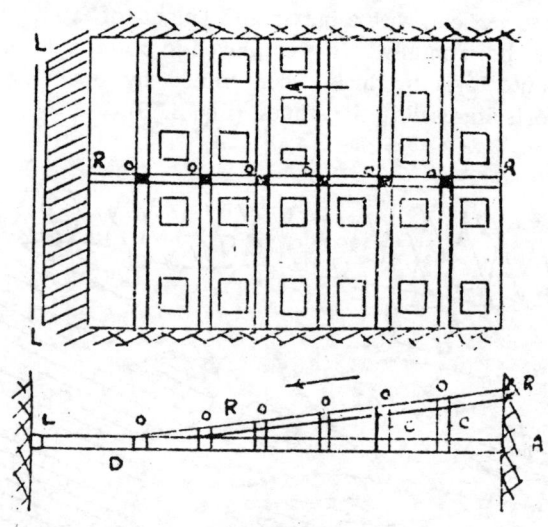

**Fig. 8.21**

When breast stoping is systematically carried out it resembles the bord and pillar or pillar and stall method, in coal mining. Even though mining, of bedded salt deposits, is a stoping method, it is often called Bord and Pillar, e.g., U.S.A. District Michigan, Kalabagh, Pakistan etc.

The following are some of the main advantages of breast stoping :
1.  The mining costs are low.
2.  It is possible to carry out sorting in the stopes.
3.  It is possible to mechanise ore loading, from the stopes into mine cars.

4. Larger output per man, leads to the reduction of men employed.

The chief disadvantages are :

1. Considerable loss of ore during mining.

2. Generally more risky than the filled stopes, or supported stopes.

3. Since the method is chiefly employed in deposits, where the grade of ore varies greatly from place to place, the working is less organised and systematic.

3. *Open underhand stoping*: Underhand stoping is best suited to narrow, steeply dipping veins, with strong walls, and where the ore requires no sorting. This method was practised in Cornwall by the ancient tin miners, and also in the Kolar Gold mines, at shallower depths. (See definitions of mining terms for overhand and underhand stoping.)

From the upper level L stopes are developed through the winze. When, only one side, of a winze, is developed into stopes, S,S,S, it is called single stope. Fig. 8.22, in the upper portion, shows the development of double stopes. Sometimes, where underhand stoping is practised, the raise or winze is connected to the level L below, so that the broken ore falls down, and it is loaded directly in wagons, or cars. The lower portion of the figures shows a overhand stope being developed simul· taneously.

As a protection to the level, a pillar is often left, in the position P. As far as practicable, the mouth of the winze, at P, is not widened, as automatic loading into mine cars will then become difficult.

Where the stope appears dangerous, stulls or props are employed, for

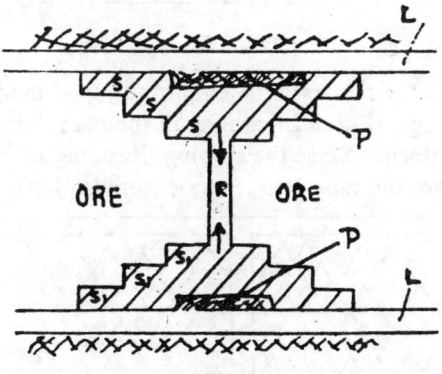

Fig. 8.22

support. Waste packing behind lagging, or pillars are utilised for this purpose. Since most of the broken ore remains at the working face, after blasting, in the stopes, waiting to be lifted out, the underhand method is not popular, in the larger mines, as compared to overhand stoping.

Underhand stoping has the following advantages :
1. The miner, standing on firm solid ore, works with greater efficiency.
2. Where the dips are steep and greater than 45°, overhand stoping involves erection of platform, for miners to stand. This is not required in the case of underhand stoping.
3. The method (comparatively) requires little use of supports.
4. Losses of fine ore are less than in overhand stoping.

With modifications, underhand stoping can also be adopted for mining thicker veins or ore bodies. This method was adopted in some of the old mines, e.g., Kolar, Dariba (Alwar).

4. (1) *Open overhand stoping* (Fig. 8.23): In this method, stoping is started from a raise (R), in the lower level L-L, and progresses upwards towards the next higher level. Stope S, is started and widened, and then advanced, in steps $S_2$, $S_3$ etc. The broken ore gravitates down, and the

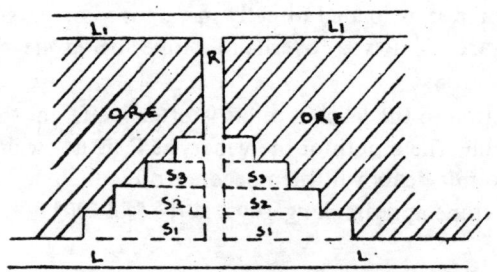

Fig. 8.23

working face is free, for miners to be employed. A modification of this method is the rill stope (Fig. 8.24) where, in the plan, along the vein, the stope shows a V-pattern. Since the stoping is done in inclined slices, S, $SS_2$ etc. starting from the raise R. the stope can also have a flat back, as

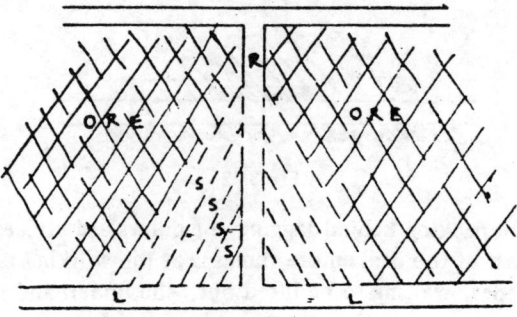

Fig. 8.24

shown in Fig. 8.25.

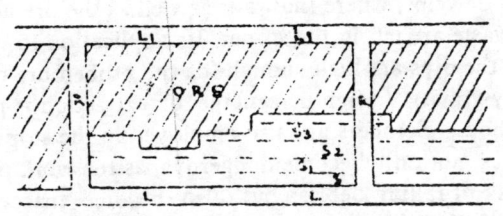

Fig. 8.25

*Open overhand stoping* is more advantageous in the following respects as compared to underhand stoping (See definitions of Mining terms
for overhand and underhand stoping).

1.  Greater distance between levels can be maintained, i.e., the back
    of the stopes are high.
2.  Sorting of the ore in stopes is possible.
3.  Rejected ore or waste can be left in the stope.
4.  Gravity can be utilised for loading ore, even when the dip is low.
5.  Working allows for safety, and the method is applicable to a wider
    variety of conditions.

5. *Underground Glory hole or Milling or Mill hole:* The underground glory hole, Fig. 8.26 is similar to the surface "glory hole", and
comprises funnel shaped excavations. From the foot wall shaft (2), cross
cuts (1), are drawn to intersect the ore body. Raises (3), are made to connect the cross cuts. The top of the raises are widened into the glory

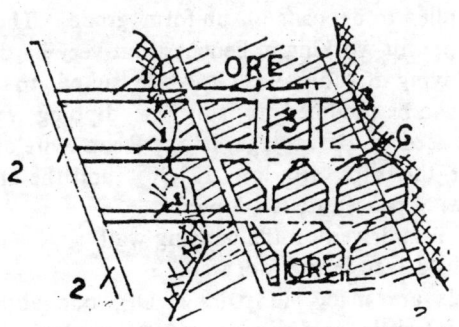

Fig. 8.26

holes G. This method can be employed advantageously, in working large ore bodies or wide veins, where the walls as well as the ore are strong. Low grade ore and waste are left in the stopes. Its application to narrow veins is possible, where the dips are high enough for the broken ore to slide down the foot wall, by gravity. High percentage of extraction is possible by the use of this method. Scrapers are also employed in the stope, for handling ore. Glory holes, not only become dangerous, as the work progresses, due to increasing risk of falling debris, but also become inaccessible, at the same time (Fig. 8.27).

6. *Pillar and Chamber method, or Pillar and Stall, or Room and Pillar* : This method, in its typical form, has particularly become obsolete

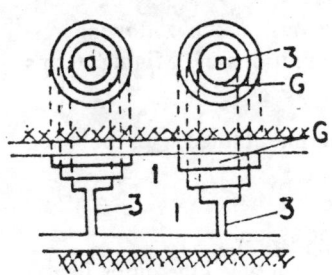

in metal mines. The chambers may be opened up either as underhand stopes, glory holes, or as overhand stopes. Support for chambers is affected by debris, or by square sets, or occasionally by waste. The method, in general, is wasteful and the recovery is only 60% while much of the ore is locked up as pillars. It is applicable, when the ore body is large, and the ore is massive and strong. The walls should be sound and the mineral should be cheap, while the deposit is of uniform character. Some of these methods are practised in mining stratified or bedded deposits, e.g., coal, salt, iron ore etc. and these are discussed later on (see under coal mining methods).

Fig. 8.27

7. *Sub level stoping*: As distinguished from sub level caving to be described later on, was developed in the iron ore area of Michigan, U.S.A.

It can be applied to deposits of uniform grade. The ore should be strong enough to permit working in benches, but very hard ore may make it costly for the driving of sublevels. It is best suited to steeply dipping ore bodies, but it can be modified to suit low dipping veins also. The method has been used to advantage in the Roan Antelope mines, North Rhodesia. Where the dips are low 12°-30°, and the ore body is wide enough, the following procedure is adopted :

1. A raise R is driven on the hanging wall.
2. A raise is driven in the foot wall.
3. Sublevels are made, in pairs (DD$_1$), one above the other, at regular intervals, near the foot wall D$_1$, and also just below the hanging wall D. These are known as hanging-sublevels D and foot-sublevels D$_1$.

4.  A slice is made, by working both the
    hanging wall raise  R and
    foot wall raise.

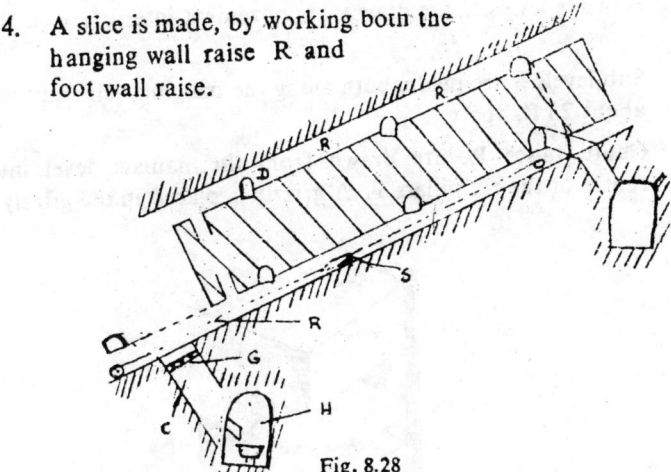

Fig. 8.28

5.  The  broken  ore  falls into the foot  wall  raise into  the  grizzly
    G (or tough screen) and into the  chute C, and  finally into the
    haulage levels (H).   Scraper S is used if necessary.

Where the dips are steeper 45° or more then, modification is made as
shown below (Figs. 8.29 and 8.29a) :

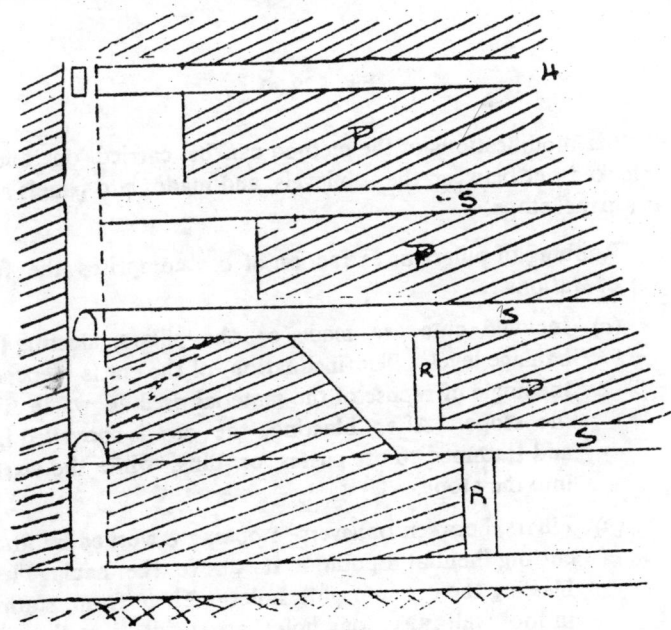

Fig. 8.29

1.  Haulage levels H are drawn, at convenient intervals, i.e., 150 ft. apart.

2.  Sub-levels S are made, both along the hanging wall and foot-wall about 25 ft. apart.

3.  Chute raises R, are drawn from the haulage level, into the centre of the rib pillars P. A grizzly is placed in the grizzly level.

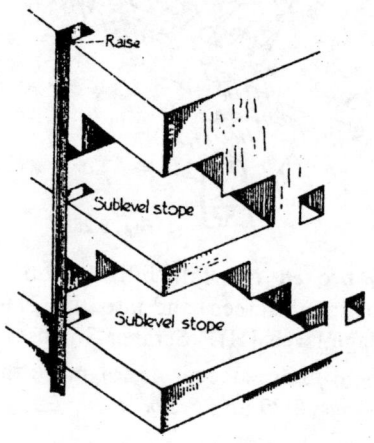

Fig. 8.29 (a)

Further modifications of the method can be carried out and the ore can be blocked out between the sub-levels, and made into panels and developed as a panel stope.

1.  Robbing of pillars or extraction of ore comprises the following operations :

   (a) Inclined cuts are made, on the pillars, starting from the haulage level. The inclination of the cut is dependent on the angle of repose of the material used as waste filling, in the stope. After blasting, the ore is shovelled into cars, and trammed to ore passes or bins, while the waste is run into the stope.

   (b) Pillars, between transverse stopes, are worked by first under cutting them at a point 50 ft. above the haulage level, and blasting the ore into pull holes. Then from stations, cut in foot wall raise, long holes are drilled into the pillar to a maximum depth of 120 ft. The holes are charged with 60% gelatine dynamite, and fired, with cordeau fuse, about 1/10 lb. of powder is used per ton of ore.

(c) Floor pillars are mined, by drilling long slabbing holes, from stations in the foot wall, and blasting the ore into the pull holes, until the pillar fails and falls in pieces.

The sub level stoping method has the following advantages :

1.  It is economic from the point of view of consumption of explosives.
2.  The miners are more efficient in breaking ground.
3.  Where the dip is steep, the ore is loaded by gravity. Scrapers are employed in case of low dips.
4.  Only a small amount of timber is consumed.
5.  Ventilation is more effective; and
6.  There is lesser danger from the occurrence of underground fires.

The method has certain disadvantages such as :

1.  The operation, of the method, involves a considerable amount of development work.
2.  Losses may be appreciable, due to dilution of ore.
3.  It is often necessary to blast ore on grizzlies and ore chutes.
4.  There is no possibility of carrying out sorting in stopes.
5.  The large unsupported open stopes may cause trouble with increasing depths.

**B. (a) Supported stopes**

1.  *Overhand stoping method with supports*: In this overhand method

Fig. 8.30

(Fig. 8.30) a certain amount of supports is used, in the open stopes, but this is limited to props P, and some waste filling (B), (Fig. 8.30).

2. *Timbered stopes (Square set method)*: This method is used where the ore body, and walls, are weak, due to the presence of fractures and faults. The general pattern for forming square sets, with timber, is given in Figs. 8.31 (a, b and c). It will be noticed that the timber is fitted in a manner so as to form skeletal cubes which appear in the vertical direction as well as in the horizontal plane.

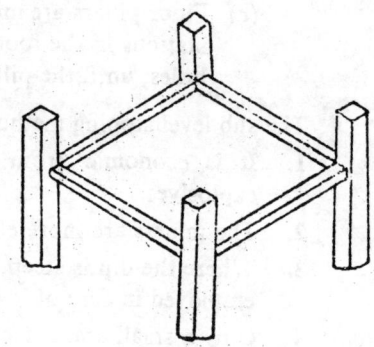

Fig. 8.31 (a)

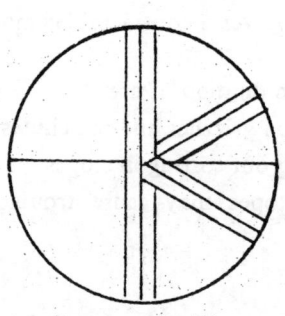

Fig. 8.31 (b)

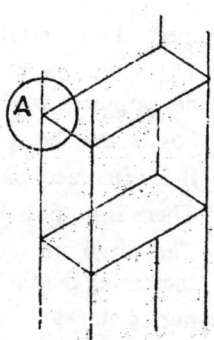

Fig. 8.31 (c)

### 3. Square set method—general

The sets are fitted in position, as soon as possible, after the extraction of the ore. Not more than two floors are opened, at one time, and the waste is run into the stope, through a raise, from the level above, and distributed. "Inclined" square sets are used, in mining horizontal floor pillars. In this way, the ore is removed from the stope, and replaced by waste. The stope is started from a raise in the centre of the pillar. Four rows of sets, are mined, across the full width of the pillar. Then two rows of sets are mined, in the floor, before the filling is run into the stope. In this way, the desired slope, of the filling, is obtained. Once the slope is established, a new cut is started, at the bottom, and timbering is placed, before the filling is run in. Very little effort is required, to work the filling under the caps. Care must be taken, to block the sets well in place, since filling is kept at an angle, which is slightly steeper than its angle of repose, and consequently it exerts a pressure on the timber sets. Usually, while one side of the stope is filled, the other side is mined.

The following are the variations, in the method of development, with square-set timbering.

(*i*) Flat backed or stepped face overhand stopes.

(*ii*) Domed or pyramid stopes.

(*iii*) Rill stopes.

(*iv*) Vertical face stopes, and

(*v*) Underhand square set stopes.

(*i*) *Flat backed* stoping is used in Anaconda, Copper mine, Butte, Montana, Zinc. The conditions at Butte, where square sets are applicable, are:

1. The ore body comprises complicated vein systems.
2. Dips are steep.
3. Numerous faults cause the walls to collapse.
4. Country rock is usually granite (hard).
5. Gangue comprises quartz with pyrite and zinc blende.

The stoping widths vary between 4-100 ft. but usually between 10-30 ft. The development of flat backed or stepped face, overhand stopes with filling is illustrated in Fig. 8.32.

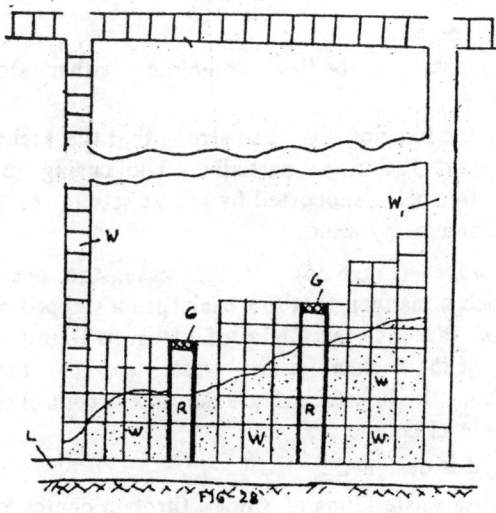

Fig. 8.32

L and L are levels,

R and R are raises,

$W_1$ and $W_1$ are raises for manholes.

G and G are grizzly.

Where enough waste is not available for packing, then the hanging wall is blasted down, to make good this deficiency.

(*ii*)  *Domed or pyramid stopes* :   The stope is dome shaped or pyramidal in section (Fig. 8.33).   This method of stoping is used :

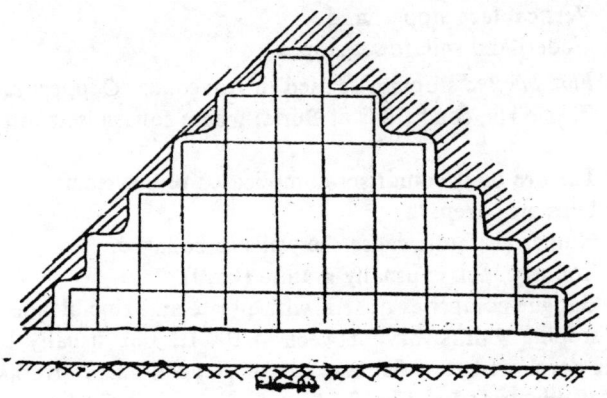

Fig. 8.33

(*a*)  in massive ore bodies, containing either strong or weak ore, and

(*b*)  where the hanging wall is so strong that the arched back is able to support the stope partially. The caving in of the hanging wall, after being supported by square sets, gives the appearance of a dome or pyramid.

(*iii*) *Rill stope*—(Fig. 8.34).  In this case, the overhand stope is developed, in such a manner, that the back has a stepped appearance and he gradient, of the inverted flight of steps, is slightly in excess of the ngle of repose, of the broken ore and waste, so that the broken rock noves down, under the influence of gravity, into a central chute, and falls nto the haulage level below.

The method is designed :

1.   to utilise waste filling of stopes, through chutes, and

2.   for filling the stope, with waste by gravity. The chutes are laid together.

This method was employed in the Kolar Gold mines in India, but with increasing depth, and with increase in cost of timber, it has since been given up.

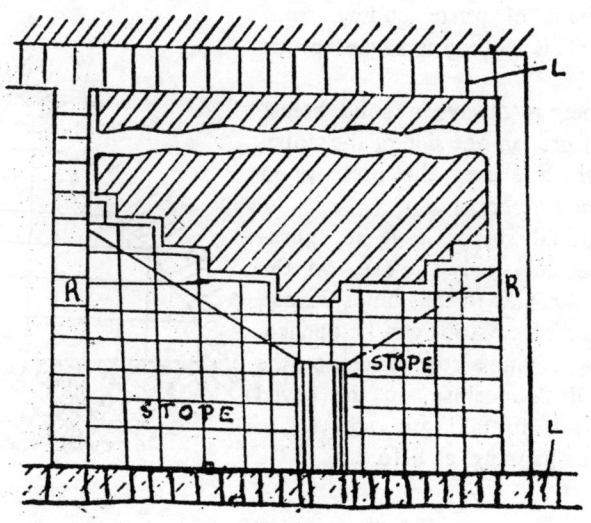

Fig. 8.34

(iv)  *Vertical face stope*—(Fig. 8.35 a and b). The stoping is initiated,

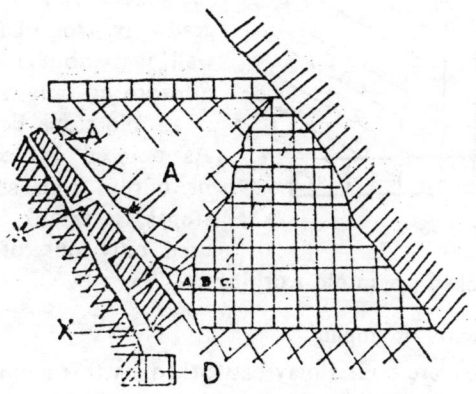

Fig. 8.35 (a)

in the ore body, by placing the "lead sets", A, close to the hanging wall. The stope is advanced, and the set B is placed, while sets C are also inserted, after stoping. By this method, the ore body is worked and the face is kept vertical or nearly so. The broken ore is drawn into the foot wall drive D, through a chute X, while waste filling is accomplished by means of waste chutes, from the surface. This method is advantageous in that :

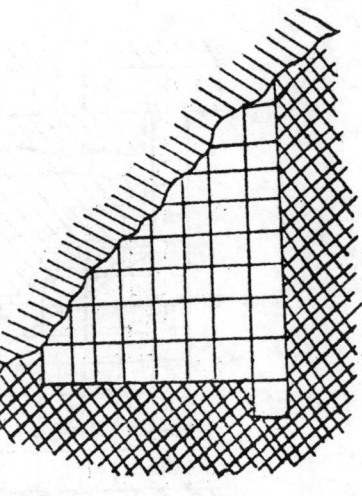

(a) there is always a solid breast of ore, on one side of the stope, which relieves the pressure on the timbers, and

(b) should stopes cave in, unexpectedly, only the ore, on the floors and in the chutes, is lost, and a new stope can be opened, by driving a row of sets, on the sill floor close to the caved stope, for taking it right up to the hanging, as before.

Fig. 8.35 (b)

(v) *Underhand square sets* : This is rather an unusual method (Fig. 8.36). This method is practised, where the ore body is nearly vertical,

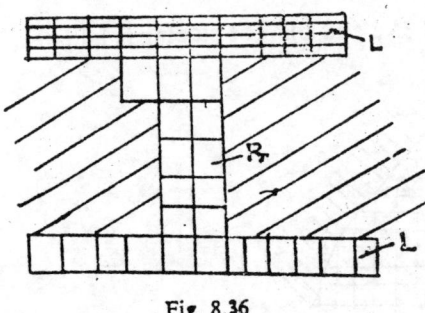

Fig. 8.36

and has considerable width. Stoping starts by making vertical slices between sets, and extended gradually from hanging to foot wall. It is adopted, with the object of exposing only small sections, of an old stope floor at a time. As pointed out earlier, raisings or output from such underhand stopes, is small. The method is used only out of necessity, in reclaiming ore locked up, in old workings.

### Advantages of square-set stoping

1. Irregular ore bodies may be worked by this method. Horses or lenses of lean ore or waste, can be left, in the place.

2. Waste may be sorted and left in the stopes.

3. It is easy to control the grade of ore, sent out of the stope, since each new face can be sampled and assayed, before the

ore is drilled. The flexibility of square set stoping is important, in mines, where the ore varies greatly in value, from set to set.

4.  By fitting the sets, as soon as possible, only a small space is exposed at a time.

### Disadvantages of square-set stoping

1.  The cost of mining is high.
2.  Extraction of the ore is slow, and a small tonnage is mined per man shift.
3.  A large amount of timber is required, and this constitutes a fire hazard.
4.  Square-set stoping has the highest, accident rate of any underground stoping method.

### B. (2) Filled stopes

(i) *Cut and fill methods of stoping or filled stopes* : In this method, the stope is quickly supported, either by stowing, or packing, or by running in waste etc. For the successful operation of this method, it is preferable that the material of the ore body is strong, while the walls are weaker. The method has several modifications depending on the situation obtaining in a particular mine. Steep dips are an added advantage, to the operation of the method. The ore is usually loaded into mine cars by gravity, but scrapers may also be employed, and dilution and contamination are prevented, by laying a layer of planks over the waste filling. As waste filling is let into the stope, the chutes are raised upwards, by adding additional sets of timber. Occasionally, pillars are also left in the stopes, for added support, and these may be mined later on by using square set stoping.

The waste required for filling is obtained by :

1.  blasting the walls, where the wall rock is strong, or
2.  by driving cross-cuts into the walls or using waste from cross cuts, driven for exploratory purposes, or
3.  by obtaining waste from the surface. This may comprise mill tailings, waste rock, or sand etc. or
4.  Waste may also be got by opening glory holes, in the barren ground, or
5.  "Caving in" for waste is also done. This is done by running in the overburden, by placing workings quite close to the surface, so as to cause subsidence.

(i) *Filled flat back* and stepped face overhand stopes, or horizontal cut and fill stopes—for Narrow Veins (Fig. 8.37).

At first, the back of the stope is opened up, and a narrow slice is cut behind the timbering, above the lower level (P). As stoping progresses, chutes or mill holes (C), or ore passes are built at regular intervals. As the stope is filled with waste, the upper end of the chutes are built up, and the chutes increase in length. The chutes also contain man ways. Entry, into the stope, can also be had at the point X [Fig. 8.37(a)].

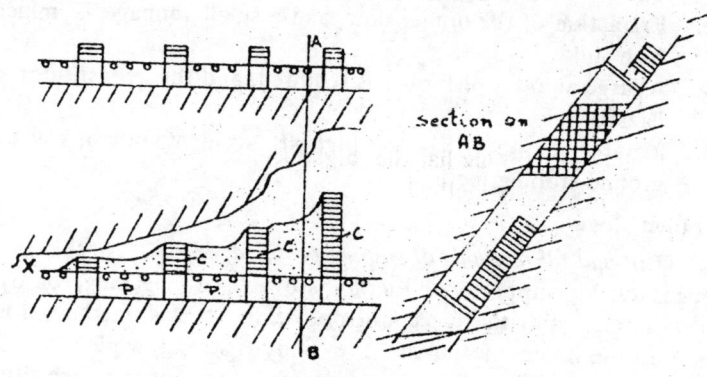

Fig. 8.37 (a)

Sorting is possible in these stopes, but within limits.

(ii) *Filled flat back* stopes or horizontal cut and fill stopes—for wide ore bodies [Fig 8.37 (b)].

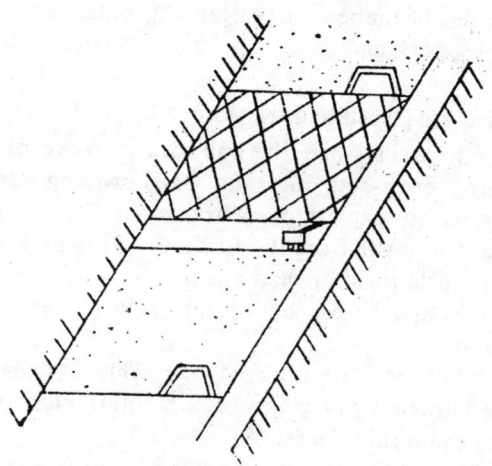

Fig. 8.37 (b)

Favourable conditions :
1. The ore should stand unsupported, at the back of the stope. No supports, other than posts or cribs, resting on filling are employed.
2. Strong walls are not required.
3. General dip should be steep, except for local variations.
4. Filling in is employed, in case of weak ore.

This method is essentially the same as that employed in the case of veins, except that :
 (a) filling materials, in the case of wide stopes, have to be obtained from outside,
 (b) transport of ore may have to be arranged within the stope, and
 (c) a chute or raise, for conveying waste, for filling in, is to be driven in the foot wall. This method is employed, in some of the mica mines of Hazaribagh, Bihar, with some modifications.

 (iii)  *Baltic dry wall method—filled flat backed stopes—for wide ore bodies* (Fig 8.38).

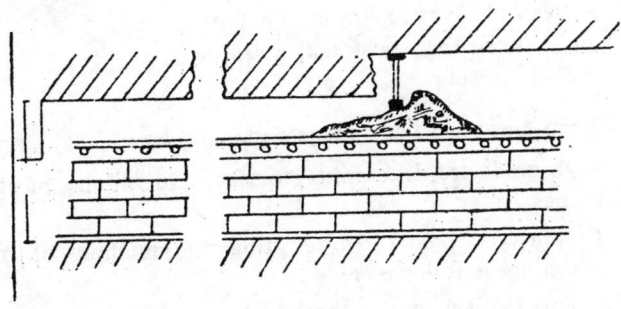

Fig. 8.38

In the dry wall method, no timber is used, except for temporary supports, and only dry stone pack walls are used, to support tunnels which go through waste filling, and also for supporting drives and chutes. From the well supported level drive the top lagging is removed, and a narrow cut is made. The cut is then widened into the flat backed stope. The ore is blasted by drilling holes and the blasted ore is sorted, and dropped through the lagging for loading into wagons. Waste filling is also done side by side.

In Kolar gold mines, complete stone packing of the stopes is the common practice. With increase in depth, it has been found that only dressed granite blocks are satisfactory for supporting stopes.

(*iv*) *Resuing or stripping* is used in mining narrow veins, or pay streaks, where the ore occurs as a distinct band, which can separate easily from the wall rocks (Fig. 8.39).

High grade ore occurs only in the narrow portion XY, and this is clearly separated from the barren rock PQ, by gouge or crushed rock G.

A flat back stope is opened from a lower level. As the stope advances, the cribs or chutes (C) are also built up. At first, the barren portion is blasted down, and it fills up the vacant portions, of the open stope, all around the chutes (C), then the pay streak XY, is broken, and the clean ore drops down into the chute, and is loaded directly into the wagons (W).

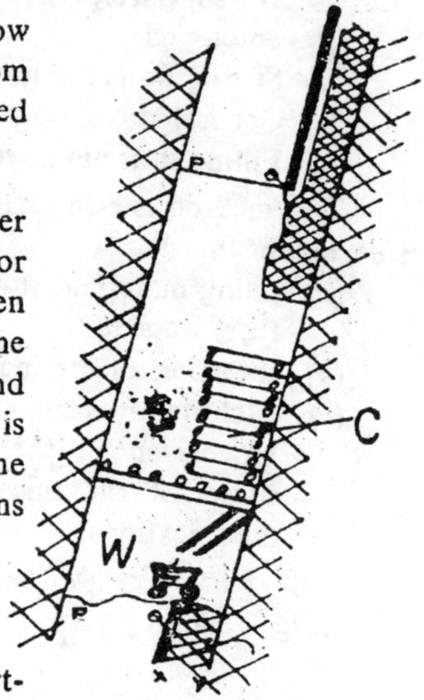

Favourable conditions :

(*a*) A clean plane of separation or parting between the pay streak, and the walls ;

(*b*) Steep dips are an advantage ; and

Fig. 8.39

(*c*) High grade ore, is essential, for economic mining of narrow ore bodies.

(*v*) *Cross-cut method* : In adopting this method of mining the following procedure is followed :

(1) A shaft (S) is made, on the hanging wall.
(2) Cross cuts C & C, are driven, at suitable level intervals.
(3) The foot wall drives D, are made in the ore, from the cross cuts.
(4) Raises footwall (R), are developed from the cross cuts, to connect with the cross cuts, at the higher level.
(5) Stoping starts by removing slices of ore, along the strike $S_1$, $S_2$. $S_3$ etc. The foot wall drive is taken as the stoping level (Fig. 8.40).
(6) As the stoping of the slice proceeds, filling is also carried out with waste, which is dropped in, through the foot wall raises, from the level above.
(7) After filling is completed, in a slice, slicing is extended horizontally and laterally.

(8) Then slicing is repeated in the same manner ; slices 1, 2, 3, 4, 5, 6 etc. are removed, in order, in the upper levels.

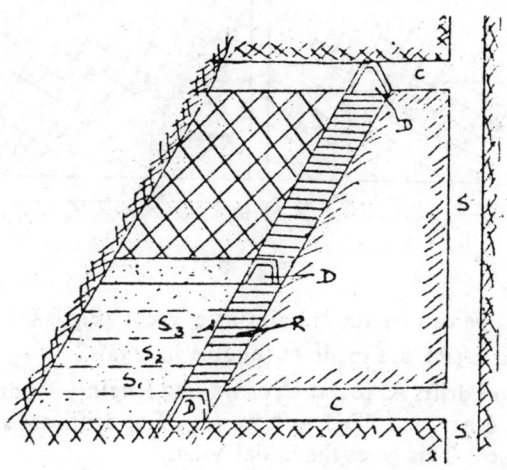

Fig. 8.40

Slices are sometimes mined, in the descending order, as well. The cross cut method allows for the safe, and complete extraction of large and weak ore bodies, of almost any shape, with no serious surface subsidence. The cost of development is high. Timber consumption is, however, low, if the ore is strong enough to stand, until filling is completed. Even otherwise, the cost of timber is low, generally, lower as compared to most of the other methods.

(vi) *Inclined cut and fill stopes or filled rill stopes* : This method is applied, where dips are steep.

In this method (Fig. 8.41),

(1) Winzes W, are driven first, at regular intervals.

(2) An opening is made, at the back of the level, called the cut out stope (S).

(3) Chutes (C), are placed between the winzes, at convenient points (Fig. 8.41).

(4) The slope, of the stope face, is kept at about 40°, so as to facilitate rilling, or flow of broken ore.

(5) The broken ore falls through the chutes, into cars standing on the level below (L).

(6)  Waste filling is used to maintain the gradient required, in the stope, for effecting rilling of ore.

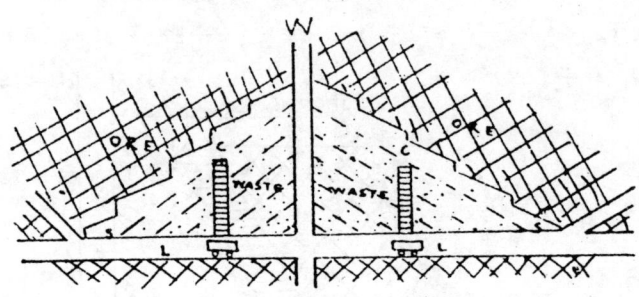

Fig. 8.41

A rill stope can be started also as a V-cut (Fig.8.42), in which

(1)  Raises (R), are made at suitable intervals.

(2)  Three drifts A, B and C, of varying lengths, are driven, one above the other, and filled with waste. The drifts are started, from the raise. This gives the initial V-cut.

(3)  The stope P, is started on the slope afforded by the superposed drifts A, B and C.

(4)  After completing slice P, slice S is taken up, and thus the V-cut of the stope advances, when slices $P_1$ and $S_1$ advance (Fig. 8.42).

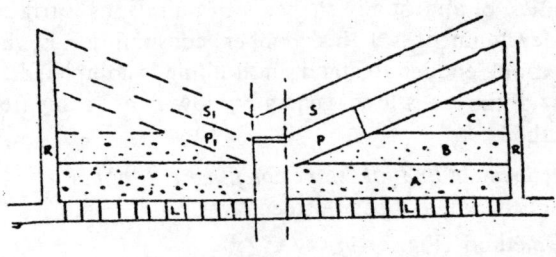

Fig. 8.42

Rill stopes are commonly used to various extents, in practically all metal mines, wherever conditions are favourable. Rill stoping is practised where the gradient is favourable, e g., Kolar in Mysore.

Advantage of cut and fill methods in general :

1.  Weak walls, are supported, and the general stability of mine is maintained.

2.  High recovery is possible (upto 100%).

3.  Any irregular extension, of the ore body, can be easily mined.

4. Waste can be sorted, in the stope, and pockets of gangue can be left behind.
5. There is less dilution of ore, than in shrinkage stoping.
6. Secondary breaking can be done, and the choking up of the chutes can be prevented.
7. Accident rate is less, in cut and fill than in shrinkage or square set methods.
8. Ore is removed as fast as it is broken, hence, capital is not locked up.
9. Stopes are easily ventilated.

Disadvantages :

1. The output, from stopes, is limited or else irregular, due to intermittent breaking of ore.
2. Filling must be provided, as and when required.
3. Cost of mining is generally greater, than for shrinkage stoping, but less than that of, the square set method.

### C. Shrinkage stopes

In this method of mining, overhand stopes (S), are developed and the broken ore is not removed from the stope, but utilised to support the stope (Fig. 8.43). Soon after each blast, the broken ore fills up a considerable part of the stope, and the stope appears to shrink or become smaller in size. Hence the method is named shrinkage stoping.

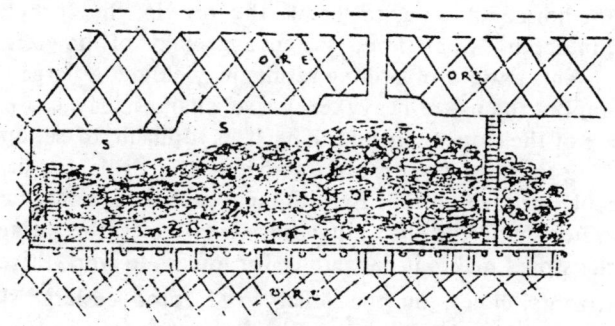

Fig. 8.43

Some amount of ore has, necessarily, to be removed from the stope, as broken rock, occupies 30-50% more volume than solid rock. Waste filling is used for supporting after the ore is withdrawn from the stope.

This method is used, in steeply dipping lodes, with strong walls.

Narrow veins are amenable to complete extraction, but level pillars may be necessary, for preventing premature collapses.

In the case of wide massive deposits, shrinkage stoping is not advisable, as the size, of the unsupported stopes, is limited, by the nature of the ore and wall rock.

Thus large pillars may be required for supporting stopes. After mining by shrinkage stoping is over, the pillars may be extracted. by caving methods.

Complete recovery, may necessitate the use of square sets, or the use of top slicing methods.

Shrinkage stoping is practised in most metal mines, when the conditions are favourable, e.g., within limits in the K.G F. at Ghatsila etc.

**Shrinkage stoping procedure**

In starting a stope, two methods are used, in supporting the floor, at the back of the level.

1. The sill floor may be entirely cut out, and may be left unlined. This is commonly practiced, in narrow stopes.

2. The roof may be supported by pillars. These pillars afford sufficient support, and no great tonnage is lost, by leaving such pillars.

With drills mounted on vertical stulls, placed between broken ore at the back of the stope, holes are made, either horizontally or at a slight angle to the horizontal, so as to break the ore in the form of benches, 6 ft.-8 ft. in depth. Sometimes, the arrangement of holes is fan-shaped or radial. This pattern of drilling results in yielding a large tonnage of broken ore, but the use of heavy explosive charges may cause shattering, of the back of the stope. In addition, it is difficult to set up the tripod, for attaching the large stopers or large pneumatic drills, for drilling vertical or steeply inclined holes. Even though, such drill holes, when blasted have a greater shattering effect, than horizontal holes, the weakening of the back of the stopes makes it dangerous, for miners to work there.

Horizontal holes, on the other hand, produce large slabs, which require breaking and the use of grizzlys i.e., screens made of iron bars, or rails for sizing.

Inspection of the stope, is essential, after blasting. Pillars left are extracted either by (1) the under cut method or (2) by using square-sets.

**Shrinkage stoping compared with other methods**

In mining deposits with weak walls, much stoping cannot be done, without the risk of dilution. Hence supports are essential.

If the ore is strong, then the cut and fill method is advantageous. Waste filling serves to support the weak walls, but not the ore. If both the ore and the walls are weak, then the alternative method, of mining is top slicing, or a method using square sets. Mining by square sets gives a high percentage extraction, but it is costly, and hence it is only suitable for high grade ores.

Top slicing is a good method, for soft ores, with weak walls. Since it is a caving method, the surface is necessarily affected or destroyed. In many ways, sublevel stoping is applicable to the same type of deposits, to which shrinkage stoping is, also, applicable. Sublevel stoping, however, requires greater amount of development work, but it gives larger production, on account of the greater number of faces. In narrow ore bodies, sublevel stoping will probably be more costly than shrinkage stoping, because of the large amount of development work necessary. It is also to be pointed out that in shrinkage stopes, the ore is liable to oxidation, clogging, and spontaneous combustion.

**Advantages of shrinkage stopes**

1. If the ore is strong, and the walls are firm, a stope can be quickly developed to produce ore, since the amount of development work is not large. This feature is of advantage to small mining companies, with limited capital.

2. Broken ore, which is left in the stopes, serves to support the walls, and thus largely eliminates the use of timber. Occasionally props and stulls are required, to support loose ground.

3. As the miners work on a solid floor, they obtain a firm footing and work with greater efficiency.

4. A large number of men can work, with advantage in one stope.

5. Long ore passes, from level to level, are not required, since the ore is drawn from the bottom of the stope, through a number of chutes.

6. It is not necessary to tram the ore (i.e., transport the ore along the levels).

7 Large reserves of broken ore are maintained, in the mine, and these can be drawn when required, without much loss of effort or time

8. Large blocks of ore may be left, in the stope, if so desired, instead of blasting, for delivering into the chute.

9. Good conditions, for ventilation of workings, are obtained.

10. The method is cheap when applied, under suitable conditions.

**Disadvantages**

1. Although rapid development is possible, only 1/3 of the broken ore, from a stope, is available at hand. The rest is locked up in the stope.
2. Broken ore lends support, but dilution of ore is a consequent danger.
3. Chutes must be closely spaced, if the ore is to be drawn out easily, from the stope.
4. Capital is locked up, since a large quantity of ore is left in the stope.
5. It is difficult to maintain man ways, through the broken ore.
6. There is little chance for sorting, in the stope, and the broken ore hides any traces of the lode, or ore shoots, and causes some inconvenience in planning further development.
7. Oxidation may lead to fires. Oxidation of the ore may also cause the ore to become less amenable to flotation (i.e., beneficiation).
8. To run the waste packing, into the stope (when the ore is finally drained out) is a difficult task.
9. It is difficult to travel over broken ore, when the drawing out of the ore is irregular and uneven. Drills must be removed before blasting.
10. Collapse may occur during drawing of ore, from the stopes.
11. The air, in the stopes, may be poor in oxygen, unless care is taken.

**Comparison between sublevel and shrinkage stoping**

1. Sublevel stoping is safer, since the men are protected, by firm ground.
2. Less secondary breaking is required, in sublevel stoping.
3. A larger daily output is possible, since the drawing can be done at a few chutes, thus increasing the efficiency of drawing and hauling.
4. A larger percentage recovery is possible than in shrinkage stoping, in which certain amount of ore is locked up', in tightly folded areas of the foot wall.
5. Lean ore may be left as pillars, but this is not possible in shrinkage stoping.

**D. Mitchell slicing system**

Mitchell slicing system is a special modification of the square set stoping method, which is applied under certain conditions :

1. When the ore is (generally) flat bedded, i.e., with little or no dip.
2. When the hanging wall is well defined, and can be well supported.
3. Where the ore bodies are relatively free from inclusions, or patches of waste.
4. Where the thickness of the ore body is about 50-60 ft.

The general plan is to form a flat back stope, near the hanging wall, and to support the hanging wall, by square set timbering. The ore block, under the square sets, is termed a pillar. By continuing the square set supports, the pillar is mined, using the underhand stoping method. The method is in fact, underhand square set stoping.

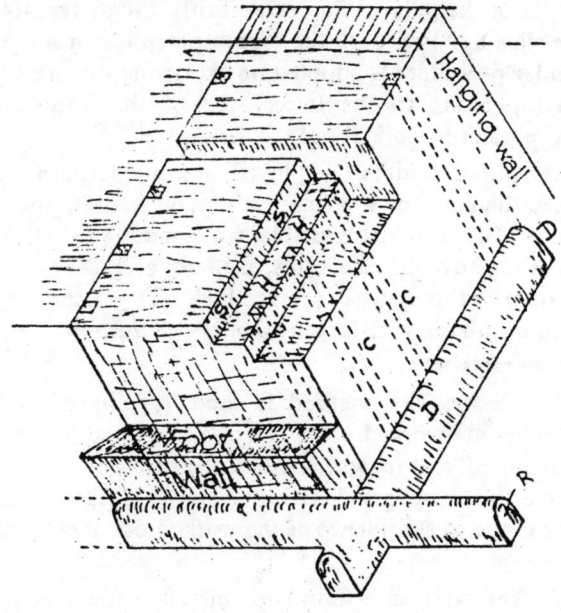

Fig. 8.44

The following is the sequence of operations involved :
1. The main working drift R, is driven, along one of the walls, say, in the foot-wall.
2. From the working drift R, cross cuts D are driven 20 ft. apart, from foot to the hanging wall. These are called the "slice leads" and these are timbered with sill-floor sets.
3. A raise (not shown in the diagram) connecting the lower or mining level, is made to the upper level, and this pass, which is termed "man way", is used as waste filling chute later on.

4. The sills, above the slice leads D, are removed, and chutes H fixed.

5. Two adjacent slice leads, in an upper level, are developed as flat backed square sets stopes S, and stoping is carried on, until the hanging wall is reached. Such stopes are called lead stopes, as broken ore falls into the wagons, from the lead stopes, through the chutes.

6. Slicing of the pillar is started at the topmost step, of the lead stope, by drifting across, to meet the corresponding step of the lead stope, on the other side. The ore rolls down, by gravity, through an inclined drift, connecting, each lead stope to the centre of the pillar. Succesive drifts are driven, to remove the ore slice by slice, until the topmost step is stoped out.

7. Similar procedure is adopted, in extracting the next lower slice, corresponding to the lower step of the lead stope, until the entire pillar is worked out.

Ordinary pillar widths employed, vary between 25 ft.-50 ft. Maximum safe height, for working is about 60 ft. even though in some cases, a height of 100 ft. is also possible, when mining in two lifts of 50 ft. each. Some of the advantages of this method are, (1) saving in labour, and time as compared to square set method, (2) greater safety, (3) increase in speed of mining with larger tonnages of output.

E. Caving methods

1. (a) *Top slicing*: The method is used (a) where wide veins or massive ore bodies are worked, (b) when clean ore, and high percentage of recovery are required, and (c) when neither the roof nor the ground surface are not required to be supported. In other words this method is applicable, where caving in, due to subsidence of the worked out area, is of no consequence.

*Method* : The ore is mined, in horizontal floors, comprising $S_1$, $S_1$, $S_2$ slices. The slices are made in the descending order. When each slice is completed, the roof is allowed to cave in, before the adjacent slice is taken up. Mining is started from the hanging H, and foot walls W, and the last slice to be removed is near the raise (Fig. 8.45).

Sometimes the slices are inclined, and the method is called inclined top slicing. If the inclination of the slice is small, then shovelling is required, for handling ore, in the stopes.

Advantages of top slicing

1. Top slicing is a safe method of mining in heavy ground, which is not amenable to any other stoping method. It is economical, where timber is plentiful, and available at reasonable cost,

2. A very high % of extraction of ore is possible, with practically no dilution from the capping and walls.

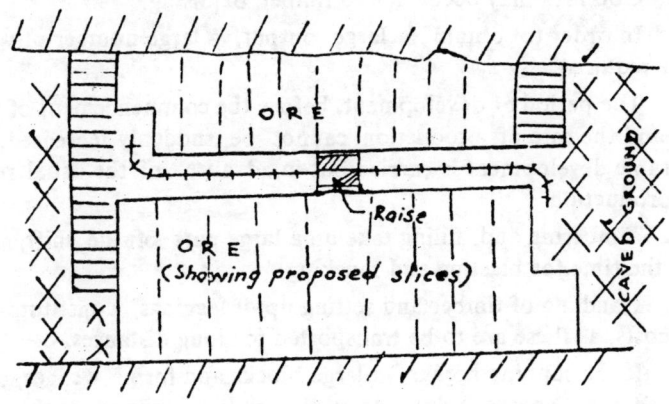

Fig. 8.45

3. Under proper supervision, the method is safe.

4. Where the market conditions are such, that the mine is faced with occasional shut downs, the stopes or slices may be blasted down, so that the ore body remains in good condition, ready for resumption of operations at short notice, after idle periods of considerable duration. Development headings may, however, require considerable repair work, which will have to be carried out, in close proximity to the caved ground.

5. Although much development is required, considerble tonnage can be got.

Top slicing can be employed even where the overburden is sandy, or is made up of unconsolidated materials, and where the risk of contamination is not so much as in the case of sublevel caving.

**Disadvantages of top slicing**

1. Supply of timber and lagging must be adequate, and the cost also should be low.

2. It is more expensive than some other methods, but is comparatively free from the risk of dilution.

3. Where the surface is to be protected, this method is not employed. Thin lodes occurring at depth can, however, be worked with advantage.

4 Ventilation is not so easy. Sulphide ores, while oxidising

generate gases.  Ventilation is important.  Booster fans are used under such conditions.

5.  Gob fires may occur due to timber oxidising.

6.  In order to obtain a large output, a large number of working faces are required.

7.  The period of development, before the commencement of slicing, is long, and the rate of production cannot be suddenly increased, unless considerable development is carried out in advance, of the usual required rate of production.

8.  Timbering and filling take up a large part of the shift, and this reduces the time for blasting and mucking.

9.  Handling of timber and setting up of laggings, constitute a large part of costs, as these are to be transported for long distances.

10.  If the capping breaks, in large blocks and forms wedges, so as to leave large open spaces below, then the sudden collapse of the hanging wall presents a serious danger.

11.  This method is not adopted, where sorting of waste, in stopes, is required..

(*b*) *Sub level caving:*  This method (Fig.8.46) can be used where the ore body is wide, and comprises soft or loose material.  The wall rock' or superincumbent rock should cave readily, and give rise to doming or arching, which, in its turn will lend temporary support, to the openings.

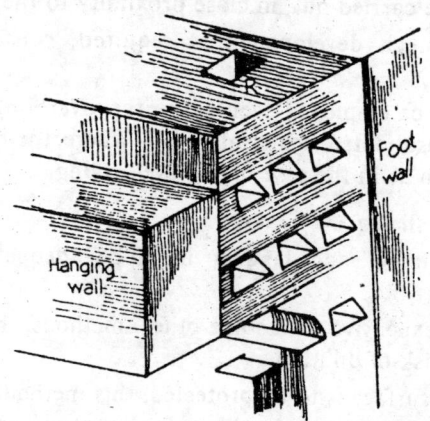

Fig. 8.46

In this method of mining (1) levels L are driven, and cross cuts are made between hanging and foot walls so as to form level pillars.

(2) In between the levels, $S_1$, $S_2$, $S_3$, $S_4$ $S_5$ are driven, and sub level cross cuts are similarly driven between the hanging and foot walls forming sub level pillars. The development of the sub levels is mainly carried out through raises R, driven upwards, from the levels. Mining starts with the pillars, in top most sub level, where the side lagging is removed to expose the ore, on the sides, and also on the roof of sub level. Holes are drilled radially and the ore is blasted down. As the broken ore runs into the sub level, it is loaded, conveyed and tipped into the raise R. The timber used in supporting the top of the sub levels, also caves in and forms a mat, which rests over the lower sub level.

**Advantages :**

1. The cost of mining heavy ground (soft rocks) by this method, is comparatively low.

2. The cost of timber is lower, and less development work is required, than in top slicing.

3. The ore is mined rapidly.

4. Under favourable conditions the percentage of extraction is high, with less of dilution.

5. The method can be applied to soft, and sticky ore, which is not suitable for block caving.

**Disadvantages :**

1. There is, usually, more dilution in this method, than in top slicing, and square set stoping.

2. There is practically no possibility of sorting ore, in the stope.

3. Stopes are difficult to ventilate.

### C. Block caving

Block caving is the ultimate development, in the caving methods of mining. Pillars are left, as in the case of room and pillar mining. A block of ore B, which is supported by pillars, and which has been isolated by raises R, on all the four sides is ready for caving. This is done by driving tunnels T, at right angles to each other, as in the board and pillar method. The spacing and size of the pillars, supporting the block. depends on the nature of the ore. Block caving can be used, in low grade ore. The blocks caved can be 200 ft. to 250 ft. long, from 100 ft to 125 ft. high and can be extended to the full width, of the ore body (Fig. 8.47). After the block is undercut in this way, the supporting pillars are blasted, simultaneously, so that the block crashes and caves in. The broken ore takes several weeks to settle, and flow in. It continues to crush itself and in some cases, it takes about 6 or 8 months for about 80% of the ore to be extracted. When sufficiently crushed, a closely timbered cross cut, can be driven through the broken material by using fore poling.

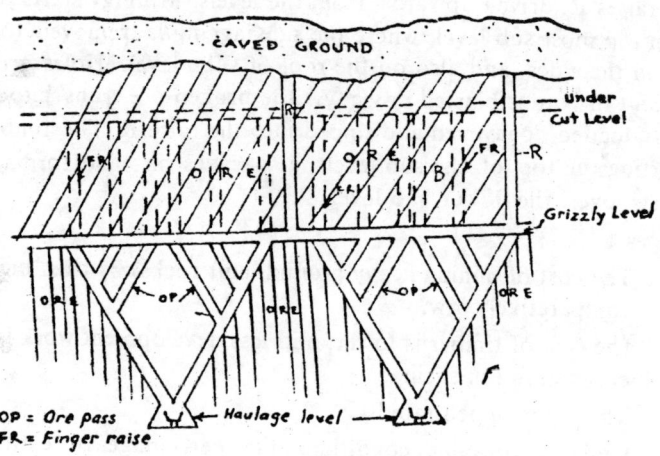

Fig. 8.47

Drifts may be run, from the cross cut to the boundary of the block, and the ore drawn in at the ends of the drift and shovelled into cars. As soon as the waste capping is reached, a timber set or two, from the roof is removed further back, so that more ore is drawn in. By this method a large volume of ore can be mined, after a small amount of development work has been done, but the ore is to be loaded into cars, by manual labour.

Block caving is adopted at Ohio Copper Mine (See Fig. 8.47).

Large blocks of ore could be caved with advantage,

1. If the blocks are properly undercut.

2. If the drawing off, of the ore, is well controlled, under proper supervision.

3. If the chutes, into which the ore caves in are evenly distributed.

4. There is some danger of fire, from timber, accummulating from successive caving operations.

5. The lower grade of ore, occurring near cappings or overburden, and close to the lease boundary, is lost.

**Conditions**

The method is applicable to wide veins, or thick beds, and to massive homogenous ore bodies overlain by ground, which readily caves in.

There are three main variations, of the method.

1.  Dividing a level area into either rectangular blocks, or often into square blocks, and drawing evenly over the entire area, this procedure helps to maintain an approximately horizontal plane of contract between the broken ore and the caved capping rocks.

2.  Dividing a level area, into panels, either crosswise or lengthwise, in the ore body, and extracting one panel after another, in such a manner that the plane of contact, between the caved capping and broken ore, is inclined.

3.  The ore body is not divided into panels or pillars, in any level, but under cutting is done, from wall to wall, retreating as mining progresses, from one end of the ore body to the other, and keeping an inclined plane of contact, between the broken ore, and the caved capping.

# PART IX


# COAL MINING METHODS

# COAL MINING METHODS

The typical or characteristic methods of underground mining of coal can be classified under the following. heads :

- (a) (i) Board and pillar or pillar and stall or room and pillar method
  (ii) Board and pillar with panels
- (b) Longwall advancing
- (c) Longwall retreating
- (d) Horizon mining
- (e) Miscellaneous
  (i) Underground hydraulic mining
  (ii) Strip Mining

In practice, however, these water tight compartments can be given up, and under suitable conditions a happy combination, of methods of working, may have to be adopted.

*Board and pillar* is a simple method of mining and comprises in driving tunnels or galleries, within the seam, right from the shaft bottom in two main directions. The two directions of the galleries, under ideal conditions, are the dip direction and a direction nearly at right angles to it, i.e., approximating to the strike. Thus, soon a pattern of intersecting tunnels is produced with intervening rectangular or square pillars of coal. In India, the strike galleries are termed level galleries and the dip galleries are known as either dip or rise galleries, depending on the location with reference to the shaft, strike direction etc. The mine plan thus comprises a miniature net work of rectilinear tunnels.

The optimum size of pillars, to be left, unless otherwise specified, by the Coal Mines Regulations, will depend on the following :

1. The depth of the seam from the surface.
2. The nature of the coal.
3. The mechanical and physical characteristics, of the roof and floor.
4. The time interval, between the development stage and the depillaring stage. That is the interval of time for which the pillar will be required to stand.

5. Geological factors such as the presence of faults, dykes, etc.

6. The percentage of coal which is to be won, during the development stage.

There are thumb rules, devised, for arriving at the size of pillars :

(a) The area of the pillar in obtained, by allowing one square yard for each foot of depth. The area thus obtained is so adjusted, that in the rectangular pillar, one of the sides does not exceed the other by 100%. i.e., if one side is X yards, the other side should not be more than 2 X yards.

(b) Another method, of calculation, is based on the volume per cent of extraction. In this method, for a depth of 500 yds., the extraction is fixed at 20%. With the decrease in depth by 100 yds. an additional recovery of 3% is allowed.

Thus if the depth of the seam is 300 yds. the percentage extraction (20+3+3) or 26% or 25%, Assuming X ft. to be the length of a square pillar and the width of galleries to be 6 ft. then each side of a square pillar can be computed. Depth = 300 × 3 = 900 ft.

according to (a) Suppose, Y = area of pillar = 900 × 9 = 8100 sq.ft.

= 8100 sq ft. as obtained by the first method, or each side will be (90 — 6) ft. = 84 ft. allowing 6 ft. for galleries.

According to (b) if the area of pillar = 8100 ft., 25% of 8100 sq ft. is extracted.

∴ area of pillar will be 8100 — 2025 = 6075 sq ft. assuming a square pillar each side will be

$$\sqrt{6075} \text{ or } 77.9 \text{ ft or } 78 \text{ ft.}$$

*Bord and pillar* : Methods of extraction of pillars involve one of the two principles;

(a) inducing the roof to break diagonally, in respect to the side of the pillars and consequently also diagonally to the cleavages and cleats. This line of break is called the "fracture line".

(b) controlling the rate at which the fracture develops, across the pillar or panels. The quicker the fracture develops the better, so that the load on the pillars is removed. Inspite of precautions this system of working involves losses which may amount to as much as 15% to 20% during extraction. Whatever the shape or size of a pillar, depillaring or pillar extraction is done in "lifts" or "Judds" or slices.

The maximum thickness of the seam which can be worked in one section is controlled by the Coal Mines Regulations 1957—i.e. 10 ft.

The board and pillar method is somewhat modified in the American practice, since machines are employed to cut coal. Due to machine cutting, ribs of coal, with a general saw tooth pattern, are produced as each 'lift' or "slice" is removed. Each unit in this saw tooth rib is called a 'fender'. Thus the rib comprises a chain of fenders, which serve to eliminate contamination of coal with stone during depillaring.

In the first method of depillaring (Fig. 9.1), the pillar is divided into "Slices" or lifts. Two of the slices are started at the diagonally opposite

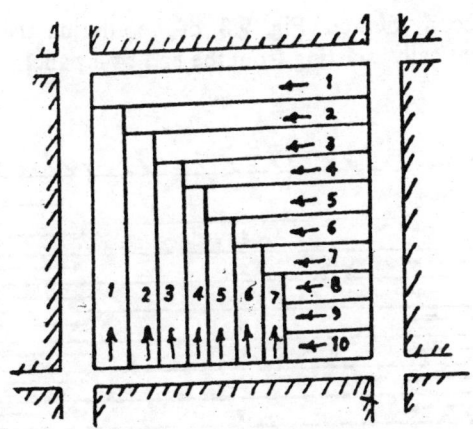

Fig. 9.1

corners of the pillar, and advance at right angles to each other. Thus slices (1—1), (2—2), (3—3) etc. are driven successively leaving a stook or "Chowkidar". It may be indicated that as the slices are cut, timbering in the form of stulls or props or cogs is employed to support the roof.

The arrow indicates the direction in which the Judd or "lift" is worked out.

Figure 9.2 shows a modification of the first method. In this case the lifts or slices are cut by machine, and these have a serrated margin, known as fender. The arrows show the direction in which the lift is removed.

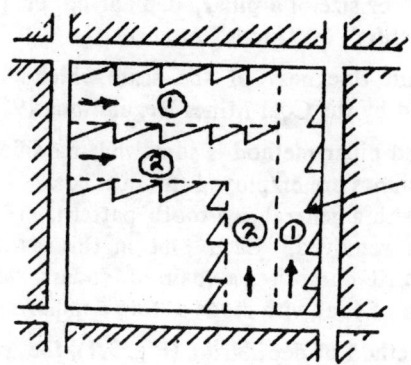

Fig. 9.2

In the second method Fig. 9.3 of depillaring the lifts or slices are removed systematically, starting from the end of a panel.  The slice drifts

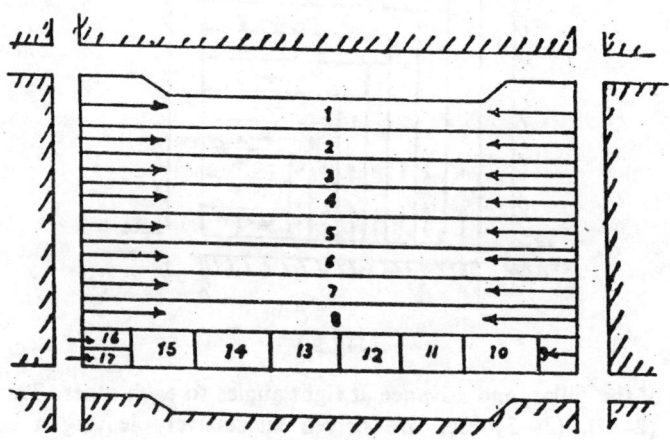

Fig. 9.3

advance toward each other.  Finally only a rib of coal is left.  This is also removed by blocks starting from one end of the pillar.

In the third method of depillaring, Fig. 9.4, the pillar is divided by driving a Jenkin or split.  The method otherwise is similar to the second method.  The advantage here is, that instead of having two working faces at a time for each "lift", four working faces may be arranged simultaneously.

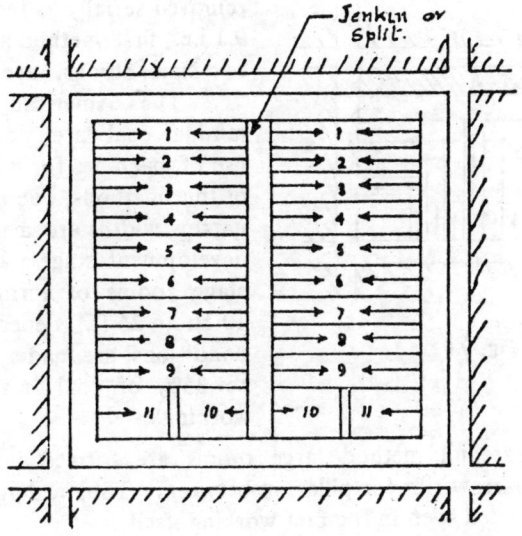

Fig. 9.4

*Fourth method* of pillar extraction is as follows (Fig. 9.5) :

(a) The pillar is split into four quarters by cross Jenkins.

(b) (2) and (4) are mined simultaneously and removed.

(c) (1) and (3) are taken also up together. The method of extraction is similar to the first method.

(d) Timbering in the form of props is generally employed and also pig styes or cogs.

(e) The stook or chowkidar is attacked quickly and

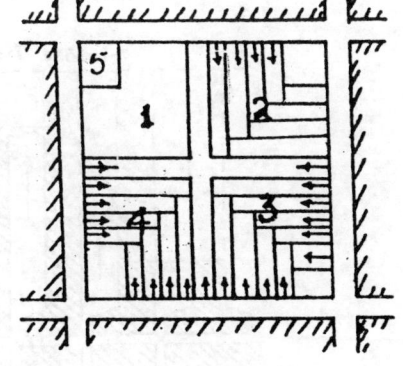

Fig. 9.5

actually robbed leaving as little coal, behind, as possible.

*Fifth method* : In the fifth method (Fig. 9.6) T-shaped Jenkins are driven first and then slices are removed serially as indicated in Fig. 9.1 i.e., first method and finally the two small stooks, are also removed.

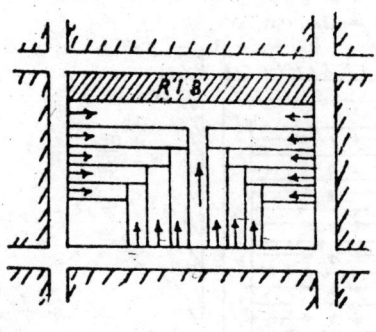

Fig. 9.6

The American methods of mining coal are designed for the use of machines for winning. In one of the methods, rooms or stalls of narrow widths are made during the development stage. ` In the winning stage rooms of normal width, i.e., 10 ft. to 25 ft., depending on local conditions, are made. Thus only 15 to 35% of coal is won in the first working.

In the second method large rooms are formed in the first stage, leaving very narrow ribs for pillar, in between. Thus a larger percentage of coal is extracted even in the first working itself.

A combination of both the systems of mining (Fig. 9.7) is also practised. This lay out is, however, bet*er suited for machine mining or mechanised mining as done in America.

In the American system of mining (Fig. 9.7) a larger number of "entries" or galleries are made, even at the development stage. Where locomotives are used, the ripping of the roof is necessary for gaining head

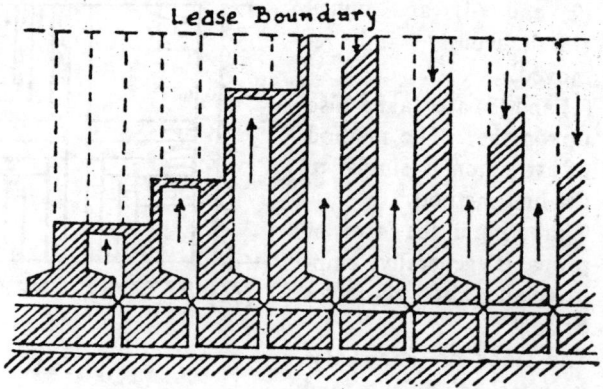

Fig. 9.7

room. The waste so produced is stowed away in the extra galleries, as also waste obtained by working dirt bands in the coal. Moreover, more openings mean an increase in the output during the development stage. Where conveyors are employed, the height of galleries is reduced and so ripping the roof may not be needed. The entries of galleries are usually 10ft. to 25 ft. in width.

### (a) Panel System

In the panel system, the development is so arranged that groups of pillars, in a mine, are blocked out and worked out as a separate unit, with separate ventilation arrangements. Further the panel can be excluded or isolated in the eventuality of a mine fire, or flood etc. Fig. (9.8).

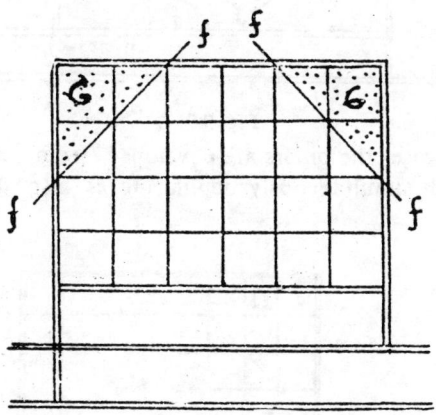

Fig. 9.8

G is Goaf ; f, f are lines of fracture. Each panel may be as large as 30 acres in extent, and is separated from the adjacent panel by a barrier of coal. The double lines show the main ventilation system for the panel while single lines show the pillars.

In the panel system depillaring or extraction of pillars can be made either in the beginning of the development stage (Fig. 9.8) or after the development is completed. In both the cases, the extraction of the pillars is so designed that the line of break is diagonal, so that the pillars support the roof, well. The two lines of break are thus complementary.

Depillaring can also be done even after the partial development of the panel. Here pillars do not suffer from crushing as they do not have to bear the roof weight for a long time. In both the cases, Fig. 9.8 and Fig. 9.9 since two lines of fracture (f f) are developed, roof control becomes difficult. In Fig. 9.9, PQRS, is the area to be developed. The area bounded by the two lines of fracture and by the level gallery PQ is the area developed already.

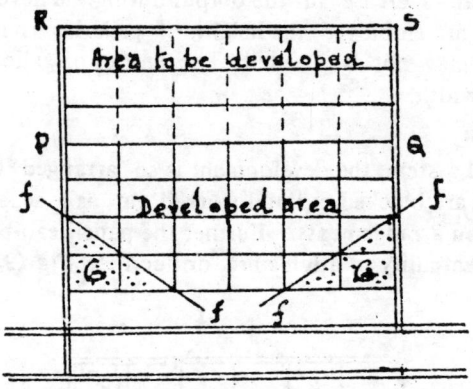

Fig. 9.9

In certain cases the pillars are developed from one of the corners of the district, and simultaneously depillaring is also done. In this case,

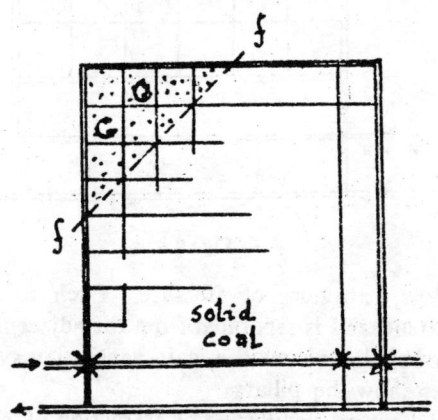

Fig. 9.10

Fig. 9.10 only one line of fracture is to be contended with and it is hence safer, even though the output is likely to be comparatively less.

(b) **Longwall**

*Advancing method* : As the name "longwall" implies, the working face, in this method, is in the form of a long and continuous wall extending for a considerable distance, which may be of the order of a few hundred yards. The vertical face appears like a wall confronting the workers and hence "longwall," is aptly descriptive of the conditions.

In the longwall advancing method, a shaft pillar, of solid coal, is left. The working face, which appears immediately outside the pillar, A,B, C,D as a wall of coal, is pushed backwards towards the lease boundaries, as the

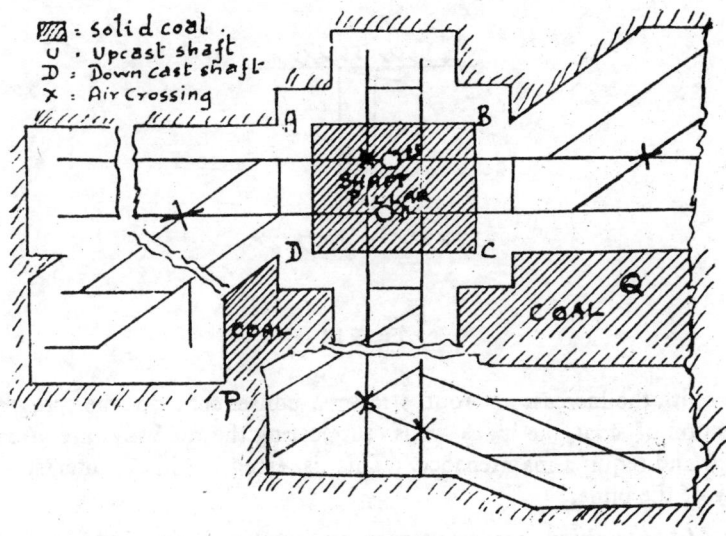

Fig. 9.11

coal is extracted (Fig. 9·11). In this manner the entire coal in the property is removed practically in a single operation.

In the Fig 9.11, U is the up cast shaft, D is the down cast shaft, X are air crossings.

In general, the tendency is to maintain straight and long working faces, eventhough, it is possible to have curved faces. The straight face has a decided advantage—where a mechanical conveying system, such as a conveyer belt, is used along the face. By this arrangement large outputs are possible. In the case of hand got coal, the output increases as more "faces" are made available. As the working face advances, away from the shaft pillars, the roof in the worked out areas (gob) caves down. Since the seams are generally thin, the broken stone from the caving of the roof, fills up the voids due to increase in volume and supports the "goaf" or "gob". It is however essential to maintain the two main roads leading to the working face (Fig. 9.11). The roads are essential for (1) communication, (2) transport, (3) ventilation, (4) drainage etc. Thus in order to support the road, stone from the roof is ripped and pack walls are built on either side of the roadways. A section across a longwall advancing face will appear as shown in Fig. 9.14.

G is Goaf, R are Roadways. Roof is ripped here for obtaining packing materials.

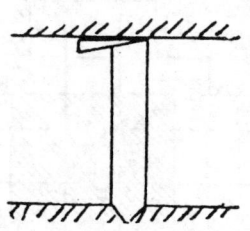

Fig. 9.12

With the increase in roof stressess, consequent to the progressive extraction of coal, the pack walls supporting the roadways are often disturbed and require maintenance. This is essential in the interests of the safety of the mine.

At the working face, temporary supports are employed in the form of "props" or "chocks" Fig. 9.12, "cribs" or "pig stys" Fig. 9.13.

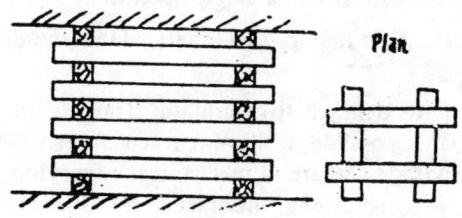

Plan

Fig. 9.13

The longwall method is generally applicable to thin seams. Even seams upto 12″ can be worked depending on the quality and nature of the coal, the nature of the roof and floor etc. Coal cutting machines can be used in conjunction with conveyors. A thick seam, say about 8 ft. thick, may be troublesome to work by the longwall method, if the roof conditions are poor. The maximum thickness of a seam that can be worked conveniently by these methods is about 4 ft. Longwall method can be applied to deep mining say a depth of about 1400 yds. or so. The roadways in the longwall system are known as gates.) The roadways at right angles to the longwall face are known as stall gates and the roadways driven at an angle to the stall gates are known as cross gates. The typical arrangemet is shown in Fig 9.14.

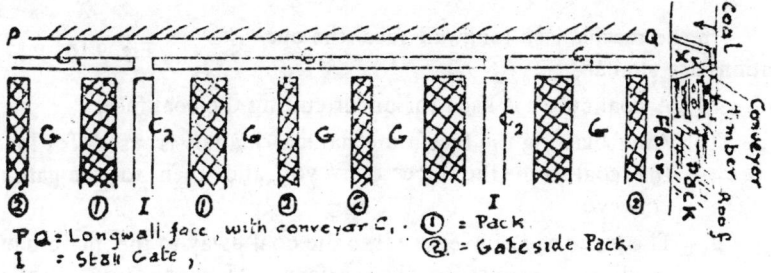

PQ = Longwall face, with conveyor $C_1$ . ① = Pack
I = Stall Gate ,    ② = Gate side Pack.

Fig. 9.14

In Fig. 9.14 PQ is the longwall face, with conveyor C. I are stall gates. (1) is intermediate packwall and (2) Gate side pack. GG are cross gates. SG are stall gates (Fig. 9.15).

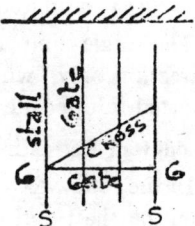

Fig. 9.15

In general the pattern of a longwall layout is as follows :

At the coal face PQ, a long belt conveyor $C_1$ is laid (Fig. 9.14) which carries the broken coal to the gates from where the gate conveyor $C_2$ transports the coal to the loading points I. At times the coal is loaded straight into tubs or mine cars, at the stall gate for being transported to the surface.

( Longwall faces worked by hand (manual labour) are known as hand got faces.) When worked by machines the face is said be mechanised.

The operations involved in working by hand are :

1. Spragging or supporting the face by props (Fig. 9 16).
2. Undercutting the coal i.e., making a channel at the lower part of the seam. (1) is an undercut, (2) is a sprag.

3.  Coal getting and setting up of props
    or cogs as may be necessary.

4.  Ripping of roof for building packs.

5.  As the coal face advances the face
    conveyor also should be shifted.

6.  Withdrawing of timber from areas
    which fall in the goaf.

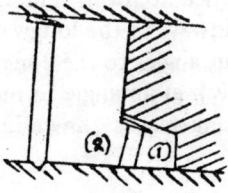

In the mechanised longwall face the operations are similar :

Fig. 9.16

1.  A coal cutter is used for under cutting the coal face.

2.  After blasting the face a mechanical loader is used for loading
    the coal into the face conveyor and then to the gate road
    conveyor.

3.  The gate road conveyor takes the coal away to the main haulage
    road for transport to the surface. Mechanisation is affected
    either as :

    (a)  single unit or

    (b)  double unit.

The single unit transport system works only "one-way", i.e., in
one direction only, whereas in the double unit layout the coal can be
transported in opposite direction also

(c) **Longwall Retreating Working**

In the longwall retreating system, the gates or roadways are developed from the shaft bottom, right upto the boundaries of the coal property.
Then the longwall face is opened and coal is won ; such that the working
faces "retreat" from the boundary and come nearer and nearer to the shaft
bottom.

This method is costly during the development stage and capital is
locked up until production starts. It is thus not popular. It is less
adaptable for mechanisation, It is, however, cheaper in respect of
maintenance of roadways. Goafs may have to be sealed of by pack walls
and clay mortar especially where the seam is liable to spontaneous
combustion.

On comparing the three main methods viz. Bord and Pillar, Longwall advancing and Longwall retreat, the following points may be
noted :

1.  That the bord and pillar method can be used with advantage
    where the seam is liable to spontaneous combustion. Panels can
    be arranged with facility for sealing off fires, When crushing

of pillars takes place, it is difficult to prevent the formation of leaks and this is likely to aggravate the situation.

In the case of advancing longwall, fires caused by oxidation of coal in the workings, are practically excluded. Sealing off an area is more difficult than in the bord and pillar system. Special methods of goaf pack sealing are required.

In the case of the longwall retreat, fires can be brought under control quickly by sealing. Fires are also liable in the solid coal.

2. Working in virgin areas.

The bord and pillar is satisfactory. The longwall advance is largely unsatisfactory while the longwall retreat is generally also unsatisfactory.

3. Working more than one seam.

The bord and pillar method is satisfactory.

The longwall advance is better suited.

The longwall retreating method may cause some amount of difficulty in driving headings.

4. Output and concentration of workings.

Working faces are small in bord and pillar and hence for large outputs concentration is not possible, due to increase in the number of working faces.

In longwall advance method a high degree of concentration of output is possible.

Concentrated work is possible in longwall retreat.

5. Percentage extraction.

In the pillar and stall method, there is a possibility of loosing more coal due to crushing (say 10-15%).

In the longwall advance method, the recovery is practically 100%.

In the longwall retreat the recovery can be as much as 90-95%.

6. Capital investment in development.

In the bord and pillar method the investment is generally excessive, and it increases with the narrowing of galleries. Some recovery of capital is possible if the panel system is introduced.

In the longwall advance method, capital is not locked up as production starts immediately.

In the longwall retreat locking up of capital cannot be avoided.

7.  Labour and organisation.

In the bord and pillar method, the small working faces, permit employing small groups of labour. Thus output per man can be controlled. The method, however, requires an increase in the strength of the supervisory staff, if the labour is not well organised.

In the longwall advance method large labour teams are required, especially in the hand got faces, and there is no means of checking efficiency in smaller units—the overall efficiency may thus fall. In the longwall retreat method, there is the possibility of employing smaller teams of labour.

8.  Effect of seam thickness on size of workings.

In the pillar and stall method, manual labour is employed in seams of 5 ft. and above in thickness, and thick seams can also be worked, subject to conditions laid down in the coal mines regulations. Mechanisation is possible. Thinner seams can yield larger production by this.

The longwall advance method can be practiced in thin seams upto 15 inches thick, but ripping of the floor or roof will be necessary for making roadways.

In the case of longwall retreat, ripping can be avoided by laying roads from suitable points from the boundary.

9.  Depth of seam.

In the bord and pillar method, the size of pillars increases with depth, and at greater depth the percentage recovery may become uneconomical. Limiting economic depth is about 2000 ft. or thereabouts.

There is no practical depth limitation to the application of the longwall advance method. At very shallow depths the effect of subsidence may probably militate against its use.

10.  The influence of dip.

The pillar and stall method is suitable, if modified, and can be used in very steeply dipping seams.

11.  Nature of coal.

For the application of the bord and pillar method hard compact coal is preferable.

In the longwall advance method the more friable coals give more output.

In the longwall retreat method hard compact coals are advantageous.

12. Arrangements for ventilation.

It is rather easy and simple in the pillar and stall method.
Ventilation requires careful planning especially in gassy mines in the case of longwall retreat ; the same difficulty arises as in longwall advancing method.

13. Stone bands and shale intercalations.

In the bord and pillar method the disposal of these present no major difficulty.
In the longwall advance dirt can be used in making pack walls.
Stone bands are a problem in longwall retreat.

14. Creep or squeezing of floor.

Presents a problem in maintaining roadways in the bord and pillar method.
Creep can be controlled by regulating the rate of development in the longwall advance method.

### (d) Horizon Mining

This practice should be adopted, where it is felt that subsidence, caused by mining at lower levels, is likely to affect the upper seam to such an extent, that its recovery may be impaired.

In the case of very highly disturbed areas where the coal measures have been folded, and faulted, and no definite method can be adopted, mining may require a combination of two or more methods. In general however, the practice has been to adopt a system in which vertical shaft 1 and 2 are used for gaining access to the seams and also for ventilation. From the shafts $S_1$ and $S_2$, cross cuts are driven at fixed intervals (levels) $L_1$, $L_2$, $L_3$ (Figs. 9.17 to 9.19) and the seams are worked level by level as indicated in the methods described earlier.

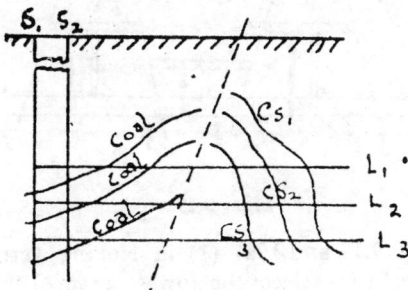

Fig. 9.17

The seam (CS) can be worked by shafts $S_1$ & $S_2$ from cross cuts (drifts) L, $L_1$, $L_2$ etc. (Fig. 9.17 and Fig. 9.18)

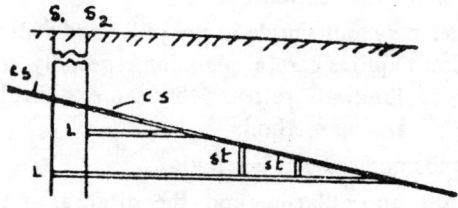

Fig.  9.18

It may not be possible to adopt the general pattern of mining in all cases.  If the main haulage is at the  intermediate  level,  sufficient precautions, in the form of pillars etc., should be provided  in  order to maintain its stability. This is so as this level will be very soon surrounded by worked out areas.   In order to avoid this situation, in larger mines, the  practice is to drive the main cross cut well below the level of  the  lowest  seam to be worked. Staple pits or blind shafts are employed for affording interconnection with the workings in all the levels.

This method, which is  called  horizon  mining is adopted in the disturbed coal measures of the Kuznetsk basin of the U.S.S.R., in  Poland, as well as on the continent, in the Ruhr coal fields  in  Germany and  also in Holland.

This method of mining can be  employed  with advantage  in  highly disturbed areas, i.e., folded and faulted ; and where a number of seams are known to occur.

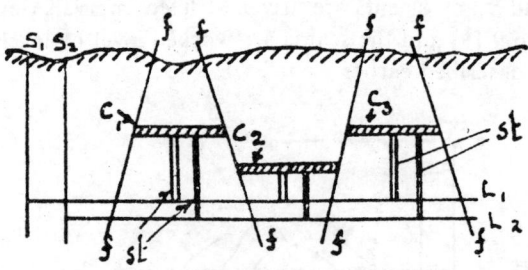

Fig.  9.19

It consists (Fig. 9.19 and 19a) (1) in sinking  vertical shafts, $S_1$ $S_2$ (at least 2 in number) to intersect the lowest seam, (2) in driving drifts (cross cuts) at various levels from the  shafts to connect the  various seams, and (3) sometimes in sinking down staple shafts to connect intermediate points between levels (Fig. 9.19). $C_1$, $C_2$, $C_3$ are coal seams, $S_1$ and $S_2$ are shafts $L_1$, $L_2$, are levels. St are staple shafts. In India, horizon mining is to be practised in general in the Sudamdih Colliery of the National Coal Development Corporation. The mine is being developed under Polish collaboration.

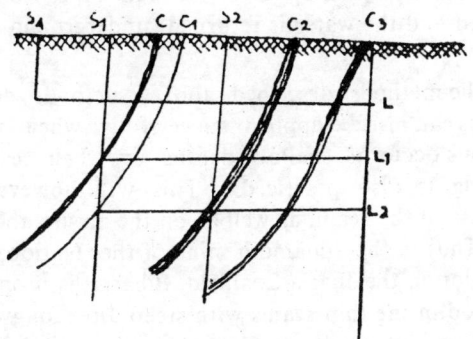

Fig. 9.19a

### (e) Miscellaneous

### Underground Hydraulic Mining

In certain areas of the Donets and Kuznetsk coal basins of the U.S.S.R., hydraulic mining, of coal underground, has been practiced with success. Here the seams are very susceptible to spontaneous combustion. The dip is of the order of about 70° and the thickness is about 35 ft. The development is very similar to the sub-level stoping method used in metal mining but the mine is divided into panels. The mining proceeds from the top to the lower levels. Fig. 9.20. The panel selected for mining is developed by the paired sub-level drives (1) which are made at intervals of about 20 ft. from cross cuts (2) driven from the main winzes (3) which are spaced about 200 ft. apart. Stoping starts from the upper most sub-level and the stopes are maintained in a steplike arrangement. Stopes are filled and supported by packs, since roof collapse will tend to contaminate the coal. Even during hydraulicking (using monitors) the floor is covered

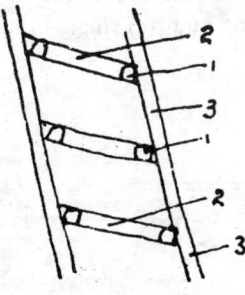

Fig. 9.20

over by a wire netting so that the stone from the floor does not contaminate the washed coal.

The general arrangement consists of monitors from which powerful jets are made to play upon the coal face. Sluice boxes as well as riffle boxes are used to recover lump coal. A hydraulic elevator and hydraulic lift are used to convey the lump coal through pipes to the surface. The coal is separated and the water is returned underground for re-use in the monitors.

Besides the methods described above, methods described as metal mining methods can also be applied successfully when conditions permit. When coal seams occur at shallow depths or when seams are exposed, opencast mining is also practiced. This will however depend on the nature, thickness of the seam, as well as on the nature and thickness of the overburden. Thus in the Sudamdih mine of the National Coal Development Corporation in the Jharia Coalfield, Bihar, it is proposed to adopt the 'horizon' method in the thin seams with steep dips, longwall method when the seams are thin with moderate dips and caving (sub level) method where the dips are steep and the seams are thick.

## Strip Mining

Strip mining is commonly practiced in the U.S.A. It is one of the methods of opencast mining with a high degree of mechanisation employing either power shovels or a combination of dragline and power shovels for stripping the overburden. The overburden to coal ratio may be as low as 1 : 12 upto a depth of 30 or 35 ft. or 1 : 15 where the overburden is about 90 ft. If the overburden is very thin, being only a few feet, and conditions are favourable then the stripping can be managed with "bulldozers". It may be pointed out that this method destroys the countryside, alters the drainage and the surface is marred.

The more common layouts are (a) single stripping shovel, (b) Tandem stripping and (c) single dragline.

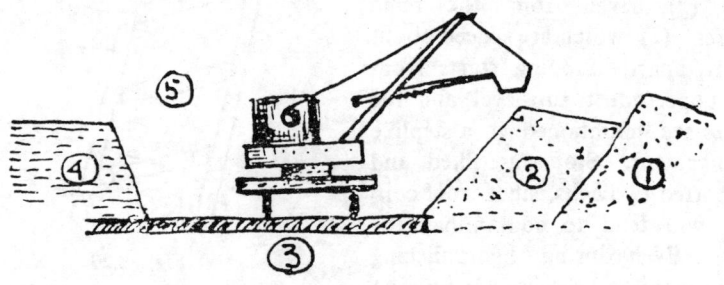

Fig. 9.21

In the single stripping shovel layout (Fig 9.21), the shovel removes the in situ cover (shale or clay) at (4) and dumps the same at (2). The dump formed earlier is designated (1). After the coal is exposed, the machine moves in one direction so as to form a long trench, or cut, in which a "strip" of coal is exposed. The strip is excavated by conventional methods. On the return trip the shovel exposes another strip and so on.

In the tanden stripping layout a dragline, in combination with a power shovel, operates simultaneously for stripping greater thickness of over-burden. The dragline is employed advantageously for removing the softer material from the top, forming the top bench, while the power shovel digs up the harder formations below, so as to expose the coal. The dragline, because of its longer boom, dumps the spoil beyond the range of the shovel, and the spoil from the shovel is dumped nearer the exposed strip of coal. The plan of working is in other respects similar to the single stripping shovel layout.

In the layout where a single drag line (Fig. 9.22) is employed, the drag line strips the overburden in two stages. First bench (A) is formed and a thin bench (B) is made exposing the coal C. When making the benches, the dragline operates on position A and then on position B. It thus proceeds forming a long open cut, stripping the overburden.

Fig. 9.22

# PART X

A) MANAGEMENT ECONOMICS
B) ELEMENTARY TECHNIQUES IN
   PLANNING/PROGRAMMING
   1. PERT AND CPM
   2. USE OF THE COMPUTER

## (A) MANAGEMENT ECONOMICS

In any business undertaking there are four aspects :

(1) Stewardship which takes into account profit and loss accounts and balance-sheet. i.e., it dischagres the reponsibility vested by the share-holders.

(2) Decision accounting aspect relates to the estimates of cost and revenues based on alternatives, which may pertain to investments or technologies, etc.

(3) Control accounting aspect provides the information required for management for maximising efficiency.

Stewardship is mainly interested in financial accounting, so much so that it brings out the cash position at the end of a period annually or bi-anually, in respect of the business. While financial accounting is essential for understanding how the business has behaved during the period under review, it may not help directly in decision accounting for management purpose. It only helps to look back in retrospect as to how the company has prospered and analyses any failure or losses. That is why it is important to know the elements of book keeping. This subject has however been omitted in the present volume for want of space.

Control accounting, in business, is of prime importance since it exercises control on performance. Comparisons are made between the planned performance and actual achievements and these guide further planning and progress. Hence for planning the basis is control accountting. It does not merely correct the past mistakes, but also directs both current and future operations so as to enable realisation of the physical and financial targets. Hence, the three aspects involved are, (a) communication which involves passing of information in respect of proposals, (b) suggestions for modifications if needed for achieving the planned targets ; and (c) reporting, comprising monitoring and reporting on performance.

The above three activities are inter-related. Planning is the basis of control, information affords the guides while the monitoring evaluates performance and aids in the same. Thus in control accounting there are various aspects, such as, (1) financial accounting as brought out by the book keeping records, (2) judging and measuring profitability, (3) financial planning for long range objectives, (4) appraisal of capital expenditure, and (5) short-term planning and budgetary control.

Basic financial accounting has already been dealt with earlier under 'book keeping' and no further clarifications are required for the preparation of balance-sheets or profit and loss accounts, which constitute the issues of financial accounts. Basic cost accounting has also been dealt with, elsewhere in an earlier chapter, and the reader may refer to the same. It is, hence, proposed to deal entirely with control accounting which is a major factor in management economics.

**Classification of Costs**

Costs can be classified on the basis of functions, such as :

1. production,
2. selling,
3. distribution,
4. research and development, and
5. administrative.

The elements of cost comprise :

(a) material cost,
(b) labour cost,
(c) expenses, such as, services, infra-structure,
(d) particular costs relating to assets and capital besides depreciation, sunk capital, etc.

Costs accounts can be related either to direct costs or indirect costs. Direct costs are associated with the production and activity. Indirect costs are not associated with production, hence these have to be apportioned to the cost operating centres. Indirect costs of production, selling cost, distribution cost, R and D cost as well as the administrative expenses are called 'overheads' and overheads have to be absorbed at a definite rate under direct costs, by apportioning either to labour hour rate, or man hour rate etc.

Costs are further divided into fixed costs which are constant in the short run, such as, rent, taxes, salaries, insurance, etc. With time, these can also change. Variable costs are those which vary with the volume of output but which cannot be allocated to the cost unit and have to be absorbed.

**Cost Systems**

Cost systems can be classified as ;

(a) job costing,

(b) process costing,

(c) historical costing, and

(d) standard costing.

*Job costing,* is employed either for obtaining costs for a specified quantity of product, equipment, or for a quantum of repair or service. Hence in this system, individual jobs or batches of jobs are separately identified. This system is employed in industries, such as, construction, heavy engineering, etc. Process costing, on the other hand, is used in industries where large quantities of identical products are manufactured continuously by mass-production methods. In *process costing,* emphasis is on the accumulation of costs for all the work units, during a specified period and at the end of each period, the average cost per unit is determined. This method of costing is applicable to industries producing a single product, such as, cement, sugar or to a variety of products utilising the basic facilities such as, tiles, bricks, pipes in the clay industry or a variety of products using separate facilities as in dairy plant—producing butter, cheese etc.

While the process costing system has the advantages of easy maintenance plus easy clerical effort. It has also some disadvantages.

**Historical Costing**

Job costing and process costing can be of two categories, viz., historical costing or standard costing. In *historical costing,* actual costs are determined after the operations have been completed, and the costs incurred, whereas in standard costing, the costs are calculated ahead, even before they are actually incurred. The disadvantages of historical costing are :

1. costs obtained are of limited value and cannot be projected for any length of time,

2. interpretation of cost becomes difficult because of the unknowns,

3. there is the lack of a standard or norm for evaluating the efficiency of the operations. and

4. because of delays being inevitable, inefficiencies are likely to be exaggerated.

Thus historical costs are informative but do not provide a means for cost control; and standard costing has been adopted for this purpose.

In *standard costing,* on the other hand, costs are pre-determined carefully and compared with actual costs incurred so that cost control can be affected. The differences between the actual cost and pre-determined standard costs are called variances, and variances can be analysed in respect of causes and inefficiencies. Variances can be in respect of material costs, and variances in material costs can be price variances and quantity variances. Variances can also apply to wages and wage variances may be related to rate variances or efficiency variances.

The following equations will simplify what has been said above:

Material cost variances = actual cost − standard cost

i.e.,            standard cost = standard quantity × standard price.

Material cost variance = material price variance + material usage variance.

Material price variance = actual quantity × actual price − actual quantity × standard price.

Material usage variance = actual quantity × standard price − standard quantity × standard price.

In addition to standard costing/variances, *ratio analysis* is also widely accepted as a technique for providing a basis for cost control. *Standard ratios* employed include:

1. Net profit/sales.
2. Gross profit/sales.
3. Net profits/total assets.
4. Sales/total assets.
5. Production cost/cost of sales.
6. Selling cost/cost of sales.

## Pricing

In managerial decision making, setting the price and formulating price policies, is of primary importance. Profit or revenue is dependent on price. It is also a device for expanding markets. Hence the objectives of pricing should be clear :

1. whether it is maximisation of profits for the entire line of products,

2. whether it is for the promotion of long range welfare of the firm so as to curb/discourage competition,

3. whether it is to suit competitive situations faced by various products,

4. whether it is to introduce flexibility which would meet the changing economic conditions experienced by consumer industries, and

5. whether pricing policy is for stabilising the price and margin of profit.

In setting the price one has not only to consider cost, but other factors also such as, demand. Thus with an increasing demand, it may be possible to increase the price without an increase in costs. Often it is possible to determine the cost after fixing the price. Hence with rising costs, there is a decline in quality, if the price remains constant.

The factors governing the price are either external or internal. External factors are related to the elasticity of supply and demand, the goodwill of the company, nature of competition and market trend, purchasing power of buyers and governmental policy towards pricing. The internal factors are costs, management policy regarding profit, margin and sales turn-over.

With regard to demand elasticity and pricing, this is clear, if we take for example two products, one for which the demand is slightly elastic and for the other the demand is highly elastic, it will be noticed that a reduction in price, for expanding sales, would yield better results for the latter product, with a highly elastic demand than for the product with a slightly elastic demand. Further, if the demand is inelastic, a reduction in price will be less profitable than the existing price.

## Methods of Pricing

Methods of pricing may be classified as :

(a) cost plus or full cost pricing, and

(b) rate of return pricing.

The cost plus pricing is done so as to cover all costs including material, labour and over-heads, as well as the pre-determined percentage of profit. In the rate of return pricing, the emphasis is laid on market conditions and the price policy has to adjust to the general pricing structure of the industry. This method is particularly useful, in cases where costs are difficult of appraisal. The formula for the rate of return pricing is :

$$\frac{\text{Capital employed}}{\text{Total annual cost}} \times \frac{\text{Profit}}{\text{Capital employed}}$$
$$= \frac{\text{Profit}}{\text{Total annual cost}}.$$

The technique adopted for cost plus method, is known as absorption pricing. Total costs including the convention of absorbing indirect cost will not give a correct answer to the questions, as these do not represent incremental cost or the additional cost resulting from change in volume, while these include fixed costs which are historic sunk costs, in relation to the changes in volume.

Since both the above methods have certain disadvantages in the short run, the third method known as 'the margin pricing' is sometimes employed. In 'this, the producer fixes the price so as to maximise the total contribution to fixed costs and profits. This method is particularly useful in secondary pricing decision, for example, in tendering, by products, unusual work, export orders etc.

### Profit—measurement—profitability

Profit is the residual revenue and after defraying cost. The net profit is what remains after meeting costs of business including contractual outlays. Definition of profit is, however, different to the economist. Profit is not merely the residual amount after deducting not only the explicit costs but also imputed costs, such as, salary of managers in self-owned factories, rental of self-owned land, interest on self-owned capital etc. Whereas to the accountant profit is related to the explicit or actual costs. Taking the accounting version, items of revenue or expenditure are :

1. depreciation,
2. valuation of stock,
3. allocation of expenses over a period including deferred expenditure, and
4. capital gain and loss.

### 1. Depreciation

In any industry, equipment, machines, buildings are subjected to wear and tear due to use and time. Hence their value is lowered. Since all these form part of capital, the value of capital decreases with the passage of time. Hence a suitable deduction has to be made for this purpose from the capital, so much so, that after a certain period an asset has to be completely written off from the books and its value, ultimately becomes zero. This, however, need not be true, even though the book value is zero, because the assets may yet have a sale value or scrap value. An asset, though subjected to depreciation, can at the same time appreciate some value, if properly maintained. Hence while depreciation is deducted from the original value of the asset, cost of maintenance appreciates the asset and adds to its cost.

### Method of calculating depreciation

The methods of calculating depreciation are :

1. straight line method,

2. declining balance method, and

3. sum of digits method.

Of these methods, the straight line method is the most common. It is based on the assumption that wear and tear of the capital asset is evenly distributed throughout its normal life. Take for instance, a piece of machinery. If it is worth Rs. 5,000/- and taking the life of 10 years, its scrap worth may be just Rs. 500. Hence depreciation charge will be as follows :

$$\frac{Rs. 5000 - Rs. 500}{10} = Rs. 450 \text{ per annum.}$$

Sometimes depreciation is also calculated from the rated working hours. Taking the cost of the same machine again, it is found that if its rate life is 43,000 hours then the depreciation per hour will be,

$$\frac{5000 - 500}{43000} = \frac{4500}{43000} = \text{Approx. } 0.10 \text{ p.}$$

In the declining balance method, depreciation is calculated on the written down value of the asset at the beginning of the year. Taking the case of the machine valued at Rs. 5,000, if 10% is its depreciation, the depreciated cost at the end of the first year will be Rs. 5000 − 500 = Rs. 4500 and at the end of second year the depreciated cost will be Rs. 4500 − 450 = Rs. 4050. Similarly the calculations can be made for the remaining period of 10 years. The formula for use in the declining balance method is :

$$d = 100\left(1 - \sqrt[N]{\frac{s}{c}}\right)$$

The sum of digit method is similar to the declining balance method. But the amount of the depreciation charges per year is progressively lower with the passage of time as shown in the example below :

In this method it will be noticed that the final book value remains the same while the rate of depreciation varies from year to year. Taking the example :

| Cost of equipment | Rs. 5000/- |
| Scrap value | Rs. 500/- |
| Life of equipment | 5 years |

At the beginning of 1st year the equipment's expected life is 5 years.
      "      " 2nd "     "     "     " 4 years.
      "      " 3rd "     "     "     " 3 years.
and this proceeds year after year and the expected life diminishes from

5 to 1. Thus the total anticipated period is $5+4+3+2+1=15$ years. The distribution of depreciation is done as follows :

| Year of Use of Asset | Depreciation Rate | Annual depreciation amount | Commulative depreciation | Book value of asset |
|---|---|---|---|---|
| 1 | 5/15 | 1500/- | 1500 | 3500 |
| 2 | 4/15 | 1200/- | 2700 | 2300 |
| 3 | 3/15 | 900/- | 3600 | 1400 |
| 4 | 2/15 | 600/- | 4200 | 800 |
| 5 | 1/15 | 300/- | 4500 | 500 |

## 2. Valuation of stock

Three common methods are often adopted :

(a) FIFO (First In First Out). In this method, the material acquired first is also issued first. Consequently, the inventory comprises only the most recent goods.

(b) LIFO (Last In First Out). In this method, the last units to be acquired are issued first, so much so that the inventory would consist of goods purchased earliest.

(c) Weighted average method. In this method, the goods purchased later are mixed with the goods already available, so much so that the cost of the one cannot be separated from the cost of the other. Hence when these are issued, the overall cost of each purchase, taking into account the quantity, is charged.

## 3. Allocation of expenses over a period of time

These include intangible fixed assets with a limited life, e.g., patents, copyrights, leasehold, licences and permits. There are other intangible assets such as trade marks, goodwill, etc. which endure for an unlimited period.

Those with limited life are generally written off during the life or even earlier ; whereas intangible assets which are of a perpetual nature writing off depends on individual discretion.

## 4. Capital gains and losses

There is a considerable variation in the treatment of capital gains.

This is sometimes included in current profits, while capital losses are often written off out of retained earnings. Thus these affect the amount of profit.

In the case of unrealised capital gains, they are not taken into account. A gain due to revaluation of property is usually transferred to capital reserve.

Profitability in business is primary, and is a yardstick for efficiency in competitive business. Thus, the rate of return on capital is taken as a standard measure. The rate of return on capital employed is the percentage profit in relation to capital employed and includes :

1. percentage profit in relation to sales, and
2. the relation of sales to capital employed.

Thus,

$$\frac{\text{Profit}}{\text{Capital employed}} = \frac{\text{Profit}}{\text{Sales}} \times \frac{\text{Sales}}{\text{Capital employed}}.$$

The relations between the factors affecting the return on capital employed, is shown in the following table :

RETURN ON CAPITAL EMPLOYED

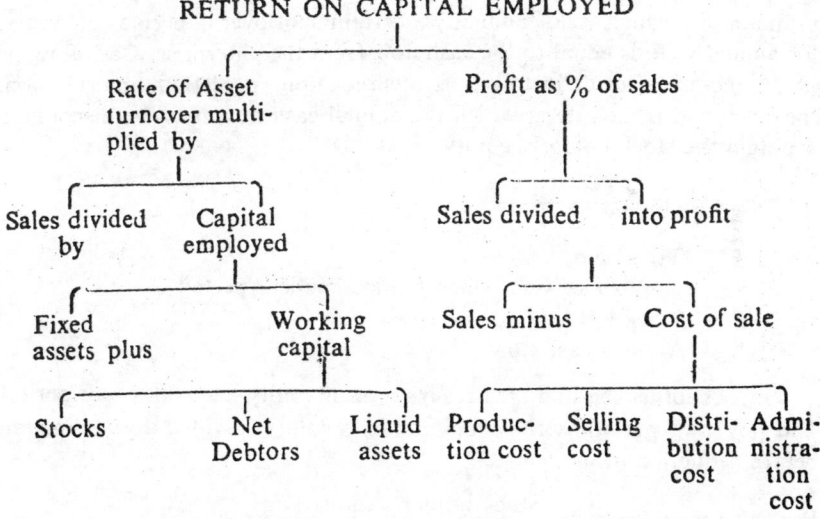

For calculating profit, cost of the capital employed, is an important factor. Certain ratios have been used for measuring profitability.

(a) Comprehensive ratio, i.e.,

$$\frac{\text{Profit after corporate tax, but before payment of loan and debenture interest}}{\text{Total net assets employed irrespective of the source of finance.}}$$

(b) Profit after corporate tax and interest and after deduction of profit
attributed to outside shareholders and preference share holders
—————————————————————————————————————————————————
Equity capital

This Ratio measures the return on equity or shareholders' investment.

(c)  Control  ratio$= \dfrac{\text{Trading profit}}{\text{Net trading assets.}}$

A. Profitability of American mining companies is measured by the "Book yield" or the rate of return. The rate of return is the ratio of net income, after deducting income tax, to the stock holder's equity. The stock holder's equity is called the book value.

Net Assets = Stock holder's equity or book value = Total assets — All liabilities.

The stock holder's equity is the total value of the preferred stock at par value, common stock at par value, capital surplus, and earned surplus. Rate of return (book yield) = net income (after deduction for I. ɪ., depreciation, depletion, and amortisation charges).

B. Another method of calculating profitability is true yield (discounted cash flow or D.C.F.).

In this concept the stock holder's equity is taken to be the present value of an annuity which yields annually a certain sum over a period of years, this annual yield is equal to the cash flow from the enterprise. Cash flow is the net income after deducting I. T., depreciation, depletion, amortisation. The true yield is the rate at which the annual cash flow must be discounted to obtain the stock holder's equity.

True yield is $\dfrac{E - [1 - I]}{I} - \dfrac{E}{A}$

where  I — True yield
N — Number of years when the income is expected
E — Stock holder's equity
A — Annual cash flow.

C. Another method of measuring profitability is the pay out period. The pay out period is the stock holder's equity divided by net income before deduction of income tax.

Pay out period $= \dfrac{\text{Stock holder's equity}}{\text{Net income before deduction of I.T.}}$

## Capital Investment Evaluation

In evaluating capital investment, one of the problems relates to the future investment as apart from the measurement of past performance.

## Capital Budgeting

Since investment is necessary, if capital is to increase, scrutiny of investment is also equally essential. In many cases, it is possible that there is more than one project to be considered at the same time, but then priority will have to be assigned, after taking into consideration, (1) which of the projects will pay back quickly, (2) which of the projects will yield greater return on investment, and (3) on which of the projects, capital invested can be quickly recovered. Hence, in the long term view, the decision making is tied up with capital budgeting. Projects, such as, expansion, or replacement of equipment can fall within this category.

Capital required for investment is acquired from two main sources which may be described as (1) internal and (2) external. The internal sources are further sub-divided into (a) depreciation charges and (b) retained cash flows, whereas the external sources are, (a) shares which may be equity or preference and (b) debenture, either from financial institutions or banks etc. which may be in the form of loans. Hence capital can be broadly divided into (a) debt capital and (b) share capital comprising preference equity shares. Retained earnings would form a separate category. The influence of each of these types of capital, on the overall financial picture of an undertaking, is very significant, hence the cost of each type of capital should be separately evaluated.

1. *Debt capital* : This capital is raised through banks, financial institutions in the form of loans repayable after a definite period say, 10 years or 15 years depending on the type and nature of the project. It is generally agreed that the debenture amount will be repaid at a definite rate of interest which may vary from $7\frac{1}{2}$ to 8 per cent. Hence the cost of debt capital may be stated as follows :

$$C_d = \frac{i + \dfrac{MV - NP}{n}}{\dfrac{MV + NP}{2}}$$

where,

$C_d$ = Cost of debt capital (before tax).

$i$ = Annual interest payment.

$MV$ = Value payable on maturity.

$NP$ = Net proceeds.

$n$ = Number of years after which the debenture is to be paid.

$C_d$(after tax) = $C_d$(before tax) × (1 − tax rate)

2. Since preference is a redeemable share, it is paid back within a

fixed period at a rate of interest varying between 9 to 10% and may be treated in the same way as debt capital. The cost of preference capital is given by the formula :

$$C_p = \frac{PD + \dfrac{MV-NP}{n}}{\dfrac{MV + NP}{2}}$$

where

$C_p$ = Cost of preference shares (after tax).
PD = Amount of annual preference dividend.
MV = Value payable on maturity.
NP = Net proceeds.
n = Number of years after which debenture is to be paid.

$$C_p(\text{before tax}) = C_p(\text{after tax}) \times \left( \frac{1}{1 - \text{Tax rate}} \right)$$

3. In the case of equity capital if it is to be increased, then its return should not fall below that earned by the existing equity shares. Thus new equity shares when invested should earn at least as much as existing shares say, 12%, but this may also be as high as 20%.

$$C_e = \frac{EPS}{NP} \times 100$$

where,

$C_e$ = Cost of equity capital (after tax).
EPS = Minimum earnings which a new share must earn.
NP = Net proceed per share.

$$C_e(\text{before tax}) = C_e(\text{after tax}) \times \left( \frac{1}{1 - \text{Tax rate}} \right).$$

4. The cost of retained earnings is to be calculated separately since they form part of the total earnings payable to shareholders and have an "opportunity cost", capable of being invested elsewhere. If no income tax were payable by share holders, the formula for cost of retained earnings will be the same as the cost of equity capital say about 12%. Since income-tax is payable, the formula is modified as follows :

$$C_{re} = \frac{EPS(1-t)}{MP} \times 100$$

where

$C_{re}$ = Cost of retained earnings (after tax).
ESP = Earnings per share.
t = Average tax rate applicable to the shareholders.
MP = Market price per share.

$$C_{re}(\text{before tax}) = C_{re} (\text{after tax}) \times \left( \frac{1}{1 - \text{Tax rate}} \right).$$

## Leverage or Trading on Equity and Cost of Capital

Capital structure of any project depends on the cost of capital employed and by a careful adjustment it is possible to minimise the cost of capital. Such an analysis would involve the working out of a combination of equity and debt capitals, in such a manner, that the earnings per share is the maximum as shown in the example below :

Leverage and its effects can be exemplified as given below in which the total capital, viz., whether equity or equity and debt is the same.

It is seen that the ESP for case III is the highest, since it depends largely on debt capital ; while in case I the EPS is lowest. Case III shows that the leverage is largely afforded by the smaller amount of tax. When expansion is affected by loans, the condition is known as "trading on equity".

|    |                                        | Case I   | Case II  | Case III |
|----|----------------------------------------|----------|----------|----------|
|    |                                        | Rs.      | Rs.      | Rs.      |
| 1. | Equity capital                         | 500,000  | 300,0(0  | 200,000  |
| 2. | Debt capital                           | —        | 200,000  | 300,000  |
| 3. | CBIT (cost before interest and tax)    | 75,000   | 75,000   | 75,000   |
| 4. | Interest at 10%                        | —        | 20,000   | 30,000   |
| 5. | Profit before tax                      | 75,000   | 55,000   | 45,000   |
| 6. | Income tax at 50%                      | 37,500   | 27,500   | 22,500   |
| 7. | Profit after Tax                       | 37,500   | 27,500   | 23,500   |
| 8. | ESP (earning per share)                | 7.5      | 9.2      | 11.8     |

## Average Cost of Capital

The average cost of capital can be worked out, if the proportions of the various types of capital invested, are known. It is worked out in the following manner :

If in the capital structure the % of equity, preference, debit and retained earning is $P_1 + P_2 + P_3 + P_4$ then $P_1 + P_2 + P_3 + P_4 = 100\%$. If the cost of capital before tax, consisting of equity, preference, debit and retained earning is $X_1$, $X_2$, $X_3$, $X_4$, then the average cost of capital before tax is

$$= \frac{X_1P_1 + X_2P_2 + X_3P_3 + X_4P_4}{P_1 + P_2 + P_3 + P_4}$$

$$= \frac{X_1P_1 + X_2P_2 + X_3P_3 + X_4P_4}{100}$$

The average cost of capital after tax=
    cost of capital before tax X (1−tax rate)

### Evaluation of Project Profitability

The standard method of evaluating the profitability of an investment are as follows :

1. Pay back method.
2. Accounting method.
3. Discounting cash flow method.
4. Present value index method.

*Pay back method* : As a rough and ready means, this method is very convenient. The pay back period, is the period required for recovering the amount invested by annual cash flow resulting from the investment :

$$P = \frac{i}{ACF}$$

where

$P$=period,
$i$=Initial investment.
$ACF$=Annual cash flow.

This is equal to net profits after tax+depreciation

This criterion is useful in evaluating three or four projects with varying investments and varying annual cash flows.

*Accounting method* : Accounting method is also simple and this can be calculated in the following manner :

In the form of percentage, i.e., $\dfrac{\text{net estimated profits}}{\text{capital employed}} \times 100.$

The accounting method, however, has the disadvantage that it does not take into account the time value of money, and considers money to have the same value whether received now or later.

*Discounted cash flow* : This method is designed, basically to evaluate money and time together. Thus Rs. 100 received today is worth much more than Rs. 100 received 10 years hence. The method consists in adjusting the amounts received, at a future date, with a view to arrive at the present value, by imputing a rate of interest which is known as discount. In effect, discounting is the reverse of compounding interest. The method of discounting cash flows, at the present value, is shown in the example below :

Alternative to the D.C.F., is the net present value (NPV) method. This is in fact the converse of the DCF rate of return. By this method, the rate of return is selected and taken to be the minimum expected and this percentage rate is employed when discounting separate cash in-flows and outflows to the net present value i.e. after deducting for depreciation and taxes. In case the total present value of the inflows so obtained exceeds the total of the out flows, the difference is taken as the net present value with a surplus. The net present worth would be a deficit, if the net present value of the outflows exceeds the total of the inflows, and the project will run at a loss. where the algebraic sum of the discounted net amount of cash flows is positive, the project will be profitable. If the algebraic sum of the discounted cash net annual cash flows is zero, then the project will just break even.

DCF method can be applied to both cases where

(a) the cash flows at the end of each year are variable and

(b) the cash flows at the end of each year are uniform.

For example if the investment on a marble mill is Rs. 25,000/- and the life of the capital is 5 years, and the anticipated cash flows for the five years are as follows :

|  |  |  |
|---|---|---|
| 1st year | — | Rs. 5000 |
| 2nd year | — | Rs. 6000 |
| 3rd year | — | Rs. 10000 |
| 4th year | — | Rs. 9000 |
| 5th year | — | Rs. 7000 |

Taking the anticipated rate of interest or internal rate of return to be 15%, the present value of the cash flows after discounting may be calculated as shown below:

| Year | Annual cash flow Rs. | Present value discounting factor at 15% (from tables) | Present value Rs. |
|------|------|------|------|
| 1 | 5000 | 0.9286 | 4643.0 |
| 2 | 6000 | 0.7793 | 4675.8 |
| 3 | 10,000 | 0.6879 | 6879.0 |
| 4 | 9000 | 0.5921 | 5328.9 |
| 5 | 7000 | 0.5096 | 3567.2 |
|   |   |   | 25093.9 |

Hence the net present value is Rs. 25093.9
           present investment is Rs. 25000.0

                Excess Rs.      93.9

Thus the anticipated 15 % rate of return can be obtained.

Suppose, if the anticipated rate of return was 20% then the net present value after discounting will be determined as follows :

| Year | Annual cash flow Rs. | Present value discounting factor at 20% (from tables) | Present value Rs. |
|------|------|------|------|
| 1 | 5000 | 0.9063 | 4531.5 |
| 2 | 6000 | 0.7421 | 4452.6 |
| 3 | 10,000 | 0.6075 | 6075.0 |
| 4 | 9000 | 0.4974 | 4476.6 |
| 5 | 7000 | 0.4072 | 2850.4 |
|   |   |   | 22,386.1 |

Since the net present value of the cash flows comes to only Rs. 22,436.1, this shows that the rate of a return from the anticipated or actual cash flows given in column 2 cannot be 20%. It is only about 15%.

*Present worth Index methods*

It is thus clear that the maximum rate of return from the project can be only about 15 per cent. It will be seen that the net present value method (N.P.V.) is also a method of discounting. But in this case the discounting is done at the same rate in order to evaluate the relative financial benefits of the projects.

In the example given below two projects No. 1 and No. 2 have been taken and for each the N.P.V. has been computed with a view to illustrating the above principle. It is assumed that the outlay for each of the projects is Rs. 4000. The duration is 3 years, and the expected rate of return is 12 per cent per annum.

| | | Project No. 1 | | | Project No. 2 | |
| --- | --- | --- | --- | --- | --- | --- |
| | Cash Flow Rs. | Discount Factor 12 Per cent | Present Value Rs. | Cash Flow Rs. | Discount Factor 12 Per cent | Present Value Rs. |
| Now | -- 4000 | 1.0000 | -- 4000 | -- 4000 | 1.0000 | --4000.00 |
| YR 1 | + 1600 | .8929 | + 1428.64 | + 1200 | .8929 | +1071.48 |
| YR 2 | + 2400 | .7972 | + 1913.28 | + 2800 | .7972 | +2232.16 |
| YR 3 | + 1200 | .7118 | + 854.16 | + 1200 | .7118 | + 854.16 |
| | | | + 196.08 | | | + 157.80 |

It is seen that both the projects yield return of more than 12 per cent per annum, but the Project No. 1 has a higher N.P.V. as compared to Project No. 2.

While computing the optimum economic life of a project, it may be pointed out that unless the incremental financial return is equal to the cost of capital, the expansion of production, with consequent shortening of the project life, will present serious hurdles in the way of decision making. Mere evaluation by the D. C. F. method may not yield a positive answer. Thus it has been suggested that the net present value (N.P.V.) of cash flows discounted at the cost of capital, be employed. The optimum is the highest N.P.V., and this determines the economic life as well as the rate of production. Since the D.C.F. rate of return, due to the last expansion, will be equal to the cost of the capital, weighed against the costs internally and externally generated funds, including both the minimum acceptable return on equity as well as the expected interest on debt.

## Demand Forecasting

Demand forecasting is a must in countries which are highly industrialised, if forward planning is to be done and investment decisions are to be taken. In backward economies, demand is subordinate to supply.

This may be especially true in the case of construction materials, e.g., steel, in addition to equipment for power generation or transmission etc. Idle capacity in certain sectors may be available, but then there are other causes than lack of demand, in the normal run.

The following are the main considerations in forecasting :

1. The period or time duration for the forecast, e.g., 5 years or 10 years etc. or 1 year.

2. The level of forecasting whether

   (a) National levels, e.g., planning for aluminium for the country's overall industrial requirements,

   (b) Industrial level i.e., development of the aluminium industry in respect of the various inputs such as, technology or expertise, raw materials such as bauxite, cryolite, petroleum coke, aluminium fluoride, electric power, caustic soda etc., and

   (c) At the company's level—where capital, staff selection, site selection, contracts for various supplies, arrangements for knowhow etc. are major items.

3. Forecasting may be done with regard to one aspect or requirement, e.g., cryolite or petroleum coke or bauxite, in aluminium manufacture.

4. Forecasting may be product oriented, electric grade aluminium. or aluminium alloys or structural products.

5. Forecasting of product may be categorised as :

   (a) Capital goods machinery of the medium and heavy types which are productive, e.g., tractors, lathes, vehicles for transport, mining machineries, etc.

   (b) Consumer goods—cosmetics, soaps, paper, explosives, petroleum products, fans (light domestic), sewing machines etc.

   (c) Services—for knowhow or expertise, consultancy.

6. Last but not the least, the nature of the market may be studied, any possibility or innovations or substitutions coming up, will have to be given due regard.

Examples : In opencast mining, gelatine slurry explosives may be preferred to AN/FO types or EIMCO rocker type loaders may be preferred to slusher hoists or rotating arm (or joy) type machine- aluminium, may be used instead of copper in electrical machines.

## Methods of Forecasting Demand

1. By taking various economic indicators, the demand can be forecast from the micro plan.

   (a) For example the demand for coal in relation to proposed thermal stations, taking 0.3 million tonnes per annum per 100 Mega watt, we may say that if 2000 MW is proposed to be generated the coal demand will be $\dfrac{2000}{100} \times 0.3$

   $$= 6 \text{ million tonnes/annum.}$$

   (b) Similarly taking the power demand for 1 tonne of aluminium metal to be 20,000 k wh, the requirement of power for 1000,000 tonnes p.a. will be 2000,000 kwh.

   (c) So also if the demand for fertiliser is known, e.g., phosphatic, nitrogenous or potassic, the forecast for the production of rock phosphate, ammonia etc. can be worked out.

2. By taking the per capital consumption as the basis, the demand growth, corresponding to the increase in population (and in keeping with the increase in standard of living) can be forecast. Say if the per capital consumption of non-coking coal is 1.5 tonnes per annum in 1970, the production can be assumed to be 90 million tonnes in 1980 taking the population to be 60 million. So also the per capita consumption of steel per annum, or fertilizer per annum per capita or copper etc. can be computed.

3. Another common method, is to identify the trends in growth of demand. If these are plotted as percentage increase over a base figure on a semi logarithmic paper in a linear equation can often be identified, for the curve so obtained.

The following example brings out the steps involved.

| Year | Index of in-come from mineral in-dustries X | Sale oj mining equipment Y | $X_1$ | $Y_1$ | $X_1Y_1$ | $X_1^2$ |
|------|------|------|------|------|------|------|
| 1967 | 100 | 110 | 10 | 11 | 110 | 100 |
| 1968 | 120 | 140 | 12 | 14 | 168 | 144 |
| 1969 | 140 | 160 | 14 | 16 | 224 | 196 |
| 1970 | 160 | 180 | 16 | 18 | 288 | 256 |
| 1971 | 200 | 200 | 20 | 20 | 400 | 400 |

n=5       $\Sigma X_1 = 72$, $\Sigma Y_1 = 79$, $\Sigma X_1 Y_1 = 1190$, $\Sigma X_1^2 = 5184$

The simultaneous equations used in the Least Square method are given below:

$$\Sigma Y_1 = n\,a + b\,\Sigma X_1 \qquad \text{...(1)}$$
$$\Sigma X_1 Y_1 = a\,\Sigma X_1 + b\,\Sigma X^2. \qquad \text{...(2)}$$

Substituting the values of $X_1$ and $Y_1$ from the above table:

Equation (1) is 79 = 5a + 72b.       ...(1)
Equation (2) is 1190 = 72a + 5184b       ...(2)
Multiplying eq. (1) by 72, 5688 = 360a + 5184b       ...(3)

Subtracting eq. 3 from 2, 4498 = 288a
a = 15.597 or 15.6
Substiuting value of a in eq. (1) 79 = 78 + 72b.
b = 0.014
The standard equation for a straight line is

$$Y = bX + a.$$

Hence the values of the variable Y (sale of mining equipment can be got for any yeay, since X, the income from mineral industries is known.
If the value of X = 220, then

$$Y = 15.8 + 0.014 \times 220$$
$$= 19.$$

*Profit, planning and control* : One of the techniques commonly used for this purpose is known as the break-even analysis. In this method, the break-even point, i.e., the point at which the cost of production is equal to cost of sales, and the net income is thus equal to zero.

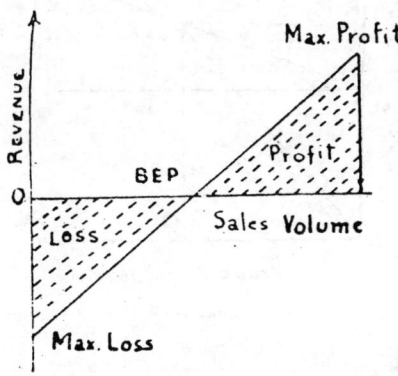

Fig. 10.1

The break-even point (BEP) is determined as shown below

$$BEP = \frac{\text{Fixed cost}}{\text{contribution of margin per unit.}}$$

Figure 10.1 illustrates the case.

In making the break-even analysis, the following conditions are taken for granted :

(a) The costs, both fixed and variable, behave uniformly.

(b) That the revenue is directly proportional to turnover.

(c) That the sales are equal to production.

Despite these limitations, the BEP does bring out certain essential features which can be utilised for economic management.

1. Profit value relationship as shown in Fig. 10.2.

2. Safety margin $= \dfrac{(\text{Sales} - \text{BEP}) \times 100}{\text{Sales}}$

3. Value required to reach targetted profit level

Target sales volume $= \dfrac{\text{Fixed cost} + \text{Target profit}}{\text{Contribution margin/unit}}$

4. Change in price $Q = \dfrac{FC + P}{SP - VC}$

Q = New sales volume
FC = Fixed cost

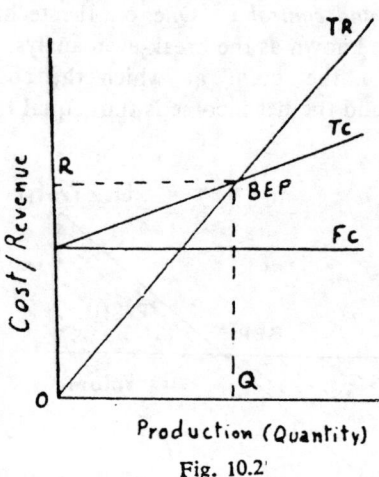

Fig. 10.2

P = Profit
SP = New selling price
VC = Variable cost/unit

One of the technique commonly used in management is the ABC analysis, which is a versatile tool. It is employed in (a) cost control, (b) profitability analysis, (c) wage analysis, (d) machine utilisation, (e) time studies, since time means money and (f) inventory control.

Examples in respect of each are given below :

(a)  Cost control

In a certain industrial operation the break up of cost, itemwise, is as given in table :

| No. | Item | % of total cost |
|-----|------|-----------------|
| 1. | Raw materials | 36 |
| 2. | Ancilliary materials | 15 |
| 3. | Electricity (power) | 6 |
| 4. | Fuel | 7 |
| 5. | Water | 7 |
| 6. | Rent | 3 |

| 7. | Salaries of staff and wages | 2 |
|---|---|---|
| 8. | Salaries of supervisors | 10 |
| 9. | Depreciation on building | 8 |
| 10. | Depreciation on machine | 5.0 |
| 11. | Packing and forwarding | 4.5 |
| 12. | Stationery | 0.5 |
| 13. | Conveyance/travelling allowances | 0.4 |
| 14. | Overtime | 0.6 |
| | | 100.00 |

In the above statement even though raw ancilliary material may constitute only about 14% by marginal value, they form 51% of the cost. Hence they will form Category (A). Similarly, labour, electricity, fuel. water make up 34% of the costs. They will be categorised as (B). Since the remaining items make up only 15% of the cost, they will be termed Category (C).

From the above analysis it is clear that any cost control system should aim at economy in category (A) primarily then turn its attention to category (B) and later to category (C).

(b) Profitability analysis

In this case again a break up of the total profit in terms of the products will help in maximisation. The following is an example.

| Products | a | b | c | d | e | f | g | h | i | |
|---|---|---|---|---|---|---|---|---|---|---|
| (a) Net profit earned (Rs. lakhs) | 2.5 | 0.3 | 0.4 | 0.6 | 0.5 | 3.8 | 0.8 | 1.0 | 0.1 | Total profit Rs. 10 lakhs. |
| (b) % of total profit | 25 | 3 | 4 | 6 | 5 | 38 | 8 | 1 | 1 | |

The analysis shows that the products labelled *a* and *f* are most remunerative and contribute to 63% of the profits.

(*c*) Wage analysis

The following example shows how wage analysis can be made.

| Item No. | Category of staff | % of total annual wage bill | Percentage strength of staff categories |
|----------|-------------------|------------------------------|------------------------------------------|
| 1. | High supervisory | 25 | 10 |
| 2. | Middle and lower supervisory | 35 | 15 |
| 3. | Mechanics | 15 | 30 |
| 4. | Assistants (clerks) | 10 | 15 |
| 5. | Stenos | 12 | 20 |
| 6. | Watchmen | 3 | 10 |
| | | 100 | 100 |

From the analysis it is seen that items Nos. (1) and (2) comprise categories A and B respectively, whereas the rest will fall under category C.

It is seen that though the number of lathes, milling machines and presses is only 55%, the losses incurred on account of these machines is 67% of the total or nearly 2/3. Hence this will form the 'A' category. In the 'B' category, shapers and drilling machines will be included, while the rest will constitute the 'C' category.

(d) Machine utilisation

| 1<br>Machine | 2<br>Numbers | 3<br>% of Idle machines | 4<br>No. of surplus machines | 5<br>Cost per machine in Rs. | 6<br>Depreciation at 5% | 7<br>Loss % of total |
|---|---|---|---|---|---|---|
| 1. Lathe | 32 | 20 | 6.0 | 30,000 | $6 \times 30000 \times 0.05 = 9000$ | 24.2 |
| 2. Milling machine | 20 | 15 | 3.5 | 50,000 | $3.5 \times 50000 \times 0.05 = 8750$ | 23.8 |
| 3. Shaper | 15 | 30 | 5.5 | 10,000 | $5.5 \times 10000 \times 0.05 = 2750$ | 7.5 |
| 4. Drilling machines | 12 | 50 | 5.7 | 15,000 | $5.7 \times 15000 \times 0.05 = 4275$ | 11.7 |
| 5. Grinding machine | 10 | 60 | 7.0 | 5,000 | $7 \times 5000 \times 0.05 = 1750$ | 4.8 |
| 6. Power saw | 4 | 80 | 3.5 | 7,000 | $3.5 \times 7000 \times 0.05 = 1225$ | 3.4 |
| 7. Press (Hydraulic) | 3 | 75 | 2.0 | 70,000 | $2 \times 70000 \times 0.05 = 7000$ | 19.2 |
| 8. Gas cutting and welding | 4 | 60 | 2.5 | 15,000 | $2.5 \times 15000 \times 0.05 = 1875$ | 5.2 |
| | 100 | | 35.7 | | 36,625 | 100.0 |

*(e)* Time study

Since time lost is capital lost/idle, an account of time consumed and the manner are important. Thus for evaluating manpower requirements, for executing a certain job, the following data will be necessary ; *(a)* time taken in each operation where more than one operation is involved, *(b)* frequency for each such operation, i.e., hourly, daily, etc. The product of *(a)* and *(b)* gives the total time required by a certain operation, in a week, or day or an hour, and this may be tabulated as shown below. It will be found that the number of a certain operation, may be only 20% of the total number of operations but the estimated time taken by that operation will be say 70%. This then will fall in category 'A'. Group 'B' and 'C' can be identified likewise. The data may be conveniently placed in a tabular form for enabling easy identification.

| *Name of operation* | *Time taken in hour, minutes etc.* | *Frequency* | $T \times F$ | $\dfrac{T \times F}{ET \times F} \times 100$ *i.e.* % |
|---|---|---|---|---|

*(f)* Inventory control

In a large industry whether manufacturing or any other where large number of machines are involved, it becomes essential to maintain sufficient stock of spares and accessories, so that breakdowns do not mean a complete shut down. Manufacturers of many machines do advise the purchase of spares/accessories required for the next two years. While accepting such advice one should also be able to visualise the rationale behind such advice. In fact the A, B, C categorisation is a handy tool in this regard.

A, B, C categorisation should be applied to each machine separately. Taking a tractor identify the parts which are subjected to *(a)* great stresses, *(b)* wear and tear due to higher speeds, and *(c)* greater abrasion. These will be classified under category 'A'. Similar criteria can be used to identify categories 'B' and 'C'.

**Stock Reduction and Control**

In store keeping inventory is a serious problem. Shortage creates problems for the manufacturing side or production wing whereas excess creates problems for the administration/management e.g. capital is locked up, stores get obsolete, storage problem increases etc. Hence methods of arriving at the optimum stock have been thought about. The following formula is suggested for determining the economic lot and order quantity.

$$\frac{\text{Economic lot}}{\text{Order quantity}} = \sqrt{\frac{\text{annual demand} \times \text{ordering cost per order}}{\text{inventory carrying cost per cent}}}$$

The following example explains the use of the formula

Annual demand = 18,000 Nos.

Ordering cost per order = Rs.2/-

Inventory carrying cost per cent = 20

$$EOQ = \sqrt{\frac{2 \times 18000 \times 2}{20}} = \sqrt{3600}$$
$$= 60 \text{ Nos.}$$

## (B) PERT AND CPM. TECHNIQUES

Part is an acronym for Programme Evalutation Review Technique whereas CPM stands for Critical Path Method. While both these techniques involve the use of net-works, the emphasis in pert is more on the minimising of the execution time of the project and monitoring the various inputs other than the financial. On the other hand, CPM is more concerned with the financial aspect of the project, with the view to minimising the overall costs within the project execution time.

Pert in a way supercedes the older technique known as the Gant Chart. For example, in the Gant Chart certain overlapping activities, or activities that can occur concurrently cannot be displayed with clarity. This is not so in the case of the Pert net-work in which every individual item of work can be accommodated and displayed in the logical sequence. These basic differences can be illustrated by an example. If a drag line is to be procured then the various activities or operations or steps to be taken are as follows :

1. Inviting tenders and finalising and placing the order, time taken is 6 weeks
2. Dilivery time, say, 5 weeks.
3. Installation at site, say, 4 weeks.
4. Training of operator, say, 6 weeks. (including recruitment).
5. Trial run, say, 4 weeks.

It will be seen from the fig. 10.3 and 10.4 the such operations items 4 and 5 which can go concurrently are better depicted in the Pert net.

Even though the use of Pert for major projects is a difficult job, it is necessary that the principles be understood so that its utility can be appreciated. It is clear that large operations will require the services of an expert for coordination. With the view to understanding the pert net-work and for preparing smaller net-work for minor projects, it is necessary to know the basic notations employed. The net-work comprises circles in the

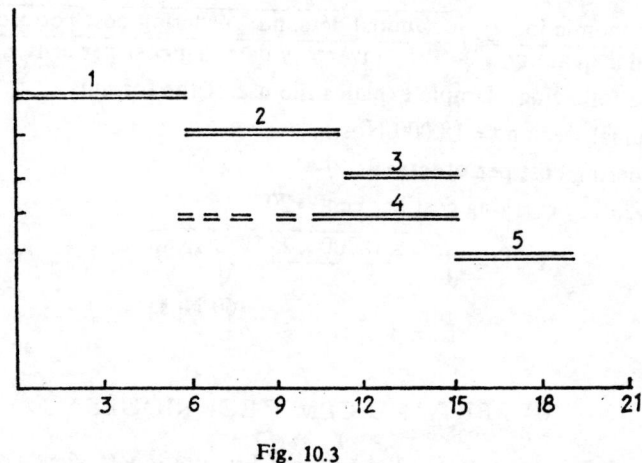

Fig. 10.3

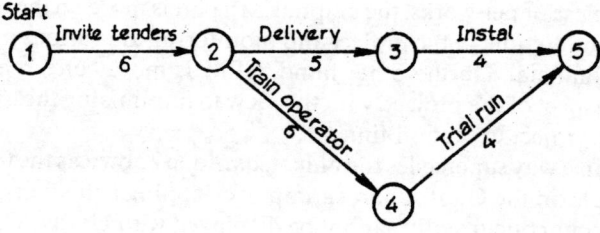

Fig. 10.4

middle of which a number is inscribed. The circle denotes an activity and the number its sequence. The circles are joined by arrows which stand for the activity. Alongside the arrow is inscribed a number which indicated the duration of the activity which are either, days, months, weeks etc. The tail of the arrow is attached to the preceeding event, while the head of the arrow shows the suceeding event. It is also to be pointed out that while the activities consume time, the events on the contrary do not. Further, an event is taken to be completed when all the activities preceeding and connected to it have been completed. It also clears that the event at the tail of an activity will have a number which is smaller than the one at the head.

Further elaboration of the net-work is in the introduction of the computation of the 'earliest start time' (E.S.) and the latest finish time (LF) of an activity. The earliest finish time is calculated after the completion of the net work when the duration of all activities have been incorporated. In order to arrive at the earliest start time the time at the initiation of the project is assumed to be zero. The lower half of the event circle is made into two quadrants. In the quadrant to the right zero is inserted. The next step is to determine the earliest starting time of all the subsequent activities, in relation to the zero time. The earliest start-times obtained are recorded in the lower right hand quadrant of the respective activity circle. Hence it will be noticed that the earliest start time of an event is the maximum value of the earliest finish time of all the activities ending with that event. So also the earliest start time of the final event of the project is its earliest finish time.

In order to find out the latest finish time of events, a procedure similar to that employed in the computation of the earliest start time is adopted. In this case however, instead of proceeding forward the computation is started from the last event of the project. In short, it may be stated that the earliest start time is got by using the 'forward pass' procedure, and the latest finish time is got by using the 'backward pass' procedure. The value of the latest finish time of each event is recorded in the event circle in the left hand quadrant. Thus, the event circle will contain three figures. The number of the event will be on the top half, while the bottom half will contain the value of the earliest start in the lower right quadrant, and the value of the latest finish in the lower quadrant. The simple net-work fig. 10.5 shows the methodology

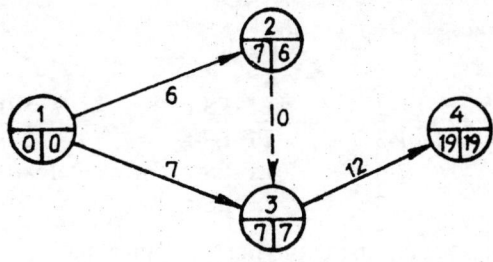

Fig. 10.5

of the computations as well as the method of recording. What has been stated in the above paragraphs can also be set out in the form of formulae, as shown below :

1.  Earliest finish time of an activity=Earliest start duration.
    $$or \quad EF=ES+t.$$
2.  The latest start of an activity=Latest finish duration,
    $$LS=LF-t$$
3.  Tatal float=LF−ES−t.

The total float or slack is the difference between the earliest and the latest starting times for an activity. In other words it is the difference between the available time and the time required for an activity. It is necessary to know which of the activities have floats/slacks, with a view to rescheduling, keeping in view the available resources, by adjusting the starting time. Such

flexibility is not possible if the activity has no float. Activities with no float
are known as critical activities. When such critical activities form a
continuous chain right from the start of the project to the completion then
the chain of events consitute a Critical path. Hence, in a net-work in which
there are several chains of events branching from the start and terminating
at the final event, it is possible that there are more than one critical paths.
The fig. 10.6 shows the activities and events of a project. The method of
determining the critical path is given in the tabular statement.

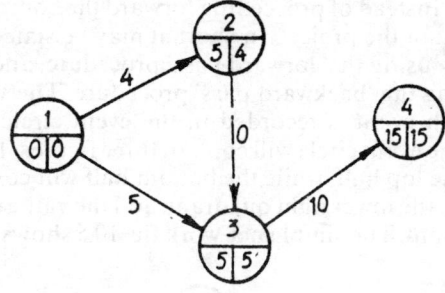

Fig. 10.6

For activity        1-2        LF-ES-t  =        5-0-4        or 1
                    1-3        LF-ES-t  =        5-0-5        or 0
                    3-4        LF-ES-t  =       15-10-5       or 0
                    2-3        LF-ES-t  =        5-4-0        or 1

Total float is usefull in identifying the Critical path/paths. In addition,
it is also essential to determine the free float. For obtaining the value of the
free float, it is assumed that all activities start at the earliest. Hence, in fig. 10.7
the free float for activity 2−4 is 24−4−6=14. The free float for activity 2−3 is
12−6−6=0 and that for activity 1−2 is 6−6−0=0. Hence these last
mentioned activities become the critical ones. Identification and utility of
the Critical Path/paths.

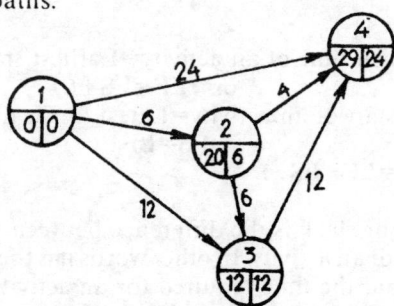

Fig. 10.7

1. All the events on the critical path are critical.
2. There can be more than one critical path.
3. The project duration is obtained by adding the duration times of the activities lying in the critical path.
4. Any delay either in the start or finish of the critical event will increase the project duration by an equal amount.
5. The reduction of the project duration by allocating additional resources for reducing the duration is known as 'Crashing'.
6. The allotment of resources, for the events on the critical path, take priority.

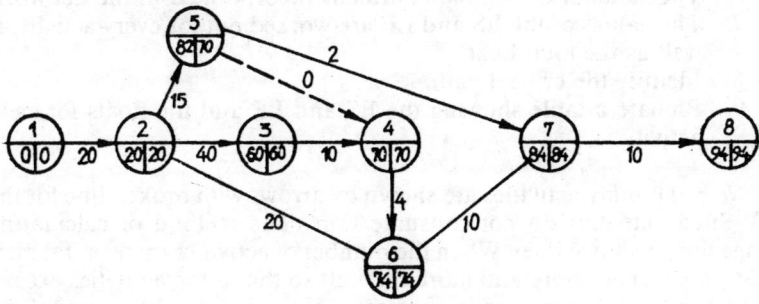

Fig. 10.8

Consider fig. 10.8. The float or slack can be determined as shown in the table given below :

### Slack or Float is LF−ES−t.

| Activity | Duration | Earliest start | Earliest finish | Latest start | Latest finish | Slack |
|---|---|---|---|---|---|---|
| 1-2 | 20 | 0 | 20 | 0 | 20 | 0 |
| 2-3 | 40 | 20 | 60 | 20 | 60 | 0 |
| 2-5 | 15 | 20 | 35 | 67 | 82 | 47 |
| 2-6 | 20 | 20 | 40 | 64 | 74 | 64 |
| 5-7 | 2 | 35 | 37 | 80 | 82 | 45 |
| 6-7 | 10 | 74 | 94 | 74 | 84 | 0 |
| 7-8 | 10 | 84 | 104 | 84 | 94 | 0 |
| 4-6 | 4 | 70 | 78 | 66 | 74 | 0 |
| 3-4 | 10 | 60 | 80 | 50 | 70 | 0 |

From the above table it is seen that the activities 2-6,, 2-5 and 5-7 have excessive float and that these become cretical and so transfer of resources is needed to tighten the slack and keep up the schedule.

Procedure adopted in applying PERT.

1. Obtain a clear picture of the objectives as well as the stages in the development.
2. Identify the key events, their sequence, and assign them to the respective departments responsible for the execution.
3. Take out a list of the activities in logical sequence.
4. Construct the net-work.
5. Compute the 3-time estimates for every activity and obtain the mean value.
6. The value of the estimated time is incorporated in the net-work.
7. The values of the ES and LF are worked out for every activity as well as the total float.
8. Identify the critical path.
9. Prepare a table showing the ES and LF and the floats for each activity.

*Note :* Dummy activities are shown by arrows with broken line for the shaft. Such activities do not consume time. The method of calculating average time is shown later. When the number of activities increase manual operation becomes more and more difficult so that if the activities exceed 300 the use of the computer is necessary. Net-work can be prepared for different levels of management in which the information to the extent needed for that particular management level is incorporated. hence, sub net-works can be prepared for civil works, or for the electrical division, or mechanical division etc. Such net-works can be prepared for the crushing unit, or the grinding unit or for the blending unit etc. Projects with a large number of subnet-works are troublesome for 'monitoring' with the view to 'updating' and for the reallocation of resources.

### The Critical Path Method (CPM)

Resources allocated to a project can fall under either of the five categories (a) men (b) money (c) machines (d) materials (e) time. In any project, money is of primary importance and its utilisation requires all possible care. If, for instance, machines or materials arrive earlier than scheduled it will lead to the blocking up of the capital which in turn will delay critical events, increase the gestation period etc. Hence, a money-oriented net-work technique has been devised which is the CPM. In this method, duration of the project is related to the cost such that the time and the cost are minimal.

If efficiency is to be maintained then the requirement of men should not fluctuate as shown in fig. 10.9 but follow the trend shown in fig. 10.10. Further, critical events cannot be altered, and it is only the non-critical ones, having floats, that can be started at a suitable time within the available float. Thus the allocation of resources to such jobs will tend to level the

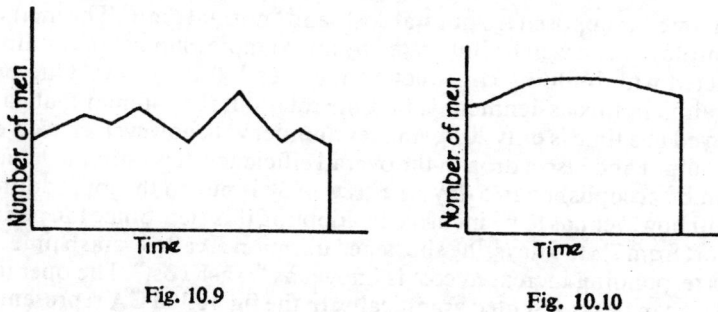

Fig. 10.9                    Fig. 10.10

requirements of man power at that point of time. Hence, the total float of any job is the criterion for levelling costs. There are three methods of levelling (a) variable (b) fixed and (c) mixed.

A variable levelling assumes that the supply of that a particular resource such as or certain raw material, etc. is unlimited. Hence, variable amounts of that resource is inducted from time to time. For example, 5 carpenters are recruited to-day, 10 carpenters taken in the next day, then only 3 are retained on the third day and so on. The scheduling should be regulated so that the demand for personnel is built up gradually and the demobilisation is also gradual, so that on a graph the schedule forms a smooth convex curve.

In the case of fixed levelling it so happens that the supply of the resource is fixed due to shortage. In such a case the activities have to be taken up by introducing overtime or by keeping standby personnel.

Mixed levelling is accomplished by a combination of the above two methods. A term commonly used for describing the resource allocation to a project is Multiple-resource Allocation Procedure which is referred to as MAP.

In order to obtain the estimated time or duration of an activity in a network the following formula is employed :

$$te = \frac{a-4m-b}{b}$$

a is the optimistic estimate.
b is pessimistic estimate.
m is the most likely estimate.
te is the expected time.
The formula is based on the normal statistical distribution.

## Costing Systems in PERT

One of the techniques used in PERT for decreasing the overall costs and minimising the project duration is the induction of "crash cost" and "crash time as apposed to "normal cost" and "normal time". The methodology employed can well be illustrated by an example. Suppose a certain job, connected with building construction requires 150 man days. One of the constraints in this assignment is that the total number of men that can be employed at a time is only 20. A smaller number will, however, cause a drop in the output and also a drop in the overall efficiency. It is estimated that the job can be accoplished in 5 days if a crew of 30 is put on the job, additional cost will however has to be incurred in adopting this step. Since the period is cut short from 7½ to 5 days, the shortened duration is called "crash time" and the corresponding increased cost is known as "crash cost". The operations involved can be represented graphically. In the fig. 10.10a OA represents the crash cost, and OB is the normal cost. Normal time is the time within which

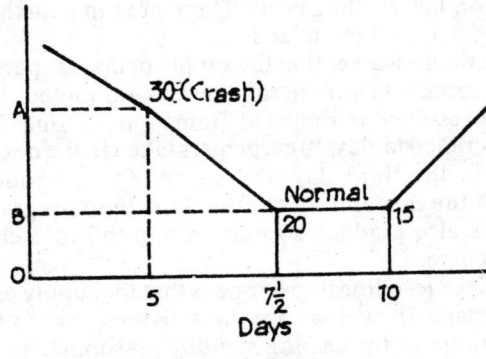

Fig. 10.10a

the job can be performed at the minimum cost, while the crash time is the time which has minimum value. From the above 4 variables a factor known as the "job slope" is derived.

$$\text{Job-Slope} = \frac{\text{Crash cost} - \text{normal cost}}{\text{Normal time} - \text{crash time}}$$

Considering the example given above, if the normal rate is Rs. 20/- per man day then the normal cost will be Rs. 20×20×7½=Rs. 3000/- If it is decided to do a crash job by reducing the time to 5 days with 30 men working overtime, the overtime rate being 10% then the crash cost will amount to 30×5×20×1.1=Rs. 3300/-

$$\text{Job Slope} = \frac{3300 - 3000}{7½ - 5} = \frac{300}{2½} = 120 \text{ per day}$$

Thus, by the use of the above formula in which there are 5 variables in all, it is possible to determine one missing variable if the other 4 are known. By the use of this formula, the 5 variables for every job or activity in a network can be determined.

## THE COMPUTER—ITS USE

A computer can be described as an electric data processor. It is a machine which can accept data which is stored in various forms, such as punched cards or paper, etc. Since, the computer has a wide range of applications, it is essential that the basic elements are known. Its use in business, in science and technological fields are well known. In the more recent years, organisations involved in geological and mining operations have also taken to the computer.

The Computer has four principal components : 1. Input 2. Storage—both internal and external 3. Processing which includes the arithemetic and the control units 4. Output.

1. The computer takes in data from one or more input sources. The data are processed and the results are produced on one or more output sources. The input and output devices are an integral part of the computer. Checking of the input is also carried out by an in-built device. The commonly used input and output devices with the computer are : punched cards. magnetic tapes, magnetic discs. floppies etc.

1. In the punched cards the information is transferred to the card in the form of holes made in the 80 columns contained in the card. Accessory of the computer or a perepheral known as the card reader interprets the information and it is processed and the result is again delivered on a punched card by an accessory known as the card punch. The reading and the punching speeds vary with the machine. Generally, the reading speed is much greater than the punching speed.

2. Magnetic tapes are made of plastic material, coated with a metallic oxide on one side. The metallic side can be magnetised readily and so the data is recorded as magnetised spots in the metallic oxide in seven parallel tracks or channels. The tape is mounted on reels and the information contained in the tape is several times larger than that contained in a stack of punched cards or is equal to 5,00,000 cards fully punched.

3. Magnetic discs are thin metal discs thich are coated with magnetic oxide on both the surfaces and floppies are similar but they are made of very thin plastic material and hence, they are flexible.

The Central Processing Unit (CPU) controls and directs the entire computer system, and it is responsible for the arithmetic and logic operations on the data. This unit includes the two sections (a) the arithmetic/logical units, which is called the brain of the computer and (b) the control section which comprises the nerve centre of the computer. The control section coordinates the several operations such as the input, processing, output as well as the storage.

The arithmetic section performs the mathematical functions, including, addition, subtraction, multiplication, division, and also carries out the comparison of the arithmetical data fed. The above mentioned five operations comprise the logical counterpart of the processing unit of the computer.

The control unit of the computer carries out the desired operations on the stored programme in a specified logical sequence.

The control unit records the instruction from the storage, examines it, and sets up the electronic circuit, which is needed to perform the operation. When an operation is completed, the control unit reads the next instruction from the storage and then repeats the same steps. The operation on an instruction is carried out in 2 stages. (a) the instruction phase or I phase and (b) the execution phase or 2 phase.

The computer can have two types of storage elements. The internal storage which forms an integral part of the machine and the external storage or secondary storage which may comprise magnetic tape, disc, or floppy etc. The secondary storage is also accessible to the primary storage and to its input/output devices.

The primary storage of the computer consists of thousands of individual units or bits. The bit is represented in the storage portion of the computer by a magnetic core. This is a tiny ring of ferromagnetic material, a few hundredths of an inch in diameter and this can be readily magnetised or demagnetised. The primary storage is made of two groups of such bits which is called a cell. In the binary system, a character is represented by a 5-bit cell. In the binary coded decimal system (BCD), a character is represented by two separate cells each of which contain 4 bits.

The programme which is stored in the internal storage is accessible and the computer can modify.

The data fed into the computer is placed in a specific location in the memory section which has a specific number for its identification, and this is called the address, i.e. a number such as 1519 can be the address and if this number is operated, the computer will instaneously provide the display or even print out the information contained at that address. Commonly, the instruction format for an operation comprises a figure with 6 digits. The first two digits known as the operation code, direct the machine as to what it should do i.e. divide or add etc. The succeeding four digits are the address and refer to the location in the memory.

Thus, for example, if a certain data is, say, Rs. 1200/- and it is located at address 1000, and if another data is Rs. 1500/- placed in address 1100 and if these two are to be added and then placed at a new address 1219, the instruction to the computer will be conveyed in the manner shown below :

| Operation code | Operation code abbreviation | Address |
|---|---|---|
| 10 | clear add | 1000 |
| 11 | add | 1100 |
| 15 | store accumulate | 1219 |

The first instruction is in the form of a numeral, 101000 rep e ̤nting a code. This code, if expanded, will read as "clean the accumulator and add the number from address 1000 to the accumulator. Similarly, the second instruction is to add the number in address 110 to the number already in the accumulator. Thus, Rs. 1200/- has been added to Rs. 1500/-. The third instruction is an order to the accumulator to place the final result in location 1219. There are various types of instructions for various operations to be performed by the computer. Thus, "sub" will mean subtract, "Acs-plus-jump" will denote that after performing an operation as instructed the result in the accumulator will jump to another address avoiding the intermediate addresses. So also the instruction "Unjump" signifies an unconditional movement to the location specified in the instruction.

When dealing with computers certain terms are often used which may be defined :

1. On line means an operation of the peripheral equipment under the control of the central processor.
2. Off line signifies an operation of the peripheral equipment independent of the central processor.
3. Access time is the time required to get a number from the memory into the arithmetic or control circuit after the central circuit has called for it.
4. Accumulator is a portion of the memory where the arithmetical and logical operations are performed.
5. Peripheral equipment are those which do not actually form an integral part of the computer. These are accessories which operate in conjunction with the central processor e.g. printer, card reader, punches, magnetic tape, floppy, etc.

*Language of a computer* : As in the case of the well known Morse code, which is used universally by the telegraph departments all over the world, for transmitting messages by using only two signs namely, the dot and the dash, so also in the case of the computer, a binary code has been devised taking advantage of the state of the bits which constitute the memory. As mentioned earlier, the core of the bit is a tiny ferromagnetic ring, which can be magnetised or demagnetised with a change in the polarity. It can, hence, be positive or negative. Hence, when the core is positively charged it is designated as "1" and when it is negatively charged it is designed as "0". Thus, a binary system has been devised for framing a language for the computer. The numerals from 0 to 9 can be expressed in the binary language and so also the alphabets from A to Z. In addition, the language also includes punctuation marks as well as other symbols. In expressing numerals, the binary system fits in readily, especially if one is familiar with logarithms to base 2, or even natural logarithms, to which this system can be correlated. For example, the figure 2468 or 1111 in natural logarithms will mean, starting from the left, $10^0$ plus $10^1$ plus $10^2$ plus $10^3$ So in the case of base two 1111 will denote, $2^0$ plus $2^1$ plus $2^2$ plus $2^3$ that is it will amount to 1

plus 2 plus 4 plus 8 or 15. If the figure includes a 0, the value of that digit is nil
e.g. 11101, this will be 1 plus 0 plus $2^2$ plus $2^3$ plus $2^4$ will add up to 29.

The binary system also lends itself to symbolise the alphabets. The
binary language is transmitted to the computer through, either punched
cards or magnetic tape etc. to which the instructions have been transferred.

*Assembly programme/Assembler*—The computer can modulate or
understand only the binary language which has been in-built in the memory
circuit. Hence, a programme written in the spoken language has to be
transformed into the 'machine language'. For this purpose, a device known
as the 'assembler' is employed. This device thus acts like a dictionary. The
complier also uses synonyms.

In short, a programme, which is the input, written in the spoken
language, called the "Source Programme", when fed into the computer will
be transformed into the machine language with the help of the Assembly
programme or Assembler. The transformed programme is known as the
"Object Programme".

Programming is a specialised operation and requires experience, if the
job is to be done quickly and accurately. In most organisations where
computers are employed extensively, persons called programmers are
employed for the purpose. A programme may be defined as a series of
instructions or commands given to the computer for carrying out the job.
These commands are to be in the language of the computer. If this is not so
then the source programme has to be converted into the object programme.
To-day, computers operate on various languages, such as Basic, Cobol,
Fortran, known as high level languages. Each step in a programme contains
instructions and these instructions are expressed in a coded language. For
example, a certain numeral, say, 11 may mean the location of an address in
the memory of the computer or another numberal, say, 10 may mean add.
Hence, the instructions and commands are in a logical sequence and the
codes are used to translate the instructions into the computer language.

To cite an example as how a programme is operated, say, the
programmer instructs the control circuit to look to address 1209 in which
either a number or a computer word is stored. This number is read into the
control circuit which has been instructed to add this number to one already
at another address 1223. This address is the accumulator (AC), which is that
part of the memory which does the arithmetic operations.

The memory of a computer has a very strong resemblance to the
human brain. For example, a driver of a motor car has stored in his memory
two important bits. The first bit is "I will stop my car if the traffic signal is
red." In the second bit the contents is "if the traffic signal is green I will
proceed". Thus, as long as the optic nerve perceives red in the traffic light, the
driver automatically jams the brakes and if the optic nerve senses the green
in the traffic lights, the driver proceeds.

The decisions made by the computer may also be compared to a
person who after knowing the locality wants to locate a particular door
number. He makes a few enquiries and finally reaches his destination. In a

computer which has stored-programmes, the programmes stored in the magnetic core of the memory, is a package of sequential instructions which can be located in a particular memory plane in particular addresses numbers. Thus, in the Hewitt-Packard computer which is combined with the ABEM=Terrameter, standard resistivity curves for 2 or 3 layers have been stored in the memory. The field data can also be transferred to the memory and a comparison made with the sheaf of standard curves until a suitable matching fit is got with one of the standard curves comprising the package stored in the memory.

Even though the computer is a complex machine, it can be used by one who can identify what inputs are available and what output is required. In this context, it neither neccessary to know the machine language or any of the high level languages. This part is that of the PROGRAMMER and it should be left to him. Hence, it remains for the user to state the problem in such a manner, the steps should be in the correct sequence and the logic should be clear. The best way of acheiving this is by constructing a flow diagram. A few examples of flow diagrams are given below, with the view to clarifying the method and procedure.

*Programming* : In so far as the geologist is concerned, it is advisable that he confines himself to the drawing up of the flow chart and leave all further manipulations to a qualified programmer. It is not essential that the geologist learns and familiarises himself with the use of the high level languages such as fortran, basic etc. Since, the geologist has all the data and is fully aware of the inputs as well as what is required, he is qualified to play the role of the systems analyst. For example, it can be a mining problem where the mining geologist is required to match the output of the machines to obtain the optimum output. The operations involved include blasting both primary and secondary, loading, transporting, unloading, crushing etc. All these operations can be taken to constitute a single system or these can be split into subsystems also. Taking the case of a limestone quarry, where the input data are available, such as the capacity of the blast hole drill/drills, and the rate of drilling which is taken to be 12 m/hr. per machine. The depth of holes taking the height of bench to be 10 m. is 10½ m. The burden and the spacing are taken to be 4 m and 5 m. respectively. From these data, the rate of blasting can be calculated. If the loading is to be done by shovels, say, either 2,5 or 3m. details regarding the digging time, swell factor swing time and dumping time should be available. The number of passes for loading a dumper of, say, 10 or 20 tonne capacity, and the time consumed in loading. In regard to the transport part, the time required for positioning the dumper for loading, and the travel time up and down, taking into the road factor should also be calculated.

If the output required is taken to be 3,000 tonnes per day, the computer can be used for decision making to afford the best. combination of the available machinery. On the other hand, if machinery is to be acquired, then the computer can be used to decide on the most economical combination of machinery. The flow diagram fig. 10.11 shows the various stages in decision making, in relation to the former aspect. In fig. 10.11a, the logic of the programme is depicted. (See Annexure)

Another area where the computer can be used with advantage is in the estimation of reserves, in an ore or mineral deposit which has been blocked out in a grid pattern, the grid interval being, say, 100 m. The input data regarding the thickness of the bed and the depth of overburden is available. Supposing there are 100 blocks in the grid the computer can find out which of the blocks had the minimum reserves or maximum reserves etc. The flow diagram fig. 10.12 illustrates the steps. An alternative programme is given in fig. 10.12a.

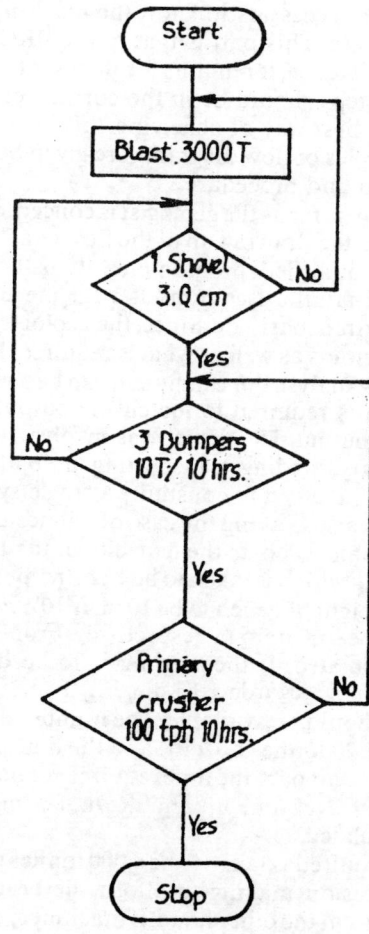

Fig. 10.11

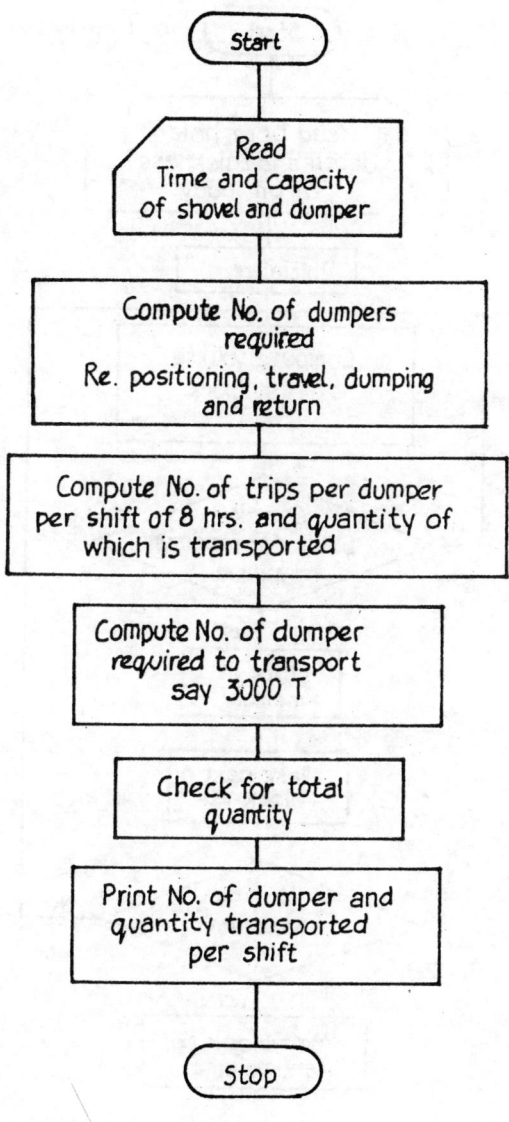

Start

Read
Time and capacity
of shovel and dumper

Compute No. of dumpers
required
Re. positioning, travel, dumping
and return

Compute No. of trips per dumper
per shift of 8 hrs. and quantity of
which is transported

Compute No. of dumper
required to transport
say 3000 T

Check for total
quantity

Print No. of dumper and
quantity transported
per shift

Stop

Fig. 10.11a

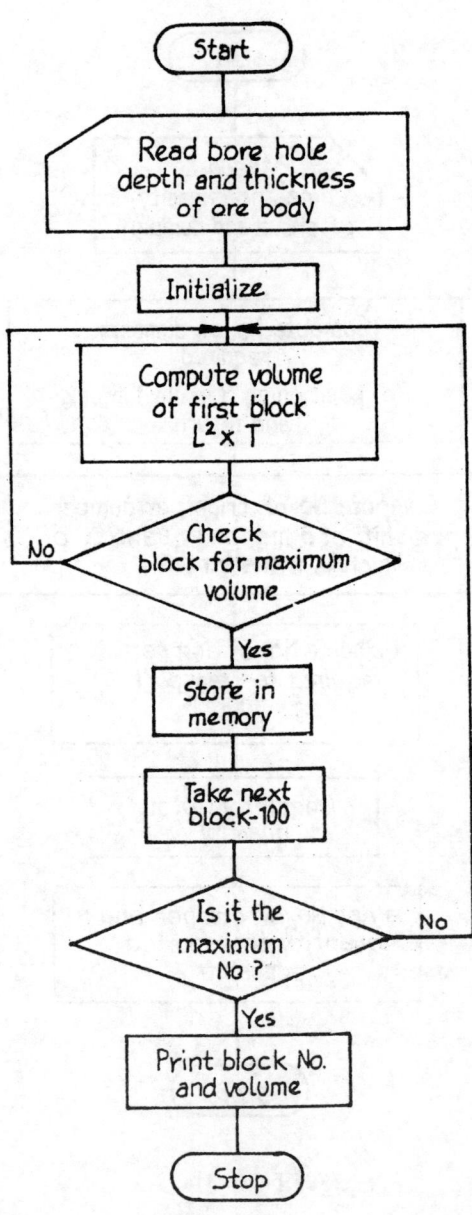

Fig. 10.12

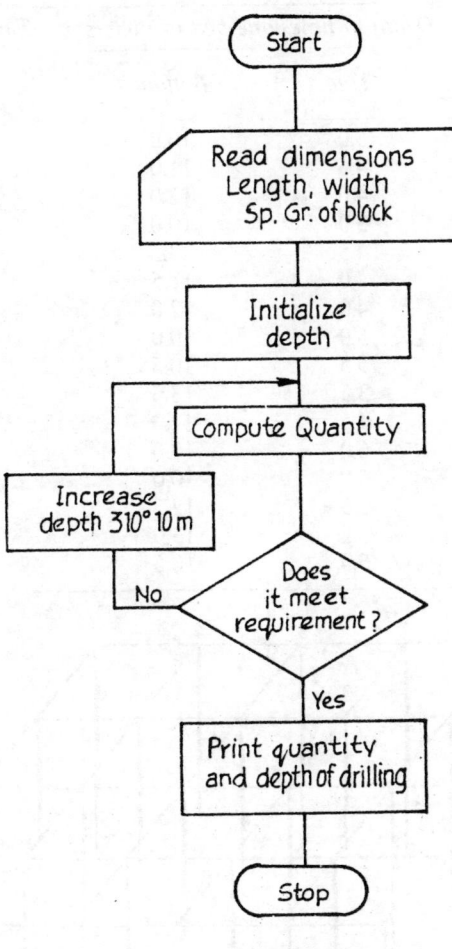

Fig. 10.12a

*Computer Graphics* : Another areas where the computer can be of immense assistance to the mining geologist, is in the elucidation of the disposition and structure of an ore body which has been drilled systematically and logged. It is preferable if the drilling had been carried out in the grid pattern with regular intervals. Taking the example of a limestone deposit which is more or less tabular, with varying thickness, and where the drilling had been done in the grid pattern, the following data had been obtained and tabulated as shown below : and the fig 10.13 gives the particulars.

| Coordinates of bore hole | Depth to limestone bed in meters | | Thickness in m. |
| :---: | :---: | :---: | :---: |
| | Top | Bottom | |
| (1, 1) | 5.0 | 10.0 | 5.0 |
| (1, 2) | 4.5 | 11.0 | 6.5 |
| (1, 3) | 4.0 | 12.0 | 8.0 |
| (1, 4) | 5.0 | 10.0 | 5.0 |
| (2, 1) | 4.5 | 9.5 | 5.0 |
| (2, 2) | 4.0 | 11.5 | 7.5 |
| (2, 3) | 4.0 | 12.0 | 8.0 |
| (2,4) | 5.0 | 10.0 | 5.0 |
| (3, 1) | 5.5 | 10.5 | 5.0 |
| (3, 2) | 3.5 | 13.5 | 10.0 |
| (3, 3) | 4.5 | 12.0 | 7.5 |
| (3, 4) | 6.0 | 10.0 | 4.0 |
| (4, 1) | 6.0 | 10.0 | 4.0 |
| (4, 2) | 5.5 | 12.5 | 7.0 |
| (4, 3) | 5.5 | 13.0 | 7.5 |
| (4, 4) | 6.0 | 10.5 | 4.5 |

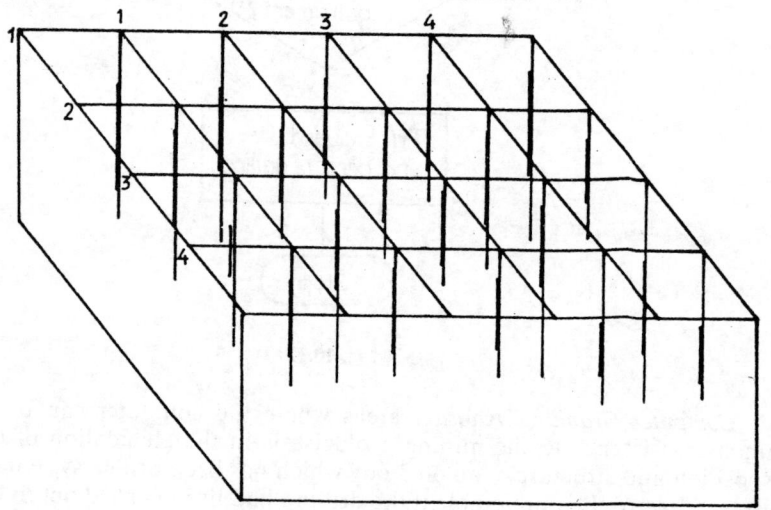

Fig. 10.13

The material contained in the tabular statement constitutes the basic data for the "input" to the computer. When the data is fed into it, the computer can be instructed not only to reproduce the data graphically, but also be instructed to produce the cross sections of the deposit in any desired direction.

Further, since the levels of the top and the bottom of the limestone bed, as encountered in the bore hole, are also known, it is possible to reproduce mathematically the surface trends of both the top and bottom of the limestone bed, either by using the surface trend technique, or Lagrang's method, or Fourier analysis. FORTRAN language has been specially designed for solving such problems.

PART XI


EXAMINATION OF MINERAL PROPERTIES

# SAMPLING AND STATISTICAL COMPUTATIONS

The method so far followed in computing average width and average assay assumes that the average assay of material, in the intermediate position, between two samples in a drive winze, is the arithmetic mean.

ABCD is the plan of an ore body, I, II and III are the sampling positions. The widths are $W_1$, $W_2$ and $W_3$ and the assay values are $a_1$, $a_2$ and $a_3$ at these positions.

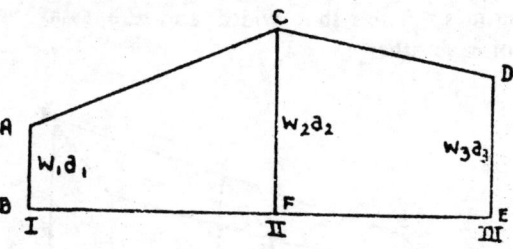

Fig. 11.1

Taking the section between I and II sampling positions

The average value $= \dfrac{W_1 a_1 + W_2 a_2}{2}$

The average width $= \dfrac{W_1 + W_2}{2}$

$\therefore$ Average assay $= \dfrac{W_1 a_1 + W_2 a_2}{W_1 + W_2}$

Adopting the same procedure to the section between II and III sampling positions—

The average values $= \dfrac{W_2 a_2 + W_3 a_3}{2}$

The average width $= \dfrac{W_2 + W_3}{2}$

$\therefore$ Average assay $= \dfrac{W_2 a_2 + W_3 a_3}{W_2 + W_3}$

Thus for the entire length between sampling positions I and III, it is seen that the total width sampled is

$(W_1+W_2)+(W_2+W_3)$, while the assay values are $(W_1 a_1 + W_2 a_2) + (W_2 a_2 + W_3 a_3)$

Hence the overall average assay value

$$= \frac{W_1 a_1 + 2W_2 a_2 + W_3 a_3}{W_1 + 2W_2 + W_3}$$

In this method, the intermediate values enter twice in the computations while the end values occur only once.

Thus the end values have to be averaged once more by extending the sampling beyond positions I and III, so that these also occur twice in the computations and the arithmetic mean gives the average assay. In sampling winzes and drifts, this precaution should be taken.

It is sometimes possible that width and the assay value may vary independently of each other.

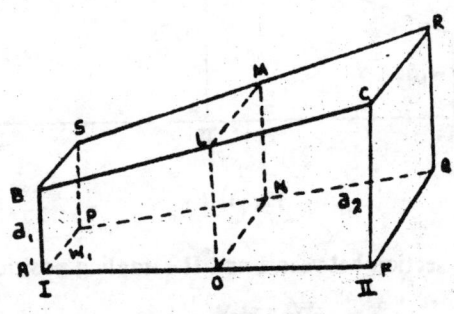

Fig. 11.2

On this basis, the average assay value is not related to the area of the trapezium ABCF but controlled by the assay widths $W_1$ and $W_2$, shown in the prismoid ABCFQRSP. The rectangles ABSP and CFQR control the assay values.

The volume of the prismoid

$$= \frac{AF \times \text{Area of ABSP} + 4(\text{Area of LMNO} + \text{Area of CRQF})}{6}$$

$$= \frac{AF \times W_1 a_1 + 4\dfrac{(W_1+W_2) \times (a_1+a_2)}{2} + W_2 a_2}{6}$$

$$= \frac{AF}{6} W_1 a_1 + 4\left(\frac{W_1+W_2}{2}\right) \times \left(\frac{a_1+a_2}{2}\right) + W_2 a_2$$

It will be seen, that the second factor represents the average relative assay values, in respect of the two cross sectional areas.

Thus, if this is divided by the average width $\dfrac{W_1+W_2}{2}$ the average assay value can be obtained.

Average assay value

$$= \frac{W_1 a_1 + 4\left(\dfrac{W_1+W_2}{2}\right) \times \left(\dfrac{a_1+a_2}{2}\right) + W_2 a_2}{\dfrac{W_1+W_2}{2}}$$

$$= \frac{W_1 a_1 + W_2 a_2}{W_1+W_2} + \frac{(W_1+W_2)(a_2+a_1)}{3(W_1+W_2)}$$

The first part of the formula represents the arithmetical average, while the second part$=0$, when either $W_1=W_2$, or $a_2=a_1$.

Where narrow widths are associated with high values, the prismoidal methods yield higher values than the arithmetical average.

**Examples.** The width of samples from a narrow vein are 50 inches and 30 inches at either end, and the assay values are 15 dwts/ton and 6 dwts/ton respectively. Calculate the average width and average assay value by the arithmetic mean, and prismoidal methods.

1. Arithmetic mean :

| Width W(inches) | Assay A(dwts/ton) | $W \times A$ |
|---|---|---|
| 50 | 15 | 750 |
| 30 | 6 | 180 |
| 80 | | 930 |

Average width $= \dfrac{80}{2} = 40$ inches

Average assay $= \dfrac{W \times A}{\text{(average width)}} = \dfrac{930}{40} = 23.25$ dwts/tons

**2. Prismoidal method :**

| Width W(inches) | | Assay A(dwts/tons) | | Au content |
|---|---|---|---|---|
| Ends Mid ×2 | | Ends Mid ×2 | | |
| 50 | ... | 15 | ... | 750 |
| ... | 80 | ... | 21 | 1680 |
| 30 | ... | 6 | ... | 180 |
| 80 | 80 | | | 2610 |

$$\text{Average width} = \frac{80}{2} = 40 \text{ inches}$$

$$\text{Average prismoid assay} = \frac{2610}{6} = 435 \text{ dwts.}$$

$$\therefore \text{ Average assay} = \frac{435}{40} = 10.9 \text{ dwt/tons.}$$

*Note :* The prismoidal formula is not applied extensively, i.e., in estimated reserves in large mines, but it is often applied to winzes or drives.

**Other methods**

**Application of Statistical Techniques to Sampling Problems—and Mineral Economics**

The application of statistical techniques, to various problems in geology and mining, is assuming greater importance; hence a knowledge of the fundamental concepts is essential for appreciating the situations. Statistical analysis helps in solving the problems which arise in day-to-day sampling and assaying by providing the mining geologist with deductions which otherwise would have been lost in the volume of data. One of the most common problems, where statistical analysis has immediate application, is stope sampling. Suppose there are 150 analyses of gold ore from a stope, in which the values range from 0 dwts/ton to 23 dwts/ton, it is essential to know many of these fall within the economic limits (leaving aside mining width etc.). In order to arrive at an answer, the analyses grouped into classes or categories 0–2 dwts., 3–5 dwts., 6–8 dwts., 9–11 dwts., 12–14 dwts., 15–17 dwts. etc. (per ton). Classification is essential, and then the number of samples, falling in each category or class, is next obtained.

This is the frequency of the group or class. For this purpose, a suitable class-interval is selected. This can be illustrated by an example, where 150 samples were collected, and classified, with 2 dwt/ton intervals.

| Class Interval Dwts/ton. | Frequency No. of samples (of the class) |
|---|---|
| 0-2 | 14 |
| 3-5 | 22 |
| 6-8 | 28 |
| 9-11 | 22 |
| 12-14 | 20 |
| 15-17 | 17 |
| 18-20 | 15 |
| 21-23 | 12 |
| | 150 |

From the tabular statement it is clear, which group of values are relatively more frequent, e.g., 6-8 dwts/ton has the highest frequency.

With a view to facilitate the use of these values, graphical methods are employed. The more common methods are (a) the frequency polygon, (b) the histogram and (c) the frequency curve.

For constructing the frequency polygon and histogram, the middle values of each class are represented on the abscissa (X-axis), while the frequency of the class is represented on the ordinate (Y-axis). The mid-values and the respective frequencies of the above example can be tabulated as follows :

| (X-axis) | Mid values | 1 | 4 | 7 | 10 | 13 | 16 | 19 | 28 |
|---|---|---|---|---|---|---|---|---|---|
| (Y-axis) | Frequency | 14 | 22 | 28 | 22 | 20 | 17 | 15 | 12 |

When the points, so obtained, are joined by straight lines, a polygon (frequency polygon) is obtained. Polygon, depicting the cumulative frequency distribution, is known as an ogive.

The histogram comprises a series of rectangles, standing side by side on the abscissa or X-axis. One of the sides, of the individual rectangles, is equal to the mid value of class interval, while the other side is equal to the frequency of the class.

In the histogram frequencies are represented by areas, while in the frequency polygon the frequencies are represented by linear measurements. In the frequency polygon, the mid values are prominent, whereas in the histogram the range of each class is prominent.

The frequency curve is, on the other hand, obtained by joining the points, of the frequency polygon, by a smooth curve. The frequency curve is characteristic for a particular distribution of values, in the sample, and it may be said to be representative of the frequency curve of the original population.

While classification and frequency describe a sample to some extent, it is necessary to obtain certain other parameters for defining the sample. e.g., (1) average and (2) standard deviation (3) mean (4) mode etc.

Average is a quantity which locates the central value of the distribution, and is in a way the typical or central tendency value. There are three averages in common use (a) the arithmetic mean, (b) median and (c) the mode.

The arithmetic mean is the value most commonly employed as the average. If the series of values is $x_1, x_2, x_3, x_4 \ldots \ldots x_n$ and $n$ is the number of terms in the series, then (the arithmetic mean)

$$\bar{x} = \frac{x_1 + x_2 + x_3 \ldots \ldots \ldots x_n}{n}$$

If however, the value $x_1, x_2, x_3 \ldots \ldots x_n$ can be grouped and the frequency of the values are $f_1, f_2, f_3 \ldots \ldots f_n$ respectively, then, the value of the arithmetic mean,

$$x = \frac{x_1 f_1 + x_2 f_2 + x_3 f_3 \ldots \ldots \ldots \ldots \ldots x_n f_n}{f_1 + f_2 + f_3 \ldots \ldots \ldots \ldots f_n}$$

$$= \frac{\Sigma fx}{\Sigma f}$$

Where $x_1, x_2, x_3$ etc., represent numbers, or where class-intervals are available together with frequencies, then the mid-values, for each class interval, are represented by x.

The following example illustrates the principle involved in the computation :

| Class interval | Frequency (f) | Mid-value (x) | (fx) |
|---|---|---|---|
| 0-2 | 14 | 1 | 14 |
| 3-5 | 22 | 4 | 88 |
| 6-8 | 28 | 7 | 196 |
| 9-11 | 22 | 10 | 220 |
| 12-14 median class | 20 | 13 | 260 |
| 15-17 | 17 | 16 | 272 |
| 18-20 | 15 | 19 | 285 |
| 21-23 | 12 | 22 | 264 |
| | $\Sigma f = 150$ | | $\Sigma fx = 1599$ |

$\therefore$ Arithmetic mean $x = \dfrac{\Sigma fx}{\Sigma f} = \dfrac{1599}{150} = 10.67$

The arithmetic mean can also be computed by using an arbitary origin. If A is the number, in the series, which is taken as origin, then $\overline{x}$ (arithmetic mean)

$$= A + \frac{\Sigma fd}{\Sigma f}$$

where f is the frequency and d is deviation of the individual term from A (an assumed mean).

The example given below illustrates the procedure.

F = Cumulative frequency.

| Class interval | Frequency (f) | F | Mid-value (x) | d=deviation from A=10 | fd |
|---|---|---|---|---|---|
| 0-2 | 14 | | 1 | −9 | −126 |
| 3-5 | 22 | 36 | 4 | −6 | −132 |
| 6-8 | 28 | 64 | 7 | −3 | − 84 |
| 9-11 | 22 | m = 86 | A=10 | 0 | 0 |
| 12-14 median class | 20 | 106 | 13 | 3 | 60 |
| 15-17 | 17 | 123 | 16 | 6 | 102 |
| 18-20 | 15 | 138 | 19 | 9 | 135 |
| 21-23 | 12 | 150 | 22 | 12 | 144 |
| | $\Sigma f = 150$ | | | $\Sigma fd = 441$ | 342 |

$$\overline{x} = 10 + \frac{441-342}{150}$$

$$= 10 + \frac{33}{50} = 10 \frac{33}{50} = 10.66$$

(b) The median is that value of the variate, at which the cumulative .requency is $\dfrac{F}{2} = 75$ i. e. its value should lie between 6 and 11 of the class interval.

1. When the series has an *even number* of terms, the median is the number which is the average of the *two middle* terms.

2. When the series has an odd number of terms, the median is the number which occupies the mid position.

$$\text{The median} = L + \frac{\frac{F}{2} - m}{f_m} \ C$$

where L is the lower boundary of the medial class, $f_m$ is the frequency of the same medial class.

C is the class interval.

m is the cumulative frequency upto the lower boundary of the medial class, e.g., the medial class is (9-11).

F is the cumulative frequency.

$$L = 6, \ m = 14 + 22 + 28 = 64, \ f_m = 28, \ C = 3, \ \frac{F}{2} = \frac{150}{2} = 75$$

$$\text{Median} = 6 + \frac{75 - 64}{28} \times 3 = 6 + \frac{11 \times 3}{28} = 9 + 12 = 7.2$$

(c) The *mode* defines the variate when the frequency is maximum. In the previous example, the frequency of the modal class is 28 i.e. where the value of the variate is 6 to 8.

$$\text{Mode} = L + \frac{f}{f_1 + f_2} \ C$$

L is the lower boundary of the modal class.

Where $f_1$ and $f_2$ are the frequencies, of the classes immediately before and after the modal classes respectively and C is the class interval.

Another measure, which is used in computations, is the geometric mean. The advantage of the geometric mean are (a) its rigid definition and (b) its property of not being affected by drastic variations between extreme values, as can be seen from the example worked out. It is useful in (1) averaging ratios, (2) computing average rate of change and (2) computing average for a series with logarithmic distribution.

The data presented may be either (a) ungrouped or (b) grouped.

(a) Taking ungrouped data, the following example indicates the method, of computing the geometrical mean.

| Item | Cost | Logarithm |
|------|------|-----------|
| 1 | Rs. 12.20 | 1.0864 |
| 2 | Rs. 3.50 | 0.5441 |
| 3 | Rs. 2.70 | 0.4314 |
| 4 | Rs. 2.20 | 0 4324 |
| 5 | Rs. 1.75 | 0.2430 |

$$\frac{2.7373}{5} = 0.5455$$

The geometric mean = Rs. 3.512

The procedure is as follows :

1. Logarithm of the value of each item is found out.
2. The sum of the logarithms is determined.
3. The sum of the logarithms is divided by the number of items.
4. The antilog of the result obtained in 3 above, is found out, and this gives the geometric mean.

(b) Taking grouped data, the geometric mean can be determined as shown in the example below. From the formula log geometric

$$\text{mean} = \frac{\Sigma f \log x}{\Sigma f}$$

| Values in Cu % class intervals) | mid. values x | log x | frequency (f) | fx log x |
|------|------|------|------|------|
| 0-2 | 1 | 0.0000 | 50 | 0.0000 |
| 3-5 | 4 | 0.6021 | 78 | 46.9638 |
| 6-8 | 7 | 0.8451 | 27 | 22.8177 |
| 9-11 | 10 | 1.0000 | 12 | 12.0000 |
| 12-14 | 13 | 1.1139 | 3 | 3.3417 |

$\Sigma f=170$   $\Sigma f \log x=85.1232$

$$=\frac{85.12}{170}=0.5\% \text{ cu.}$$

*Deviation*: The mean is a single number which is used to represent the average of a group of measurements. Thus it is possible that two or more groups of numbers, in which the items differ very much from one another, and yet have the same mean. Cónsider the following groups:

No doubt the average deviations reveal the basic differences in the dispersion, found in the two groups of variables A and B. But the most useful and popular measure of the dispersion is called the standard deviation. In order to obtain the value of the standard deviation, the following method is adopted (1) The deviations D of the variable X from the arithmetic $\overline{X}$ are worked out. (2) The values of these deviations are squared $10^2$. (3) The values of the square of the deviations, $10^2$ are added up together $\Sigma D^2$. (4) This sum $\Sigma D^2$ is divided by the number of observations $\Sigma N$, to get $[D^2]$ $\Sigma N$. (5) The square root of $\Sigma[D^2]/\Sigma N$ gives the value of the standard deviation i.e. $\gamma = \sqrt{\dfrac{\Sigma D^2}{\Sigma N}}$ where $\gamma$ is the standard deviation.

This method is time consuming since the arithmetic mean has to be found out first, and so a simpler formula is commonly used

$$\gamma = \sqrt{\frac{\Sigma(fd^2)}{\Sigma N} - \frac{\Sigma(fd)^2}{\Sigma N}}$$

In the above formula D is the deviation from any convenient origin A in the group of variables. The methods adopted for obtaining the standard deviations for non-grouped and grouped variables are shown in the examples given below.

| Item | 1 | 2 | 3 Deviation from mean | |
|---|---|---|---|---|
| | Frequency | | | |
| | Af | Bf | $d_A$ | $d_B$ |
| 1 | 4 | 2 | 9 | 11 |
| 2 | 7 | 4 | 6 | 9 |
| 3 | 10 | 13 | 3 | 0 |
| 4 | 13 | 16 | 0 | 3 |
| 5 | 16 | 17 | 3 | 4 |
| 6 | 19 | 18 | 6 | 5 |
| 7 | 22 | 21 | 9 | 8 |
| | $\Sigma Af = 91$ | $\Sigma Bf = 91$ | $\Sigma dA = 36$ | $\Sigma dB = 40$ |

$$xA = 13.0 \qquad\qquad xB = 13.0$$

$$\text{deviation } d = (x - \bar{x})$$

It is noticed that the variations, in group B, are greater than in Group A. Further, the mean is the 4th item in group A, while it is found as the 3rd item in Group B. A measure of this variability, which can be readily subjected to statistical analysis, is the deviation, d, which is the difference between the variable and the arithmetic mean. The average deviation is given by the following formula:

$$\text{Average deviation for A } = \frac{\Sigma dA}{f} = \frac{36}{7} = 5.14$$

$$\text{Average deviation for B } = \frac{\Sigma dB}{f} = \frac{40}{7} = 5.71$$

|  | F | D = (Z − X) | D² |
|---|---|---|---|
| 1. | 3 | −7 | 49 |
| 2. | 9 | −1 | 1 |
| 3. | 12 | 2 | 4 |
| 4. | 10 | 0 | 0 |
| 5. | 13 | 3 | 9 |
| 6. | 11 | 2 | 1 |
| 7. | 12 | 2 | 4 |
| $\Sigma N = 7$ | $\Sigma F = \frac{70}{7} = 10$ |  | $\Sigma D^2 = 68$ |

$\bar{x}$ = arithmetic mean.

$$\sigma = \sqrt{\frac{68}{7}} \qquad \sigma = 3.114$$

| Variable grouping | Group mid-values | F=frequency | d=deviation of x from A in class units | fd | fd² |
|---|---|---|---|---|---|
| 0.5 | 2.5 | 14 | −2 | −28 | 56 |
| 6 — 11 | 8.5 | 22 | −1 | −22 | 22 |
| 12 — 17 | 14.5 | 28 | 0 | 0 | 0 |
| 18 — 23 | 20.5 | 20 | 1 | 20 | 20 |
| 24 — 29 | 26.5 | 17 | 2 | 34 | 68 |
| | | N = 101 | | $\Sigma fd = 54 - 50 = 4$ | $\Sigma fd^2 = 166$ |

$$\Sigma(fd) = 4N = 101$$
$$\Sigma(fd^2) = 166 \qquad \text{Class unit} = 6$$

$$\text{Standard deviation } \gamma = \sqrt{\frac{\Sigma(fd^2)}{N} - \frac{\Sigma(fd)^2}{N}}$$

$$= \sqrt{\frac{166}{101} - \frac{4}{101}} = \sqrt{\frac{162}{101}}$$

$$= 1.266 \times 6 = 7.596$$

| Item Serial No. | Class interval in dwt/ton | Kind values X | Frequency (f) | Deviation (d) from mean A = 3 | fd | fd² |
|---|---|---|---|---|---|---|
| 1. | 0.5 to 1.5 | 1 | 15 | −2 | −30 | 60 |
| 2. | 1.5 to 2.4 | 2 | 40 | −1 | −40 | 40 |
| 3. | 2.5 to 3.5 | 3=A | 50 | 0 | 0 | 0 |
| 4. | 3.5 to 4.5 | 4 | 30 | 1 | 30 | 30 |
| 5. | 4.5 to 5.5 | 5 | 20 | +2 | +40 | 80 |
| 6. | 5.5 to 6.5 | 6 | 5 | +3 | +15 | 45 |
| | | | N=169 | | $\Sigma fd = 155$ | $\Sigma fd^2 = 255$ |

$$\gamma = \text{Standard deviation} = \sqrt{\frac{255}{160} - \frac{155}{160}} = \sqrt{\frac{100}{160}}$$

$$= 0.7907$$

### Pearson's measure of skewness

If the distribution is symmetrical then the arithmetic mean=median =mode. Where these three averages are in the ascending order of magnitude, the skewness is positive, and if the magnitude is in the descending order the skewness is negative.

Since skewness is a ratio, it is independent of the units of measurement and thus it is useful in comparing two different distributions.

$$\text{Skewness} = \frac{\text{mean} - \text{mode}}{\text{standard deviation}}$$

$$\text{or} = \frac{3(\text{mean} - \text{median})}{\text{standard deviation}}$$

## Time Series Trend

When quantitative statistical data are assembled with a time sequence, the result is a time series. Examples, of time series, are the monthly production of coal, annual L.M.E. prices of copper etc. Various factors influence time series, some of which are operative continuously, known as secular, others are periodic (with a variable frequency), known as cyclic, while some may be erratic, known as random. Such tendencies can be readily depicted graphically. Fig. 11.3 shows periodic of cyclic tendency (A), and erratic tendency (B). Deciphering and isolation of such factors, is known as time series analysis.

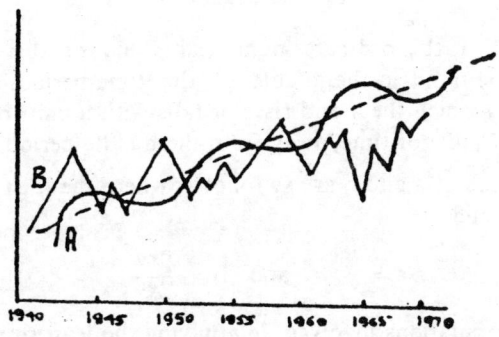

Fig. 11.3

Such an analysis is of particular importance in industry, in general, as well as in the evaluation of mineral economics. In industry, an analysis of the time series, will show its behaviour, and in predicting the production and sales. In economics trends can be forecast from a study of the time series.

Once the data have been screened, they can be plotted on a graph for facilitating visual identification of the trend. Sometimes the secular trend is so evident that a line can be drawn freehand. This is termed as manual trend fitting. Another technique, which is commonly used, is known as the moving averages method. In this method, the values of the series are averaged for a sufficiently long periods, so as to eliminate the influence of the short term fluctuations. Taking business cycles, one or more cycles are taken, in averaging, with a view to minimise local or short term distortions. This method has already been discussed earlier (vide statistics in the chapter on Cost Accounting).

The moving averages method is advantageous, as it is easily understood and minimises the influence of the extremes. However, the disadvantages are : (1) It cannot be brought upto date. Hence it is of little use in current period analysis, and for projections. (2) Even though the effects of cyclic variations are sought to be eliminated, the aim is scarcely attained, since cycles may not have the same frequency. (3) When the amplitudes of the cycles vary greatly, their effect remains, even though the period is constant. (4) Where the trend is non-linear, the moving average lies either above or below the general trend line obtained.

Another method for determining the values of the secular trend, is to obtain what is known in mathematics, *the least square line*. In its simplest form this is a straight line which satisfies the equation

$$Y_e = a + bx.$$

Where $Y_e$ is the ordinate numerical value of the trend, a is the ordinate of the trend at the middle of the time period, when $X = 0$, b is the amount by which the trend rises or falls with a unit of time, and X is the number of units of time, away from the middle period.

The values of a and b are as follows, where the least squares show a straight-line trend.

$$a = \frac{\Sigma y}{N} \quad \text{and} \quad b = \frac{\Sigma xy}{\Sigma x^2}$$

The computations involved, in applying the least square method for finding out the trend in the sales of iron ore or barytes from a small concern, are shown below :

(a) Taking an odd number of years, say 7 years.

(b) Taking an even number of years, say 6 years.

The following example shows the procedure for applying the least square method to the case where the period covers an odd number of years. Taking for example, a small barytes mine, the production trend can be worked out as shown below :

| Items $Xa$ | Tons per year $Y$ | $X$ | $X^2$ | $XY$ | $Y_e$ |
|---|---|---|---|---|---|
| 1961 | 535 | $-2$ | 4 | $-1070$ | 545.6 |
| 1962 | 593 | $-1$ | 1 | $-593$ | 584.8 |
| 1963 | 631 | 0 | 0 | 0 | 624.0 |
| 1964 | 667 | 1 | 1 | 667 | 663.2 |
| 1965 | 694 | 2 | 4 | 1388 | 702.4 |

$$\frac{\Sigma Y}{N} = \frac{3120}{5} = a. \qquad \Sigma X^2 = 10 \qquad \Sigma XY = 392$$

$$a = 624 \qquad\qquad b = \frac{392}{10} = 39.2.$$

$$Y_e = a + bx$$
$$„ \ = 624 - 78.4 = 545.6$$
$$„ \ = 624 - 39.2 = 584.8$$
$$„ \ = 624 + 0 \ \ = 624.0$$
$$„ \ = 624 + 39.2 = 663.2$$
$$„ \ = 624 + 78.4 = 702.4$$

*Note* : In the above example the values of X, i.e., years are plotted on the abscissa (x-axis), and the corresponding values of Y are plotted on the ordinate (Y-axis), on a graph. This will give the actual production trend. On the same graph, the corresponding values of $Y_e$ are also plotted, and the curve so obtained will give the trend, which can be used for projection, beyond the period indicated, by the available production figurss.

In the case, where an even number of years are involved in the period, the middle term is wanting ; and the centre lies between the 31st December of one year and the 1st January of the succeeding year. In order to avoid this difficulty, X is taken to represent half-yearly intervals, instead of one year. Thus as shown, in the example below, the values of X, for the two middle six month, is $-1$ and 1, and for the subsequent and preceding yeais 2 or $-2$, are used instead of 1 or $-1$, as shown in the earlier example. The interval is two, since each year is represented by two six-monthly periods. The following example, is the case of the barytes production from a small mine.

| Years | Production per year Y | Period X | 6-month $X^2$ | XY | $Y_e$ |
|-------|-----------------------|----------|---------------|------|-------|
| 1961 | 535 | −5 | 25 | −2675 | 549.97 |
| 1962 | 593 | −3 | 9 | −1779 | 585.98 |
| 1963 | 631 | −1 | 1 | − 631 | 621.99 |
| 1964 | 667 | 1 | 1 | 667 | 658.01 |
| 1965 | 694 | 3 | 9 | 2082 | 694.02 |
| 1966 | 720 | 5 | 25 | 3600 | 730.03 |
| | $\Sigma\dfrac{Y}{N} = \dfrac{3840}{6} = a$ | | 70 | $\Sigma XY = 1264$ | |

$$a = 640 \qquad\qquad b = \frac{1264}{70} = 18.006$$

$$Y_e = 640 - 5 \times 18.006 = 549.97$$
$$,, \ = 640 - 3 \times 18.006 = 585.98$$
$$,, \ = 640 - \quad 18.006 = 621.99$$
$$,, \ = 640 + \quad 18.006 = 658.01$$
$$,, \ = 640 + 3 \times 18.006 = 694.02$$
$$,, \ = 640 + 5 \times 18.006 = 730.03$$

In the straight line trend, the changes, per year, are constant. There are other trend patterns, where the constant change is related to percentages rather than to absolute amounts per annum. Plotting of such data, on normal graph paper, will give rise to a curve ; but if the data are plotted on semilogarithmic paper, the result will be a straight line.

This is called the *geometric straight line trend*, given by the formula $\log Y_e = a + bx$.

| Year | Production of Copper '000 tons Y | Log Y | X | $X^2$ | X log Y | Log $Y_c$ | $Y_c$ |
|------|------|------|------|------|------|------|------|
| 1961 | 5960 | 3.7753 | −3 | 9 | −11.3259 | 3.7547 | 8777.7 |
| 1962 | 6550 | 3.8162 | −2 | 4 | − 7.6324 | 3.8160 | 911.96 |
| 1963 | 7190 | 3.8567 | −1 | 1 | − 3.8567 | 3.8373 | 9228.8 |
| 1964 | 7910 | 3.8982 | 0 | 0 | 0.0000 | 3.8786 | 9437.9 |
| 1965 | 8700 | 3.9395 | 1 | 1 | 3.9395 | 3.9199 | 9637.4 |
| 1966 | 9580 | 3.9814 | 2 | 4 | 7.9628 | 3.9612 | 9828.1 |
| 1967 | 10540 | 4.0228 | 3 | 9 | 12.0684 | 4.0023 | 13617.3 |
| Totals : | | 27.2901 | | 29 | 1.1557 | | |

$$a = \frac{\Sigma \log Y}{N} = \frac{27.2901}{7} = 3.8786$$

$$b = \frac{\Sigma X \log Y}{\Sigma X^2} = \frac{1.1557}{28} = 0.0413$$

$$\log Y_c = a + bX$$
$$= 3.8786 + 0.0413X$$

where the values of X are given in column 4 of the table.

**Index Numbers**

Index numbers represent a special type of averages, which provide a yard-stick, for evaluating relative changes in price or quantity consumed or produced etc. either from time to time or from place to place. Index numbers are advantageous in that unwieldy statistical data are reduced to percentage, so as to make them amenable to comparison. Further, if industry is reluctant to release actual figures (of prices or quantities of sales etc.) it can be persuaded to furnish at least index numbers, so that the performance, of the industry, can yet be evaluated, in general terms.

There are four basic techniques employed in the calculation of index numbers.

(A) The first method is relative of Aggregates Price Index. The computations involve the following steps :

1. Addition of the prices of the materials or articles included in the index for each year or place.

2. Dividing each of the totals obtained, by the total for the year selected as base year.

The method is illustrated by the following example.

Base year is 1954=100

| Type of coal (sold at a particular place) | Price per ton in 1954 | Price per ton in 1969 |
|---|---|---|
| Raniganj | Rs. 22.00 | Rs. 34.00 |
| Jharia | Rs. 25.00 | Rs. 40.00 |
| Coking coal | Rs. 30.00 | Rs. 45.00 |
| Totals : | Rs. 77.00 | Rs. 119.00 |
| Index number | 100 | 154.5 |

(B) The second method is the Average of Relative Price Index. The steps involved, in the computations, are as follows :

1. To find the relative price of each commodity or article by dividing the price of the year in question, by the price of the base year.

2. To add up the relative price or prices of each item and obtain the average, by dividing by the number of items.

Base year in 1954 = 100

| Type of Coal | Price per ton in 1954 | Relative price in 1954 | Price per ton in 1969 | Relative price in 1969 |
|---|---|---|---|---|
| | Rs. | | Rs. | |
| Raniganj | 22.00 | 100 | 34.00 | 155 |
| Jharia | 25.00 | 100 | 40.00 | 160 |
| Coking | 30.00 | 100 | 45.00 | 150 |
| Total : | | 300 | | 465 |
| Index numbers | | 100 | | 155 |

(C) The third method is Relative of Weighted Aggregates Price Index.

One of the main disadvantages of the methods described earlier, is that the items taken are not given any weightage, in accordance with their relative importance, e.g., taking coal. Surely coking coal requires a greater weightage than non-coking coal. But the shortcoming, in the average relative method, is that the price, of each item, is taken as being relative to its importance or utility. On the other hand, the relative of weighted aggregate method assumes that the price of an item in terms of units, is a measure of the importance of each item. This need not necessarily be true. In the case of consumer goods, a pair of shoes may cost less than a radio, but the need for shoes is obviously greater than that for a radio.

The following method is used in obtaining relative of weighted aggregates price index.

1. To determine the weight price of each item by multiplying the price prevailing in each year by the selected weight.
2. Add these products to obtain the weighted aggregates.
3. Divide each of these totals by the weighted aggregate of the year selected as base.

**Index Numbers :**

| Items | Unit | 1954 | | | 1969 | | |
|---|---|---|---|---|---|---|---|
| | | Quantity consumed | Average price | Weighted price | Quantity consumed | Average price | Weighted price |
| 1 | 2 | 3 | 4 | 5 | 6 | 7 | 8 |
| | | | Rs. | | | Rs. | |
| Coal | Ton | 50,000,000 | 22.00 | 1100,000,000 | 70,000,000 | 34 | 2380,000,000 |
| Tin | Quintal | 40,000 | 2000.00 | 80,000,000 | 20,000 | 6,000 | 120,000,000 |
| Crude oil | Barrel | 6,000,000 | 80.00 | 480,000,000 | 14,000,000 | 93,000 | 1302,000,000 |
| | Total | | | 1660,000,000 | | | 3802,000,000 |
| | Index numbers | | | 100 | | | 229 |

(*D*) The fourth method—Average of weighted Relative Price Index.

In this method, the relation between the change in price of each commodity is brought out, and relative weightage is also assigned to each according to its importance.

The following steps are adopted, in order to arrive at the index number :

1. Compute the relative value for each commodity by dividing the price, for the given year, by the price of the base year.
2. Select prices and quantities for an item for a particular year. The product of these figures (prices/quantities) gives the value of the item sold during the year.
3. The relative weightage, of each item, is obtained by multiplying the relative value by the value weightage for that commodity.
4. Divide the total relative weightage for each year by the total of the value weightage.

| Item/Commodity | Unit | Quantity consumed | 1954 | | | 1969 | | |
|---|---|---|---|---|---|---|---|---|
| | | | Average price | Weighted value | Relative value | Average price | Weighted value | Relative value |
| 1 | 2 | 3 | 4 | 5 | 6 | 7 | 8 | 9 |
| Coal | Ton | 50,000,000 | Rs. 22.00 | 1100,000,000 | 100 | Rs. 34.00 | 2380,000,000 | 155 |
| Tin | Quintal | 10,000 | 2000.00 | 80,000,000 | 100 | 6000.00 | 120,000,000 | 300 |
| Crude oil | Barrel | 6,000,000 | 80.00 | 480,000,000 | 100 | 93.00 | 1302,000,000 | 116 |
| Total | | | | 1166,000,000 | | | 3802,000,000 | |
| Index number | | | | 100 | | | | 235 |

Some of the points to be considered, in computing index numbers, are the following :

1. To find out the purpose for which the data are required.

2. To make a proper choice of commodities or items to be taken into consideration.

3. To find out whether the commodities selected be assigned the same weightage.

4. To collect actual prices of the commodities, selected, and obtain data on the relative importance for each in a weighted series.

5. To make a choice as to whether the index number be represented as average variation of prices or variations of a sum of actual prices.

6. To choose the base for computing relative prices when average variations are shown.

7. To choose the form of the average when averages are used.

Geostatistics or the application of the statistical concepts, as a means to aid in solving problems in geology and mining, has been employed successfully in ceratin areas. Hence, it is of interest to know the basic principles. The use of the frequency, averages, deviation and dispersion have been dealth with earlier. In dealing with ore and mineral deposits, it is have been dealt with earlier. In dealing with ore and mineral deposits, it is found that most often than not, variations, either in the width, or depth or quality etc. are ubiquitous, and even the experienced geologist is sometimes at a loss in understanding. Statistics can help in finding out whether "there is a method in this madness". The basic data employed in the technique comprise all particulars regarding the deposit, such as the chemical compotion, thickness, or thickness of overburden or topographical variations etc. One of the methods, that most commonly employed, is the surface trend analysis which is used for defining certain surfaces such as geological boundaries or quarry levels. In the simple cases, this technique involves the use of the "least square method". However, statistical data can also be used for evaluating inter-relationships either by means of cross correlation, or spatial correlation by assessing the degree of variance either within a population or between several populations. The variances can be depicted graphically by a "variogram". Hence, by the application of statistics, isopleth maps as well as block plans can be made, which will form the basis for mine design.

The procedure adopted includes : 1. the determination and analysis of all the statistical variables 2. the construction of the variogram and the spatial correlation of any particular variable. If, for example, the grade is considered, then the variogram shows how the grade differs vectorially from the average in relation to the sample interval. In case the sample intervals are small, then the squared distances are also small and hence the correlation is apparent. With increasing sample intervals, correlation becomes progressively difficult. The variogram affords a means of evaluating the amount of error involved. The variogram is a function of the expected mean squared differences between any two values of the variable, such as z (x) and z(x+h) which are separated by a *vector* h and this may be represented in the form

$$rh = \frac{1}{2N} \left[ z\ (x) - z\ (x-h) \right]^2$$

In short, the variogram shows the variations in the grade in relation to the interval of separation h in a specified direction. Thus, if *variograms* constructed from the same starting point, but in different directions, are identical then the ore body is isotropic, and vice versa.

The statistical technique called KRIGING is also an extension of the principle of variance. It is a technique used to estimate the value of a variable occurring either at a particular point, or within a particular volume of material, contained within an area surrounded by a set of sample points. In *kriging*, again there are two modifications : 1. point kriging and 2. block kriging. Where *kriging* is applied, the area under consideration is divided into grids. Hence, an area of 1 sq. km. will contain 6561 grid points if the spacing adopted is 12.5 m. In kriging, the value of the specific variable is determined at every grid intersection, and for instance the value may be designated Z*. The procedure is illustrated in fig. 11.4. The values of Z* obtained are used in preparing the isopleth maps. Fig. 11.5 illustrates the procedure for block kriging. In this case to the value of Z*, the estimated

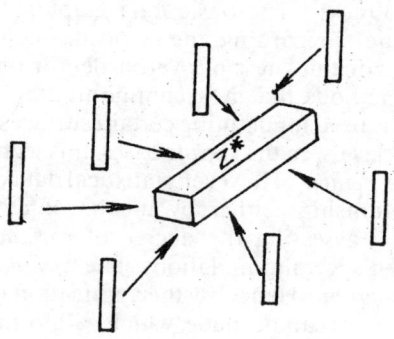

Fig. 11.4

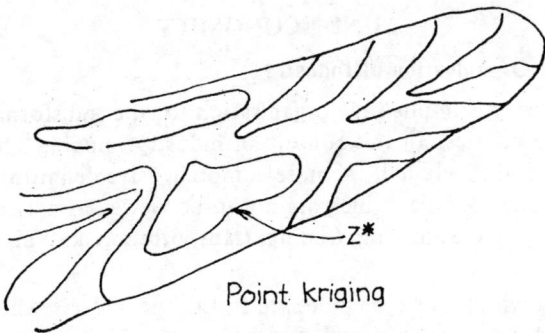

Point kriging

**Fig. 11.5**

value of the variance E is added, so as to obtain the kringed value of the particular block of the grid. In this manner the kriged values are worked out for all the blocks of the upper level. Then, with the bore hole data available, block kriging is carried out for the next lower level. When this is done, the variations in the tenor of the ore body, with depth, can be identified, and the same can be demarcated on the block diagram. Fig. 11.6

Block kriging

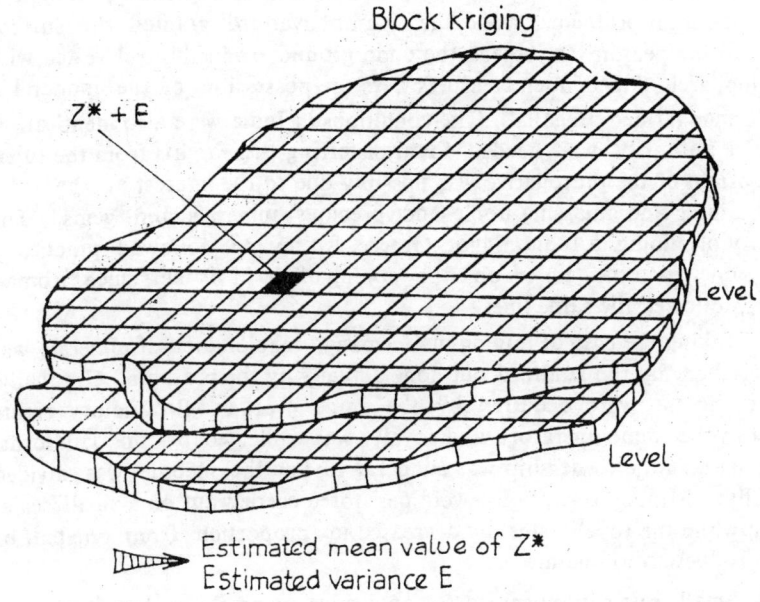

$Z^* + E$

Level

Level

⮕ Estimated mean value of $Z^*$
Estimated variance E

**Fig. 11.6**

# MINE ECONOMICS

## (a) Economic Organisation of Industry

By industry is implied the organisation for the transformation of raw materials into utilities. In this concept of industry, mining and extractive operations are not included. A more comprehensive definition of industry is, "all economic activities which are not to be viewed as agricultural". It is to this category that mining, trading, transportation and all such operations are assigned.

To start with mining, as practised in Europe and elsewhere, was not organised and it consisted in surface diggings of coal, iron ore, or copper. Alluvial mining, for gold, was also of ancient origin. Primitive open cast mining of minerals, eventually gave rise to elementary underground methods, and later on to systematic exploitation. Hazards of mining and the attendent risks not only to life, but also in working and predicting the behaviour of ore bodies, which depended on the nature of mineralisation and the type of structure, soon displaced the individual miner and led to formation of "cooperatives". Initially these groups comprised small numbers of individual miners, which grew in size with the passage of time.

Expansion of mining activities, both in scale and intensity, brought in its wake, legal problems, while the king or overlord granted the surface right to the peasant or subject, the underground ownership yet vested with the monarch. All "hidden treasures were the possession of the monarch". and hence termed "regalia". The conditions in India were also identical. A special authority was needed for transferring these rights from the ruler. The origin of the prerogative was possibly due to the interest of the ruler in obtaining and amassing noble and precious minerals and gems. The history of mining in India, Egypt, Greece, Rome, etc. followed practically the same pattern. Later on, the mining of tin in Britain, since Roman times, followed the same lines.

Taking the case of mining in Germany, exploitation of minerals was undertaken by the landlord but done actually by the peasants. Gradually, more rights were acquired by the mine worker. Proprietory rights went to the mine operators, and royalty was collected by the landlords. At this stage the ownership was cooperative, and the income was divided equally. Mining operations were per force carried out on a small scale. Meanwhile the royalty steadily decreased in proportion from one-half of the production to one ninth.

Small scale capitalism started and 'mine owners' employed workers on a domestic system, and for short periods profits accrued. Large increases in mining expenditure, on ventilation pumping and transport, required larger investment and more capital was drawn in. Development received a fillip when mineral trade grew, and the mineral trader's status grew in importance.

A stage was reached where two separate organisations developed, (1) the minor organisation for the owners of shares and (2) a workers organisation. Soon the wide fluctuations in productivity, of various mines, caused the formation of a guild to share profits as well as the losses equally. Here again, miner's income varied considerably and shares were sold out or surrendered, especially by non-workers. This led to the increase in capitalistic tendencies, and consequent expansion in mining activities. Thus mining rights came to be vested with share holders and those who had capital. As a consequence, the union of owners sold the output and distributed the proceeds.

The original unions had restricted the number of shares to three per share holder. This condition was lifted, but mining became unsystematic. Landlords again interfered with the miners or with the guilds, and they were joined by the miners or workers. Capital increased and developments followed. The workers' guild, however, remained, and soon a union of landlords or regale was born, which employed the workers and managed the production.

Metallurgical workers also grew in numbers. Since charcoal was essential, owner of forests, i.e:, feudal landlords and monastries became the first operators of early smelters. English monastries down to the 14th century dominated the field of small scale smelting operations, and larger works were also sponsored by monastries.

Monastries also started coal mines, including those of New Castle in Britain and Saar in Germany. When coal replaced charcoal in iron smelting, mining of coal leaped forward, and with the help of the newly invented steam engine, deepening of mines, with the use of pumps, was undertaken.

A word may be said about inventions. In precapitalistic times, inventions did not have industrial application, as a direct objective. When mining became established, inventions were purposeful and deliberate.

Mining grew unfettered in the West, with complete freedom for internal and external economy, leading to expansion of foreign trade, and this was followed by the breaking down of the barrier between internal and external economy, and the introduction of the commercial aspect into industry. Labour was also organised on this basis with the entrepreneurs taking the lead.

It will be noticed, that the portion dealing with general concepts in economics is dealt with earlier. The reason for adopting this procedure is not to emphasise the importance, but by taking up at that stage, the study will indicate how general economics affects or influences mining industry or any other industry. That such a background is essential for the fuller appreciation of the problem of mine valuation, is beyond doubt, but whether one should know the general principles of economics before undertaking a study of mine economics or not, is more a matter of choice.

Mining geologists, employed in a working mine, as well as those operating mineral prospects, are often required to estimate the reserves and sometimes work out the value of the property. An assignment of this type involves considerable pains and effort, and in addition calls for understanding of general economic principles.

Thus even though the subject is strictly neither mining or geology, yet an understanding of the principles is essential for every mining geologist. In order to facilitate its treatment, a broad five-fold division has been adopted.

Viz :   (A) Sampling
        (B) Estimation of reserves
        (C) Evaluation of assets
        (D) Mineral Economics
        (E) Miscellaneous computations

### (b) Sampling

Before taking up the actual processes of sampling, it is necessary to define a sample. Very often, novices to the art are unable to differentiate between a "sample" and a "specimen", and consider the terms to be synonymous.

When used by a mining geologist, the word 'specimen' refers to material collected, in order to represent a rock type, ore type, or formation and the term has only a qualitative sense. In other words, the specimen is defined by two vectors, (a) location and (b) composition, i.e. mineral and chemical also, and does not go beyond that. In short it is mainly representative of itself. Thus a granite specimen does not represent the granite body, from which it was collected or anything beyond what it contains.

On the other hand, the term 'sample' is used for the material collected to represent a rock type, or a formation or an ore body in the quantitative sense. Thus, a sample while being defined by (a) location and (b) composition, in its turn defines an ore body, formation, etc. with regard to quality and/or quantity, very often both. Thus a granite sample is intended to represent the whole body of the granite, in chemical, mechanical, and physical properties, and can be employed in computations involving estimates of reserves etc.

Sampling may thus be defined as a technique by which "a part is collected to represent the whole". In actual practice, however, an elaborate procedure is followed to eliminate various errors that may creep in, and vitiate the results, despite the honest endeavour made by the mining geologist (a) to fix the boundaries, in the three dimensions of the forma- tion, ore body etc. lying within the area under examination, (b) to decide on the interval of sampling and (c) to arrive at the quantity of sample to be collected. Ultimately his prudence, judgement, and experience are the sole guides. The actual process of sample collection is a mechanical or manual operation, and it has been shown that, if the quantities collected and the spacing have been of the right order, the average so obtained corresponds, very closely, to actual values, obtaining in the parent material.

Thus the mining geologist has to shoulder the responsibility, and for this purpose, a basic knowledge, of other allied subjects is often called for, in order to make concrete recommendations. He should (a) know the reason why sampling is undertaken and (b) acquire knowledge of the various aspects of the project envisaged. (1) If, for example, the sampling is required for a mining undertaking, the knowledge of the geology, struc- ture, etc. in addition to a working knowledge of the mining methods and ore dressing methods is essential. (2) If the sampling is required in connec- tion with the valuation of a mine, then methods of accounting, valuation etc. will be required in addition to mining methods, ore dressing methods, etc.

A good knowledge of mineral economics, variation in prices of minerals, cost per ton of various grades of ore, penalties to be paid for the presence of deleterious components and premia paid for higher values etc. is essential.

Sampling of limestones, either for cement or for fluxing in steel furnaces, is a case to the point. The type of limestone, used in steel furnaces, is different to that which may be acceptable to the cement industry. Limestone used in steel making is fine grained compact and massive. The magnesia can go up to 6%, while the silica must be low, not more than 6%, in the good grades, and may be 10% in the blast furnace grades. Sulphur and phosphorous are deleterious constituents.

Again in the case of complex sulphide ores, the geologist will have to device his own methods, with the ultimate object of obtaining a sample which will represent the "run of mine ore"

In many cases, a knowledge of the subsequent treatment, to which the ore is to be subjected, is also of considerable value. This is true of many minerals. China clay sampling is a technique in itself. Each type of clay, occurring in certain deposits, is to be sampled separately as each may require a different set up for beneficiation. This is the case with the deposits near Chaibasa, in Bihar.

Sampling should be done with the utmost care and precision. Hence it is essential that a systematic and uniform procedure be adopted, as far as practicable. (1) The sample location should be indicated with reference to fixed points wherever possible, (2) the depth and width of the groove or channel should be uniform in every band, in case of a banded or bedded deposit, (3) the width of each sample should be recorded and (4) surface of the bed, lode or veins should be cleaned by chipping where weathering is suspected. In the case of an auriferous vein, it is safer to clean the surface with a wire brush and water jet, in order to eliminate the effects of salting.

It is needless to emphasise the importance of recording the width of sample and the manner in which the same was collected. For example high values may be obtained, on analysis of a sample, but if the width is negligible, then the sample may not be considered in calculating the ore reserves. In all cases, the true width is what is required, and where the true width, of the bed or vein, is not readily obtained then the projected width can be calculated. This method is adopted when sampling in tunnels, cross cuts, stopes etc. where the lode or vein is exposed on a curved surface. In the case of low grade deposits, greater control is necessary in obtaining the quality of the ore. Hence in addition to the width, the specific gravity of the material collected should also be noted.

Sometimes it may be necessary to restrict the length of the channel, so as to reduce the bulk of the material, in a sample. The bulk can also be reduced by coning and quartering the total amount of material, collected from a groove or channel. As is more often the case, the length of a channel sample is restricted, because of the nature of the material to be sampled. For example, a coal seam 10 metres or 12 metres thick, may appear to be of uniform quality from visual inspection. Metre by metre sampling, if carried out, is likely to throw light on the variation in coking properties or volatiles or moisture content etc. It is a good practice, where large scale channel sampling is undertaken, that at least from a fair percentage of channels, samples be taken for shorter lengths.

In certain areas, where an ore body has been exploited over a considerable period of time, methods of collecting and analysing samples generally get standardised. Practical experience has shown, that the methods adopted, are the best suited under the conditions. In such cases, variation or modification, of the procedures of sampling and analysis, should be done with the utmost caution and with full justification. The geologist in charge of an investigation should, in all cases, try to explain every step taken when laying down procedures for sampling. Check sampling is a necessary step and will prove to be of considerable assistance in this respect.

Modifications, in the methods of collecting samples, are necessitated by variations in (1) the nature of the mineral deposit, and (2) the purpose of the sampling, i.e., whether for preliminary examination or for calculating of reserves or for checking. For preliminary examination of mineral deposits, it may be sufficient, at times, to collect "grab samples". This consists in collecting pieces of ore at random from various parts of a deposit, and reducing the same in bulk as well as in respect of the sizes of individual pieces, by the method of "coning and quartering" as described later. As an example, in examining a deposit of diaspore clay, in which diaspore-rich lenses occur, preliminary sampling is required, more to establish the presence of diaspore rather than for determining the quality or quantity. In such a case, it will be sufficient if only the nodular material, presumably rich in diaspore, is collected from various parts of the deposit and analysed. This method can also be applied to certain bauxite deposits, especially those associated with the gneisses (Khondalites and charnockites), in which nodular bauxite occurs in kaolinic and ferruginous clays.

One of the common techniques adopted in sampling, is channelling or grooving. This is best suited to bedded, banded, and vein type mineral deposits, in which materials showing distinct variations in colour and/or grain size and presumably also in composition representing various grades, can be visually separated. The banding, may, at times, represent a variation in mineral assemblage as well. This is true of certain crystalline limestone deposits, found in Hazaribagh district of Bihar, and in the Ramanathpuram, Salem and Coimbatore districts of Tamil Nadu.

Grab sampling consists in picking pieces of ore at random to make up a sample. This method can be adopted when run of mine ore is to be sampled. In this case pieces of ore are collected at random as the mine cars or tubs are passing by. All the material collected from a train is coned and quartered to get the final sample.

When sampling deposits, in which materials with considerable variation in hardness occur, great care should be exercised or else there is every possibility that only material of lesser hardness will be collected. An example of this type of material is the hard nodular bauxite associated with the clayey material derived from the weathering of charnockitic rocks in South India. In such cases, it is best to collect bulk samples by pitting into the deposit and working out the recovery in kilogramme per cubic metre. If this factor is worked out, the reserves can be estimated for the blocked out area.

Certain other types of deposits also require the use of special methods, of sampling and estimation of reserves. One such case is that of barytes which occurs as "box work" or reticulate veins in limestone, or shales. In such a deposit, sampling by grooving and channelling may not give the correct picture. It is safer to pit or trench and collect a bulk materials by sorting out the various grades of baryte and obtain the quantity of barytes in kilogramme per cubic metre for each grade. This factor will enable the estimation of reserves. In the case of open cuts or quarries, where large faces are available for examination, a closer approximation of the factor (tonnage per cubic metre) can be obtained by inspections of the various faces, and then working out the average concentration for the fares examined. This will afford a better control.

The same technique can be applied in sampling anastomosing magnesite veins associated with altered dunite; as well as reticulate veins of asbestos occurring in the serpentinised ultrabasic rocks.

In the case of disseminated ore bodies, the sampling technique should be such, that variations in tenor can be successfully counteracted or anulled. In such cases, deeper groves $(6'' \times 4'')$ can be made, in various parts of the exposures to be sampled, and then the averages worked out for obtaining a better estimate of the quality, within reasonable limits of tolerance.

A similar problem arises when sampling and estimating reserves of Kaolin, in an area where the deposit is patchy. Pitting, should invariably be resorted to for sampling the various grades and for the estimation of reserves. Where pits and quarries are available, careful inspection of the faces, often provides a fair approximation of the factor for quality and quantity. In such cases, routine grooving and channelling may lead to erratic and erroneous results. Samples should be collected separately either on the basis of criteria, e.g. colour or grit content in order to facilitate the gradewise estimation of reserves. If samples are not properly separated then a workable deposit may appear inferior for economic exploitation. Sampling of lateritic manganese ores is a rather tricky proposition. A geologist requires considerable familiarity, with the various grades, in order to sample and estimate reserves. Pitting is the only safe method of collecting samples, but sorting the material so obtained, into grades, requires considerable experience and familiarity. To start with, in work-ing areas, assistance of labour, with experience, may be sought for visual estimation of the grades until sufficient familiarity is acquired.

It was pointed out earlier, that the actual or true thickness of the vein or lode or band carrying mine-ralisation should be obtained. The measurement of the actual thick-ness of the ore body exposed in the underground working such as stopes, cross cuts, and drives re-quires certain precautions.

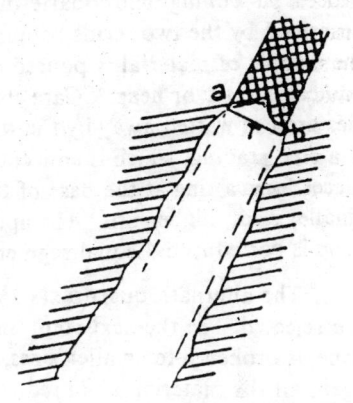

In figure 11.7 the ore body is exposed at the back of a stope. The points a and b lie on a straight line across the strike. i.e., true dip. The actual thickness is T, and this can be measured out with a tape

Fig. 11.7

to determine the width of the sample. It should be remembered that suffi-cient quantity of sample should be collected and this quantity will depend on the type and number of analyses to be conducted. Normally at least 1 kg of sample should be extracted.

In the case of channel/groove sampling, the true width of each band sampled should be recorded giving complete details regarding the location as such distance and direction from a fixed point, the sample number, as noted on the map on mine plan, the name of the material, copper/zinc ore colour, lustre, grain size, and wherever possible a visual estimate of the metal content if it is an ore. The sampling form as given below will serve as a guide.

Form of Sample Slips

Name and nature of deposit: Banded crystalline limestone.

Location: 5 miles N. of...Village/Coordinates/Type of sample grab/groove/ bulk.

| Serial No. | Description and width of sample | Remarks |
|---|---|---|
| 1. | Dark grey, coarse grained, hard 1 1/2 metre wide, sample 90 cm. only | 2 kg. sample See sketch |

## Coning and quartering

Large samples, amounting to as much as 500 kgs., will have to be reduced, before sending for chemical analysis. One of the methods employed, for obtaining a laboratory sample from the field sample, is the process of coning and quartering. The operations involved are aptly described by the two terms "coning" and "quartering". In this method the sample of material is poured out from a container so that it forms a conical mound or heap. Care should be taken to see that the finer particles are not wafted away by the wind. In practice, it will be seen that there is a size grading, which is concentric about the apex of the cone. The large pieces form a ring at the base of the cone, and this is followed by relatively smaller sizes, higher up. The apex of the cone is flattened out and the cone is parted into 4 equal segments as shown in figure 11.8.

The alternate quardrants 1 and 3 are taken, while quadrants 2 and 4 are rejected. In the next step, material, from segments 1 and 3 of the cone, is broken into smaller sizes, e.g., say from 1″ sizes to ½″ sizes. Once again all the material is poured, as before, to form a conical heap, and the quartering process is carried out. Coning, as well as the crushing, are repeated, alternately whereby reduction in quantity of the sample together with the concomittant reduction in the size, of individual particles in the sample is achieved. Thus no particular particle has any undue influence on the final sample.

In reducing the bulk of the sample, by coning and quartering, it is considered satisfactory, if the weight of the heaviest individual particle, in the sample, is 1/1000 of the total weight i.e., in an 1000 gm (1 kg) sample no piece should be more than 1 gm in weight.

The following formula may be used to arrive at the size of the screen opening (d), which may be used, in sampling; where the ratio of allowable error is (E) :

$$d = \sqrt[3]{EW/G}$$

where  d = Screen opening in cms (where the particles are cubes).
W = Weight of sample in grams.
E = Ratio of allowable error.
G = grams of metal in 1 c.c. of richest ore.

*Salting* : The term implies means "adding salt". This term is applied to describe samples which have been tampered with in such a manner, after collection by interested parties, so as to increase the grade. Taking the example of gold, after samples have been collected from a mine,

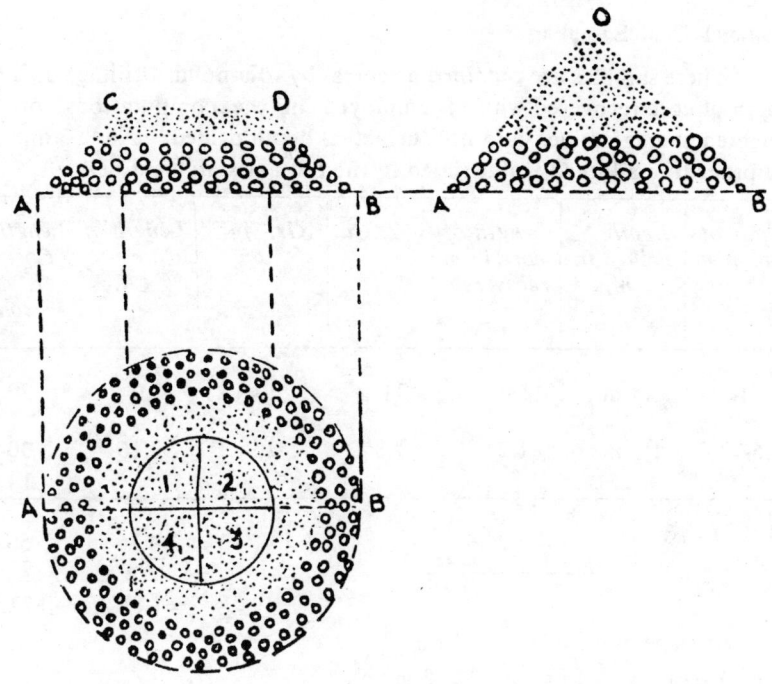

Fig. 11.8

interested parties, commonly inject or drop, into the samples, solution of gold chloride or silver nitrate so that the sample, when analysed, will show very high values of Ag and Au. Sometimes, especially designed pistols are used to fire gold or silver pellets. The pellets penetrate through the sample container and get dispersed. Salting can also be done before samples are drawn from any mine working. In this case, the solution of either gold chloride or silver nitrate is sprayed over the area to be sampled, so that when a sample is collected it will naturally show high values.

As a precaution against face salting, it is a good practice to wash all sampling faces with a jet of clean water and wire brush before starting operations. Samples collected should be carefully sealed and kept in the custody of trust-worthy persons. It is necessary that the geologist refrain from any strong alcoholic drinks, which may cause him to neglect the samples collected, and afford opportunities for salting. In any case, the mining geologist should be vigilant and never neglect the samples for fear of salting. Remember, salting can ruin a sample and waste all the hard labour, spent in mapping and toiling in the underground workings.

## Diamond Drill Sampling

Where samples are obtained as cores by diamond drilling, it is the length of core recovered which is employed in the computations, of the weighted average values, and not the actual depth drilled. An example of the procedure adopted is illustrated in the tabular statement below :

| Depth of hole in m. | Depth drilled in m. | Length of core in m. recovered. | % Cu | Gr. Ag/ ton | Length of Core × % Cu. | Length of Core × gr. Ag/ ton |
|---|---|---|---|---|---|---|
| 70-71.5 | 1.5 m | 1.0 | 2.0 | 20 | 2.0 | 20 |
| 71.5-73.0 | 1.5 m | 1.2 | 1.8 | 25 | 2.16 | 30 |
| | | 2.2 | | | $\frac{4.16}{2.2}$ =1.89% | $\frac{50}{2.2}$ =22.73 |

Average grade of Cu = 1.89%

Average grade of Ag = 22.73 gr/ton.

Even though it is common practice to recover the sludge and use the same in averaging the values of the ore, the values so obtained should be regarded, at least, only as very approximate ; unless supported by analysis of solid cores. However, where core losses are very heavy and the recovery is of the order of 30% or less, there is no other alternative than to save the sludge for analysis. As a word of caution, it may be pointed out that the sludge saving equipment should be so designed and located that all the sludge is recovered. It is well to remember that negligence, in this respect, may lead to loss of either the heavy portion or of the light fractions, in the sludge.

## Sampling placers

In sampling alluvial and placer deposits by the banka drill or by means of the churn drill, using the sand pump, it is essential to estimate the quantities of material recovered, and the values recovered. Normally, a 6″ drive pipe, with a 7.5 in. cutting shoe, is employed. Since the pipe, is driven into the ground, the material gets consolidated below the pipe, so that the material entering the pipe expands. Further, large pebbles and boulders are pushed aside. Thus the material within the pipe represents, a larger length of core than the depth of penetration. Thus normally, every foot

drilled is taken as 1.5 ft. of core.   It may also happen that the flush  water
or sand pump may pick up material from below the cutting shoe.    To
prevent this, the hole is not flushed or cleaned to the full depth  right upto
the drive shoe.   A short stub is retained at the bottom, when  flushing  is
stopped, to prevent contamination by side wash.

In order to allow for these discrepancies, and facilitate the computation of the volume of core, a factor is used  which when  multiplied (by
the length of pipe), i.e., depth drilled gives the volume in cu. yds. or cu.m.
Thus for a 6-inch diameter pipe with a 7.5 inch diameter cutting  shoe, the
factor is  0.3 sq. ft. For the sake of convenience this figure is taken as
0.27 sq. ft. so that for 100 ft. length, the volume works out to 1 cu. yd.
(i.e. $\dfrac{0.27 \times 100}{9 \times 3}$). The assay values may be calculated in the  manner
shown below:

(a) Use of pipe factor in computing values

| Depth | Wt. of Au in mgm. |
|-------|-------------------|
| 10—11 | 3 |
| 11—25 | 30 |
| 25—28 | 50 |

The pipe factor is 0.27 sq. ft.
Vol. drilled$=(10 \times 0.27)$ cu. ft.

∴   Assay value of the section
$$= \frac{27 \times 83}{18 \times 0.27}$$
$=461$ mgm/cu. yd. of Au.

Computations of assay values using the volumes recovered.

(b) In this case *the core ratio* is used i.e. $\dfrac{\text{Volume of material recovered}}{\text{Volume of material cored}}$
(using the pipe factor)

| Depth ft. | Measured Vol. in cu. ft. | Core ratio | Wt. in m. gm. (Au) | Corrected | | Remarks |
|-----------|--------------------------|------------|--------------------|-----------|---------|---------|
| | | | | Vol. cu. ft. | Wt. mgm. | |
| 10—11 | 0.30 | 1.11 | 3 | 0.27 | 3 | (100/111) × 3 = 2.70 |
| 11—25 | 5.04 | 1.33 | 30 | 3.78 | 23 | (100/133) × 30 = 22.56 |
| 25—28 | 0.72 | 0.89 | 50 | 0.81 | 56 | (100/89) × 50 = 56.18 |
| | | | | 4.86 | 82 | |

Value is $\dfrac{82}{4.85}$ m. gm/cu. ft.

$\dfrac{82}{4.86} \times 27 = 455$ m. gm./cu. yd. of Au

(c) Computations of assay value, using weights of material recovered can be made in the following manner.

| Depth ft. | Length of run (L) ft. | Measured vol. in c. ft. | Weight m. gm. (Au) | Values per cu. yd. mgm (V) | Relative contents L×V |
|-----------|----------------------|------------------------|--------------------|----------------------------|------------------------|
| 10—11 | 1 | 0.30 | 3 | 270 | 270 |
| 11—25 | 14 | 5.04 | 30 | 161 | 2254 |
| 25—28 | 3 | 0.72 | 50 | 1875 | 5625 |
|  | 18 |  | 83 |  | 8149 |

Value is $\dfrac{8149}{18} = 453$ mgm/cu. yd.

Thus the computations given are in close agreement, i.e., 460, 455, and 453 mgm/cu. yd.

## (C) CALCULATION OF RESERVES ETC.

In order to illustrate the variations in sampling procedure and computations involved, a few illustrative examples have been selected. The procedures indicated are subject to variations and modifications. Sometimes a combination of these may be the better suited to the case on hand.

*Problem 1 :*   (a)  To find out the average thickness of a bed.

(b)  To find out the average assay value per cu. yd. or cu. metre.

The simple case is that of a flat lying deposit, e.g. placer or a bed of iron ore etc. Assuming that such a deposit has been drilled on a square pattern grid (Fig 11.6) and that the following data are available :

(a)  the holes are 50 m. apart and
(b)  the values obtained are as given in the columns 1, 2, 3, of the table below :

| 1 | 2 | 3 | 4 |
|---|---|---|---|
| B. H. No. | Thickness of Placer in m. | Assay value/ dwt. per cu.m. | Metre Assay Product |
| 1 | 3.0 | 4 | 12 |
| 2 | 5.0 | 3 | 15 |
| 3 | 5.0 | 3 | 15 |
| 4 | 4.0 | 4 | 16 |
| 5 | 3.5 | 2 | 7 |
| 6 | 3.0 | 2 | 6 |
| 7 | 4.0 | 4 | 16 |
| 8 | 5.0 | 3 | 15 |
| 9 | 3.5 | 6 | 21 |
| 10 | 4.0 | 2 | 8 |
| 11 | 4.0 | 4 | 16 |
| 12 | 3.0 | 3 | 9 |
| 13 | 3.5 | 4 | 14 |
| 14 | 4.5 | 2 | 9 |
| 15 | 5.0 | 5 | 25 |
| 16 | 3.0 | 3 | 9 |
|  | 63.0 |  | 193 |

$\therefore$ Average assay value per cu. m. is $\dfrac{213}{63.0} = 3.38$ dwt./cu. m.

and the average thickness is $\dfrac{63}{16} = 3.94$ metres.

Assuming that the influence of every bore hole extends for half the interval between each, then the area in which the deposit can be expected to occur, will be 200 m × 200 m.

Since the average thickness is 3.9375 m,

The volume of gravel = 3.9375 = 157500 cu.m.

The gross values available = $157500 \times \dfrac{213}{63} = 22188.0$ ozs.

[dwts = pennyweight, i.e., 24 pennyweights = 1 ounce (oz)]

In the above problem, the weightage given to all samples is uniform. This is not so in actual practice, as it will be seen from figure 11.6. the area of influence of samples depends on the location of the sample in respect of the deposit. Normally 'the area of influence', of a sample, may be taken to be represented by a plane figure, obtained by joining the mid points of the lines joining the sample points, to the adjacent sample points. In the figure 11.6 the area shaded is the area of influence of sample point A.

The weightage for areas of influence varies with the location of the sample point in respect of others. Samples, B, C, D, E and F, have three linkages or valencies each, i.e., taking B, it influences the lines BC, BF and BA. Sample A on the other hand, has 5 linkages or valencies viz. AB, AC, AD, AE, and AF.

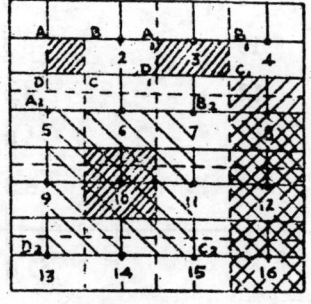

Thus in the example given earlier, the weightages of samples vary from one, for those located near the four corners of the area, to 4 for those located in the interior. The intermediate samples have a weightage of 2 only. This is clear from the areas shaded round each sample location, viz., ABCD, $A_1B_1C_1D_1$, $A_2B_2C_2D_2$. (Fig. 11.9).

Fig. 11.9

Thus the calculations will have to be modified in the manner indicated, if the weightages are also to be considered.

| B. H. No. | Thickness of placers in (M) | Assay value in dwt per cu.m. (A) | Metre Assay product (M)×(A) | Weight-age of sample | M× Weightage | Weightage × Metre Assay product |
|---|---|---|---|---|---|---|
| 1 | 3 | 4 | 12 | 1 | 3 | 12 |
| 2 | 5 | 3 | 15 | 2 | 10 | 30 |
| 3 | 5 | 3 | 15 | 2 | 10 | 30 |
| 4 | 4 | 4 | 16 | 1 | 4 | 16 |
| 5 | 3.5 | 2 | 7.0 | 2 | 7 | 14 |
| 6 | 3.0 | 2 | 6.0 | 4 | 12 | 24 |
| 7 | 4.0 | 4 | 16.0 | 4 | 16 | 64 |
| 8 | 5.0 | 3 | 15.0 | 2 | 10 | 30 |
| 9 | 3.5 | 6 | 21.0 | 2 | 7 | 42 |
| 10 | 4.0 | 2 | 8.0 | 4 | 16 | 32 |
| 11 | 4.0 | 4 | 16.0 | 4 | 16 | 64 |
| 12 | 3.0 | 3 | 9.0 | 2 | 6 | 18 |
| 13 | 3.5 | 4 | 14.0 | 1 | 3.5 | 14 |
| 14 | 4.5 | 2 | 9.0 | 2 | 9.0 | 18 |
| 15 | 5.0 | 5 | 25.0 | 2 | 10.0 | 50 |
| 16 | 3.0 | 3 | 9.0 | 1 | 3.0 | 9 |
| | | | | 36 | 142.5 | 467 |

$$\therefore \text{ Average thickness} = \frac{142.5}{36} = 3.96 \text{ m.}$$

$$\therefore \text{ Average value} = \frac{467}{142.5} = 3.28 \text{ dwts.}$$

Where the bed has a dip of considerable magnitude, the relevant correction factor is to be introduced or else the alternative method given below can be adopted for obtaining greater accuracy. It is based on the principle of subdividing the area into smaller units so as to localise the effect of variables.

| 1 | 2 | 3 | 4 | 5 | 6 | 7 | 8 |
|---|---|---|---|---|---|---|---|
| Sub area square | Sample No. | Thick- ness in metres (M) | Assay value % Cu | M-Assay product (3×4) | Average thick- ness | Average assay value | Volume assay product (Columns 6×7) |
| 1 | 1 | 3.0 | 4 | 12 | $\dfrac{14.5}{4}$ $=3.63$ | $\dfrac{40}{14.5}$ $=2.76$ | 10.00 |
|   | 2 | 5.0 | 3 | 15 |   |   |   |
|   | 5 | 3.5 | 2 | 7 |   |   |   |
|   | 6 | $\dfrac{3.0}{14.5}$ | 2 | $\dfrac{6}{40}$ |   |   |   |
| 2 | 2 | 5.0 | 3 | 15 | $\dfrac{17.0}{4}$ $=4.25$ | $\dfrac{52}{17.0}$ $=3.0$ | 13.00 |
|   | 3 | 5.0 | 3 | 15 |   |   |   |
|   | 6 | 3.0 | 2 | 6 |   |   |   |
|   | 7 | $\dfrac{4.0}{17.0}$ | 4 | $\dfrac{16}{52}$ |   |   |   |
| 3 | 3 | 5.0 | 3 | 15 | $\dfrac{18.0}{4}$ $=4.50$ | $\dfrac{62}{18.0}$ $=3.4$ | 15.30 |
|   | 4 | 4.0 | 4 | 16 |   |   |   |
|   | 7 | 4.0 | 4 | 16 |   |   |   |
|   | 8 | $\dfrac{5.0}{18.0}$ | 3 | $\dfrac{15}{62}$ |   |   |   |
| 4 | 5 | 3.5 | 2 | 7 | $\dfrac{14}{4}$ $=3.50$ | $\dfrac{42}{14}$ $=3.0$ | 10.50 |
|   | 6 | 3.0 | 2 | 6 |   |   |   |
|   | 9 | 3.5 | 6 | 21 |   |   |   |
|   | 10 | $\dfrac{4.0}{14.0}$ | 2 | $\dfrac{8}{42}$ |   |   |   |

| | | | | | | | |
|---|---|---|---|---|---|---|---|
| 5 | 6 | 3·0 | 2 | 6 | $\dfrac{15}{4}$ | $\dfrac{46}{15}$ | |
| | 7 | 4.0 | 4 | 16 | $=3.75$ | $=3.0$ | 11.25 |
| | 10 | 4.0 | 2 | 8 | | | |
| | 11 | $\dfrac{4.0}{15.0}$ | 4 | $\dfrac{16}{46}$ | | | |
| 6 | 7 | 4.0 | 4 | 16 | $\dfrac{16}{4}$ | $\dfrac{56}{16}$ | |
| | 8 | 5.0 | 3 | 15 | $=4.00$ | $=3.5$ | 14.00 |
| | 11 | 4.0 | 4 | 16 | | | |
| | 12 | $\dfrac{3.0}{16.0}$ | 3 | $\dfrac{9}{56}$ | | | |
| 7 | 9 | 3.5 | 6 | 21 | $\dfrac{15.5}{4}$ | $\dfrac{52}{15.5}$ | |
| | 10 | 4.0 | 2 | 8 | $=3.88$ | $=3.3$ | 13.00 |
| | 13 | 3.5 | 4 | 14 | | | |
| | 14 | $\dfrac{4.5}{15.5}$ | 2 | $\dfrac{9}{52}$ | | | |
| 8 | 10 | 4.0 | 2 | 8 | $\dfrac{17.5}{4}$ | $\dfrac{58}{17.5}$ | |
| | 11 | 4.0 | 4 | 16 | $=4.38$ | $=3.3$ | 14.50 |
| | 14 | 4.5 | 2 | 9 | | | |
| | 15 | $\dfrac{5.0}{17.5}$ | 5 | $\dfrac{25}{58}$ | | | |
| 9 | 11 | 4.0 | 4 | 16 | $\dfrac{15}{4}$ | $\dfrac{59}{15}$ | |

| 12 | 3.0  | 3 | 9  | =3.75 | =3.9 | 14.75 |
|----|------|---|----|-------|------|-------|
| 15 | 5.0  | 5 | 25 |       |      |       |
| 16 | 3.0  | 3 | 9  |       |      |       |
|    | 15.0 |   | 59 |       |      |       |

$\therefore$  Average thickness $= \dfrac{35.64}{9} = 3.96$ m.

$\therefore$  Average assay value $\dfrac{116.15}{35.64} = 3.26\%$ cu.m.

For further details of bore hole sampling, the chapter on drilling may be consulted.

Where the programme of exploration comprises irregularly spaced holes, with no set pattern, the procedure given below may be followed :

1    The area of influence of each hole (Fig. 11.10 is calculated. This is obtained by connecting any one bore hole say A to the surrounding bore holes B, C, D, E and F. Normals ab, bc, cd, de, ef, fa are drawn at the mid points of these radial lines. Polygon a b c d e f is formed by the intersection of the normals. The areas of influence, in the figure, are given by the polygons a b c d e f and d g h i j e.

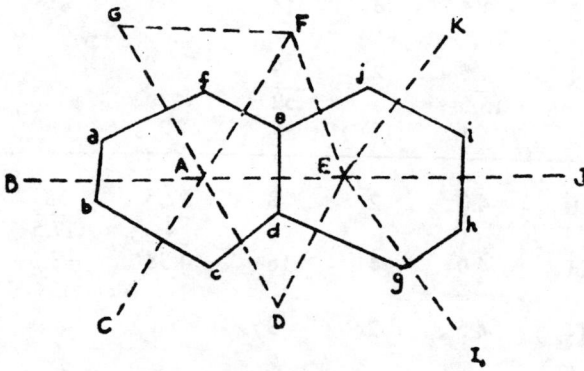

Fig. 11.10

2.   The area of each of the polygons is measured by any method, either by counting squares or by planimeter etc.

3. Knowing the thickness of the deposit, the volume of material is computed.

4. Knowing the assay value, the volume-assay product is calculated.

Thus $$\frac{\text{Volume of material}}{\text{Area of influence}} = \text{Average thickness.}$$

$$\frac{\text{Volume—assay product}}{\text{Volume of material}} = \text{Average grade or value.}$$

If the specific gravity of the ore is also known, then the tonnage can also be computed.

Again, in the case of irregularly spaced bore holes, it is possible to subdivide the area into triangles. In this case, the following procedure is adopted :

1. The area of each triangle is computed.

2. The average thickness, for each triangle, is obtained.

3. The volume of material, in each triangle, is computed.

The rest of the procedure is similar to that indicated above.

Where sufficient drilling data are available, it may be possible to construct isopach maps. By considering one of the surfaces to be plane, the other undulating is divided into smaller areas, having nearly plane surfaces. The volumes of prims so formed, can be computed.

### Sampling underground workings

In sampling underground excavations, e.g., stopes, cross cuts, winzes, raises etc., the normal practice is to carve out channels at regular intervals, just as in the case of bore hole sampling where the average assay value is obtained by adopting the procedure indicated.

viz. : $$\frac{\text{Widths} \times \text{Assay (Assay-metre product)}}{\text{Width (of samples)}}$$

In a raise, channel samples are collected regularly at 3 m intervals, for the entire width, and the following data as obtained. The sample interval is, of course, subject to variation depending on local requirements.

The following example shows the procedure adopted in calculating average assay value for a raise.

| Width in m. | Assay value in dwts/ton | Assay metre product |
|---|---|---|
| 3.0 | 4.0 | 12 |
| 2.5 | 5.0 | 12.5 |
| 4.0 | 3.5 | 14.0 |
| 3.5 | 6.0 | 21.0 |
| 3.0 | 5.0 | 15.0 |
| 16.0 | | 74.5 |

$$\therefore \text{ Average assay value} = \frac{74.5}{16} = 4.66.$$

Where the sample intervals are irregular, then averaging of 2 adjacent sample distances is adopted, in order to obtain 'the distance of influence'. Thus if the sample intervals are 4, 6.5, and 7 metres, the distance of influence will be 5, 5.5 and 6 metres for the 3 intervals respectively. Since the width of each is known, the area of influence can be computed.

Since the assay value of each sample is known, the area-assay product, and so the area-assay can be obtained.

$$\therefore \text{ Average assay value} = \frac{\text{area assay product}}{\text{area of influence}}.$$

*Stoping width* : Rich metalliferous lodes are often narrow. Thus during the course of stoping or working, it may be found necessary to excavate some portions of either the hanging or foot wall or both, so that the stope is sufficiently wide, to afford enough working room, for men or for machines to operate with maximum efficiency.

For example a vein, as exposed at the back of a stope, is measured at regular intervals of 3 m and sampled. The following are the data obtained. Assuming the stoping width to be 1.5 m, it is required to find :

(1) the average assay value of the stope face,

(2) the average width of the stoping, and

(3) the area of the stope face.

The width of stope, the width of the vein, and the assay value are given in the table in column 1, 2 and 3 below :

|   | 1 | 2 | 3 | 4 |
|---|---|---|---|---|
|   | Width of stope in m. | Width of vein in m. | Assay value dwt/ton | Assay×m product |
| 1 | 2.5 | 2.5 | 3 | 7.5 |
| 2 | 1.5 | 1.3 | 5 | 6.5 |
| 3 | 2.2 | 2.2 | 5 | 11.0 |
| 4 | 3.5 | 3.5 | 4 | 14.0 |
| 5 | 3.0 | 3.0 | 6 | 18.0 |
| 6 | 1.5 | 1.0 | 6 | 6.0 |
| 7 | 1.5 | 1.2 | 5 | 6.0 |
| 8 | 2.3 | 2.3 | 5 | 11.5 |
|   | 18.0 |   |   | 80.5 |

$\therefore$ Average assay value $= \dfrac{80.5}{18} = 4.47$ dwts.

Average width of stoping $= \dfrac{18.0}{8} = 2.25$ m.

$\therefore$ Area of the stope face $= 3 \times 8 \times 2.25 = 54.00$ sq. m.

Where an ore body is exposed in two adjacent raises, and in the interconnecting levels, it is possible to calculate (1) the average assay value of the entire block, and (2) where the specific gravity of the ore is known, the reserves also.

In essence, the computations are identical upto the stage where the face area of a stope is obtained, in the earlier example.

If in Fig. 11.11 AB and CD are two levels connected by raises AD and BC, the levels are 16 m. apart, whereas the raises are 12 m apart, A uniform sampling interval of 4 m was adopted.

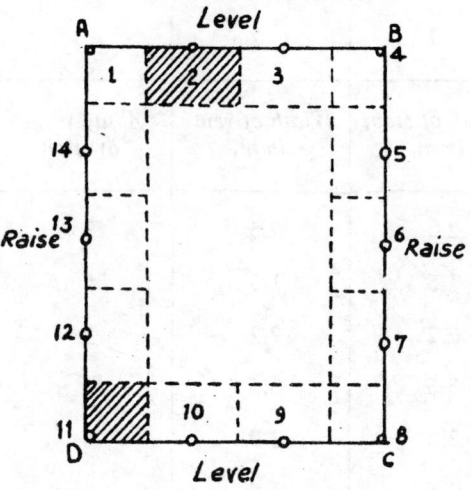

**Fig. 11.11**

Since the corner samples have only half the areas of influence of the intermediate ones, their respective weightage will be 1 : 2.

The table below gives the field data and the necessary computations.

| | | 1 | 2 | 3 | 4 | 5 | 6 | 7 |
|---|---|---|---|---|---|---|---|---|
| | | Sample No. | Width in m. | Assay value in dwts/ton | M × assay product Col. 2 × 3 | Weight-age | Weight-age × m. product Col. 2 × 5 | Weightage × m × assay product Col 4 × 5 |
| Level AB | 1 | 1.5 | 4 | 6.0 | 1 | 1.5 | 6.0 |
| | 2 | 2.0 | 5 | 10.0 | 2 | 4.0 | 20.0 |
| | 3 | 3.0 | 4 | 12.0 | 2 | 6.0 | 24.0 |
| | 4 | 2.5 | 3 | 7.5 | 1 | 2.5 | 7.5 |
| | | 9.0 | 16 | 35.5 | 6.0 | 14.0 | 57.5 |
| Raise BC | 5 | 2.0 | 6 | 12.0 | 2 | 4.0 | 24.0 |
| | 6 | 2.5 | 7 | 17.5 | 2 | 5.0 | 35.0 |
| | 7 | 3.0 | 5 | 15.0 | 2 | 6.0 | 30.4 |
| | | 7.5 | 18 | 44.5 | 6.0 | 15.0 | 89.0 |

| | | | | | | | |
|---|---|---|---|---|---|---|---|
| Level CD | 8 | 3.0 | 8 | 24.0 | 1 | 3.0 | 24.0 |
| | 9 | 1.5 | 5 | 7.5 | 2 | 3.0 | 15.0 |
| | 10 | 3.0 | 6 | 18.0 | 2 | 6.0 | 36.0 |
| | 11 | 3.5 | 5 | 17.5 | 1 | 3.5 | 17.5 |
| | | 11.0 | 24.0 | 67.0 | 6.0 | 15.5 | 92.5 |
| Raise DA | 12 | 2.0 | 4 | 8.0 | 2 | 4.0 | 16.0 |
| | 13 | 2.5 | 7 | 17.5 | 2 | 5.0 | 35.0 |
| | 14 | 1.5 | 7 | 10.5 | 2 | 3.0 | 21.0 |
| | | 6.0 | 18.0 | 36.0 | 6.0 | 12.0 | 72.0 |

Level AB : Average assay value $=\dfrac{57.50}{14.0}=4.11$ dwts.

$\qquad$ Average width $=\dfrac{14.0}{6}=2.33$ m.

$\qquad$ Area of exposure $=2.33\times16=37.28$ sq.m.

Raise BC : Average assay value $=\dfrac{89.0}{15}=5.93$ dwts.

$\qquad$ Average width $=\dfrac{15}{6}=2.50$ m.

$\qquad$ Area of exposure $=2.5\times12=30$ sq.m.

Level CD : Average assay value $=\dfrac{92.5}{15.5}=5.95$ dwts.

$\qquad$ Average width $=\dfrac{15.5}{6}=2.58$ m.

$\qquad$ Area of exposure $=2.58\times16=41.28$ sqm.

Raise DA : Average assay value $=\dfrac{72}{12}=6$ dwts.

$\qquad$ Average width $=\dfrac{12}{6}=2$m.

$\qquad$ Areas of exposure $=2\times12=24$ sq.m,

Area assay product of exposure AB$=4.11\times37.28=153.22$

,,    ,,    ,,    ,,    ,,    BC$=5.93\times30$    $=177.90$

,,    ,,    ,,    ,,    ,,    CD $=5.95\times41.28=245.62$

,,    ,,    ,,    ,,    ,,    DA$=6\times24=144$

Total area of face $=(37.28+30.00+41.28+24.00)=132.56$

$\therefore$ Average of assay value of faces is $=\dfrac{720.74}{132.72}=5.44$ dwt.

Volume of block $=12\times16\times\dfrac{(2.33+2.50+2.58+2.00)}{4}$

$$=12\times4\times9.41=451.68 \text{ cu.m.}$$

when the tonnage factor or specific gravity of the ore is known, the reserves can be computed.

In general, the influence of weightage may be ignored and the computation can be simplified. The sample width and assay value being known, the m$\times$assay product is computed, in the manner indicated in the table below :

| Sample No. | | Width in dwt/ton | Assay in dwts/ton | m$\times$assay product |
|---|---|---|---|---|
| Level AB | 1 | 1.5 | 4 | 6.0 |
| | 2 | 2.0 | 5 | 10.0 |
| | 3 | 3.0 | 4 | 12.0 |
| | 4 | 2.5 | 3 | 7.5 |
| | | 9.0 | | 35.5 |
| Raise BC | 5 | 2.0 | 6 | 12.0 |
| | 6 | 2.5 | 7 | 17.5 |
| | 7 | 3.0 | 5 | 15.0 |
| | | 7.5 | | 44.5 |

| | | | |
|---|---|---|---|
| Raise CD 8 | 3.0 | 8 | 24.0 |
| 9 | 1.5 | 5 | 7.5 |
| 10 | 3.0 | 6 | 18.0 |
| 11 | 3.5 | 5 | 17.5 |
| | 11.0 | | 67.0 |
| Raise DA 12 | 2.0 | 4 | 8.0 |
| 13 | 2.5 | 7 | 17.5 |
| 14 | 1.5 | 7 | 10.5 |
| | 6.0 | | 36.0 |

Level AB : Average assay value $=\dfrac{35.5}{9}=3.93$ dwts.

Average width$=\dfrac{9}{4}=2.25$ m.

Area of face $=2.25 \times 16=36.00$ sq.m.

Raise BC : Average assay value $=\dfrac{44\ 5}{7.5}=5.93$ dwts.

Average width$=\dfrac{7.5}{3}=2.5$ m.

Area of face $=2.5 \times 12=30$ sq.m.

Level CD : Average assay value $=\dfrac{67.0}{11}=6.09$ dwts.

Average width$=\dfrac{11}{4}=2.75$ m.

Area of face $=2.75 \times 16=44.00$ sq. m.

Raise DA : Average assay value $=\dfrac{36.0}{6}=6$ dwts.

Average width$=\dfrac{6}{3}=2$ m.

Area of face $=2 \times 12=24$ sq. m.

Assay product of level AB = $3.94 \times 36 = 141.84$

   ,,    ,,    ,, Raise BC $= 5.93 \times 30 = 177.90$

   ,,    ,,    ,, Level CD $= 6.09 \times 44 = 267.96$

   ,,    ,,    ,, Raise DA $= 6 \quad \times 24 = 144.00$

Total assay product            $= 731.70$

Total face area             $= 134$

Average assay value of block     $= \dfrac{731.70}{134} = 5.46$ dwts/ton

This value is close to the value obtained earlier, i.e., 5.44 dwt/tonne

The average width of the block $= \dfrac{2.25 + 2.5 + 2.75 + 2.0}{4}$

$$= \frac{9.5}{4} = 2.375 \text{ m.}$$

The volume of the block $= 2.375 \times 12 \times 16 = 456.00$ cu.m. This value, however, differs slightly from that obtained earlier.

Where the ore is disseminated, and of low grade, a greater control is required in sampling. In practice, larger quantities of samples are collected by making wider and deeper channels, such that about ½ kg. per 30 cms. of channel is obtained. If in addition, the specific gravity of each sample is also determined, then the computations are tabulated in the following manner :

| Sample | Width in m. | %Cu | Gr. Ag/ tonne | $W \times$ %Cu | $W \times$ gr. Ag |
|---|---|---|---|---|---|
| 1 | 1.5 | 1.5 | 10.0 gr. | 2.25 | 15.0 |
| 2 | 1.2 | 1.8 | 30.0 gr. | 2.16 | 36.0 |
| 3 | 1.3 | 1.7 | 45.0 gr. | 2.21 | 58.5 |
| 4 | 2.0 | 1·5 | 15.0 gr. | 3.00 | 30.0 |
| | 6.0 | | | 9.62 | 39 5 |

Average width $= \dfrac{6.0}{4} = 1.5$ m., Average %Cu $= \dfrac{9.62}{6.00}$, $= 1.60\%$ Cu

Average value of Ag. $= \dfrac{139.5}{6} = 23.25$ gr.Ag./ton.

The significant difference in the values, for concentration of the metallic elements, is seen, when the additional variable viz. specific gravity, present in the ore, is introduced into the computations.    The table given below brings out this feature :

| Sample No. | Width (W) | %Cu | Ag. gr./ ton | Sp. gr. | $W \times sp.$ gr. | $W \times sp.$ gr. $\times$ %Cu. | $W \times sp.$ gr. $\times$ gr. Ag./ton |
|---|---|---|---|---|---|---|---|
| 1 | 1.5 | 1.5 | 10 | 3.0 | 4.50 | 6.75 | 45.0 |
| 2 | 1.2 | 1.8 | 30 | 3.2 | 3.84 | 6.91 | 115.2 |
| 3 | 1.3 | 1.7 | 45 | 3.4 | 4.42 | 7.51 | 198.9 |
| 3 | 2.0 | 1.5 | 15 | 3.1 | 6.20 | 9.30 | 93.0 |
| Total | 6.0 | | | | 18.96 | 30.47 | 452.1 |

$$\text{Average \% Cu} = \frac{30.47}{18.96} = 1.61\% \text{ Cu. and gr. Ag./ton} = \frac{452.10}{18.96}$$
$$= 23.84 \text{ gr.Ag./ton.}$$

### Factors influencing assay width and mining width

In computing assay width, minor streaks and bands of gangue or waste, which cannot be sorted, are to be included in the ore.

An additional width, of waste' or poor grade of ore may be included in the mining width computations, in order to permit sufficient (minimum) working room, even though such a step may cause perceptible dilution or decrease in the average grade of the ore.    The assay plan in a case like this will yet show only the width and grade of the ore, irrespective of the width of the workings.    These figures thus provide the basis of the estimation of reserves.   On the other hand, mining width averages (of the value) are always lower than those for the corresponding assay widths, but the ratio of the $\left( \dfrac{\text{mining width average values}}{\text{assay width average values}} \right)$ helps in assessing the economics of actual mining operations.

In working out the assay widths and mining widths, a few thumb rules may be of assistance.

(1)   All bands of gangue less than 30 cms. are included in the sample;

(2)   Where the bands of gangue vary in width between 30-90 cms. only half the width of such bands is included.

(3) Where the width of bands exceeds 90 cms, these are totally excluded.

The examples given below show the procedure to be adopted, with certain modifications, as required in the calculation of assay width average and mining width average, in cases where polymetallic deposits are involved.

### Example 1

| Band | Assay width averages | | | Mining width averages | | | Remarks |
|------|-------------|-------|---------------|-------------|------|----------------|---------|
| | Width in m. | % Cu. | gm. Ag./ton. | Width in m. | %Cu | gm. Ag./ton. | |
| I | 1.25 | 3.0 | 45 | 1.25 | 3.0 | 45 | According to rule 1, band III is included in the sample and considered in computing assay width and mining width. |
| II | 1.00 | 2.3 | 15 | 1.00 | 2.3 | 15 | |
| III | 0.25 | 1.0 | 10 | 0.25 | 1.0 | 10 | |
| IV | 1.50 | 1.8 | 4 | 1.50 | 1.0 | 4 | |
| | 4.00 | 2.25 | 19.94 | 4.00 | 2.25 | 19.94 | |

### Example 2

| Band | Width in m. | % Cu. | gm. Ag./ton. | Width in m. | %Cu | gm. Ag./ton. | Remarks |
|------|-------------|-------|---------------|-------------|------|----------------|---------|
| I | 1.25 | 3.5 | 30 | 1.25 | 3.5 | 30 | According to rule 2, only half the width of band II is included in the assay width average. |
| II | 0.40 | 0.8 | 20 | 0.40 | 0.8 | 20 | |
| III | 1.75 | 2.2 | 25 | 1.75 | 2.2 | 25 | |
| IV | 2.00 | 2.0 | 15 | 2.00 | 2.0 | 15 | |
| | 5.20 | 2.38 | 22.1 | 5.40 | 2.31 | 22.11 | |

### Example 3

| Band | Width in m. | % Cu. | gm. Ag./ton. | Width in m. | %Cu | gm. Ag./ton. | Remarks |
|------|-------------|-------|---------------|-------------|------|----------------|---------|
| I | 3.50 | 0.5 | 10 | 3.50 | 0.5 | 10 | Since band I is adjacent to the wall, it can be excluded in the computation of assay width. By rule 3, band III can be excluded. Both I and III are, however, to be included in the calculations of mining width. |
| II | 2.50 | 3.5 | 20 | 2.50 | 3.5 | 20 | |
| III | 2.00 | 0.5 | 10 | 2.00 | 0.5 | 10 | |
| IV | 1.50 | 2.0 | 40 | 1.50 | 2.0 | 40 | |
| | 4.00 | 2.94 | 27.5 | 9.50 | 1.52 | 17.4 | |

The computation of assay values, in the case of complex ores, containing three or four valuable minerals, is more a matter of practice, but as a rough guide, the following data may be employed for the purpose.

In working polymetallic ores, the economics is depended on the total metal content as given in the following example :

If 1.5% is the cut off value for copper, and 4.0% is the cut off value for zinc, it becomes necessary to fix the equivalent economic values for the other valuable metallic contents in the ore. Thus bands assaying less than 4.0% Zn and 1.5% Cu may also be classed under workable "Ore", when they are found to carry sufficient value of either lead and/or silver. In adopting this procedure, a band with certain silver content, may be considered either as being equivalent to a band of zinc ore or copper ore. For example (depending on current market rates) 60 gms. silver/tonne may be taken as equivalent to 1% zinc and 1% lead may be taken as being equivalent to 0.8% zinc.

**Example 4**

| Sample No. | Width in m. | % Zn | % Pb | gr. Ag | Remarks |
|---|---|---|---|---|---|
| 1 | 1.5 | 20 | 3.0 | 90 | % equivalent of Zn= 20+2.4+1.5=23.9 |
| 2 | 1.0 | 4 | 3.0 | 30 | 4+2.4+0.5=6.9 |
| 3 | 2.0 | 5 | 4 | 40 | 5+3.2+0.7=8.9 |
| 4 | 2.5* | 0.1 | 0.2 | 1 | 0.1+0.16+0.003 =unworkable |
| 5 | 3.0* | 0.2 | 0.1 | 2.0 | 0.2+0.08+0.008 =unworkable |
| Assay width average=4.5 | | 9.8 | 3.5 | 54.4 | |
| Mining width average=10.0 | | 4.5 | 1.6 | 25.4 | * will be omitted in assay width calculation. |

Thus when mining a width of 4.5 m ; the grade of ore, converted to zinc equivalent, will be $\dfrac{23.9 + 6.9 + 8.9}{45} \cdot \dfrac{39.7}{45}$ or 8.8%

During the course of regular stoping operations, it is possible to obtain an additional control by means of a factor, known as the "length-height" (L.H.) factor. This is obtained by dividing actual volume of material stoped by the mining width.

$$\text{L.H. factor} = \frac{V \quad (cu.m.)}{M.W. \quad (m.)} = \frac{\text{Volume of stope}}{\text{Mining width}}$$

Certain ore deposits, once considered to be the source of a single metal, may turn out to be polymetallic, at a certain stage in the development or during the course of mining operations. This change in metal content is likely to affect the economics also. Thus a low grade copper ore, may just constitute a marginal deposit, but if it is found to be associated with lead, zinc, nickel, or molybdenum etc. the economic picture will change largely. A fair assessment, of the percentages of the various constituents, will be required in the first stage. Secondly, the equivalents of the other components, in terms of copper percentage, will have to be computed. In the third stage, an assessment of the capital investment required for the recovery, of the secondary mineral components, will have to be made before any scheme for their recovery can be formulated and recommended. The overall economics will depend on the above calculations. A position similar to this, is that of Khetri in Rajasthan, and the Rakha project in Singhbhum district of Bihar, where the average grade of copper is 1.5%, in the latter case but preliminary examination has shown that the ore contains appreciable amounts of nickel, molybdenum, etc.

### Disseminated ores

Such ores are also sampled by drilling—either diamond or churn. A grid with 200 ft. intervals, is commonly used in drilling. When the concentration is uniform, samples are collected at fixed intervals—covering every 5 ft. at a time. Where definite bands of disseminated ore are present then samples are drawn from the ore bands. A tonnage factor of 12¼ cu. ft./ton is used in computation of reserves. This may be altered slightly to suit local conditions. (It may be noted that when the metric system is used the tonnages of ore can be obtained by multiplying the specific gravity of the ore by volume in cu. meters.)

Where churn drilling is adopted, the sludge is collected by using a dart bailer or sludge pump, and a sample splitter is recommended. After drying, the residue, obtained from the sampled sludge, is put through a riffle sample splitter and the final sample which may weigh about 5 kgs. to ¼ kg. is collected for analysis.

Samples, collected from blast holes, can also be employed for obtaining estimates of the grade. Samples can be collected from the sedimentation box of the holmaster type drill or from the sludge of a churn drill. It will be seen that the values so obtained tally very closely with the actual values of the run of mine ores.

Sampling requires a good understanding of the main factors, and where large tonnages are involved, as in the case of iron ore, or coal or limestone, sampling may involve special techniques. Not only do the physical and chemical properties affect the method of sampling, but also the time available.

Hence in order to exercise greater control on the quality, it may be necessary to carry out repeated sampling of (a) the broken ore, (b) ore as loaded in wagons (c) and ore as stocked at the mill site.

Sampling is of paramount importance since it forms the basis for (1) selling the ore, (2) for planning exploration and development and (3) for controlling the grade.

In sampling, material obtained by the following methods may be utilised : (1) Diamond drill core, (2) Churn drill (sludge samples), (3) Test pit, channel or groove or bulk samples, and (4) Drive samples from workings underground. Here also grooving can be done at regular intervals, across the width of the vein or ore body, (5) Stope or working face sample, while (6) Car sample or wagon sample is drawn by scooping out equal amounts from each car, at definite points, e.g. at four corners, and at the centre or some other pattern, depending on the size of the wagon. Samples collected, for the determination of moisture, are required to be sealed in air tight box to prevent losses.

Reserves may be computed as volume of ore available at sight, but the more common practice, is to work out the reserves in tons or tonnes. Thus if the volume of ore is known and the weight of ore per unit volume, i.e., specific gravity, or weight per cu. ft. or cu. metre, then the tonnage can be determined. Before proceeding any further with this topic, it is well to remember that the usual advice, of presenting conservative estimates as being safe, is not without its dangers and disadvantages. To present a realistic and true picture should be the motto; since a conservative estimate may place the reserves at such a low level as to make a good deposit appear uneconomic.

The weight, per unit volume of ore, depends on its specific gravity and porosity.

1. The specific gravity of a mineral aggregate can be obtained, by working out the volume percentage of each mineral species in the rock, and then by employing the value of the specific gravity of each mineral species to obtain the average.

2. The powder method, using a specific gravity bottle can also be employed, and average value of the specific gravity obtained after making six or more determinations.

From the maps and sections, the volume of ore available can be computed either by planimeter or by the use of Simpson's rule, and this figure can be used, in conjunction with the specific gravity value, for computing the tonnage available.

Ores can be monometallic, bi-metallic or polymetallic. Very often the ore has considerable pore space and moisture. Hence it will be necessary to determine the true specific gravity of the ore, including voids and moisture. In order to determine the true specific gravity, the following procedure may be adopted.

1. Calculate the specific gravity of the ore mineral. This is a simple operation in the case of a monominerallic ore. Where the ore is polyminerallic, then the computation are a little more involved, a weighed arithmetical mean, taking into account the total volume of ore, and the respective volumes of each of the ore species and their specific gravities, will have to be determined.

Suppose the volume of ore is V and the vols. of the three individual (ore) mineral components are $(a+b+c) = V$ in scc.

If the sp. gr. of $a = G_1$, $b = G_2$, $c = G_3$ (in gms/cc.)

$$\frac{aG_1 + bG_2 + cG_3}{V} = \text{Average sp.gr. of mineral (ore)}$$

Using a polished section on an ore microscope, the volume percentages of ore minerals will have to be determined by modal analysis, using an integrating stage or a point counter technique and specific gravities determined by the powder method with a specific gravity bottle or pycnometer.

2. Where the porosity (i.e., Vol.% of pores and Vol.% of moisture) of the ore (mineral) is known from other determinations, then if the weight of ore specimen is (W) gm., the volume is V cc., and the weight of dry ore (at $100°$-$110°C$.) is (Wd) gm.

Then the actual vol. occupied by the ore ($V_e$) after allowing for voids and occluded moisture, may be expressed as follows :

$$V_c = V - (V_a + V_m),$$

where $V_a = \%$ vol. of pores and $V_{on} = $ vol. of moisture

Since the weight (W) includes also that of moisture ($W_m$),

Actual weight $= (W - W_m) = $ Wd

Hence the specific gravity $= \dfrac{Wd}{V_c} = \dfrac{Wd}{V - (V_a + V_m)}$

## Example

A piece of ore weighs 1600 gms. and its volume is 400 cc. After drying its weight is 1550 gm. When the specimen is saturated with moisture (after the surface is wiped of excess water) its weight is 1650 gms.

Vol. of moisture ($V_m$) = 1600 — 1550 = 50 cc.

Vol. of air ($V_a$) = 1650 — 1600 = 50 cc.

Dry weight of ore (Wd) = 1500 gm.

Hence sp. gr. $= \dfrac{Wd}{V - (V_a + V_m)} = \dfrac{1500}{400 (50 + 50)}$

$$= \dfrac{1500}{300}$$

Sp. gr. = 5.0

The term, tonnage factor, has fallen out of use very much in this country since the introduction of the decimal system.

Tonnage factor is the number of cu. ft. of ore (may be in situ or broken) per ton long (i.e., 2240 lbs.). Thus a tonnage factor of 10 will mean that 10 cu.ft. of ore make a ton. For broken limestone the tonnage factor will be about 18.

In the decimal system, the relation between volume, sp. gr., and weight are so simple that the importance of tonnage factor is not as great as the conversions are simple.

If an ore has a sp. gr. of 5, the weight of one cubic metre will be

$$\dfrac{5 \times 100 \times 100 \times 100}{1000} = 5000 \text{ kgs. or 5 tonnes.}$$

In giving the estimate of the reserves, one is often at a loss as to how best to describe, or what epithets to use in the definitions. This problem has confronted economic geologists for a long time, and finally the U.S. Bureau of Mines and the U.S.G.S. have accepted the use of the following terminology in defining ore reserves.

(1) The term "measured ore" is employed, when reserves have been computed on the basis of the dimensions, obtained from measurements made on outcrops and exposures in workings, as well as in the excavations made during prospecting operations. The grade should have been computed from bore hole data, as well as from detailed sampling. On the whole, it is accepted, that the interval used, in these measurements should be so close, that it will afford a reliable picture of the structure, size, nature, and behaviour of the ore body. Hence the estimates of the tonnage and grade may not differ from the actual figures by a maximum $\pm 20\%$.

(2) The term "indicated ore", is employed in the case where the estimates, for the tonnage and grade, have been based partly on computations and partly on measurements as in the former case, i.e., from data obtained in exposures, excavations drilling etc. as well as from the general structure, so that the behaviour of the ore body can be predicted with reasonable accuracy on the basis of detailed geologic mapping. This may be taken to include and synonymous with "drill indicated ore."

(3) The term "inferred ore" is applied when the estimates of reserves are based mainly on very general and broad observations of a qualitative nature. In this, the continuity of the ore body is generally assumed on purely geological evidence, and the behaviour of the mineralisation inferred from observation made on ore bodies of a similar nature in nearby areas. The figures may include estimates of reserves of possible ore bodies, as revealed by geological evidence. It is, however, necessary to state the basis on which the computations have been made. The term geological evidence may include strong anomalies obtained by the application of geochemical and geophysical methods of prospecting.

## (d) Procedure for the estimation of reserves

*In estimating the reserves in a virgin area,* where in addition to surface geological examination a limited extent of drilling has been done, considerable prudence is called for in exercising judgement. It may be pointed out that the conservative attitude of "Playing Safe" should not go to the extent of lowering the reserves to an absurd minimum, such that a valuable deposit is lost to posterity. On the other hand, it is unbecoming and dangerous to exaggerate the figures where the reserves are poor. Hence it is the considered opinion that the deposit warrants further examination, then the estimates of reserves should be in keeping with that view, so that the proposition is attractive for an entrepreneur to take an investment decision, and proceed with further systematic exploration.

Useful aids to making preliminary estimates of reserves are certain thumb rules which have been tested over the years, in several areas. One of these states that the depth extension of an ore body or ore shoot can be taken to be approximately equal to half its exposed strike length, as seen on the surface, or as exposed in the workings underground, additional support should also be obtained from observations made in the vicinity where possible. Another empirical rule which also requires confirmation likewise, is that the maximum width of an ore body or an ore shoot is located midway down the dip. This will be true in the ideal case where the ore body is an ellipsoid.

Knowing that even in the case of a tabular ore body mineralisation is nearly always confined to a zone comprising ore patches known as ore shoots, and that these ore shoots very often bear a constant relation to the strike, known as the pitch, the study and the identification of the pitch is of prime consequence. Pitch is defined as the angle between the longer axis of the ore shoot and the strike of the ore body, in the plane of the longitudinal section. (Fig. 11.12). This should not be confused with rake or plunge which is the apparent dip of the ore shoot. It can so happen that the longer axis of the ore shoot may show various degrees of inclination which can vary from 0 degree to 90 degrees.

Since the most probable extension of the ore body is in the direction of the pitch and since the ore shoots occur either 'En echelon' along the

Fig. 11.12

same strike or either with sinistral or dextral parallel shifts, invariably preserving the same pitch, a few of the guides to ascertaining the pitch are indicated below:

1. The trend of the boundaries of the ore shoot. This may not be clear in the early stages of exploration.

2. Observation on the limits of the mineralisation from level to level.

3. Pitch of other ore shoots, if known.

4. Pitch of ore shoots in the adjoining area, if controlled by the same major structure.

5. Plunge of the structural elements in the country rock, in case of

structural control or of replacement type deposits.

In addition to the above, the following structural elements may also serve as aids to identifying the pitch.

1. Plunge of the drag folds.

2. Lineations such as (a) intersection of cleavage and bedding, (b) of two sets of cleavages. (c) mineral lineations etc.

3. Pitch of the intersection of veins/and or of faults, or of the bedding and veins etc.

But the reserves estimated by the use of these methods must be placed under the category of 'possible'.

The tonnage of material available, in a deposit, can be readily computed, if the volume as well as the specific gravity are known or if the volume and the tonnage factor are known. This will, however, be purely of theoretical interest, if the economic bias is absent. Hence, it is the general practice to establish reserves of the marketable material or ore. For this purpose, the "cut off grade" is to be ascertained, and this figure, for the tenor of the ore, will form the dividing line between ore and waste during mining operations. It is a good practice to estimate reserves even for a grade lower than that of the "cut off", since improvements in mining methods and/or in ore dressing practice etc. can affect the economics, by attaining higher efficiency.

Thus the estimates of reserves will include :

(1) The total tonnage of material found in the ore body,

(2) The total tonnage of ore, of a grade above the cut off grade, as well as quantity falling within the cut off grade.

(3) Tonnage of material of a grade slightly lower than the cut off grade.

(4) The tonnage of ore recoverable, after allowing for losses in mining, transporting etc.

In practice, it has been found that the preliminary estimate of reserves will be subjected to revision after development, and the reserves computed at the later stage are always greater. Estimates of reserves pertaining to bedded deposits such as coal, salt, iron ore etc. are more amenable to rigid treatment, whereas estimates of reserves, of irregular ore bodies, always have less chances for attaining a high order of accuracy. Even in the case of bedded deposits, for example limestone, the presence of solution cavities is an unpredictable factor and great caution is necessary in estimating reserves.*

## Indian Standard Procedure for Coal Reserve Estimation and Rank Classification of Coals

---

*See addendum.

The following standard procedure for coal reserve estimation is intended to provide uniform rules and definition for Geologists and Mining Engineers so that the estimates prepared, in different areas and by different persons, can be suitably correlated, and combined into all-India and international figures. Obviously, this procedure cannot be applied rigidly to all coal-bearing areas, and modifications within its broad frame work, to suit the geological features in particular areas, are permissible. Whereever such modifications are made, however, proper explanations should be given.

1. The basis of calculation of all regional coal reserves, is the geological map, of the coalfield. The degree of reliability of the geological map and the data, available from the same is primary. The latest available geological maps should, therefore, be used, in every case.

1. Reserve data shall be reported according to (A) the amount of overburden on the coal and (B) according to the depth from surface, as follows :

(A) 1. Overburden equal to the thickness of the seam or seams, where two or more seams occur in close proximity.

2. Overburden equal to the thickness of two seams, where they occur in close proximity.

3.    "    "   three   "    "    "

4.    "    "   four   "    "    "

5.    "    "   five   "    "    "

(B) 1.   0 metre to   150 metres or    0— 500 ft.

2. 150   " to   300   "    500—1000 ft.

3. 300   " to   600   "    1000—2000 ft.

4. 600   " to   900   "    2000—3000 ft.

5. 900   " to   1200   "    3000—5000 ft.

In arriving at the figures of reserves, with in the above mentioned overburden limits and depths, the dislocation, caused by faults, should be taken into account, and their effect eliminated, while demarcating the sectors, under different categories, for the purpose of calculation.

2. *Classes of reserves*

2 : 1 On the basis of the relative reliability of data coal reserves shall be classified as follows :

2 : 1 : 1  *Proved Reserves* :  In this case, the reserves are estimated from dimensions revealed in outcrops, trenches, mine workings, and bore-holes, and the extension of the same, for reasonable distance, not exceeding 200 metres (660 ft.), on geological evidence. Where little or no exploratory work has been done, and where the outcrop exceeds one kilometre (3300 ft ) in length, another line drawn roughly 200 metres (660 ft.) from outcrop will define a block of coal, that may be regarded as proved, on the basis of geological evidence.

2 : 1 : 2  *Indicated reserves* :  In the case of indicated reserves, the points of observation are 1,000 metres (3300 ft.) apart but may be 2,000 metres (6600 ft.) for beds of known geological continuity.

Thus a line drawn 1,000 to 2,000 metres (3300 to 6600 ft.), from outcrop, will demarcate the block of coal, to be regarded as indicated.

2 : 1 : 3  *Inferred Reserves* :  This refers to coal, for which quantitative estimates are based, largely, on broad knowledge of the geological character of the bed but for which there are no measurements. The estimates are based on an assumed continuity for which there is geological evidence, and more than 1,000 to 2,000 metres (3300 to 6600 ft.) from the outcrop.

2 : 2  Proved reserves shall be further divided as follows :

  (*i*) In working collieries.

    (*a*) Coal standing on pillars, and in partings, roof and floor.

    (*b*) Solid coal.

  (*ii*) In closed mines.

  (*iii*) In areas covered by mining leases, but not worked.

  (*vi*) In other areas.

3 : 1  *Thickness range* :  The average thickness, of coal seams, shall be calculated and stated in centimetres (and feet).  Partings greater than 5 centimetres (2 inches), in thickness, shall be excluded, in calculating reserves.  The burnt out portions of coal and *jhama* shall be excluded, while taking thickness of the seams, for the purpose of calculation.

3 : 2 The thickness range for individual seams for the calculation of reserves shall be as follows :

  (*a*) 0.5 metres to 1.5 metres (1.5 ft. to 4.5 ft.)

  (*b*) 1.5 metres to 3.5 metres (4.5 ft. to 10.5 ft.)

  (*c*) 3.5 metres to 5.0 metres (10.5 ft. to 15 ft.)

  (*d*) 5.0 metres to 10 metres (15 ft. to 30 ft.)

(*e*) above 10 metres (above 30 ft.)

4 : 1  The reserves, of each individual seam, in a sector or part of the coal-field, shall be given separately.

4 : 2  Reserves in virgin seams, or solid coal, may be calculated, in the case of flat seams, on the basis of area, the thickness of coal beds, and a correction applied, in the case of seams, with inclination above 5° by multiplying the figures with the secant of the dip angle. All reserves shall initially be expressed in cubic metres or cubic feet.

5.  *Coal mined and lost in mining* :  This shall be calculated by either of the two methods given below :

(*i*)  By actual quantitative measurements, in working mines. Its comparison, with coal production figures, gives a yield percentage of recovery data, which can be applied under similar conditions, in virgin areas, to determine the recoverable reserves.

(*ii*)  Where no data is available, the production figure increased by a percentage factor, of losses in mining, gives the quantity of coal worked out, in a particular coalfield. The balanee, or the available reserves, will give the recoverable reserves where no other data is available, the yield percentage factor may be taken as 50 tons, for every 100 tons of coal mined.

6.  *Coal in existing mines, closed mines and areas under mining leases*

6 : 1  The reserves of coal, in the seams which have been or are being worked, may be calculated from the pillars, partings, roof and floor in the mines, and coal standing as solid blocks. Standard stipulations, regarding overburden, thickness categories, class of reserves etc., will apply as in the case of virgin areas.

6 : 2  The possible losses in working shall be given under the following heads :

(a)  Coal likely to be lost, due to geological features.

(b)  Coal likely to be locked up, under roads, railway lines, rivers and jores.

(c)  Coal likely to be lost, in barriers.

(d)  Mining losses.

The working plans, of the collieries and the geological maps, shall form the basis of calculations.

6 : 3  The standard procedure shall be followed in every detail, in the computation of reserves, in the areas under mining leases, the latest

geological maps of the coalfields forming the basis of calculation.

6 : 4    The reserves shall be reported separately, for each seam, in the form of tables, indication thickness taken and dip to facilitate break up the figures quality-wise.

6 : 5    Both *in situ* and recoverable reserves should be estimated, in every case separately. In the case of working collieries, the recoverable reserves shall be classified as follows :

    (*a*)  Recoverable without stowing.

    (*b*)  Recoverable with stowing.

This will also help in arriving at a recovery ratio factor, applicable to the whole coalfield, and the calculation of overall recoverable reserves.

7.    *Rank classification of coals* :    The figures for reserves shall be classified as follows according to the rank of the coal :

I.    Anthracite.

II.    Bituminous.

    (*A*) *Low to Medium volatile coals or coking coals.*     Air dried, moisture upto 2%, and volatile matter usually not more than 35%, on unit coal basis.

    (*B*) *High volatile or High Moisture coals.*     Air dried moisture more than 2%, or volatile matter usually more than 35%, on unit coal basis. Sulphur less than 1% in either case.

    (*a*)  Semi coking coals.

    (*b*)  Weakly to non-coking coals.

    (*c*)  High Sulphur coals.     Sulphur more than 1%. These occur in Assam and although some of them are semi coking they are not suitable for metallurgical purposes.

III.    Lignite.

8.    The bituminous coal reserves shall be further classified, according to quality, on the basis of the analysis of seam samples, as follows :

    (*A*)  *Low to medium Volatile Coals or Coking Coals.*

Class    I—Ash not exceeding 17%.

Class    II—Ash exceeding 17%, but not exceeding 24%

Class    III—Ash exceeding 24%, but not exceeding 35%.

Class    IV—Ash exceeding 35%, but not exceeding 50%.

(B)  *High Volatile or High Moisture Coals.*

Class I — Ash + Moisture not exceeding 19%.

Class II — Ash + Moisture exceeding 19%, but not exceeding 28%.

Class III — Ash + Moisture exceeding 28%, but not exceeding 40%.

Class IV — Ash + Moisture exceeding 40%, but not exceeding 55%.

(C)  *High Sulphur Coals.*

No quality classification is needed.

9. *Specific Gravity*

Where reliable data are available, the following average specific gravity, of each class, within each category, should be used :

(A)  *Low to Medium Volatile Coals or Coking Coals.*

| | | |
|---|---|---|
| (a) Class I | — | 1.42 |
| (b) Class II | — | 1.47 |
| (c) Class III | — | 1.57 |
| Class IV | — | 1.70 |

(B)  *High Volatile or High Moisture Coals.*

| | | |
|---|---|---|
| (a) Class I | — | 1.40 |
| (b) Class II | — | 1.45 |
| (c) Class III | — | 1.55 |
| (d) Class IV | — | 1.70 |

(C)  *High Sulphur Coals.*
     *Ash Content*

| | |
|---|---|
| (a)  0 to 5% | 1.30 |
| (b)  5 to 10% | 1.34 |
| (c)  10 to 15% | 1.38 |

10. *Unclassified reserves*

Where no data are available, or where reliable data are lacking, the reserves should be placed under a category of 'unclassified' reserves, and for the calculation of the quantity, a specific gravity of 1.5 may be used. These will be split up into various categories, and classes, as and when reliable data become available.

11. *Abnormal coals*

Some coals may exhibit abnormal properties, because of their unusual petrographic composition, e.g., some coals may show low moisture content, but at the same time be non-coking. These should preferably be reported separately.*

## (E)  EVALUATION OF ASSETS

In order to find out whether a certain mineral property is going to yield sufficient profit or not, it is just not enough to estimate the reserves of the ore, and its present market value. It is essential, in such calculations, to take into account the cost involved, in extracting the ore at least upto a certain stage.

There are normally three stages to be gone through, before the metal is extracted from the ore : (1) mining, (2) ore dressing and (3) metallurgy. Examples of this type of ore are gold, copper ore, zinc ore or lead ore. In some cases, a mineral can be mined and sold directly, e.g., iron ore coal, felspar, kyanite etc. Thus every mining undertaking involves investment of capital (a) on purchase of land and (b) purchase of machinery. Hence the margin of profit should be such that the capital invested may be recovered in reasonable period and also allow payment of a reasonable rate of interest on the investments.

Machinery is required in various stages, (a) exploratory stage, (b) development stage and (c) mining or exploitation stage. A careful assessment of the essential requirements of machinery for these various stages, in the operation, must be made. Over estimation of the costs will have adverse effects on the economics, so also is the case with an under estimation. A prudent investment is what is necessary, which will mean minimum investment with maximum output. Under the head of capital, royalties to be paid on land and mineral, the cost of land the cost of building for office, quarters etc., should also be included.

During the development stage again, the capital is locked up and hence openings, excavations, etc., made during this stage of the operations, are also part of the investment. In fact, it may not be possible to demarcate the stage, when exploration stops and development begins. Thus a careful assessment should be made, and appropriate weightage given to the various variables which influence the ore reserves. Quite often, an ore body may comprise several independent ore shoots, separated by barren rock. An allowance should be made for the barren rock, in the final estimates for the reserves, as per rules enunciated earlier.

Of other considerations, to be taken into account, blending is of importance. For example, in a limestone area if there are several bands of limestone, each assaying differently, e.g., cement grade < 2% MgO,

---

*See addendum.

blast furnace (BF) grade < 10% MgO, < 6% insolubles, steel melting shop (SMS) grade < 6% MgO, < 5% insolubles. It is just possible, that with a careful blending of the cement grade with the BF grade the reserves of BF grade can be increased. In certain cases the cost per ton is likely to be affected by lease terms, e.g., where surface rights have not been acquired, and compensation is to be paid for any damage to the surface by subsidence.

No method of mining can ensure cent per cent recovery of the ore. It is to be expected that losses will occur either due to (a) blocks of ore that will have to be left as pillars to support, shafts or stopes or surface installations, (b) due to crushing of pillars, (c) difficulties in mining, such as the presence of faulted and badly fractured zones, and (d) due to excessive dilution with waste rock. It is thus prudent to take a higher tonnage factor while estimating reserves, in order to allow for such unpredictable losses.

During the course of mining, sometimes, it is possible to recover 75%, or even 100% of the estimated reserves, in a section of a mine. It may also happen, that the recovery is in excess of the estimated reserves, even in a case where there is no doubt as to the accuracy of the estimate. This can happen due to "dilution" and it will mean that the ore won must have been of a lower grade than the grade estimated. Thus, mining efficiency does not merely depend on the % recovery, but also on the extent of dilution. Hence cent per cent recovery with no dilution is the maximum efficiency.

The capital invested on development is locked up and it goes to waste, if in the long run, the deposit proves uneconomic. Hence in the renewal of mining lease, this aspect has been recognised, and if an operator has done enough development, the priority for renewals remains with him.

In addition, to the above investments, capital is required for running the mine. Routine expenditure, such as payments to labour, maintenance of machines, fuel, power supply, transportation, known as working capital has to be incurred. Such expenditure will have to be paid out of the capital, until such time the output is large enough to be able to sustain the mine. At times, it may be necessary, to stop sales of ore, if the market is down. This will produce a shortage or scarcity and consequently help recovery of the price level. It is common practice, where steel plants own colleries, to buy coal from the open market, when the market is down, and increase production from their own mines, when the market recovers and prices go up. Since steel plants are one of the main consumers of coal, with the fall in sales a slump sets in, and this is advantageous to the steel plants since they can buy coal in a buyer's market. Thus a fair price level is maintained. What is true for steel plants, is also true for railways, owning coal mines.

In the case of a property which had been abandoned, assets (investments) may have to be taken into account in assessing its value. A certain amount of useful machinery may be available, and a certain amount of development may have been completed. Thus such investment already made will affect a saving in capital. It may so happen that excessive initial outlay, in developments and machinery, may have caused the mine to close down for want of funds, but then the low sale value of the property known as "distress sale", gives an advantage to a subsequent purchaser who has a better opportunity to make a profitable business. Such instances, may be cited in mica mining, where the original owner had invested capital on development, but recovered little mica, and sold the property incurring a loss. The purchaser on the other hand, with little investment, on development, struck a bonanza, and the property became a paying proposition. Other factors, requiring evaluation, are availability of timber (such as standing trees), lands other than the mine etc., availability of water, besides sources of power and labour.

### (F) MINERAL ECONOMICS

The income from a mine is dependent not only on the reserves, effective machinisation etc., but also on the market, supply and demand and the value of the currency (rupee or pound or dollar).

(1) The value of a mineral is adversely effected by the lack of accessibility, and transport facilities. This is more so in the case of minerals of low value, like limestone, silica sand, clays, or coal etc. An example of such a deposit, is the Jharia coal field, which could not be opened up until the railways came into existence. The same is the case of several clay deposits notably those in the Singhbhum district of Bihar.

(2) Sale of the mine product is affected either (a) as raw ore, in cases like uranium, or coal, silimanite, kyanite, magnesite, (b) or as benificiated ore, i.e., either after upgrading by washing or screening, e.g.. coal, or by the use of more expensive methods like jigging, floatation, e.g., felspar, nepheline, etc. or by dressing, e.g., mica, (c) after recovery of the metal, as in the case of copper, lead, or zinc. This is the case with most minerals, except for gold and uranium for which there is ready sale.

It will thus not be sound economics, if the income from the sale is computed on the optimum output for the mine, where there is no market. Business contracts, agreements, financial position, even the names of person on the board of management are factors of importance. Sometimes, the sale of a mineral is affected by a factor which has little or no relevance to the use, e.g., in the case of mica where colour is a primary consideration, even though it may not have any bearing on the use for which mica is required. Allowance will also have to be made when, due to a slump in the market, the sale of the mineral is restricted or stopped—so that expenditure continues to be incurred on mining, taxes, royalties etc., while no returns are forthcoming. Consequently when the ore, after stock piling, is sent to the market, it will be sold at a higher price than that before stock piling, so that the expenditure incurred during the slack period can be realised from the sale, e.g., interest on capital (say $4\frac{1}{2}\%$), royalties on ore and property etc.

Mining is dependent on demand, for mineral raw materials. A few exceptions are gold, silver, and consumer goods, such as salt or coal etc. Hence, to a certain extent, it may be said that mining operations are regulated by the producer or manufacturer in keeping with the rise or fall in demand. Stock-piling by the manufacturers may cause a slump for a certain period. Mining is also affected by the supply. For instance the supply position for cement, in a particular year may be very high, due to large scale construction programmes. As a consequence there is a boom in the mining of limestone, gypsum, coal etc. If the consumption of cement falls, the production of limestone, gypsum etc. also slumps.

Mining cannot be compared with other business undertakings.

(a) A mineral property is a wasting asset. The reserves in a mine are continuously decreasing. It is not like agriculture where crops can be raised again and again on the same land.

(b) The pit mouth value of a mineral can vary considerably, from mine to mine, even in the same area, depending on the tenor of the ore, size of the deposit, depth of working, nature of the ore etc. In the case of agricultural products, the costs may not differ so widely. Sometimes a certain amount of uniformity can be expected, in minerals produced, and this is more common in bedded deposits. For example, it is well known that seams below the X seam in Jharia coal field yield coking coals and that the Dishergarh seam or Poniati seams, in the Raniganj coal field contain high grade uniform quality coal.

(c) The extension of workings, with time, increases the cost of mining, and as the saying goes, it is largely true that "ninety per cent of the profits in mining are recovered in the first 1000 ft.".

Even though minerals are essential for the continued industrial development, as well as for industries, the mine often does not last long. Exceptions to this are rare, but the industries utilising minerals, operate continuously and often outlive the mines.

The term "profit spread", is used to signify the difference between the total cost of production, and the sale price of the mineral. This is dependent on (1) the value of money as well as (2) on value of the mineral in relation to other commodities. If there is no variation in the second factor, the cost of production and the sale price may also vary in accordance with the general "price index."

The main factors influencing the production of an industry are : (1) land, (2) labour, (3) capital and (4) management. Of these land and capital are fixed assets, whereas cost of labour and management fluctuate with the price index, thus affecting the cost of production. Hence in the history of a mine, the profit in terms of money (profit spread) does not vary very much, even though the percentage of variation is markedly large.

For example if the "price index" is 150, the cost price is Rs. 5/- and the selling price is Rs 8/- the profit is Rs. 3/-. This profit is derived from an investment comprising land, capital, labour and management. Taking the royalty on land to be 10%, and the depreciation on capital invested (on equipment) to be 10%, it will mean that 20% of the costs will not be affected by the price index. A rise in price index from 150 to 200 will produce the following effects.

| Price index 150 | Price index 200 | Difference in terms of | |
|---|---|---|---|
| (1) | (2) | Money    (3) | % |
| Sale price  Rs. 8.00 | $8.0 \times 1.25 =$ Rs. 10.00 | Rs. 2.00 | 25 |
| Cost price  Rs. 5.00 | $1+(4 \times 1.25)=$ Rs.  6.00 | Rs. 1.00 | 20 |
| Profit       Rs. 3.00 | 4.00 | 1.00 | $33\frac{1}{3}$ |

Thus it is seen from columns (1) and (2) that if the sale price increases by 25%, and the cost price increases by only 20%, then instead of an apparent increase of Rs. 4/- the actual profit Rs. 1/- only.

In case the price index falls, e.g. to 120, then its effect on the cost, and price will be as follows :

| Price index 150 | Price index 120 | Difference in terms of | | |
|---|---|---|---|---|
| (1) | (2) | Money | (3) | % |
| Selling price 8.00 | $8 \times .75 = 6$ | 2 | | 25 |
| Cost price    5.00 | $1 + (4 \times .75) = 4$ | 1 | | 20 |
| Profit        3.00 | 2 | 1 | | $33\frac{1}{3}$ |

In this case also, as seen from columns (1) and (2), the selling price has fallen by Rs. 2/-, and the cost price has fallen only by Re. 1/-, and there is yet a profit of Re. 1/-

Other factors, which minimise the variation in profit are influenced by the policy adopted by the management.

(a) During a period of boom, comparative inefficiency creeps in and management is slack. There is a dearth of skilled labour, and this also contributes towards the overall decrease in output.

(b) When there is a slump, unemployment is high, and stricter measures are imposed, by the management for increasing the efficiency, which is reflected by an increase in production.

A boom affects the mining industry as well, and lower grades of ore become payable. Hence the output and gross profit increase even though the profit per tonne may be lower. Thus during the World War I, when coal was in short supply, the miners in the Jharia coalfields stepped up production, for making hay while the sun shines, with the result that even pillar robbing was resorted to. This, however, brought troubles in its wake. Crushing of pillars, roof falls etc., followed by extensive mine fires, increased mining hazards later—putting up the costs, since sand stowing had to be enforced.

When a period of slump sets in, the selling prices fall. It is economic to work only the higher grades of ore.    This leads to a high rate of depletion of the reserves of high grades.    This can be exemplified by conditions, which prevailed in the Raniganj and Jharia coalfields, soon after World War II, when economic conditions permitted the working, only of the better quality coals, from the Dishergarh and Ponlati seams, in consequence the poor quality seams, like Koithi, were not worked.

(c)  During periods of slump maintenance is poor and expenditure on this account is low.    Replacements and purchases are made during the boom periods.    A good example is again drawn from the coalfields, where prosperous mines, once well equipped, now operate with dilapidated machines, due partially to the slump as well as to depletion of reserves, which do not warrant further capital expenditure.

(d)  Minerals and metals are affected by the market rates, of other commodities as well    Even gold is no exception.    Gold pays when the price index is high. but the profit decreases with a fall in the market.    It may not be possible. to fix, with certainty, the price of ore or metal in relation to that of any other commodity.    Yet, it is possible to follow the trends in a general way. A fall in prices, of a particular metal. should be anticipated, when large reserves of ore discovered.    This happened, when suddenly Rhodesian Copper flooded the markets, and made marginal deposits uneconomic for exploitation.    Under such circumstances, occasionally Governments come to the rescue of the industry by levying protective tariff on imports, so that the indigenous industry may not be throttled.    As an example, the protective tariff levied on imported soda ash from Africa may be cited.

**Example :**

The average price of Copper per ton in 1929 was £ 81/-.

,,          ,,          ,,          ,,          in 1938 was £ 45/-.

Price index in 1929 was 180.0

,    ,,    1938 was 105.0

Taking price index in 1926 as 100

Thus the relative price of copper in 1929 as compared to

$$1926 \text{ is } 81 \times \frac{100}{180} = £ 45$$

Thus the relative price of copper in 1938 as compared to

$$1926 \text{ is } 45 \times \frac{100}{108} = £ 44.4$$

This shows that copper did not loose much of its value during this period, but the price fluctuation was relative to that of other commodities.

It is thus a good practice, to obtain the statistics for the annual fluctuation, in the prices of the particular ore, or metal for 20 years or more. This data may be plotted on a graph, together with the annual wholesale price index of commodities. Such a graph will help in deciphering the particular phase of the business cycle i.e., whether it is the boom phase or recession phase, and thus assist the prediction of the general market price trends.

The procedure for obtaining the graph will be as follows :

(1) Get the prices/tonne of ore or metal (in the case of gold and silver price/oz) for every quarter, for the whole year, i e., March, June, September, March.

(2) Assume that the general price index, for any particular year, to be 100. This is the base index.

(3) Work out the relative price index, for the ore or metal etc., for each quarter and every year, during the period, say 10 years.

(4) Plot these values on the Y-axis of the graph and the year on the X-axis.

By this method any local highs in the values can be eliminated and a more rational picture will emerge.

Sometimes it is possible to detect, an overall tendency to increase or decrease, when a time series of this type is plotted on a graph. Such a tendency may comprise a linear curve, forming the least mean squares and satisfying the equation.

$$Y_c = a + bx \qquad \qquad ...(1)$$

where a and b are constant, X and Y are the variables. This has been discussed under statistical techniques.

**Estimating future costs and profits**

"Profit spread" is the best index of the future earnings of an industrial undertaking. When it becomes necessary to estimate the costs and future sale prices, of a product or commodity the first step is to prepare an outline or frame work for the cost sheet.

If the problem is related to a mine or factory in operation, and a record of the previous history is available, the problem is relatively simple. The methods of costing, adopted earlier can be employed with modifications, if necessary, in order to arrive at the probable figure for the future costs.

Some of the points, which are to be taken into account in such computations are that

(1) A change in the pattern or quantity of output will alter the cost of production.

(2) A change in the grade or quality will affect the cost per unit, e.g., per ton of metal, or per unit.

(3) A variation in the nature of the ore (or in the nature of raw material) may require a change in the method or scale of operations. An increase in hardness of the ore is likely to increase the cost of labour, steel consumed and explosives, while there is likely to be reduction in the cost of timbering.

(4) With increasing depth a change of mining method may be required, more supports, more investments on ventilation, drainage, lighting, haulage and winding etc. and these may cause a possible fall in output.

Since the redemption of the purchase price has been provided in the formula used in mine valuation, (as shown later) there is no need to allow for depreciation, depletion or deferred charges, in the cost of production. It is, however, necessary to include cost of additional plant machinery, development etc. which may be required in future

Taxes and royalties are based on output, as well as on the profits. Hence these should be estimated at the end.

The various taxes which will have to be provided for, will include (1) royalty on land, (2) royalty on raisings, (3) cess on road improvements, (4) cess on the labour creche, (5) cess on insurance, and (6) income tax.

A variation in the cost, of the mineral per unit, is called for, when (1) the grade, of the ore mined, is different to that obtained earlier, (2) there is an increase or decrease in the amount of gangue, (3) there is a possibility of contamination and consequent dilution and (4) the economic conditions indicate a fluctuation in prices.

In computing costs, the price offered, whether it is (a) pit mouth value or (b) it is F.O.B. or F.O.R., or CIF should invariably be taken into account.

(Note : F.O.B./F.O.R. is cost after loading on steamer/railway.
C.I.F. is cost including insurance and freight.)

If from the above data, (a) the reserves are known together with (b) the profit in Rs./ton etc. and (c) output in tons/annum, then the present value of the mine, can be worked out by allowing for the speculative rate of interest on investment, as well as for the additional overplus (interest on capital), as shown later on.

# PART XII


# ORE DRESSING OR BENEFICIATION

# ORE DRESSING OR BENEFICIATION

An ore is a mineral (solid) of economic importance or value. In determining the economics, the grade and reserves of the ore deposit are no doubt important factors, but these again depend on the nature and location of the deposit. To quote an example, it may be indicated, that what was considered to be poor grade of ore, a few decades earlier, may now fall within the economically workable range, due to advancements in technology, and improvements in communication, etc., besides shortage of supply, or an emergency. The glaring example is the low grade grunerite-magnetite rocks, of the U S.A. known as taconite ; which are now being worked, when the resources, of higher grade iron ores, have been depleted.

Ore dressing or beneficiation is the process by which an ore is improved in grade, so that the product can be used in the metallurgical industry. There are several methods adopted, for the purpose, each taking advantage of some peculiar physical, mechanical or chemical property, or a combination of properties, of the ore mineral. The process may range from simple sorting, by manual methods, as in the case of chrome ore, or mere washing as in the case of iron ore, to very complicated operations, involving several stages, especially in the case of polymetallic or complex ores, containing a mixture of several metallic sulphides e.g., lead-copper-zinc etc.

Some properties of minerals, commonly utilised and the possible beneficiation process/processes, are as follows :

1. Particle size, i.e., screening, e.g., beach sands, coal, etc.
2. Cleavage or fracture, i.e., whether the grains are flat or rounded jigs, e.g., felspars, manganese ore, etc.
3. Specific gravity—sink and float,—tables (Wilfley)—Cyclones—Spirals, e.g., coal, gold, tin ore, etc.
4. Magnetic susceptibility—magnetic separation,—e.g., magnetite, tungsten, etc.
5. Surface energy—flotation, e.g., complex sulphide ores of copper, zinc, lead, etc.
6. Fluorescence, as in the case of certain minerals, e g., uranium minerals and scheelite, etc.
7. Hardness—washing, e.g , clayey iron ores, kaolin.
8. Volatility—, e.g., sulphur, also crude antimony, zinc, etc.

9.    Solubility—e.g., certain salts from evaporites, e.g., potash salts

10.   Colour—e.g., coal, manganese ores, etc. in hand picking.

11.   Electric conductivity—in electrostatic separation, e.g., graphite, beach sands, etc.

12.   Interfacial energies, e.g., diamond, greased table.

13.   Amalgamation, e.g., gold and silver with mercury.

Ore dressing comprises the following processes :

1 Crushing,    2. Screening,    3. Finer crushing or grinding, i.e., Communition,   4. Concentration,   5. Storage.

## CRUSHING

### Crushers

| | |
|---|---|
| 1. Jaw Crushers | (a) Blake |
| | (b) Dodge |
| 2. Gyratory Crushers | |
| 3. Cone Crushers | |
| 4. Sledging Rolls | (a) Single |
| | (b) Giant |
| 5. Hammer Mills | |
| 6. Stamps | |
| 7. Spring Rolls | |
| 8. Manual Methods of breaking. | (a) Sledging |
| | (b) Spalling |
| | (c) Cobbing |
| | (d) Heating and quenching |
| | (e) Blasting |

### 1. Jaw Crushers

(a)   The Blake Crusher (Figure 12.1) : The essential parts comprise :

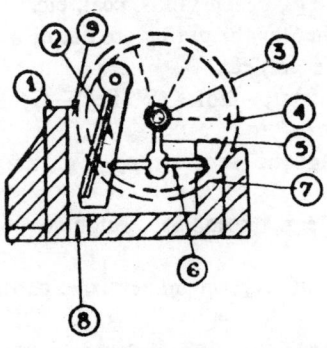

Fig. 12.1

(1) Fixed Jaw (Mn Steel)

(2) Jaws (Swing jaw) Mn Steel

(3) Eccentric

(4) Flywheel

(5) Pitman

(6) Toggle

(7) Fixed support for toggle

(8) Exit

(9) Feed

The fly wheel is actuated by a prime mover, i.e., any type of motor or engine, and as the axle rotates it moves the eccentric. The pitman attached to the eccentric also moves up and down, carrying the toggle with it. Since one end of the toggle is fixed, the moveable jaw, attached to the other end, moves back and forth, from the fixed jaw. Thus any material lodged in the throat of the *crusher* is broken up.

(*b*) The Dodge type of crusher is also similar but the swing jaw is pivotted at the bottom (Figure 12.2).

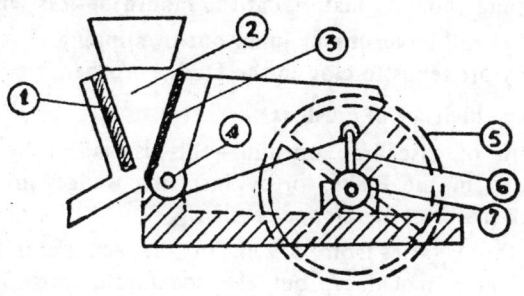

Fig. 12.2

(1) Fixed jaw (Mn steel)

(2) Mouth of Crusher

(3) Moveable jaw (Mn steel)

(4) Axle for jaw

(5) Fly wheel

(6) Pitman

(7) Eccentric

When the fly wheel (5) rotates, the pitman (6) moves up and down, on the eccentric (7). This causes the lever arm (8), attached to the moveable jaw (3), to move up and down, while the upper portion of the jaw, which is pivotted at its lower end, swings forward and backward. The feed, dropped into the throat, is delivered at X. Any material, large enough to be lodged in the throat of the crusher, is reduced in size.

Jaw plates, for crushers, are made of chilled iron, white iron, high carbon cast steel, manganese and chrome steel. The fly wheel supplies the inertia for smooth running during breaking.

Blake versus Dodge crushers—advantages and disadvantages :

1. The Blake crusher delivers a product in which the sizes are variable ; as the lower portion of the jaw opens and closes. The

Dodge crusher delivers a more uniform size of product as the opening at lower end is fixed.

2. In the Dodge crusher, breaking is done at the throat, while in the Blake crusher breaking is done near the discharge end. Thus the mechanical advantage is in favour of the Blake crusher. Thus the power consumption is more in the case of the Dodge crusher.

3. The capacity (tons/hour) is less for the Dodge crusher as compared to the Blake of same size.

4. Nipping (holding material at the mouth) is less effective in the Dodge and larger lumps jump out by slipping.

5. Sticky ore tends to clog in the Dodge crusher.

**Factors affecting efficiency of crushers**

(a) Width of discharge opening—this is varied by adjusting the toggles in the Blake, or by inserting wedges in the Mn steel cheek plates of the Dodge crusher.

(b) The throw varies from 3/8 in. in the small crushers to 1 in. in the larger machines, but the maximum throw is about three times: Brittle rocks, like quartzite, and granite, require minimum throw, whereas tough rocks like, trap, limestone, and rocks which resist compression like shale, etc., require larger throw.

(c) Speed has no marked effect on the efficiency, but at higher speeds, the efficiency drops.

(d) Size of feed depends on the gape, nip angle, and the angle of the mouth opening which is about 24°. Taking these into consideration, the largest size of the individual piece of rock, should not exceed 80 to 90% of the gape (discharge opening). Larger fragments tend to clog the mouth.

If undersize particles are screened, and then the oversize is fed in, then less clogging occurs, and consequently there is an increase in the amount of oversize crushed.

(e) Reduction ratio is the ratio of the screen aperture of the feed, to the screen aperture of the discharge. This has been defined variously e.g., limiting reduction ratio, apparent reduction ratio, working reduction ratio, mean reduction ratio, and 80% reduction ratio.

The 80% reduction ratio, is the ratio of that aperture which passes 80% of the feed, to that of an aperture which passes 80% of the product.

($f$) Capacity is affected by volume of crushed material per hour, as well as by the density of the material crushed. It is also affected by the area of the discharge opening, and the rate of travel of the material fed. These conditions again depend on the moisture content, throw, strokes per minute, size reduction, nip angle etc.

($g$) Reduction ton is a factor which involves (i) the tonnage crushed (per hour) and (ii) the reduction ratio, e.g.,

Tonnage reduction = Tonnage (per hr.) × 80% reduction ratio.

## 2. Gyratory Crusher

The working of this machine may be described briefly as follows :

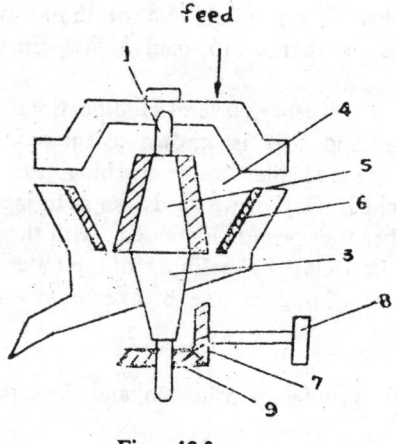

feed

Fig. 12.3

(1) is the bearing suspension for shaft. When the pulley (8) is rotated, by a belt, it actuates the bevel gear (7), and the bevel gear (9). Thus the shaft (3) is rotated, and the main conical jaw, or breaking head (5), is also rotated. ·The crushing chamber (4) is in the form of a inverted hollow truncated cone, the surface of which is lined with Mn steel plate, known as mantle (6). The breaking head (5) is also lined with Mn steel plates. The sleve (12), lined with babbit metal or bronze, imparts a small amount of eccentricity, to the breaking head, which acquires a hammer action in consequence. The action of the crusher may be likened to that of a continuous wedge.

There are three types of gyratory crushers : (1) Suspended spindle type which is a popular machine, (2) the supported spindle type which has been outmoded, and (3) is the fixed spindle type or the Telsmith breaker.

Comparison between jaw and gyratory crushers.

1. Cost of the gyratory crusher is only 29 to 72% of that of the jaw crusher of the same capacity.

2. Cost of installation of the gyratory machine is less, as foundations, for the jaw crusher, should be very substantial to stand against the severe vibrations.

3. The housing, for the gyratory machines, costs more as it occupies more space.

4. Feed arrangements for the gyratory crusher are.simpler than for
the jaw crusher.

5. The gyratory machines are more efficient than the jaw machines
and more so, when working on full load.

6. Maintenance costs are more for the gyratory machines.

7. Since the jaw crusher had a larger latitude, in the adjustment
of the throw, it functions better when crushing soft rocks. Further the jaw has a larger latitude for setting the size of the
product.

8. The gyratory crusher can work more efficiently where the material tends to break in slabs. The general rule or thumb rule is
that if *tonnage per hour* is less than 0.115, then a jaw crusher is
to be preferred.

In a later version of the gyratory crusher, several modifications have
been introduced, for finer crushing, and this is known as the reduction
crusher. Since the discharge opening is smaller in fine crushing, the convergent crushing zone is flared (belled out) downwards, so as to increase
the length of the discharge. Further the speed is increased, with the consequent increase in the rate of impacts. The lining of the outer shell
(concave) is curved, while the mantle (lining of the breaker) may also be
curved. This prevents choking.

### 3. Cone Crusher

This is very much like a coffee grinder in principle, and the essential
parts are :

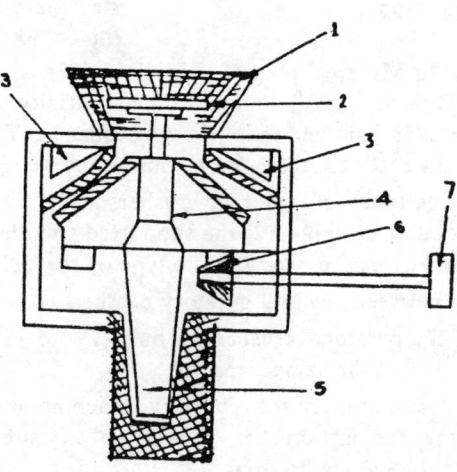

Fig.  12.4

(1) Feed hopper. (2) Feed distributor tray. (3) Fixed jaw. (4) Breaking head (eccentrically mounted) with check plates. (5) Spindle axle (eccentric). (6) Bevel gear drive for the breaking head. (7) Pulley drive.

In the cone type crusher, the breaking head (conical) is squat or flat, and the ratio of the height to diameter is 1/3. The throw of the head is also greater, i.e., a/b = 1/9 or 1/5.

It is also stronger mechanically. With these modifications, the volume of the fine crushing zone, is larger as compared to the volume of coarse crushing zone. Movement of the material is quicker, and due to the high speed of gyration, the material is acted upon rapidly, and the size of the discharge is governed by the finer setting.

This machine has also a modified version in which the head is a segment of a sphere.

The maximum size of a material, fed into the cone crushers, should be such that it easily gets into the crushing zone. Material smaller in size, than that discharged by the crusher, if fed into the machine, tends to choke up in the fine zone. Excessively·large sized material may cause "bridging", and prevent crushing.

The product, from a cone crusher, will pass through a screen 1.5 or 3 times the setting of the machine. In general the percentage of particles, which is coarser than the setting, varies from 14 to 70 and increases with the coarser settings, but on the average the percentage of product, that will pass through a mesh, equal to half the setting aperture, varies between 35 to 45.

Reduction ratio in the cone crushers is large, due to the relatively larger crushing area. The annular (concentric) corrugation of the crushing face, near the opening, in effect decreases the nip, to prevent slipping. The reduction ratios are of the order of 1 : 7 or 1 : 6 the common figures being 1 : 3 or 1 : 5.

Giant rolls (known as slugging rolls) comprise two rollers, 3 ft to 6 ft in diameter, which have knobs. They rotate inwards, i.e., towards each other and the material is broken down by the hammer action of the knobs or slugs. The size of discharge varies from 4 in. to 8 ins. The action of the slow speed machines (peripheral speed being 150 ft. per min. to 350 ft per min.) is mainly nipping and compression ; whereas the high speed machines depend largely on sledge hammer action.

## 4. Hammer Mill or Pulverator

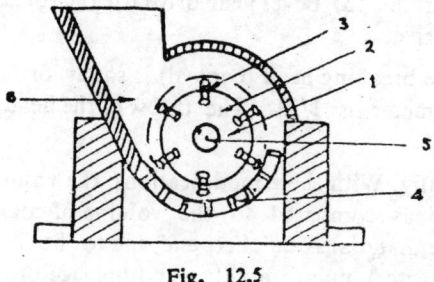

Fig. 12.5

The machine comprises a rotor (1), which consists of two circular discs (2), to which the hammers (3) are pivoted. The rotor is rotated about the axle (5). A grate or grizzly (4) is placed at the lowest position and the discharge passes through this. The hammers swing out, under the action of centrifugal forces, and the material, lying on the breaker plate (6), is struck and it disintegrates.

## 5. Stamp

Stamps (Fig. 12.5) are in principle large size pestle and mortar. The pestle is actuated mechanically—either by steam or compressed air directly. When stamp is lifted by a cam and drops down by gravity, it is called a gravity stamp.

In the gravity stamp the cam (3), on rotation, lifts the tappet (2), and this in turn lifts the pestle (1). While lifting, a rotary motion is also imparted, to the pestle, by the cam. The mortar is provided with screen (6), on one side so that only the undersize material flows out, while the coarser fraction is retained for further breaking and crushing.

Stamps are not generally used, except in small gold milling units, where amalgamation is also carried out in the stamps.

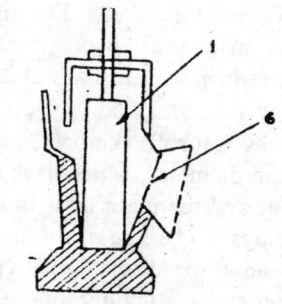

Fig. 12.6

## 6. Spring Rolls

The roll (Fig. 12.7), essentially comprises two rollers (1) and (2) [Steel cased (X)] mounted on a common frame (7). The roller (2) is rigidly fixed by bolts, while the roller (1) is mounted on a sliding block, which moves on the frame (7). The moveable roll (1), is kept in contact with roll (2) by the pull of a tension bar (3), provided with springs.

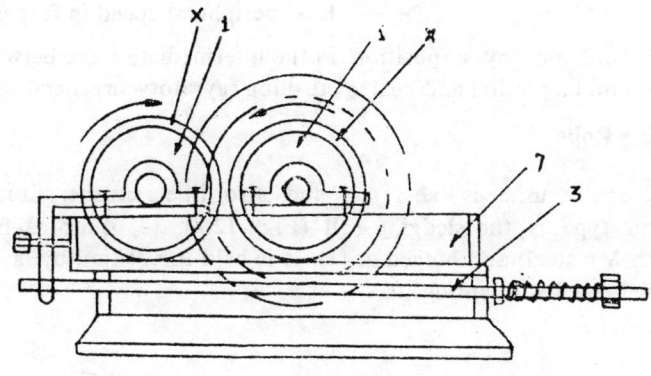

Fig. 12.7

Both the rollers are driven by pulleys (not shown), but the pulley for roller (1) is smaller than that of roller (2) (fixed roller). Thus, the speed, of the moveable roller, is greater than that of the fixed roller. A feed hopper is attached at the median position between the rollers.

The angle of nip generally is about 25°, and the diameter of rolls required, may be got from the formula

$$t = d + D/40$$

where     $t$ = the maximum thickness of feed,

$d$ = the setting of the rollers, i.e., space between them,

$D$ = diameter of rollers.

The reduction ratio varies from 1 to 2, in general, but it is slightly more for heavy duty rolls.

Capacity of a roll, $T_e = \dfrac{R \times D \times W \times d \times \text{Sp. gr.}}{293}$

where     $R$ = r.p.m., $D$ = diameter, $W$ = width of the roll,

$d$ = setting in inches,  Sp. gr. = Specific gravity.

This calculation of tonnage ($T_e$) is true only in the ideal case, where a solid flat sheet of crushed rock is being delivered, by the rolls. This, however, is normally not the case, and thus a factor ($r$), known as the ribbon factor is introduced.  Ribbon factor ($r$) is the ratio between the actual volume of rock crushed, to the theoretical volume of material passing through the rolls, i.e.. $L \times d \times W$ where $L$ is the speed of of the roller in ft. per min.

$$r = \frac{2900\ T_a}{L \times d \times W}$$    where $T_a$ = actual hourly tonnage,

$d$ = setting in inches,

$W$ = width of roll.

$L$ = peripheral speed in ft. per. min.

The rolls occupy a position in the intermediate stage between fine crushing (tumbling mills) and coarse crushing (gyratory crushers).

## 7. Sledging Rolls

Single roll crusher is like a roll and also like a crusher, in its functions. This type is the sledging roll (Fig. 12.8). (2) is a breaking plate (lined with Mn steel) and hinged at (1). It is held in position by a tension bolt (5) tightened with springs (6).

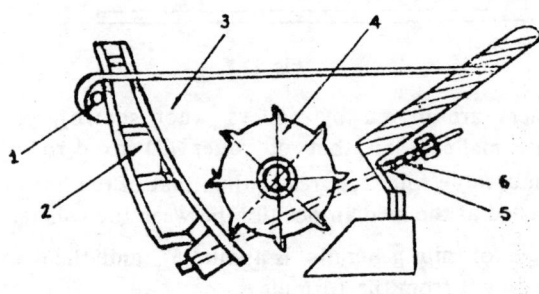

Fig. 12.8

A roller (4), provided with hard teeth, is rotated by gearing through pulley and belt arrangement. Material, which is dropped into the mouth (3), is pulverised as it gets between the breaking plate (2) and the roller (4). The roller speed is 200 to 300 ft. per min.

This type of crusher is used for fairly soft rocks, e.g., limestones, shale, clays, and the performance of these machines is superior in sticky materials of medium hardness. The capacity varies from 50 to 1500 tons per hour. On an average, the reduction ratio is about 8 to 10. For a definite setting, 30-40% passes through a screen with an aperture equal to half the setting gap.

## 8. Manual Crushing

*Breaking* or *crushing* is also carried out manually.

1.  Sledging is the process of breaking rocks with sledge hammers which are 10-16 lbs. in weight.

2. Spalling is reducing 6-8 in. lumps to 2 or 3 in. sizes with the help of 2 to 3 lb. hammers.

3. Cobbing is further reduction of sizes from 2 to 3 in to less.

4. Explosives are also used in breaking up large boulders.

5. Pile driver type of breakers were used earlier, but these are not popular now. In pile driver type of breakers, a heavy solid iron cylinder was lifted up, by a pulley block, and dropped suddenly on the rock. Shattering was produced in this manner.

6. Heating and quenching is one of the earliest methods, of breaking rock, used by man, for mining. This method was also practised, by the miners, in India. This method is reported to be efficient in liberating and breaking quartz of high purity.

## GRINDING

*Characteristics* : While crushing yields a relatively coarse product, grinding produces finer material. For this purpose the parts of the machine which are used for grinding come into contact with each other, while crusher jaws and rolls do not touch each other. Another feature of grinding is that the operation is generally continuous, while crushing is commonly intermittent.

The purpose of crushing is, generally, to reduce the size, of the run of mine product, but grinding is essential for liberating valuable minerals. from the gangue. Grinding is also necessary in order to liberate minerals, where the minerals occur as intergrowths, e.g., galena and chalcopyrite, so that these ores may be rendered amenable to ore dressing processes, e.g., flotation. In the case of non-metallics, e.g., felspar, sillimanite, limestone, coal etc. grinding is done to make them marketable. The mills for non-metallic minerals are generally designed for dry grinding. Grinding is also needed where hydrometallurgical process is practiced, e.g., leaching of low grade copper ores, nickel ore etc.

In grinding, impact and abrasion are responsible for communition. Hence larger the number of impacts, and longer the period of abrasion, the finer is the product. Fortunately, both factors are dependent on the period or time taken in the breaking zones, and this is largely controlled by the resistance of the charge, and also by the size of the screens used, at the discharge end.

The important types of grinding machines are :

(A) *Tumbling mills* : These comprise a variety of machines, e.g.,

(1) Rod mills in which steel rods are used for crushing.
(2) Ball mills in which steel balls are used for crushing. Porcelain balls are also used occasionally.
(3) Pebble mills in which flint pebble are used for crushing.
(4) Autogenous mills in which the ore itself is used as the grinding media. These are similar to rod and ball mills in operation.

The principle of operation is (1) abrasion due to grinding between the particles and also between the rolling of the grinding elements—caused by cascading or cataracting, i.e., balls, rods etc. are carried up by the inner surface of the rotating drum or cylinder, and then dropped in a continuous cataract. In the zone PQRS (Fig. 12.9) maximum crushing occurs. The arrow, in the figure, indicates the direction of rotation. (1) is the drum and (2) the course taken by the pebbles.

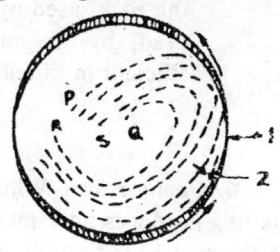

Fig 12.9

It has been found that the angle of repose, of the load (i.e., balls or rods), increases with the speed of rotation so that, at a certain speed of rotation, a truely parabolic cascade is generated, with a turbulent core C. When this stage is attained, abrasion is the chief factor for grinding. The theoretical speed, for the balls, to attain the maximum angle of repose, before the parabolic drop, is known as the critical speed or cataracting speed, and it is given by the equation, $S = \dfrac{54\ 19}{\sqrt{R}}$

where S = critical speed in r.p.m.

 R = radius of the mill, where the outer layer of balls start cascading.

The best operating speed, S (r.p.m.) = 57 to 40 log D, where D = internal diameter of mill in ft.

The tumbling mills have arrangements for continuous automatic feeding. Of the more common type of automatic feeders is (1) the scoop feeder (Fig. 12.10) which comprises a flat helical plate enclosed by two nearly semicircular discs. One of the discs is provided with a central hole with a hopper or funnel which fits on to the feed end of the mill (trunnion).

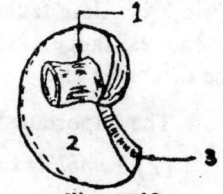

Fig. 12.10

(1)  is the spool which fits on to the trunnion of the mill.

(2)  the semicircular discs forming the side.

(3)  the digging lip.

The second form of feeder (Fig. 12.11) comprises a three-way pipe which fits on to the trunnion of the mill. It carries the feed pipe, for ore and also for balls, which are fed, as and when required. The third branch (3) is used for feeding in the required quantity of water.

The type of liners, used in these mills, are also important. It is known that the slip increases with decreases in tumbling load, speed, pulp density and hardness of the ore.

Fig. 12.11

Thus the influence of the roughness, of the liner is greater with increasing capacity. The type of liner used will thus depend on the size of the feed. Liners for rough feeds have prominent ridges, running parallel to the axis of the mill, while liners for fine feeds are smooth or gently roughened. Liners are made of manganese steel, silex or hard flint, or rubber.

Tumbling mills whether (1) Rod mills (2) Ball Mills, may either be (a) overflow type or (b) open-end type.

The essential parts of the rod mill are shown in the Fig. 11.12.

(1) Feeder (2) Inspection door or manholes (3) Body (4) Overflow (5) Trunnion or open hollow axle.

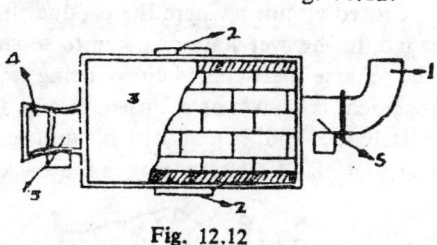

The rod mills are generally cylindrical in shape. The overflow type is supported on trunnions at both the feeder

Fig. 12.12

end and the overflow end. In the open end type, the discharge opening is closed by a heavy, tight fitting, door to prevent spillage. The door is supported independently on posts fixed to the foundations (Fig. 12.13).

(1) Discharge end, (2) roller support for discharge end, (3) door for discharge end (note the independent fit), (4) Trunnion support.

The usual size of feed is from $\frac{1}{4}''$ and this is not used in feeds unless a granular product is required, e.g., for gravity or wet magnetic separation or for fine aggregate in a concrete mixture.

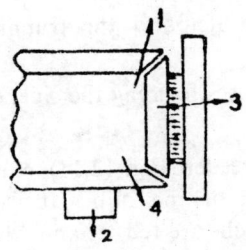

Fig. 12.13

The reduction ratios vary from 2 to 8, normally. Rod mills can be used where the final product required is between 35-mesh to 48-mesh; while ball mills are necessary for a finer product, e.g., 65 mesh and finer.

The soft ores, especially wet and sticky ores, rolls and short head cone crushers are less effective where the reduction of sizes is from $\frac{3}{4}''$ or $1\frac{1}{2}''$ (feed) to a 10-mesh (product). Rod mills operate with better efficiency in such cases.

Rolls and short-head cones are preferred for hard ores. A ball mill can operate well where a product, less than 48 mesh, is desired, irrespective of whether the material is hard or soft.

Ball mills are either cylindrical, cylindroconical or conical in shape. In the conical mill the narrow end is the discharge end.

Closed circuit : where the product from the tumbling mill is either screened, in the wet state or taken to separators or classifiers, for collecting the coarse particles, before being fed into the ball mill again, the arrangement is known as a closed circuit (Fig. 12.14). The material from the discharge end (2) of a ball mill is taken to classifier (1) and the coarser material after screening is fed back at (4) into the mill.

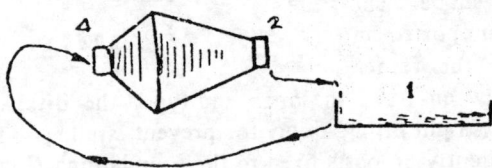

Fig. 12.14

Closed circuit improves the performance of the mill due to (1) reduction in the mean size of the feed, (2) increase in the quantity of a product which approaches the finished product rapidly, (3) removal of the finished sizes, (4) rapid movement of the material, (5) more uniform nature of the feed due to which the cataracting affects, a larger number of particles (6) increase in the quantity of material caught up in the "Vortex" of the cascade, which in effect amounts to better grinding.

The grindability of an ore is its amenability to grinding and this affects the capacity of a mill. Grindability depends on (a) hardness, (b) toughness, (c) natural grain sizes, and (d) mineral ratio or relative proportions of minerals present.

Cost of wet grinding depends on the (1) size of feed, (2) size of product, (3) grindability of the ore, (4) size of the installation and (5) the efficiency of the plant. The expenditure also depends on cost of power, steel and labour, of which power costs are about 40-60%, labour is 5 to 10%. Grinding to 100 mesh to 200 mesh costs nearly four times that of grinding for the rougher concentrates, i.e., 40 mesh.

(B) *Chillean mill* : This is very similar to the "surki" mill or "bail chakki" in principle, but it is a wet grinding process. It consists of a robust steel or cast iron tray with vertical sides (Fig. 12.15). (1) Inside the tray on the lower part of the side. is a screen. (4) There are three rollers. (3) mounted 120° apart, on a central shaft. (2) The feed is put into the mill, near the centre, together with sufficient water. As the rollers rotate the material is crushed. There is a scraper, which follows the rollers, for raking up the sediment. As the rollers move, the fines are pushed forward, in suspension.

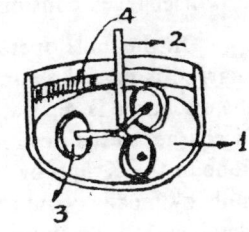

Fig. 12.15

After getting through the screen (4), the fines enter a launder or discharge channel.

(C) *The Huntington mill* : Fig. 12.16 consists of four rollers (1) which are freely suspended from a cross arm (2), attached to a vertical shaft (3). Just above the base plate is fixed a die ring (6), made of hard manganese steel, and above this is a circular screen (8). When the shaft (3) is rotated, at a high speed, by the gears (7), the suspended rollers fly apart, due to centrifugal force ; and run against the die ring (6). The material which is fed in through a hopper (9), from the top, is ground by the action of the rollers.

Choice of the method of grinding whether dry or wet depends (1) on the physical characteristics of the feed, and the use for which the product is

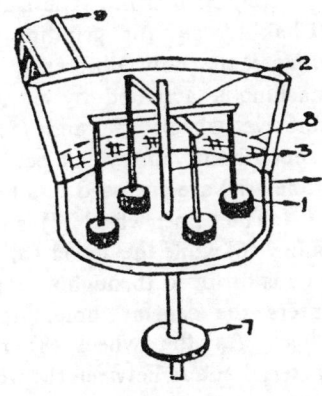

Fig. 12.16

required, (2) on the behaviour of the feed and its influence on the mill, i.e., whether it is abrasive, corrosive, stickly etc., (3) on the shape, particle size and distribution desired, (4) on cost economics, e.g., capital, expenditure, on labour, maintenance etc , (5) on the climate. (6) on the availability of water, (7) on the safety factors, and certain other factors e.g., dust, noise, vibration, speed, etc.

The efficiency of dry grinding is more or less on a par with wet grinding, but certain materials are required to be ground dry according to market requirements, e.g., coal, for being used as pulverised fuel, in boilers. Cement mixes in the dry process of cement manufacture.

Dry grinding is generally done in three ways : (a) batch or intermittent, (b) open circuit and (c) closed circuit.

In the open circuit mills, e.g., buhr and attrition mills, pans, and tube mills, the rate of feed is so regulated that the finished product is also discharged continuously, at a fixed rate.

In the mills operated on the closed circuit system, the material discharged is passed again and again into the mill. The coarse fraction of the discharge is separated and fed back until further grinding reduces the discharge. to the desired sizes. Internal separation of coarse material is done in the Krupp or Griffin mills, by means of screens, and in the roller mills by means of internal classification. In the ball and pebble mills, the classification is external.

Products of grinding may be described as (1) graded, i.e., all sizes occur (2) fine. (3) abraded, (4) granular, (5) cleaned, (6) disintegrated, (7) sharp and (8) flaky.

The principle used in grinding may be either (1) attrition (2) impact (3) jetting or ejection.

(D) *Buhr Mill* : The Buhr mill (Fig. 12.17) is similar to the ordinary "Chakki" used for grinding wheat in the villages, the only difference is that this machine is actuated by a motor. Two abrasive wheels (7) and (8), made of French buhr, emery, or pebble grit (mill stone grit) are enclosed in a housing (9).

The upper wheel (7) is rotated by a pulley (6) while the lower (8) is fixed. The feed is dropped through a hopper (1) and enters the central hole, in the upper wheel. As the wheel (8) rotates, the material enters between the wheels, and is

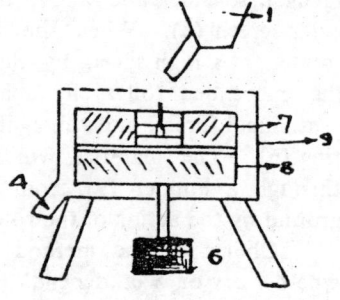

**Fig. 12.17**

crushed and discharged, all round the wheels, where it is drawn out by the opening at (4).

The Buhr mill can also be of the vertical type and a feed screw, attached to the shaft, carries the feed in between the wheels. These Buhr mills are used for grinding soft materials, e.g., clay, chalk, talc, coal, ochre, limestone and gypsum. The feed ranges between $\frac{1}{4}$ in. and the product is 20 mesh to 200 mesh. For grinding to less than 80 mesh, screens, air classifiers are employed.

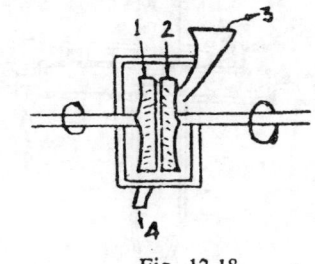

Fig. 12.18

In the attrition mill (Fig. 12.18), the feed is placed in a hopper (3). It enters between two fast rotating wheels (1) and (2). The two wheels move in opposite directions and material is thus ground and discharged at (4).

The impact mills (Fig. 12.19) are either of the hammer type or jet type. In the hammer mills the material is dropped through a hopper

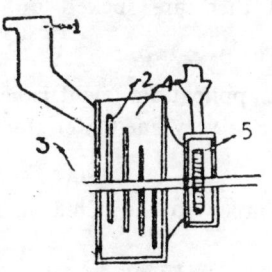

Fig. 12.19

(1), into a flat circular chamber (4). Rods or bars, radially attached (2), to a shaft (3), are placed inside the chamber and are rotated by a motor. The material is struck by the radial hammers (or bars) and is pulverised. As the correct size of the product is attained, a fan (5) is used to suck out the fines, while the coarser material is subjected to further grinding.

The cage mill (Fig. 12.20) consists of two cages (1) and (2). Each of the cage is made of iron bars fixed vertically to the periphery of a circular flat iron disc. One cage is smaller than the other, and fitted into the larger one. Both the cages are enclosed in a steel box. In this position, the two cages are rotated by a motor, but in opposite directions. When the feed is dropped in from the top, the material is disintegrated. This is also called a disintegrator or "Surki" mill.

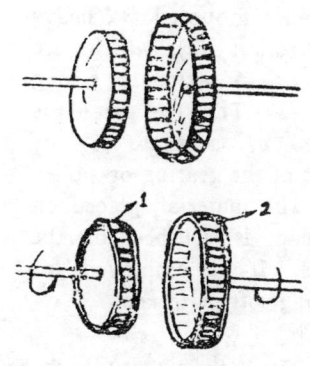

Fig. 12.20

(E) *Jet Pulveriser* (Microniser) (Fig. 12.21) : The feed 1/8ᵏ size enters through orifices (2), from a circular pipe (1), into a grinding chamber (6), which is a flat circular box. Steam or gas at high pressure, e.g., 100 lb/sq. in. or compressed air or super heated steam at 500 lbs/sq. in. is conveyed by a circular ring pipe (3) and led into the grinding chamber by ports (4).

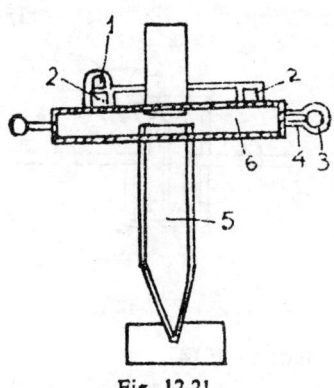

Due to sudden expansion, the gas under pressure, acquires a very high velocity, and turbulence is caused, in the material fed into the grinding chamber, which leads to the disintegration of the particles. The grinding chamber is lined with rubber to minimise wear. When grinding has proceeded sufficiently, the

Fig. 12.21

fines accumulate in the receiver (5), and then are sucked up by the exhaust pipe (7).

The microniser has a limited field of application ; as it is generally more costly hence it is best suited where a very fine material less than 325 mesh is required.

(F) *The Pan mill* (Fig. 12.22) is similar to the Chilean mill in operation. It consists of a perforated circular pan made of thick steel sheet (1), on which are mounted two heavy iron rollers (2), which are fixed on a axle, to strong posts (3). The steel pan (1) is rotated by a shaft (6), by means of the gearing or pulley (4). The material, placed on the pan, is crushed by the rollers. It is employed for grinding soft materials.

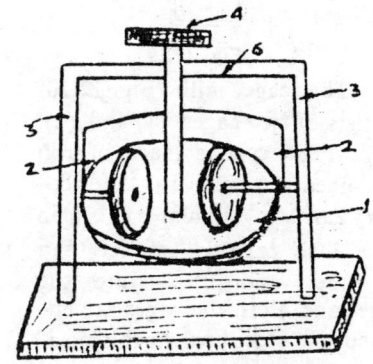

Fig. 12.22

(G) *Roller mills* (Fig. 12.23) : In one type, three free rollers (1), (2) and (3) are spaced 120° apart, within a steel ring (4). These are held in position by springs at the top (5). One of the rollers (1) is driven by gears or belt drive. This causes the ring (4) to rotate and so also the rollers (2) and (3). If the material is fed into the hopper (6), at the top near roller (1), it enters the ring, and it is subjected to grinding, by the rollers. The fine material is discharged at (7). An air classifier is generally placed in the circuit or sometimes a screen, from which the oversize is returned to the mill.

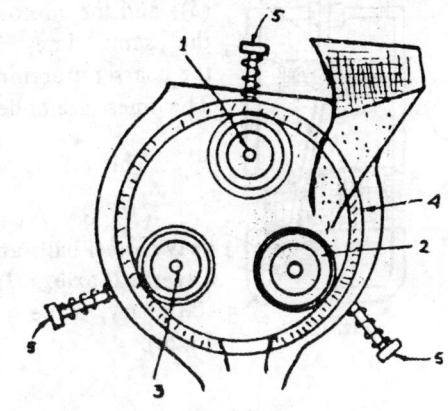

Fig. 12.23

In another design (Fig. 12.24), the ring (3) rotates, while the rolls 1 and 2 are stationary, and the capacity is 8-10 tons/hr. for barite, sizes of product varying between 1 to 40 mesh, and for marble the capacity is 3 to 4 tons/hr. for a 100 mesh product.

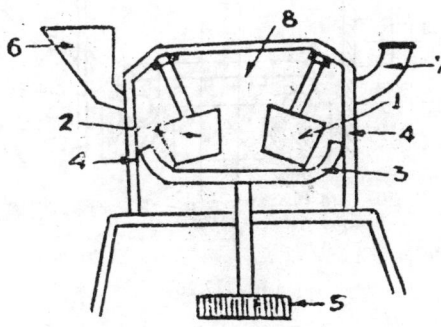

Fig. 12.24

In the Williams mill (Fig. 12.25) a single roll (1) suspended from a vertical shaft (2), is attached to the hub of a universal joint (3), at the lower side of a pulley (4). When the pulley (4), which encloses the universal joint (3), is rotated, by belt, the roll swings out, under the action of centrifugal force, and moves against the die or grinding ring (5). The fines pass, through the circular screen (6), placed above the grinding ring.

The Raymond bowl mill acts on the principle similar to the Chillean. mill. The only difference that the rolls (1) and (2) are conical and the bowl (3) is also provided with a grinding ring (4) with sloping sides. The grinding ring is rotated by gearing (5), and the motor also actuates a suction fan at the same time. The material is classified, and the coarser fraction is returned for further grinding. The fines are collected in a cyclone.

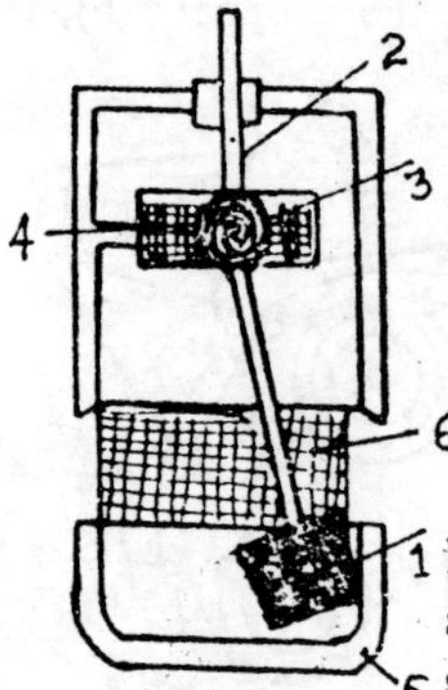

This mill is popular for grinding coal.

(*H*) *Ball bearing pulverisers* : In these machines heavy steel balls are set in circular races, around a vertical axle. In some machines the races are stationary, while in others the races rotate (Fig. 12.26).

**Fig. 12.25**

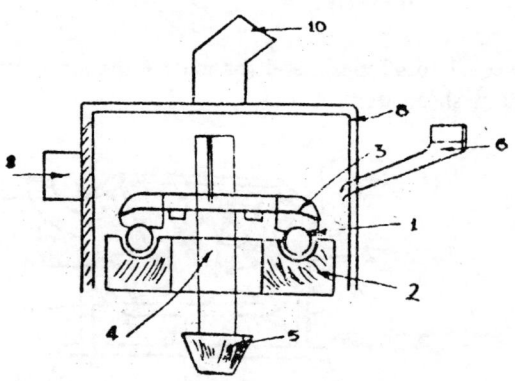

Fig. 12.26

The four heavy steel balls (1) are housed in a heavy concave ring (2). The races (3) are known as pushers and are attached by a vertical shaft (4) to the driving gears (5). When the gear is moved, the pushers rotate the balls, which start moving in the ring. The entire assembly is housed in a steel case (8). The feed is let in by the opening (6), in the shell. A fan blows in air (hot air in some cases) through the opening at (9). Vanes, attached to the pushers, help in stirring a cloud of fines, which is exhausted through the opening (10).

Modifications of this type of mill, incorporate two or even three tiers of balls, but in each case, a classifier is included which forms an integral part of the grinding circuit.

The following table summarises the salient features and characteristics of the various types of mills, and related products.

| Particulars | Type 1 | Type 2 | Type 3 |
|---|---|---|---|
| 1 | 2 | 3 | 4 |
| 1. Mills (name) | Tumbling, e.g., ball, pebble tube, rod, batch, pan. | Rolls, ring roll, ball bearing, buhr. | Hammer, disc, cage, jet. |
| 2. Relative Speed (of grinding crushing) | Slow | Medium | Fast |
| 3. Grinding action (nature) | Attrition, compression, impact. | Compression, attrition. | Impact, Shearing attrition. |
| 4. Material suitable for grinding in each type of mill. | Hard, abrasive, friable. | Medium hard, non-abrasive, friable. | Soft and friable, tough non-abrasive, fibrous. |
| 5. Advantages | Low maintenance, efficiency increases with use, minimum contamination, easy control, little attention required, uniform character of product. | Low power consumption, less floor space. | Low Capital investment, low power consumption, less floor space, quality of product is also poor. |
| 6. Disadvantages | High initial cost, high power consumption, generally larger floor space. | High maintenance on hard materials, efficiency decreases with use, and also the quality of the product. | High or excessive maintenance costs, on hard materials efficiency falls with use, time is wasted. |
| 7. Best suited for grinding | Anthracite, asbestos, barite, bauxite, coal, glass, pyrite, talc, etc. | Asbestos, barite, clay, fuller's earth, gypsum, limestone, magnesite, phosphate, etc. | Asbestos, coal graphite, limestone sulphur, talc etc. |

## SIZING

*Sizing of the mill products and the type of operations* :

1. In batch operation a predetermined quantity of material is charged into the mill, and after a predetermined interval of operation, the mill is stopped, and the product, of the desired specification, is recovered.

2. In the open circuit operation, various types of mills are employ-ed, except tumbling mills, and the product is variable, in quality.

3. Compartment mills are those in which there are more than one compartment (1, 2 or 3), and these may be considered to comprise a series of mills. These can take a coarse feed and have a relatively high reduction ratio.

4. Compartment mills are also designed with intervening screens between each compartment. These have a higher efficiency than normal compartment mills.

5. Sizing is affected by passing a current of air through the grinding zone, as in the case of hammer mills, roller mills, ball bearing mills, and tumbling mills, which are used for grinding coal.

6. In the roller, hammer, ball bearing and ring-roll mills, separa-tion is affected by circulating a current of air, through the mill housing itself. Drying, of the product, is achieved by circulating hot air.

7. In some cases, of closed circuit, the product is lifted by a bucket elevator and screened. The coarser fraction is returned to the grinding circuit. Screen is sometimes replaced by an air separator.

8. In a compartment mill, the first compartment is placed in a closed circuit, with the addition of an elevator and screen. As the product is discharged, from this compartment, it is elevated and screened. Sometimes the second compartment is placed in the closed circuit instead of the first.

9. If both compartments, of a compartment mill, are placed in closed circuit, then the sizing is very closely controlled.

10. When external sizing is done, in conjunction with a compart-ment mill, then the oversize is returned to the discharge end, of the first compartment. The finished product is got from the second compartment. An external classifier (air) returns any coarse particles to the respective compartments, for further grinding.

11. Where a fine product is required from a fairly coarse feed, two-circuit grinding is necessary. The first is closed by a screen and elevator, and the second includes both internal and external classification.

12. Where tolerance, in sizing, is fine, then three-stage, grinding in separate units, with individual external classifiers, is employed.

13. A four stage mill is also employed in certain types of grindngi circuit.

## Sizing by Screening

The various size groups, in a crushed or ground material, are known as grades. A grade is a size group, which passes through a particular aperture or limiting screen and is retained by another aperture or retaining screen.

A short range product comprises particles in which the sizes are very near to each other, without much difference.

A long range product comprises particles of various sizes.

Seive Ratio is the ratio of the aperture of screen, belonging to a particular standard (e.g., ASTM, Tyler, IMM, BSS or USS) to the aperture of the next smaller mesh screen.

Percentage opening of a particular screen is the total area occupied by the openings, in a screen, to the area of the screen, expressed as a percentage.

Purposes of screening are (1) to remove the coarser fractions, (2) to remove the finer material from the grinding circuit, (3) to obtain commercially marketable sizes of material, e.g., sands, rock chips, (4) to obtain suitable sizes for further beneficiation, e.g., jigging and (5) to separate different minerals, which occur together, but each mineral being characterised by a particular grain size. In beach placers, ilmenite, monazite garnet, etc., can be separated as well as quartz, and these can sometimes be separated by sizing.

Requirements and conditions for proper screening are : (1) All particles should be brought to the screen opening, oriented in such a manner, and moved at such a rate, that the undersize particles will pass through freely unhampered, without rebounding, from the edges of the screen opening. (2) Ideally, every undersize particle should be at standstill and centrally placed, in respect of the aperture. (3) Larger tonnages can be obtained if the particles, of the material, move over the screen. (4) Even though screen, made of extremely fine wire or metal they are ideal for efficiency, in practice, these cannot be emploped as they are mechanically too weak.

During the course of screening, conditions are far from ideal, as (1) crowding of particles takes place, and (2) the particle may lie flat or present the surface of maximum area, when placed on the screen. Thus, many undersize particles are prevented from passing through the screen, for a certain period.

The efficiency of a screen is dependent on the factors mentioned above, the probability, of a particle passing through, may be presented, mathematically in the following form :

$$P=[(n-1)/n]^2$$

$1-\dfrac{1}{n}=$ length of a side of square mesh opening.

$1/n=$ diameter of the spherical particle.

(where n is any number larger than 1)

The probability that the particle will tough the side of the screen aperture is $(1-P)$. However, relative efficiency is employed, in practice. It is judged by the amount of oversize to the products, passing through, in both cases, i.e., in the test screen, and in the screening plant.

It is clear, that as the particle diameter (size) approaches the screen aperture, in size, the chances, of passage, become greatly minimised. Further particles, which are very much larger than the screen aperture, glide along, and the smaller particles drop down through the interspaces (between the larger ones) on to the screen. Thus these cause no great harm. Some intermediate sizes, however, may cause "wedging" and "blinding" or block up of the screen apertures, while particles, of slightly lesser diameter, pass with difficultly causing delays. These latter group, range in size between 50% larger to 25% smaller than the aperture.

When the material is placed on the screen, particles have a random orientation and distribution. If the screen is vibrated or moved, then stratification is developed, since the smaller sizes slip through the interspaces of the larger size fraction, while the coarser sizes remain above. While movement of the screen causes stratification, which is an essential factor, for efficient screening, it also prevents "blinding" or choking of the screen. Excess movement may, however, cause the particles to rebound and to slide past the apertures thus preventing stratification.

Screens may be made of cast iron, steel, or low carbon steel, manganese steel, chrome-nickel (stainless) steel, brass, phosphor bronze, monel (nickel alloy) nickel. Rubber plated screens are also made. For the finest sizes, e.g., in screening, graphite or abrasives, silk bolting cloth is employed. Selection of material, for screen fabric, is dependent on the nature of the material viz., whether corrosive or abrasive, or both. The cost also is to be considered.

Wire woven screens have either square or rectangular openings. In square mesh steel wire cloth, the aperture, of the intervening screen opening can be determined when the gauge or thickness of the wire is known.

A screen, for a particular purpose, should be selected on the following considerations :

1. Large capacity,
2. Greatest freedom from blinding or choking,
3. Clear separation of sizes.

Square screens give a sharp separation, but when inclined the capacity is reduced, due to apparent loss in aperture size. Rectangular screens of relatively small differences in sizes, have a larger capacity, and yield clean separation of grades, within the range between ½ inch and 65 mesh. With materials consisting of fairly rounded grains, the capacity of rectangular screens is increased 30 to 40% as compared with the capacity of equivalent square mesh screen. The advantages are not perceptible when the screen aperture exceeds ½ in. While square screens are necessary for use with material comprising flat particles, rectangular mesh is called for in sizing elongate, acicular or fibrous materials. The rectangular mesh is set across generally, but where blinding or choking is excessive, longitudinal setting is adopted.

Punched metal screens are also made of various metals and alloys. The holes may be round or square, in shape, while the arrangement of the holes may be linear or staggered. Further, screens may be slotted, and when fitted in a rectangular frame, the slot may be parallel to the sides or diagonally placed. The slot are commonly arranged in parallel rows, and they are either full open or burred.

Advantages of metal punched screens are that (1) due to even wear with use they have long life, (2) there is less tendency to blind, (3) the smooth surface permits the use of lesser inclination and increases the space required for discharging oversize, (4) with the screen in the horizontal position, the rate of travel is increased, (5) the smooth surface increases the tendency for stratification, and brings the finer particles in contact with the screen, with an increase in efficiency and (6) in special cases, the round punched holes enable accurate sizing. The minimum size, for round and square holes, is 0.30 mm. or 0.75 mm. for punched screens.

For coarser screening, bar screens; known as grizzley, made of discarded rails etc., fixed parallel to each other are employed. Steel rods are also used to form, screens called rod decks. Wedge wire screens, comprise wires having special cross sections (of approx. triangular section.

Haller Screen is made of stretched piano strings with no cross wires support.

Screens are generally classified as (a) fixed and (b) moving.

(a) *Fixed type* screens are represented by the grizzly. Grizzlies are coarse screens made of stout iron bars. They belong to the fixed type. Rails are also used in the larger grizzlies. Advantages are mainly, simplicity and ruggedness. The disadvantages are lesser efficiency, loss of head room, blinding and breakage of oversize. The grizzly is sometimes mounted on an eccentric with an inclination of 10°. A travelling grizzley has also been designed and it comprises a wide belt made up of bars fixed at regular intervals. When modified slightly, are called drop bar grizzly. In the roller grizzly, instead of bars, corrugated rollers are used.

(b) *Moving type* consists of shaking, vibrating, revolving and travelling belt screens. Revolving Screens, or trommels are usually attached to the discharge end of a crushing or grinding mill. Compound trommels consist of two or more screens, mounted on the same shaft. Conical trommels have the advantage that when the axes are horizontal the screens are inclined, further the wider end of the cone facilitates feeding of the material. Wet screening can be carried out with great ease. The cutting and fitting of the screen is tedious and time consuming.

Revolving Screens used in sizing stone and gravel, are different to the trommels, in that there is no central shaft; and the mounting is on gudgeon. The ends of the screen are provided with heavy steel trays, which move on rollers, as in the case of the tumbling mills. This type of screen is used in dredges and stone crushing plants.

Shaking Screens consists of elongate trays which are fitted with screens at the bottom. shaking is affected by various mechanical arrangements : (1) The screens may be suspended by chains; (2) They may be supported on vertical rods attached to the sides; and (3) The screens may be supported by small rollers, at the sides, which move on rails. The screens are, generally. given a tilt of 10°-15°. Shaking motion is communicated, to the trays, either by an eccentric, or toggle mechanism, or by an adjustment of the rails and rollers, where the screen is roller mounted, The screen may be made to climb up a bend in the rail. This provides the necessary slow backward motion, but a rapid forward motion causing the material also to move forward as the screen turns back. The adjustments to the shaking screen are for (1) frequency, (2) length of stroke, and (3) angle of slope. These should be so combined as to produce a quick forward movement,

of the material, with a view to avoid blinding and induce rapid stratification. Excessive speed and length, of stroke, cause the material to be
thrown forward and prevent stratification, whereas excessive slowing down,
on the other hand, has the opposite effect. The slope is the controlling
factor in both the cases. The strokes normally employed are 9 m.m.
length and 60-70/min, the max. being 800/min. Shaking screens are used
in coal preparation and also in sizing asbestos.

The Ferraris suspension comprises flexible steel supports, for the
screen trays. On bending of the spring supports and releasing them, the
tray acquires an upwards motion on the forward stroke. Shaking screens
comprise a shallow rectangular box fitted with a screen bottom. The
screen, is generally inclined, and a mechanical motion is imparted. The
tray is supported either by rods or suspended from chains. The length,
of the supports, determines speed of travel, of the materials. The speeds
vary, generally from 60-70 strokes/min. at 9 in. strokes or 800 strokes per
min. at 3/4 in. stroke. The capacity is 2 to 8 tons per sq. foot. per 24
hours per mm. aperture.

These screens are used in the preparation of coal, asbestos, and
phosphates but they are not preferred in treating clay materials.

One of the main disadvantages of such screens, is the high cost of
repairs, caused by shock and vibrations.

*Vibrating screens* : Consist essentially of a plane screen surface,
which is stretched tightly, and made to vibrate at a high frequency ; the
amplitude being small, such screens may contain upto 4 decks or tiers,
placed one above the other. Generally, not more than 2 decks are recommended. The vibration is produced either by electromagnets, or by the
use of mechanical devices, such as hammers, cams, eccentrics, or by the use
of an unbalanced rotor.

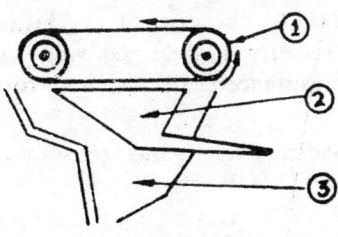

Fig. 12.27

*Travelling belt screen* (Fig. 10.27). : It consists of a long strip of
screen cloth (1) which is made into a belt by passing it over rollers. The

material is washed with water. The material, passing through the cloth belt, is collected in the chute (2), the oversize is carried by the belt, and washed down into the chute (3).

## Classification—Wet classification

The principles of wet classification can be stated as follows :

1.  For particles having the same shape and sp. gr., the relative settling velocities, vary with the size.

2.  For particles of the same shape and size, the relative settling velocity varies with the sp. gr.

3.  For particles of the same weight, the relative settling velocity depends on the shapes, e.g., tabular particles have a longer settling time.

4.  For all particles, resistance to fall, varies with the velocity. It varies directly at low velocities, i.e., (diameter)$^2$ and as the ($\frac{1}{2}$ Diam.)$^2$ at high velocities.

5.  For all particles, the resistance increases with density, of the liquid.

6.  For all particles the resistance increases with viscosity of the liquid.

The purpose of classification is to affect separation in various grades, irrespective of size, so that, in each grade, particles of various shapes, sizes, and sp. gr. are included. The largest sizes include the tabular particles of the mineral with lowest sp. gr. and the smallest particles are spherical and comprise minerals with the highest sp. gr. The minerals with intermediate sp. gr. are present in shapes, varying between the sphere and flat tabular particles.

When a body falls through any fluid medium, and the velocity is low, no considerable disturbance is set up, in the fluid, and practically the entire resistance is due to the viscosity. This resistance is the viscous resistance. When the velocity is great, eddying and turbulence set in, at this stage, the viscous resistance decreases, and turbulant resistance is of major importance.

For spherical bodies, the viscous resistance can be expressed as follows.

$$R = 3\pi D^n V$$

where    R = resistance of the fluid,

n = viscosity of the fluid,

D = diam. of spherical particle,

V = the relative velocity, in respect of the fluid.

While Newton's Law gives the value of turbulent resistance as

$$R = Kp\ D^2V^2$$

where      R = resistance of the fluid,

p = density of the fluid,

D = diameter of spherical particle, and

V = velocity, with respect to the medium.

These two equations fit in with value of resistance of bodies, moving through various media, at low and high velocities respectively, but are not in agreement with values of resistance, often encountered in experimental work. The intermediate values of fluid resistance are, however, defined better in terms of a resistance function S, known as Castleman function:

$$S = \frac{2\ g\ D\ (d-p)}{3\ p\ V}$$

g = gravitational constant, d = sp. gr. of the solid.

$$S = f\ (N)\ \text{where}\ N = \text{Reynolds number},\ N = \frac{DVp}{n}$$

$$T = N^2S = \frac{2g\ pD^3\ (d-p)}{3n^2}$$

Free settling is effected, when the body of water is very large, in relation to the size of individual particles.

When the diameter of the water column is less than 10 times that of the solid particle, the settling velocity is affected (decreased) by the wall effect. If the number of particles is so great that they collide, with each other, then, hindered setting causes the settling velocities to fall far below the free settling rates. The fall, in settling velocities, increase with the weight of particle, and on the extent of crowding ; and decrease with the increase in diameter and sphericity.

Free settling ratio, is the ratio of the diameter of a light-mineral particle, to the diameter of a heavy mineral particle, which have the same settling rate, under identical free settling conditions. In the same manner, hindered settling ratio can also be defined, except that the conditions are hindered. In the case of classifiers, in close circuit with grinding mills, conditions are akin to those of hindered settling.

## Classifiers

Classifiers may be grouped under two main types (1) horizontal current or mechanical and (2) vertical current or hydraulic classifiers. The horizontal current classifiers are mainly mechanical, and these generally help to separate two different sizes of material; whereas the hydraulic classifiers help in sorting out material according to size specification, specific gravity and shape. These machines are often called by various names, such as sand tank, spitzkasten, sand cones, or desliming cones. Where air is used instead of water for classifying, then the machines are known as pneumatic classifiers.

## (1) Mechanical classifiers

The mechanical classifiers consist of a long settling tank, the end walls of which are nearly vertical, whereas the longer sides are parallel and have vertical walls. The feed, which is led into the tank, is kept agitated, while the finer material, thus kept in suspension, overflows into a launder, at the lower end of the tank. The heavier material, which settles down, is carried upwards, along the inclined bottom of the tank, either by means of a (a) rake, (b) spiral or (c) flight conveyor. This type of classifier is generally, used in close circuit grinding, and an output, of the order of 1000 tonnes per day, can be obtained.

The machine can also be used either for desliming or for dewatering granular material, and also for washing, where scrubbing is not necessary. (a) Of the rake type classifiers, the Dorr machine is well known (Fig 12.28)

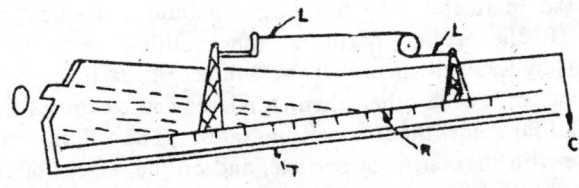

Fig. 12.28

O=Fines overflow ; R=Rake ; C=Coarse particles raked out;
T=tank; W=water level ; L=lever mechanism for
moving rake.

In the backwards journey the rake moves along the bottom of the tank 1; and when moving forward, the rake is lifted off the tank bottom by the lever mechanism.

*Operation* : The long rake consisting mild steel flats, is lifted up from the tank, and moved forward and lowered at the lower end of the tank by the lever system L. At this position, the rake is embedded in the sediment. In its upward journey, it carries the sediments to the upper end, and drops the same at C, to be discharged. The fines move downward and are discharged at the overflow O.

(b) Of the classifiers, utilising the spiral for discharging the coarse particles, the Akins machine can be taken as the type (Fig. 12.29).

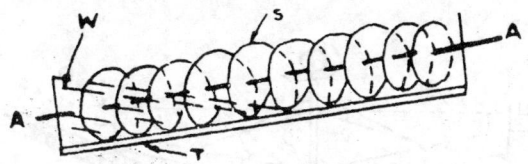

Fig. 12.29

S = spiral or helical, T = tank ; W = water level

*Note :*   There can be more than one spiral, generally two spirals.

*Operation :*   The spiral rotates about the axis A-A', within the tank. The sediments, at the lower end, are collected by the spiral, and lifted upwards. If the spiral rotates in the clockwise direction, as viewed from the top end, then the sediment will move upwards.

Of the drag type classifier, the Esparanza can be taken as being representative. The drag type machine, because of its cheapness, was popular, but now it is used only in small size plants. The modification, of the spiral classifier, is the counter current machine. In this machine, the spiral is fitted to the inside of the inclined cylindrical drum, which forms the body of the machine. The rotation of the drum cause the material to move along the internal spiral. The coarser particles which settle down are discharged at a higher end of the drum, as in the case of other classifiers, while the finer material is washed downwards and discharged through the launder. The machine is also similar, in arrangement, to the normal spiral classifier, except that the spiral is fixed to the inside of the tank.

Another variation, of the mechanical classifiers, is a group called the bowl classifier. This consists of a shallow cylindrical tank, with a flat conical bottom. The pulp, which is filled into this tank, is kept agitated by rakes, which are carried on a central arm, and fixed at the centre of the cone. Agitation is accomplished by water jets. The coarse sand, which settles down, in the centre (bottom) of the cone, is collected and discharged either by a rake or by a spiral mechanism. The overflow launder, or channel, fitted to the rim, of the cylindrical part of the tank, carries away the slime. The essential parts of this machine are shown in Fig. 12.30.

R=rakes suspended from a central rotating column C ; B=Bowl ; O=overflow for fines ; H=chamber, containing a scraper Sr., and water jets. W=water level ; G=sand discharge ; S=spiral conveyor for discharging coarse particles ; D=Drain cock.

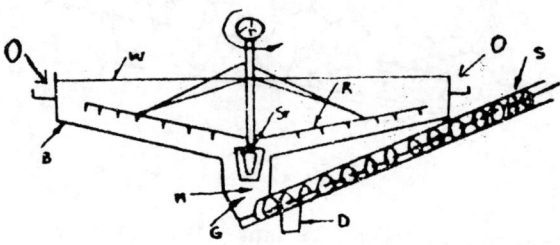

Fig. 12.30

*Operation* : The rakes are rotated about the vertical column C. The fines overflow into the launder O. while the coarser particles (sand), get into the chamber (H), and carried away by the spiral S and discharged.

Bowl rake classifiers are used when dealing with material whose size range lies between 28 and 300 mesh. The common size is the 65 mesh, and more usually 35 mesh. The machine has the advantage of flexibility, as it can be adopted to deal with fluctuations in sizes, by varying either the speed of rake or by changing the dilution etc.

### (3) Hydraulic classifiers

Hydraulic classifiers are generally divided into two groups : (a) free settling and (b) hindered settling. In the free settling classifier the cross section, of the settling column, is constant. Some models, of this type of classifier, use tanks while others use launders. The tank type classifier consists of a rather deep V-shape channel, or trough, to which the sorting column are attached. In the launder classifier, similarly, sorting columns are attached, to the launder, at convenient intervals.

Even though the efficiency of the hydraulic classifier e.g., sand cone is low, it helps in the separation of material into short (small) range groups, so that these can be more efficiently treated on shaking tables. (a) *Free settling classifiers*, known as sand cones and tanks, are non-mechanical in operation, but these are used specially in washing sand materials

L = launder ; W = water inlets ; T = troughs ; D = spigot openings.

*Operation* : The feed is let into the machine at the point F. Due to the action of the water currents, the sand is deposited at the points marked X, within the launder L, while the fines are discharged, at the spigot.

(b) *Surface current classifiers*—Spitzkasten type (Fig. 12.31).

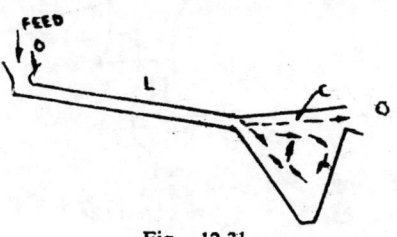

Fig. 12.31

*Operation* : In this type, the feed, from the hopper O, is led by a launder L, to the conical tank C. The arrows, shown in the figure, indicate the movement of the current, in the body of the pulp, contained in the tank C. The action is similar to that of a shallow current, and the fines are discharged at O, the overflow. The main use of this equipment is to separate fine sand from slime.

Free-settling classifier (Fig. 12.32).

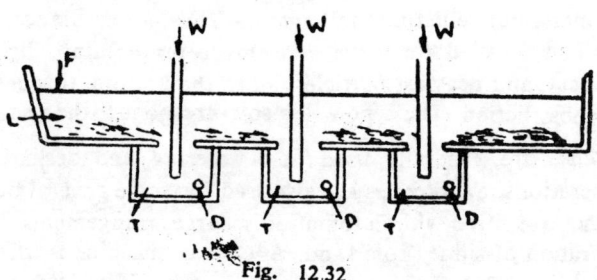

Fig. 12.32

(c) In the *hindered settling type,* the construction of sorting column is different, in design to that found in the free settling type. The cross section of the sorting column, in this type, is made by one of several means, either by the gradual reduction in the diameter to form a cone, or by the sudden construction etc.

Hindered settling classifiers (Fig. 12.33).

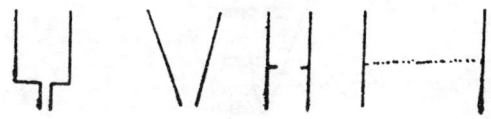

Fig. 12.33

The figures given show the various methods by which hindered settling is affected.

Launder-type of hindered settling classifier (Fig. 12.34) :

L = launder, R = reducer ;
V = Vortex.

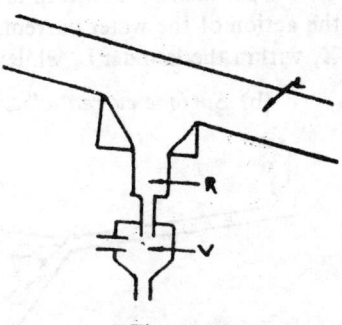

Fig. 12.34

Even in this type of classifier, the same launder or tank is used, the variation being in the type of settling column. Action of the hydraulic classifier of this type is rather interesting. Particles, falling through the large diameter portion of column, of the classifier, meet the rising current of water, from the constricted end. The increase in velocity of water, here causes higher resistance to falling particles. Hence, some particles do not drop further down and thus become suspended in the column. Particles, moving upwards to the top or to the lower pressure position, settle in a mass, but sometimes fall again. Thus, a circulation column is built up. This is called the teeter. In this teeter column, light particles appear to float, and heaviest particles fall to the bottom, through the constriction or obstruction. The type will also correspond to the Spitzluttentype.

(d) *Sand tanks* are, generally, used for dewatering and desliming in preliminary operations. Sand cones are also used for some general purposes, and at times, they are fitted with automatic discharge arrangements. The cone can affect separation of slime from sand. A typical machine is the Caldecott sand cone (Fig. 12.35). This type is known as the "Spitzlutten".

Sh = Shell of sand cone ; O = overflow for fines; F = feed pipe; R = regulating pipe outside the feed pipe, for controlling velocity of rising currents ; S = sand; C = Cone value for discharge of sand ; B = Circular baffle plate, placed across the opening of the feed pipe.

*Operation* : The feed is led into the sand cone (S) at a regulated rate. The sand settles after impinging on the baffle plate (B), while the

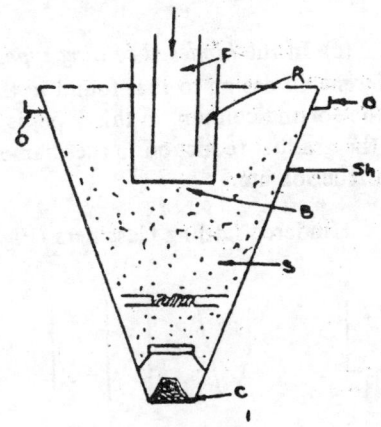

Fig. 12.35

fines are carried away by the rising current which gives rise to a suspension known as Teeter column, from where the sized particles move to the overflow (O). The conical value (C) is opened, from time to time, for drawing out the sand from the bottom of the cone.

## (e) Centrifugal classifier

This equipment makes use of the centrifugal force in classifying, and the intensity of these forces may be as much as 1000 times more than gravity. In principle, the machine is akin to the laboratory centrifuge. It is particularly useful when the material to be handled contains sizes of the order of 1 to $2\mu$ when the solid content is only about 25%. It is used in the cement plants using impure limestone. In the cases of such material, differential grinding leads to calcium carbonate being carried in the slime, whereas the shale and sand remain in the coarser fraction. Centrifugal classification helps to recover the lime from the slime discharge.

## (f) Pneumatic classifier

Classifying by means of air currents may be grouped under (1) dust collection and (2) sizing, depending on the purpose. In dust collection solid particles, suspended in the air, are removed either (a) to improve the atmosphere, e.g., in underground mines or (b) to prevent dust from getting into residential areas, e.g., spread of industrial dust into cities, (c) for cleaning industrial gas, (d) for recovering any valuable mineral present in gas and (e) to prevent disfiguring of buildings due to dust. The particle size in industrial dust varies considerably, but the common range is between $50\mu$ to $1\mu$. The equipment used comprises the following types, (1) gravity type which uses a chamber, (2) the inertial separator, which uses baffles and centrifugal action, (3) filters which use coarse fabric bags, (4) spray type in which scrubbing towers, with water jets, are used and (5) the electric type, where static electricity is used for dust precipitation.

1. In the gravity separator, dust is removed from the gas due to the fall in velocity, caused by the expansion in volume. This process is further assisted by the use of baffles and deflector plates. The settling chamber type, is mainly employed for coarse particles and, also for a rougher separation, to enable effective separation at the later stages.

2. In the inertial separator also, the settling chamber is employed together with deflector which are arranged in such a way as to reduce the velocity; as well as induce centrifugal action which cause the dust to be thrown down, in the low pressure areas or miniature air pockets.

## Air Sizing

(a) In principle, pneumatic sizing employs an apparatus, very similar to the dust collecting devices. In the gravity type classifier, (Fig. 12.36) the

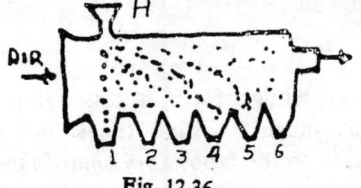

Fig. 12.36

mixture, containing gas and dust comprising various particles sizes, is fed into a long chamber, through a hopper, while, at the same time, a current of air is also forced into the chamber, along the axis of the chamber. The coarse particles fall nearer the hopper, while the finer particles which are carried further away depending on the grain size and are collected in the conical vessels 1, 2, 3... 6 and drawn out periodically. In affect, this apparatus is similar to the normal or common process of winnowing.

In the gravity type classifier, where a vertical current is used, the chamber is a tall cylinder and the feed is generally led into it at the bottom. The heavier particles fall very close to the inlet and are removed from the bottom of the chamber, whereas the lighter particles are carried away by the air current to the outlet at the top.

(b) Classifiers of the inertial type utilise the inertia, of the particles, which varies according to the size ; the velocity being constant. Thus the heavier particles will be projected for a greater distance than the higher ones. In this type of classifier, the feed is forced into a chamber together with air at high pressure. The larger particles are projected further from the inlet nozzle and the smaller ones drop closer or nearer to the point of entry. In this manner sizing is affected.

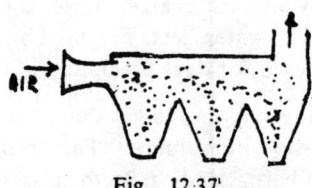

Fig. 12 37

Inertia type classifiers also take advantage of centrifugal force. In some machines the feed is projected into a conical chamber so that the current moves tangentially. Due to this gyration, the heavier particles are projected on to the shell of the chamber, whereas the lighter ones arrange themselves according to sizes, the smallest particles remain close to the axis of the conical shell while the largest particles stick to the sides of the chamber.

Commercial classifiers have, however, utilised a combination of both gravity and inertia principles, and in addition, they employ a rising air current, e.g., cyclone type classifier (Fig. 12.38). Low pressure air is introduced at the lower end of the conical chamber, in such a manner. as to induce the formation of a vortex or cyclonic movement, within the chamber. This causes the finer dust to be discharged though the central open-ing, at the top of the cone, while the coarser material drops to the bottom of the cone, from where it is discharged, whereas the intermediate size fraction gyrates near the periphery of the cone and is exhausted through another duct.

(c) Another type of centrifugal classifier employs the volute principle, and it is commonly used for roughing. The mixed feed entering the chamber, together with the gas, at high velocity, is separated into sizes, in a manner similar to that obtaining in the centrifugal inertia type classi-fier. The coarser particles are thrown towards the periphery, while the fine particles, which enter the inner most whorl, are led out through a pipe.

(d) Yet, another type of classifier, which makes use of both mechanical as well as the centrifugal means, is also of commercial importance. In this, the feed is forced on to a rotating disc and thus classification due to inertia is induced. The air current forced up into the conical shell of the machine producc a winnowing action while a fan, placed below the rotating disc, induces centrifugal separation.

(a) In the centrifugal collector or cyclone, the velocity reduction and centrifugal action again play an important role. Modern equipments employ multiple cones to increase efficiency. As a result of higher centrifugal action, and the increased drop in velocity, with increasing resistance,

there is an increase in the efficiency of separation. The dust particles, precipitated at the periphery of the conical cyclone, drop to the apex of the cone, whereas the gas, free of dust, remains in the centre or axial portion of the cyclone (Fig. 10.38). Cyclone are classified either as high or low efficiency machines. In the low efficiency cyclone, only 70% collection of dust, less than 100 mesh, is affected. These are satisfac-tory for use in the large particle size range. Low efficiency cyclone are cheap to construct. The wear is also less due to low velocity. As the pressure drop is small, the power consumption is low. These can be employed particularly for dust with high specific gravity and particle sizes of less than 200 mesh. Most industrial dusts are, however, too fine to be handled by this machine alone.

Fig. 12.38

Some disadvantages of this machine are (1) its low efficiency which, in certain cases, may be dangerous and hazardous, especially when used for dust separation. If the valuable mineral is in the finer sizes then losses may be large thus making the recovery uneconomical.

(b) A modification, of the cyclone or centrifugal separator, is the roto-clone. This consists, essentially of a centrifugal fan, comprising both stator and rotor. The narrow vanes are arranged in both the stator and the rotor, along the periphery of the rotor, parallel to the axis of the fan. The dust laiden air, entering through the inlet near the axle of the fan, is forced out at great speed, on to the stator, where it is collected, and dis-charged through a duct.

Air filters are used for collecting fine dust where the quantity of dust is small in volume. The bags, used in the filters, are made of tough fabric and they are either triangular or rectangle in shape.

Spray washers are generally of two types, viz. the vertical and hori-zontal. The principle of the spray washer is to clean dust laden gases by bringing into contact, each particle of dust with a fine drop of water. In practice, this method may be likened to shooting birds in flight, by using cartridges charged with minute lead pellets. Dust particles, brought down by water sprays, are removed by draining out the spray chamber.

Packed towers are the simplest form of spray washers. These com-prise a metal or masonry tower packed with either rock, pumice, or bricks. The gas inlet is placed at the bottom, of the tower, while the out-let is at the top. Since the gas is made to enter the tower tangentially, it acquires a gyratory motion. Spluttering and splashing, of the falling water drops, enable the entrapping of dust particles which are removed, by the water.

A modification of this design is the multiwash spray tower, in which the packing material, which induces the cascading of water, is replaced by tires of radical plates, which also act as baffles.

In the rotor spray washer, a rotating cylinder, partially immersed in water, serves to form the spray. The rotating cylinders, provided with perforated baffle plates, are placed in a long chamber. The washed mate-rial is collected in tanks, placed at the bottom, while the finer particles are discharged, at the overflow.

### Electrical Precipitation of Dust

The principle of electrical precipitation depends on the fact that if molecules of gas or particles of dust are charged electrically or in other words become ionised, they will be attracted towards a negatively charged body or a grounded electrode (Fig. 12.39). In one

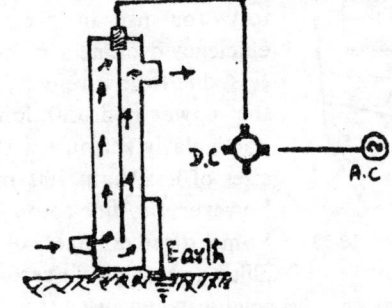

Fig. 12.39

of the designs, known as the pipe-type precipitator, an electrically charged wire is placed in the centre of a cylinder. Since the pipe is grounded or earthed it acts as the negative pole, thus the finest particles move towards it at high velocities, depending on the size, and degree of ionisation, and are thus precipitated. A number of such pipes are often used, so as to form a battery, for increasing the capacity of the precipitator.

Dust flocculation is also affected by the use of ultrasonic devices. Considerable success has been achieved by the use of frequencies in the range of 10,000 to 20,000 cycles per second.

## (A) Washing and Scrubbing Concentration

Scrubbing is affected by the use of such methods in which only relatively harder minerals are recovered, while the softer material forming the matrix is removed either by abrasion or disintegration. An example of scrubbing is the use of water for removing clay, adhering to surfaces of iron ore or limestone etc. when the pieces are rubbed together. Scrubbing may also be used for removing hard coatings, comprising decomposed material adhering to the surfaces of ores. One of the methods of scrubbing is the use of water jets. In general, monitors or hydraulic giants can also be use for the purpose.

(a) Tumbling scraper consists of a cylindrical drum arranged in more or less horizontal position. The material to be scrubbed is charged into the drum and the drum is rotated mechanically. Water is also fed into the drum for washing. This machine is used in washing of phosphatic nodules and also in cleaning nodules of manganese ore.

(b) Another machine which is also used for scrubbing, is similar to the drum, but has a number of teeth which are set vertically inside the drum. The teeth have a cutting action and are more effective in eliminating clay.

(c) Other types of scrubbers make use of a stirring mechanism for liberating clay and similar impurities. An example of such a scrubber is the pug mill which is either horizontal or vertical. In the horizontal type the stirring is done by blades mounted on a shaft. This is placed within a cylindrical drum into which the material, to be scrubbed, is charged. Water is fed at the upper end, and the clean material is removed after sometime from below. The log washer is similar in design to that used in washing alluvial materials. Here again an inclined trough is fitted with a shaft carrying blades which are arranged more or less spirally. The material, to be washed, is charged at the lower end, while water is fed at the upper end, of the trough. The washed material is collected at the upper end of the washer. The spiral washer is similar to the log washer, only that the spirally arranged blades are replaced by a helical screw.

Washing is the term applied to the separation of the particles according to sizes, specially in the case where the variation in size is very great. Screen washers employ a rotary screen on which water jets are directed. Materials, placed on the screen, are washed and the finer material goes through the screen, while the coarser material is collected from the screen. These are employed in the working of alluvial gold, as well as tin mining. Screen washers are used in working shells collected from sea bottom and from lagoons. Shaking and vibrating screens are also used in a similar way.

Washers may also take advantage of the classifier principle. In this group are (a) sand tank of the automatic type, (b) current and surface current settling tank, (c) launder or trough type classifier, (d) mechanical classifier and (e) hindered settling hydraulic classifier. The principles of these classifier have been described earlier under the relevant paragraphs. Hence descriptions of these are omitted here.

Beneficiation takes advantage of the differences in specific gravity, which come into play under the influence of forces impressed upon various particles. The common method, of making forces act on particles, is through a fluid medium viz. a liquid or gas in which the particles are more or less suspended. In this method the fluid properties also come into play, such as density etc. In general, gravity processes also induce sizing. Mineral mixtures which are amenable to gravity separation, may contain gangue and sulphide or any such combination. The methods of gravity separation are described in brief.

## (B) Jigging

The jig, is a machine utilises the differential capacities, of various mineral grains to penetrate a semi-stationary bed. This machine consists essentially of a box, with a perforated bottom. Water is generally employed, as the medium and the current induced passes through the perforated plate either (a) continuously upwards or (b) continuously downwards or (c) alternately upwards and downwards.

(1) Jigs of the fixed sieve type or the Hartz jig can be taken as an example.

This machine consists of a number of hopper shaped compartments, to the upper part of which is divided into two portions. In one of the portions (1) is a plunger, while the other portion (2) is fitted with screens (Fig. 12.40).

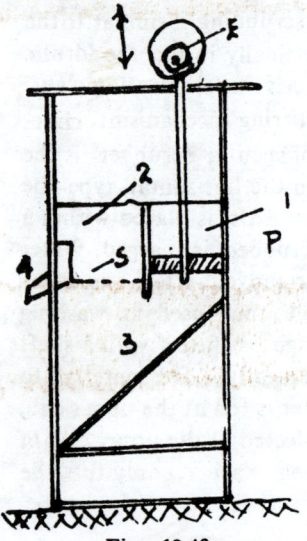

Fig. 12.40

*Note*: The figure shows only one unit, but jigs may be in groups.

When the compartments are filled with water, and the plunger is forced in, the material which is placed on the screen, (5) the heaviest particles sink and settle on to the screen, while smaller particles pass through the screen, into the lower compartment or hutch (3). The coarser particles which lie on the screen travel forward and are withdrawn through the lip or spout (4).

The lighter material, which does not pass through the screen, floats up from the first compartment, and gets into the second, and this process is repeated, till the lightest material travels further and further through the lower compartments, which are also affected by the pulsation induced by the plunger. Thus sizing and specific gravity differentiation are both affected simultaneously.

(2) In the diaphragm jigs, pulsation of the water within the machine, is caused by means of a diaphragm. The jig comprises a box in which the screen S is fitted very near the upper part. The pulsating diaphragm D is below the screen. Water is used as the medium for transmitting the pulsating motions to the materials on the screen (Fig. 12.41).

(3) The same principle, as the diaphragm jig is utilised in the American placer jigs, used for treating heavy minerals and for working gold, associated with alluvial materials.

(4) Another form of jig is the pulsator jig. In this machine, instead of a mechanically actuated piston, or diaphragm, for inducing agitation, intermittent water current is fed beneath the screen compartment, through the rotating valve. The concentrate obtained is discharged

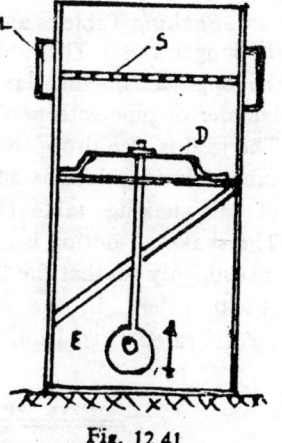

Fig. 12.41

through the gate and dam system. Water impulses can also be used, when required, to make the material loose. The normal pulsation used, is about 200 strokes per minute, but this machine uses a large quantity of water.

(5) A modification, of the pulsating jig, is the air driven machine, which is typically represented by the Baum jig. Here, instead of the plunger compartment with the plunger, compressed air is employed. Compressed air admitted into the plunger compartment, causes the necessary pulsation.

(6) Another variation of the jig is the moveable sieve jig. This consists of a tank with several compartments, having moveable sieves, and the

mechanism for moving these sieves. The material, which is placed on the screens progressively moves downwards, by small jumps, due to the reciprocating movement of the sieve frame. The heavier particles go through the sieve, and are collected nearer to the feed end, while the lighter particles travel along the screens until they can pass through the sieve. Small and lighter material travel farthest.

Mention should also be made of the hand operated jig, which is commonly used in most mining operations, as well as in prospecting work. It is frequently taken up in conjunction with vanning and panning. This consists of a wooden tray, which is about 8 inches deep and about 12 × 12 inch, fitted with a perforated plate or screen at bottom. This tray is suspended in a tank of water and moved up and down manually by means of a lever arrangement. The material for jigging is put on the tray, and if the tray is actuated, in such a way as, to move quickly downwards and slowly upwards, jigging can be achieved.

## (c) Tabling

Shaking Table is an oblong wooden table which has a slight slope across the longer axis. The table also imparts a jerky motion, in the direction of the longer axis, while, at the same time, water is fed in through a small launder or pipe with perforations, located at the upper edge of the table. The feed is also dropped from a launder located at the upper edge of the table. Thus washing is done simultaneously. Fig. 12.42 shows the action of the shaking table (1). Typical shaking table is the Wilfley Table. The shaking motion is obtained by compressing the spring S and releasing it suddenly so that the table moves forward and strikes the bumper block B with a jerk. In the other forms of the table a toggle mechanism is incorporated (Fig. 12.42).

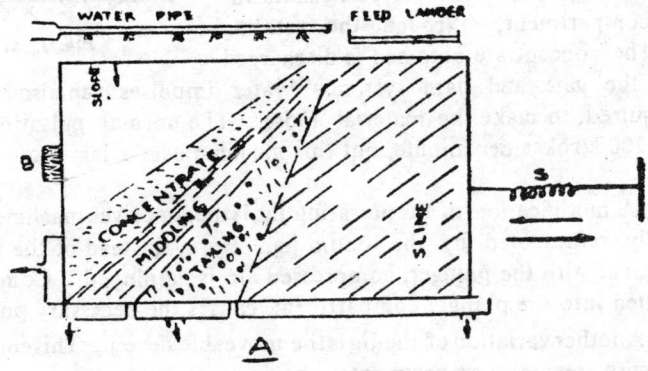

Fig. 12.42

In the Fig. 12.43 the various stages, by which separation is affected in tabling, are illustrated. The table has a number of thin wooden batons, fixed on the surface, and these are known as riffles. The feed, getting into the riffles, is subjected to the washing effect of the current of water. This is similar to the settling tank, in action.

The bumping or jerking action, of the table, produces a density, as well as size classification (Fig. 12.43 ). The heaviest particles are found at. the bottom in the layer (*a*), with the smallest sizes lower down. Similarly, the lighter material shows grading with the upper layer (*b*), with the smaller size fractions settling lower down (Fig. 12.43 ). The function of a shaking table is similar to a continuous mechanical pan, in which size stratification is affected by the gyration caused by the flow of water, and the distribution of the material is according to sp. gravity and sizes. As a consequence of this combined action four products are obtained. (a) concentrate (b) middlings (c) tailings (d) slime. The concentrate is delivered to the metallurgical plant while the middling is subjected to further grinding and recovery.

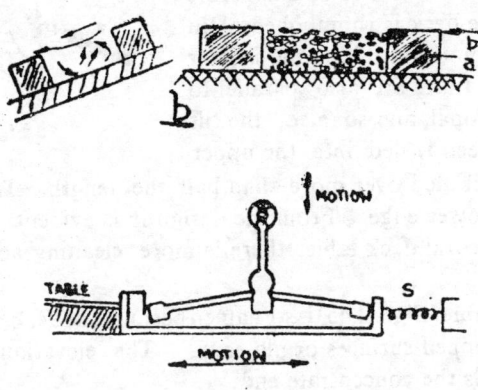

Fig. 12.43

The toggle mechanism of the Wilfley table is different to that of the blake crusher. Here the end X of the toggle is attached to the end of the table, while the other end of the toggle mechanism is attached to a moveable block Y, which slides horizontally in the frame F, called the yoke. The block Y is also attached to the spring S, while the other end of the spring attached to the fixed block B, which is independent of the table itself. When the pitman moves up and down under the action of the eccentric, the table is drawn backwards. By the toggle and part of the horizontal motion is absorbed by the compression of the spring. When the pitman moves down, then the table moves forward being pushed both by the toggles as well as by the spring.

As a result of the combined actions of the factors mentioned above four products are obtained; A. concentrate, B. middlings, C. tailings, D. slime. The concentrate forms the feed for the metallurgical process. The middling is subjected to further grinding and concentration. In the smaller plants the last two products are dumped. But in the sophisticated plants recovery from slime is also practised.

(2) Another form of shaking table is the card table. This is similar to the Wilfley table but the riffles are cut into the deck, and the shape of the riffles is trangular in section, instead of the rectangular section as in the Wilfley table. The riffles increase in depth from the head motion side upto a point, corresponding to the termination of riffles in the Wilfley table and then decrease towards the concentrate end.

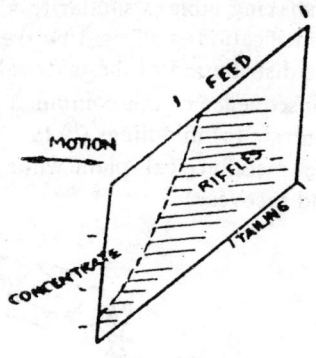

In the Deister-Overstrom table (Fig. 12.44) the deck is rhombohedral in shape, and the motion is along the shorter diagonal. The riffles are also parallel to the shorter diagonal, and so also, the tilt of the deck. Feed is led into the upper

**Fig. 12.44**

edge, and distributed over more than half the length. The tailing is discharged at the lower edge. From the design it is evident that in the case of the rhombohedral deck table, there is more cleaning action than in the Wilfley.

(3) Plate table (Fig. 12.45) is so named because the deck surface consists of 2 or more stepped surfaces or plateaux. The elevations of the steps increases towards the concentrate end.

### Shaking table-limitation

An appreciable difference in sp. gr. increases the efficiency of the shaking table ; and minimum concentration criterion of 1.25 is required, while 2.5 enables a more rapid and complete recovery. Where there is a strong contrast, in shape of the concentrate and tailing, then a concentration criterion of even 1.0. will suffice, e.g., coal and slate, the coal particles are irregular or cubical, whereas the slate particles are tabular, or in the case of beach sand and sea shell fragments.

$$\text{Concentration Criterion} = \frac{(G_1 - W)}{(G_2 - W)}$$

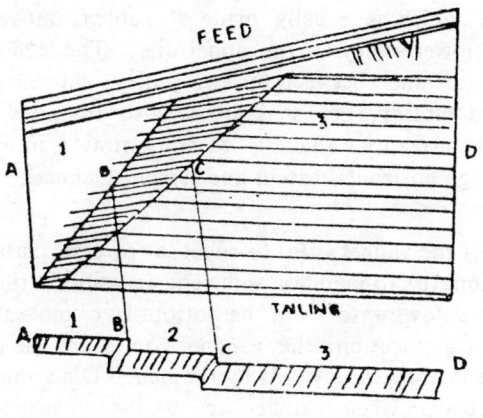

Fig. 12.45

Where $G_1 =$ Sp. gr. of the heavy mineral
$G_2 =$ Sp. gr. of the light mineral
$W =$ Sp. gr. of the liquid.

The size of the feed is also not well defined.    The velocity, of the water current employed, determines the sizes of the particles washed out. The relation between the smallest particle, of the concentrate, and the largest particle, of the tailing, is dependent on the sp. gr., shape of particles, nature of the deck, character of the feed, etc.

The smallest particles of chalcopyrite, for economic recovery on, tables, is of the order of 0.025 mm., while that for cassiterite is 0.05 mm., on the average.  As a rule, a granular or sandy material of −200 mesh (or 0.074 mm.) is easily concentrated, on a shaking table, when the gangue is not too coarse as to require a strong water current for washing.  With regard to the upper limit, for material used for tabling, it may be stated that for base metal ores, 2.5-2.0 mm. sizes are used for roughing tables.

The capacity of the full riffled tables, used for roughing, is 100-200 tons per 24 hours, employing a feed of 2 or 2.5 mm. size.  This capacity can be increased, but the recovery will fall.  In general, this capacity in the case of lead ores is higher than for copper or zinc.  In roughing tables, used for pyrite recovery, from siliceous materials, the recovery is of the same order.

### D. Vanners

Vanners are machines, employed in the concentration of fine sandy materials.  The machine comprises an endless belt which travels over two pulleys, one at either end.  The belt has a slope in the direction of width. The belt, which is usually made of rubber, moves at a slow speed, from the lower pulley to the upper one. The feed is introduced at one fourth the distance between the two pulleys but nearer the higher pulley. A second motion, i.e., shaking, is also imparted to the belt. Thus the vanner is in reality a shaking bed concentrator, in which the bed has a relatively large horizontal extent and small thickness. It is mainly employed for roughing.

The action of the vanner is (a) to effect a reverse classification and gravity stratification, (b) to maintain sufficient dialation in the bed, so that the top layer moves downwards, and the bottom layer moves upwards. As a result of reverse classification, the gangue comes to the top, while the rough concentrate is dragged upwards by the belt. Dialation of a bed, or quick sand effect, occurs when particles are sustained in suspension, by agitation caused by either rotary vibratory, mechanical or liquid flow. This is utilised in the sink-float processes. The best known of the vanners is the Frue vanner. The size feed used in vanners is 0.6-0.8 mm. on the average and the sizes are smaller than that used on the shaking tables. The pulp consistency is normally 80-90% water. Felting or sticking of the clayey slime occurs, in most vanners, if the feed is too thick. High sp. gr minerals can be concentrated with high slopes of belt, slow movement, and by the use of a larger quantity of water. An ore, with high concentration ratio, requires a slow and careful treatment.

The efficiency of vanners is low, and the recovery does not often exceed 50% even with rich galena slimes, while with copper slimes it is 20-35%, on rough copper concentrates.

### (E) Miscellaneous

Among the sink float methods, the use of heavy liquids for concentration of minerals plays a subordinate role. The liquids, used in this connection, are generally more or less concentrated chlorides. Organic liquids, of high specific gravity such as carbon tetrachloride and pentachlorethane, with specific gravity of 1.583 and 1.678 respectively, have also been used in commercial processes.

The chlorethane process was used in coal cleaning. One of the requirements, for the successful operation of this process, is perfect wetting of the solid particles, before immersion in the heavy liquid. The wetting agent used generally, is water with a little starch, acetate or tannic acid. This wetting agent prevents the absorption of chlorethane by the solid particles, and this can be readily drained away from the separated heavies, as well as the light fraction, after the recovery.

(1) *Sluice box*: Concentration by sluicing is adopted as a rough or pre-

liminary measure, where the valuable mineral occurs in the free form, in a finely divided state, with a concentration of 3.5% or greater. It is used in working alluvial gold deposits, and tin deposits (See under mining).

(2) *Long tom*: (See notes under alluvial mining)

(3) *Diamond Pan* (Fig. 12.46) consists of two concentric cylinders, the larger (1) is about 10-18 ins. high and smaller cylinder (2) is about

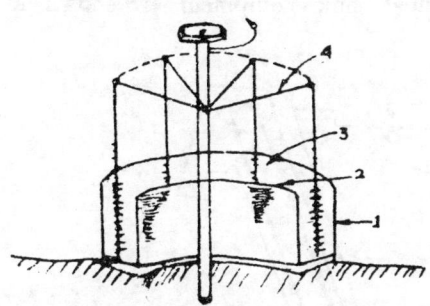

6 ins. less. The diameter of the outer cylinder (1) is 5 to 20 ft. The diamantiferous gravel is fed into the annular space (3) between the two cylinders, and it is continually raked up by knives or tines, attached to rotating radial arms (4).

The impulse, on the settled material, acts in the same

Fig. 12.46

manner as an expanding spiral. Thus the effect of the rotary agitation, is to induce a quick sand effect with an upward impulse.

(4) *Use of suspensions*: Suspensions in general, comprise particles of solids dispersed in a liquid, by using some form of agitation.

The solid particles, held in suspension, cause an increase in the apparent sp. gr. of the fluid, by an amount roughly proportional to the amount of solids present. The viscosity of the fluid, however, increases with the increase in solid content. Thus the increase in these two properties increase the specific weight, of the mixture as a whole, but a certain amount of mechanical agitation is necessary to maintain the suspension.

Contamination during use, however, causes an increase in apparent viscosity, and the converse may also be true in some cases.

Suspensions used in commercial operations comprise clay, quartz, slate, magnetite, galena, ferrosilicon, occasionally mixtures of barite, hematite, pyrite, copper, high carbon steel, and atomised lead also have been used. Water has always been preferred as the suspending medium. For the successful use of this method, the sp. gr. of the suspension must be greater than that of the mineral to be floated. The sp. gr. of the suspension is calculated thus :

$$R = \frac{W}{[p + W(1-p)]} \text{ or } p = \frac{W(1-1/R)}{(W-1)}$$

where $R$ = effective sp.gr. of suspension, which with water has a value nearly $= \dfrac{W}{2}$

W = sp.gr. of solid medium, and must be $>7$.

p = decimal fraction of solids, in the pulp by weight. This generally falls between 75-80%.

The essential parts of the float sink equipment are shown in Fig. 12.47.

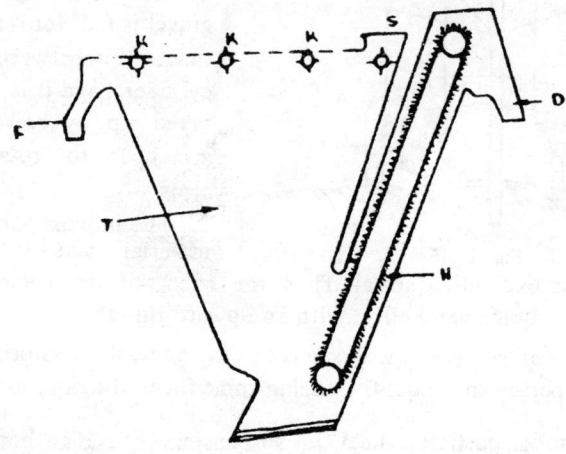

Fig.   12.47

T=tank or cone with the suspension.

K=rotary separators for removing the float.

F=float discharge.

H=elevator, with perforated, buckets for collecting heavies.

D=Heavy discharge.

S=Star feeder for affecting submergence of feed material.

Capacity : A 4 ft. cone can treat iron ore 1½-3/8 in. @ 40 tons/hour. With a 6½ ft. cone @ 110 ton/hour can be obtained, while a 7½ ft cone can treat anthracite @ 8 tons/hour.

This method is advantageous in cleaning mixtures of coarse particle (¼ or ⅜ in.), (1) where there is a strong difference in sp. gr. of the float and sink elements and (2) when the light fraction is liberated by crushing to ¼ in.  Where finer crushing is required the installation is more costly.

(5) Revolving round table

In this equipment, the table has a conical surface, the apical angle of which is very large obtuse. The feed is placed on the table which is slowly rotated. A circular launder surrounds the table and wash water is spread over the surface by a helical pipe. Due to the radially inclined surface of the conical table, particles move down towards the peripherial launder. Particles of varying specific gravity are acted differentially, by the water current, gravity, and the rotatory motion. Thus separations are affected into a concentrate. and middling, and tailing. The multiple deck table is also used to increase the capacity. This type of table has been used successfully in the separation of cassiterite, that is a mineral of high value but not amenable to flotation.

## (6) Classifier Separators

Classifier separators are used to produce rough separation, of the valuable mineral and gangue, where the former has a low price value. In this class of machines, the *Humphrey's spiral* (Fig. 12.47a) may be considered as the type. This consists of a spiral launder with a curved bottom, built up in the form of a spiral. The axis of which is held in a vertical position. When pulp of low density, after desliming, is fed into this spiral, the lighter minerals, which are held in suspension, readily acquire greater tangential velocity, as they move down and round the spiral, while the heavier minerals are less affected, tend to remain closer to the

Fig. 12.47a

centre of the launder, and to the bottom of the spiral. Coarser grains, which also acquire a higher velocity, travel towards the outer edge of the spiral. Thus separation is affected within the trough or launder of the spiral. Cocks or bleed-offs are set in the medial position of the spiral to tap the concentrate and the middlings, while the tailings flow downwards to the bottom of the spiral.

### (7) Pneumatic concentrator

These are akin to the hydraulic concentrator, excepting that air is used as the fluid. These consist of pneumatic tables and pneumatic jigs. The pneumatic jig consists of a deck covered with the porous material or thick fabric supported on an iron netted frame. The deck is provided with riffles. Below the deck is a leather bellows, from which air is forced, through the porous deck, in short impulses, through the action of an eccentric. The deck is thus subjected to jigging action. Between the riffles the heavy minerals settle, at the bottom and the light minerals settle at the top. The light minerals overflow first, from the riffles, and then the heavy minerals follow. Thus a separation is affected.

### (8) Dry rocker

Dry rocker is used in working gravel deposits which are dry and can be easily disintegrated. The disintegrated gravel is thrown into the screen and the finer particles, passing through, fall on the riffle deck on which a tough fabric or finely woven screen is stretched. Below, the riffle deck, is a bellow, with the help of which air is blown, through the riffle deck, in short impulse so as to produce a jigging action. The action is similar to that of the pneumatic jig. The pneumatic table is also very similar in its action. The deck is made from cellular metal grid with diamond shaped holes.

The deck is also riffled longitudinally, and is tilted. The action of the table is similar to that of the jig described earlier, but in addition, to the jig action, produced by the impulses of air from below, the table is also subjected to shaking, as well as to a flow of materials caused by the tilt. In the operation of the table, the feed is to be controlled, together with the fluidity of the bed. Adjustment of the time factor is also necessary for ensuring efficiency.

## (F) FLOTATION

### General principles

Flotation is also a method of gravity separation in the ultimate. In this method, advantage is taken of certain properties of minerals by which air bubbles can be attached to their surface, thereby decreasing the effective specific gravity of the mineral concerned. The apparent loss of weight, of the mineral, caused by the attachment of the bubble may not, at times, be sufficient to make the particles float, but sufficient to produce a separation on a shaking table or in a hydraulic classifier. If, however, the froth causes the mineral particles to float, the process is known as froth flotation. When the particles do not actually float, the process is termed agglomerate flotation or oil air flotation. The factors which serve to achieve flotation are (1) fine g      g to at least 35 or 48 mesh, (2) low pulp consistency, in which the sol.     utent is 15 to 35 per cent, (3) the addition of small quantities of one or more organic conditioning agents, (4) the introduction of a collector reagent, which coats the minerals to be floated with a water repellent film, (5) the addition of a frothing agent, for stabilising the bubbles, when they reach the surface, (6) aeration which may be induced either by agitation, or by air injection, for producing fine bubbles of gas and (7) separation of the froth or bubbles, containing the floating mineral particles, which were not coated by the collector regent. Often the various stages are adjusted, in such a manner, so as to save time, in the process of beneficiation. Thus, sometimes, the collector and conditioning agents are combined in the grinding circuit. The dilution, for keeping the solid content low is carried out in the classifier, while the frothing agent is added to the flotation machine, during aeration or often at the earlier stages, in the classifier or ball mill.

The method is useful to separate minerals which differ, in such a way, that one has substantial amounts of a metallic elements or acid ion which is not found in the others. It is particularly useful in the extraction of sulphide ore minerals associated with siliceous gangue. Oxidised heavy minerals can be removed, from siliceous minerals and so also, coal, graphite, and sulphur.

The principles underlying the process of flotation comprise a system which may roughly be taken to include both dynamic and chemical equilibria, in which mechanical processes as well as chemical reagents play an important part. This can be divided into four main parts: (1) conditioning, (2) collection, (3) levitation and (4) frothing.

## (1) Conditioning

Conditioning is essential in selective flotation. Conditioning, which aims at affecting collector coating, on a desired mineral group, is called activation, whereas conditioning which aims at preventing the entry of undesirable minerals, into the overflow, is termed depression. Conditioning time differ for various ores and in various treatment plants. The usual period, in the zinc-lead treatment plants, is less than 5 minutes, in the lead circuit conditioner, and about 15 to 30 minutes, in the zinc conditioner. Conditioning time is an important factor, and may affect efficient recovery, both when extended or when cut short.

## Conditioning Agents

Acids are normally used to control the pH, except hydrofluoric acid, which is used to precipitate ferrous, ferric, aluminium and earth metal ions. Hydrofluoric acid prevents re-surfacing of quartz by these ions and thus helps soap flotation of the other silicates than quartz. It also acts as a dispersant for clay and altered felspar even in acid solutions. Sulphuric acid is normally used because of its low cost. Quartz and other silicates are depressed in soap flotation by the supression of ionisation of silicic acid. Thus activation produced, by both metal and collector coating, is prevented. Depending on pH and on the nature of soap coating, acids convert metal soap coating to acid soap. At pH 7 and below, neutral soaps, of heavy metals, exist, while soaps below this range are largely acidic. Carbonate ion is used to precipitate heavy and earth metals and prevent resurfacing.

Cement is employed to depress pyrite and pyrrhotite. Chromate and dichromate are used for surface closure of galena, with selective flotation of a lead-zinc mixtures by depressing galena. Reactivation is affected by ferrous sulphide, or hydrochloric acid with or without sodium chloride, or sulphuric acid with sodium hydrosulphite, bisulphite. Copper ion is used to resurface, sphalerite, either in acid or alkaline pulp. Cyanide ion, as Na or K salt, is used to depress sphalerite, iron sulphides and arsenopyrite, when using sulphydric collectors. Lime and copper sulphate increase the activity of cyanide ion on iron minerals. Acid pulps are rarely used, in sulphide flotation. Dispersion of anionic collectors is hampered by acid pulps, while acid pulps assist the action of cationic collectors.

Acidity, however, helps to clean oxide crust and thus assists flotation. Acidity, generally, causes flocculation of non-sulphide slime, without dispersing the sulphides; thus causing slime coating on sulphides. Froth, in acid pulp, is heavy in loading, and not suitable for vigorous aeration in pneumatic machines.

Alkaline pulps are common and are effective in precipitating heavy metal ions and in complexing amphoteric metals at low pH. Such a pulp is a depressant for pyrite and other iron sulphides. Alkaline pulps are, generally, formed by the addition of lime which acts as a dispersing agent, particularly for non-sulphides at low concentration. Excess, however, affects froth adversely. Bubbling is done by weak acid. Lime is invariably used as a hydroxyl carrier, except where it has undesirable effects. Lime depresses quartz or slimes, containing talc, in the presence of sulphydric collectors. Pyrite and sphalerite may be depressed under the same conditions. Excess lime prevents galena from floating. Organic colloids such as glue, gelatin, albumin, casein containing hydrophylic groups, of both basic and acid nature, i.e , amine and carboxyl, are of the nature of proteins, and are depressants. Others are tannic acid, saponin, and carbohydrates such as dextrines and starch. The latter substances, form multimolecular aggregates, in water, and precipitate on solid surfaces. Whatever be the mechanism of coating, the surfaces resist the collector coating and levitation ; and when in excess, colloid precipitates do prevent flotation. Sodium silicate is the most common reagent for depressing quartz and silicate minerals, and also to disperse siliceous and iron oxide slime. Excess causes the formation of brittle froth. Sulphide ion, obtained from sodium sulphide, is employed in the flotation of oxidised metalliferous ores. It prevents ineffective multi coating by collectors by forming a mono film sulphide coat, which is resistant to further oxidation. Excess sulphide prevents coating and the froths produced are barren and watery.

Wetting agents are generally used for dispersion. Sodium silicate, sodium phosphate and sodium carbonate fall under this class. Since some of them give rise to excessive frothing, their addition is sometimes a disadvantage. Zinc ion together with cyanide ion is used for wetting, in the case of sphalerite.

## (2) Collection

Collection is the process by which selective water repellent coating, is produced on the mineral particles, which are to be separated from the pulp, so as to enable the attachment of gas bubbles to these alone. Organic chemicals, added to the pulp, in agitation, produce the desired effect. The collecting agent used in flotation may be divided into (1) anionic collector, (2) cationic collector and (3) oily collector. Of these groups of collectors, the anionic collectors are, by far, the most important, in the present day flotation process. They have been marketted under various trade names, by various manufacturers. These collectors can be classified, according to the active acid groups, or the hydrocarbon loading.

Hydrocarbon loading determines the solubility of the collector and the water repellent properties, while the acid group determines the extent of attachment of mineral particles.

### (a) Anionic collectors

The common collectors are of the acidic type or the carboxylic (R.COOH) and the sulphydric types. The mechanism of reaction, in such collectors may be expressed thus, X-SH, in which X is one of a general variety of loadings, which always contain hydrocarbon groups linked to sulphur, usually through carbon and occasionally through phosphorus. The sulphydric type collectors $R.SO_3H$ and $R.SO_4H$ are typical. The H, in these collectors, is acidic and as such this may be used in the form of a salt, in which H is replaced by Na or K.

In the sulphydric class, xanthates are the most important and they are used in the recovery of heavy and precious metals, sulphides and oxidised minerals, except for earth minerals. They are produced by the reaction of carbondisulphide, alcohol, and strong alkali such as NaOH in water. The effectiveness, of xanthates, increases with the molecular weight of the alcohol. Ethyl, propyl, butyl and amyl are effective in separating copper, mercury and silver minerals. Xanthates are more commonly used for the separation of sulphides, except sphalerite and pyrrhotite, and these are used even without activators. Xanthates are very powerful collectors, and are used in very small quantities, say between 0.04 to 0.08 per cent. On account of their efficiency xanthates are utilised in scavanger cells in the place of the aerofloats as collectors. The quantity of xanthates used, in flotation, is 0.2 lbs. per ton or less. Xanthates are used, perferably in weak alkaline solutions, as the bubble attachment is difficult when alkalinity increases. In acid solutions, the salt becomes acidic with a decrease in dispersion. Thus more time and more reagent are consumed for coating. More acidic solutions cause the xanthate to decompose.

Thiophosphates belong to the sulphydric class. They are produced by the reaction of phosphorus pentasulphide with various organic compounds, like phenols, alcohol, mercaptans, thioalcohols, amines and nitriles. The reaction products, of phenols and alcohols, are most commonly employed as collectors and go under the trade name of aerofloats. These compounds are highly soluble in water, and form relatively insoluble salts, with heavy metals, but not with earth metals. The consumption of aerofloat reagents is of the order of 0.25 lbs. per ton or less.

The mercaptans and thioalcohols have the structural formula R.SH where R is respectively an alkyl or aryl group hydrocarbon substituent, and the SH group is acidic. These have comparatively poor solubilities in water and acidic pulp, but are ionised as salts in the presence of alkaline compounds of sodium and potassium. They have been used for selective collection of copper and zinc minerals, in the presence of pyrite.

Diphenyl thiocarbazide has been suggested for use as collector for nickel and cobalt minerals in the presence of copper and iron minerals. Mercaptobenzothiazole is used as reagent for copper and zinc sulphide. When sulphydric collectors are used in oxidising conditions, they form organic sulphides and di-sulphides, many of which are of the nature of oily liquids at atmospheric temperatures. Dixanthogen, formed by the oxidation of xanthates, is used in minute quantities for increasing the action of typical sulphydric collectors.

The carboxylic collectors contain the carboxyl (COOX) group, in which X represents acidic hydrogen or a base. In this group of collectors are included fatty and resin acids, and their alkaline salts, known as soaps. They have been used in the recovery of base metals and earth minerals. They have, however, been substituted by sulphydric collectors to a large extent. The main use of fatty acids and soaps is in the floating of earth metal minerals when the carbon content of the collector should preferably be greater than $C_8$ or preferably $C_{12}$. The earth minerals can be selectively floated by varying the pH. Fatty acids are used as collectors for iron and manganese oxides more, especially when gold is associated with these oxides. Copper carbonate can also be floated, as well as iron sulphides, by employing differential flotation.

As fatty acids are insoluble in water and mechanical dispersion is essential, alkaline pulp and soaps are used to help emulsification and dispersion, while in the case of acid pulp. a reagent such as sulphonated alcohol is necessary. After the removal of the slime, if the pH is not too high, about 0.5 lbs. of oleate soap will be necessary for a ton.

Sulphoxy collectors are used for both sulphides and non-sulphide minerals.

### (b) Cationic collector

Cationic collectors are truly organic compounds which are readily ionised, the cationic component being represented, in the hydrocarbon and reactive groups. Such compounds have the formula $R - M - (R')_n - Y$, in which R represents the hydrocarbon group, and R' is either hydrogen or the same hydrocarbon group as R or a different hydrocarbon group. Pentavalent nitrogen or quadrivalent sulphur, are denoted by M which is equal to the valence of M, reduced by 21. Y represents a hydroxyl or an acid ion usually a halide.

Amines are cation collectors, and, their behaviour is truly so, in respect of non-metallic minerals, where ion exchange with the hydroxide or salt, takes place. In the case of metallic minerals, the amines do not react in the same manner. Amines are derivatives of ammonia obtained by the replacement of one or more hydrogen atoms by the hydrocarbon radical. According to the degree or extent of replacement, amines are designated by the prefix mono, di or tri, amine. They react in water and with acids in the same way as ammonia does to form hydroxides. Cationic collectors are essentially used in the flotation of silicate minerals including quartz, and their use can be extended in beneficiating tungstates, molybdates, vanadates, chromates, and arsenates.

Cationic collectors, in which nitrogen, sulphur and phosphorus of high valencies are coordinated between a hydrocarbon loaded cation and an anion, are designated "onium" compounds. The nitrogen equivalent is called pyridinium. The sulphur compound is sulphonium and the phosphorus compound is phosphonium.

The main application of these cationic collectors seems to be in the flotation of talc, micas, kaolin, and silicate type slime. Primary micas, zircon, kyanite are also amenable to the action of these collectors.

The quantity of cationic collectors required, is of the same order as that of soaps, except in the case of onium reagents with a long alkyl groups. Oil assists in the collection, when (low −C) cationic reagents, are employed, especially in the case of the weak ones. Oil forms a solution-coating of hydrocarbon surfaces with low water repellent properties, but oil-water selectivity is sufficient to cause oil to spread over them. Thus the surface, for bubble attachment, is formed by neutral hydrocarbons requiring no orientation. The nature of the frother used, with such collectors, is an important factor. Pine oil forms brittle froth, but it is capable of correcting the tendency for excessive frothing, due to the collector. Pine oil tends to float non-silicates, if a fatty acid is present. Excessive froth can be corrected by the use of higher alcohols Cresol, when used, should be free of carboxylic acid or neutral oils which tend to float non-silicates.

Oily collectors are, generally, good solvents for hydrocarbons which are, to a large extent, insoluble in water. Thus they tend to spread out as thin films, at solid surfaces, to produce bubble attachment, in the presence of water. These, of necessity, do not contain elements which produce reaction coatings on minerals. Examples of such oils are the neutral hydrocarbon liquids of petroleum, of wood, and coaltar, which have a boiling point between 350° to 450°F. with a low viscosity, at low temperatures and pressure, so as to enable feeding, and uniform dispersion in the pulp. Freedom from frothing is an additional important property.

Oils are used in table flotation, as anciliary reagents, to increase bubble stability. The bubbles produced do not have the tendency to spread out and survive movement, when they are mechanically pushed about. The use of oil is not indicated where differential flotation is resorted to. Oil, however, has an advantage in the flotation of metallic oxides, specially oxide minerals, for example oxidised gold ore. If the oil is sufficiently fluid, and the pulp is more or less free from slime then 0.5-1.0 lb. of oil per ton will suffice.

(a) *Activation*

Effective bubble attachment is obtained only when the coating, of the reagent has the same orientation in respect of the electric charge. This is obtained only when the reagent produces a monomolecular film. Where the reaction continues further, a thicker coating will result. Consequent to the variation in thicknesses the orientation, in each layer of the coat, will also vary, resulting in the particles having both polar and non-polar elements. Such a condition is unsatisfactory, for efficient attachment of the bubble to the surfaces. This is common in the case of relatively soluble mineral surfaces, where collectors, with small ions, are used. In such cases, the solubility of the mineral surface is reduced prior to coating (1) by using sodium sulphide, (2) by removing the extra thickness, of the coat by attrition mixing and (3) by preserving the first formed coat, with mono molecular oil film. Where slime prevents the formation of coating, then acids, alkalies, alkaline silicates, phosphates, as well as organic colloids, are employed for dispersion. In attrition mixing, the thick pulp is churned mechanically, in a tumbling mill or beater box, of the agitation-froth machine. Resurfacing is commonly practised, the best example being the use of copper sulphate to resurface sphalerite, where copper ion exchanges with zinc to produce a copper sulphide (covellite) coat. Sulphide resurfacing is also possible, within the following groups—(Hg:As) (Ag:Sb) (Cu;Pb:) Sn; (Bi:Zn) (Fe:Co) (Ni:Mn) The purpose of resurfacing is to form on the surface, of the mineral to be floated, an anion layer, which will be influenced by the collector and make it susceptible to levitation. Hence the nature of the collector and that of resurfacing obtained are inter-related. Reagents, used for resurfacing, with anionic collectors, are salts of heavy metals and earth metals, while silicate ion and metals acid ion are used with cationic collectors.

## (b)  *Depression*

The main purposes of depression are (1) to prevent activation resurfacing, due to the presence of soluble salts in the pulp, (2) to prevent collector reaction on surfaces, (3) to nullify or remove collector coatings, (4) to aid dispersion and (5) to enable resurfacing and to induce the affinity between water and mineral particles. With regard to the first property, precipitants, for heavy metals used are hydroxyl, carbonate, silicate, and sulphide ions. The earth metals require greater concentration of hydroxyl, for precipitation, while the basic carbonates, phosphates, silicates, and fluorides together with tungstates, molybdates of the earth metal, are relatively insoluble. The complexing ions are cyanide, fluoride, fluosilicide and silicate Since alkaline earth ion tends to activate quartz and alkaline silicates. In alkaline solutions, the precipitation of these tends to depress the gangue minerals, when fatty acids, and other collectors are employed. With regard to the second condition, sodium sulphide is the best example. Similar reaction can be produced by the use of cyanide or pyrite, while the addition of silicate ion causes a setback to ionisation of the surface, of the

silicate minerals, in respect of both cationic and anionic collectors. The
action of the sulphuric acid on quartz is similar.  With regard to the third
condition, the example of copper  molybdenum concentration, made with
sulphydric collectors, may be cited.  By heating, removal of only the
sulphydrate coating compounds and oily contaminants, is affected. Particles
floated by acid and alkali soaps are  also coated  and  ionisation decreases
the tendency for levitation, hence such coatings also have to be removed.

With regard to dispersion, it may  be  stated  that  soaps  of  low pH
range, both of the acid and  basic  types,  produce  dispersion.  Sulphydric
collector and neutral soap collectors, with carboxy collectors, cause flocculation while the uncoated particles  are  dispersed.   Such  a  condition  is
usually welcome,  as it affects clear  separation,  between  granular  values
(ore minerals) and  slime, and between slime, gangue and  collector  coated
particles.  The  last aspect of depression is  not  extensively  used.  The
depression of pyrite by cyanide may be taken as example, and so also  that
of galena and  barytes  by acid dischromate. Organic  colloids also produce
hydrophylic coatings, and thus cause depression.  With the addition of
small amounts of colloid, the  coating is generally selective,  and unless the
pH  is equal to that of the  general iso-electric point of the  coating, colloid
dispersion  is  not affected. At the iso-electric point,  however,  the  coated
particles  flocculate.   Excess of colloid coat produces flocculation.

### ( 3) Levitation

Levitation consists of attachment of bubbles  to  collector  coated
mineral particles, which cause mineral particles to rise up and  the  subsequent  separation  of  air mineral aggregates, from the others, by differential sedimentation and skimming of the floating  particles.

### (4) Frothing

*Froth* is an aggregate of bubbles.   The duration or life of the bubble
is an essential factor in  frothing.  The bubbles should have  life,  long
enough to enable differential draining of gangue minerals into the pulp, as
well as enable the horizontal transport of the concentrate, to the discharge
point. Large size bubbles with comparatively  thick  walls,  and sufficient
internal movement are ideally suited.

Frothing agents are substances which go  into  solution  readily  and
change the surface tension of the solvent. When the surface tension curve
is negative, the absorption is positive, and  the  reagent  is  concentrated at
the bubble surface.  When the surface tension is  negative,  the  reagent is
concentrated away from the interface.  Good  frothing  agents,  produce a
curve with an average initial slope of 1 to 2 dynes  per  centimetre,  per 10

parts per million of water.  Chemically speaking, frothing agents are polar
organic compounds, represented by the structure RA, in which R represents a hydrocarbon radical containing, generally, more than six carbon
atoms, and A is a pulp solubilising group such as hydroxyl, carboxyl,
carbonyl, amine etc.  In general, better frothers contain more powerful
solubilising groups.

Frothing agents, when absorbed produce solid films with similar
orientations.  Such films may be composed of the hydrocarbon groups
etc. in contact with the gas phase.  In the case of good frothers, with a
sufficiently high percentage of hydrocarbon bulk, the attraction of water
is resisted by the tendency of the polar radical to be enclosed by water.
Since the concentration of hydrocarbon is low, its solubility is too low
to permit molecular dispersion.  The common frothing agents are pine oil,
creosote, celic acid, eucalyptus oil, cresylic acid, liquid aerofloat mixture
of aliphatic alcohol, sulphates and sulphonates of alcohol, etc.

**Flotation machines**

Flotation machines employ one of the two principles, namely to
produce either a liquid to liquid interface, or a gas to liquid interface.
The liquid used, in flotation, may be an organic liquid, while the gas used,
is essentially one that is immiscible in the liquid.  The flotation machines
used, can be broadly divided into frothing or gravity type.  Non-frothing
machines, use various principles, which involve gravity, and a combination of gravity and mechanical devices, etc.

In present day practice, the frothing machines are of major consequence, and as such they are dealt with at length, while only a passing
reference is made to non-frothing machines.  Machines of both types can
be broadly grouped under internal or pulp body, and external or bubble
column machines.  Both, of course, use froth for affecting separation ;
but the difference is fundamental, in respect of the place of selection, as
well as in the mechanism of the selection itself.  Of the two types, of
machines, the bubble column machine plays a major part, in flotation. The
three main classes of this type of machine are (1) pneumatic, (2) cascade
and (3) subaeration  The concentration is affected by the bubble column
in all these machines.  Large volumes of air are consumed, the consump
tion being of the order of about 1000 to 2000 cu. ft. or more per cu. foot
solids floated.

The pneumatic machine (Fig. 12.48) is the simplest of the bubble
column type.  It consists of a elongated open cell or trough (1) which
is subdivided (2).  The pulp flows from one end to the other.  Air is
delivered at the bottom (5) of the trough either from a compressor. placed

outside the machine or through a blower. Air is introduced either
through pipes or through a
porous fabric placed at the
bottom of the cell or trough.

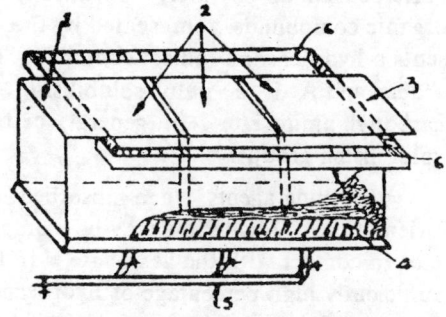

One of the machines, where
air is let in through a fabric
covered chamber, placed at the
bottom of the cell, is known as
the blanket type. The Callow
flotation cell is an example
(Fig. 12.48).

Fig. 12.48

(1) is the feed box; (2) cast iron pass of the cell; (3) tailing discharge
box; (4) porous fabric compartment; (5) compressed air manifold; and
(6) the launders.

In operation, the machines are simple and the froth, carrying the
concentrate, is skimmed away mechanically into the launder (6), to be
filtered and dried.

The typical air lift machine is the Forrester cell (Fig. 12.49). The
machine is called air lift, because it utilises the same principle as that of

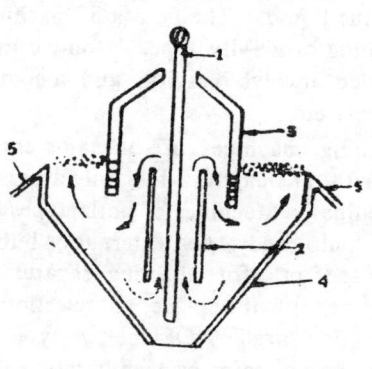

Fig. 12.49

an air lift pump. Thus when
air is injected, through the
pipes (1) placed between baffles
(2), in the central part of the
flotation cell (4), the water
enclosed forms a mixture with
a lesser specific gravity (due
to admixture with air). Thus,
this column of water, between
the baffles, rises to the top. In
this machine considerable agitation is produced in this
column, which may be termed
as air lift zone. The rising column of water produces a convection current
within the cell. Thus particles at the top are again drawn down to the
bottom, to rise again up the column, as shown by the arrows. Perforated
baffles, placed on either side of the column, preserve the floating froth
from being drawn in, by the convection currents. The froth is drained
off, through the hoppers (5). As in the case of the Callow cell, the froth,
carrying the concentrate, is collected, mechanically and conveyed to the
filter and drying sections.

In the casade type machine, aeration is produced either by a jet of water or by a jet of pulp, directed against the main body of the pulp. When such a stream is directed from a distance, slightly greater than a few times the diameter of the water column, it breaks up into a chain of drops. The number of units, in the chain of drops, increase with increasing velocity and distance. Thus a substantial volume of air is carried into the pulp and compressed. The air, thus introduced, breaks up into a train of smaller bubbles, essential for the formation of the bubble column. Cascade machines, are not very extensively employed as the control of aeration is difficult, and not entirely satisfactory.

The subaeration machine have a variety of designs, but all of them aim at maintaining a column of bubbles over the pulp in the box or cell. These can be classified under two main types. In one type (Fig. 12.50) there is an impeller (4), which is submerged in the pulp and this acts as a

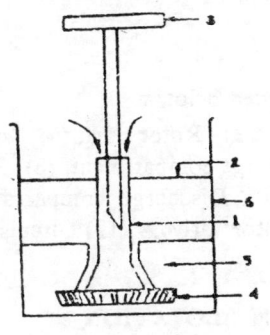

Fig. 12.50

centrifugal pump, when it is rotated by a pulley (3). In these machines, the impeller shaft is hollow and passes above the surface of the pulp Thus air is drawn into the shaft as shown by arrows and delivered at the impeller end ; where it is broken up into bubbles. Baffles (5), placed inside the cell (6), prevent gyration of the liquid as a mass. In the second type an external arrangement is required to deliver the air below the impeller. A compresser or an air pump is required for this purpose.

In some machines, low speeds are employed, while with others, double shrouded impellers are used. In these, the bubble column is pure, as the impeller merely serves for dispersing the air bubbles and distribu-ting the same under the pulp, lying over the impeller area. In the high speed machines, impellers are used to reduce cavitation effect. The preci-pitation selection is done in the impeller zone, and the selection is affected by pure ‑bubble column action. Of the subaeration machines, the Denver Sub-A is one of the best known ones. A second machine belonging to the same type is the Fagergren machine (Fig. 12.51) In this machine the rotor (4), consists of a cylindrical cage built up of rods. Cavitation is produced when the rotor is moved by a shaft (12) through the pulley. Thus air is entrapped and broken up into bubbles. In addition compressed air is intro-duced into the cell (6), by pipes and this enhances the froth formation.

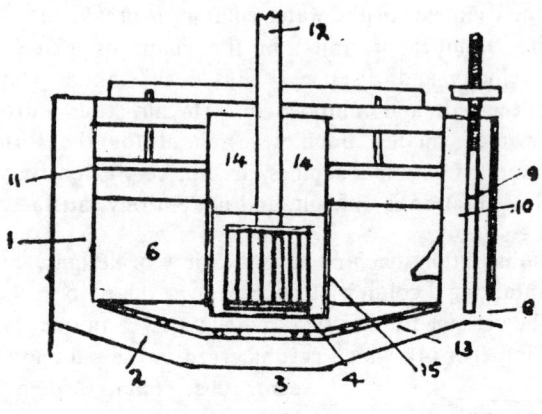

Fig. 12.51

The essential parts of the machine are listed below :

(1) Feed Compartment, (2) Feed arm, (3) Rotor port, (4) Rotor, (5) Stator, (6) Tank, (7) Baffle port to discharge compartment, (8) Sand gate, (9) Sand gate shutter with capstan, (10) Discharge compartment, (11) Skimmer, (12) Shaft coupled to motor drive, (13) Circulation ports, (14) Stand pipe.

## GENERAL OBSERVATIONS ON FLOTATION

Metallic sulphides are the most common minerals, to which flotation is applied. Where the ore comprises a single sulphide the matter is relatively simple, but when the ore is a complex assemblage of sulphides, the problem becomes involved.

In general, the problem of separation, by flotation, is to prevent the gangue from entering into the sulphide fraction. For achieving this result, the pulp is usually conditioned, with a pH of 7-9, by the addition of lime. Ethyl xanthate is employed as collector, while the frothers used consist either of pine oil or cresylic acid. The separation is clean, if the ore is not oxidised.

Micaceous materials are prevented, from entering the concentrate, if (a) the flotation is carried out with a minimum quantity of frother with little aeration, before the addition of the collector, (b) if conditioning is carried out, for depressing either with sodium silicate or with small quantities of glue, casein or dextrine or starch; and by the addition of a small quantity of frothing agent, when floating.

Clay is another member of the gangue materials, which can also be removed, by carefully controlling the pH, to obtain maximum dispersion. This takes place very close to pH 7, on the alkaline side. The addition, of sodium silicate or sodium sulphide aids a cleaner separation, and the addition of creosote affords additional assistance.

If the sulphide ores are not oxidised, then, their amenability to flotation, in the decreasing order, is copper, lead, iron and zinc. When sulphydric collectors are used, with no addition of conditioning agents, copper sulphides respond to flotation excellently, and the recovery may be as high as 95%. Normally ethyl xanthate is employed as the collector, while the higher order of xanthates and mercaptans etc. are recommended where pyrite occurs in association.

Galena is activated by copper and may acquire a coating of copper, at high concentration $OH^-$. With a pH of 10.4, it is depressed. In order to achieve this, sodium carbonate is employed, in preference to lime. A low concentration of $S^-$ also causes galena to be depressed, while dichromate has a stronger effect. Alkaline sulphites and $FeSO_4$ are used for reactivation, while HCl, NaCl in acid solution, are employed for chloridisation. The collectors used include sulphydrates, fatty acids, sulphates and sulphonates of fatty acid, alcohols. Aerofloat acts excellently after activation, but with copper sulphate the activation is superior.

In the case of iron bearing sulphides, such as pyrite, marcasite, pyrrhotite, and arsenopyrite, flotation is necessary, particularly when gold is associated. Fatty acids are employed in acid pulp, while higher xanthates are used in either acid neutral or alkaline pulps. Oxidation of sulphides is rapid, in the grinding and flotation circuits, and due to the removal of oxygen, the pulp becomes reducing. The resultant ferrous sulphate hampers flotation of other sulphides. This effect is most pronounced in the case of arsenopyrite and pyrrhotite. In order to eliminate this condition, precipitation by the addition of hydroxyl and/or carbonate ion, is resorted to. In the case of sulphides, the practice is to depress, by the use of cyanide and aminophenol for pyrite, starch for pyrrhotite, and lime and cyanide for arsenopyrite.

In the absence of activating ions, sphalerite can be floated with fatty acid, in acid pulps. By using the higher xanthates the same results can be achieved. Resurfacing can be attained by the use of As, Sb and Pb for flotation, in acid solution with cetyl xanthate ; while Ce, Pb, Cd, Cu, Ag, Hg, Bi are utilised, in activation, for flotation in neutral pulps.

Differential or selective flotation is employed to separate each mineral species, from an aggregate, and it has been used, in the case of polymetallic sulphide ores. Examples of such associated mineral separations are (a) galena and sphalerite, (b) galena and pyrite, (c) sphalerite and pyrite and (d) chalcopyrite and sphalerite.

The general procedure adopted involves (a) flotation of one of the sulphides, (b) depression of other sulphides, and (c) selecting the most suitable collector with lowest molecular weight for flotation. Conditioning, for the first mineral for flotation is accomplished, in the grinding circuit. The froth is made, as fragile as possible, to enable maximum recovery.

(a)   Galena-sphalerite flotation is carried out in the presence of lime or sodium carbonate which are added in the grinding circuit. The pH of the pulp is maintained between 8 and 9.5. Ethyl xanthate or lower aerofloats are used, in small amounts, as collectors, together with wood or coal tar creosote. Froth formation is also kept at a minimum.

Sphalerite is usually floated by the addition of copper sulphate for activation. Xanthates, of the higher order, are employed as collectors. Frothing is encouraged to the maximum. A middling comprising sphalerite rich in galena, may be separated for return either to the lead circuit or to the grinding circuit. Sufficient lime is added to keep down pyrite. In the absence of pyrite oelic acid is used, for floating sphalerite without activation of copper.

(b)   Galena pyrite combination is separated by the use of sulphydric collectors with cyanide.

(c)   Sphalerite is separated from pyrite by activation with copper sulphate and by the depression of pyrite by the use of lime.

(d)   The separation of chalcopyrite from sphalerite is most difficult and the useful practice is to add soda ash and cyanide, to the grinding circuit, to avoid resurfacing of sphalerite with copper, and then float the sphalerite, with a strong collector, by adding copper sulphate, if necessary, and lower the cyanide ion concentration below that necessary for chalcopyrite.

In affecting the separation of galena from chalcopyrite, both sulphides are recovered together in the first operation. The galena is then depressed by the use of dichromate or alternatively chalcopyrite is depressed with cyanide.

Non-metallic minerals, are also amenable to flotation.

1.    Barite is readily floated by fatty acid collectors, in the presence of sulphate and sulphonate salts, of respective alcohols, the optimum pH being 10-11. Sodium silicate is a depressant for metal-salt-silicate and also dichromate for cleaning.

2. Calcite floats from siliceous gangue, by the use of fatty acids, with sodium silicate, lignin sulphonate or dextrin at a pH of 8-9.5. This is the common technique used in cement plants when the limestone requires beneficiation. The steps taken comprise (a) desliming of the pulp, (b) flotation of carbonaceous impurities, by restricting the amount of frother to $C_7$ or $C_{10}$ monohydric alcohols or cresol, (c) dispersion of the remaining pulp with calcium lignin or dextrin and (d) flotation of calcite with emulsified fatty acid, so that mica and quartz are removed as tailings. Alternatively, mica and silicates can be flotated with cationic reagents at pH 7.5-8, so that calcite and iron minerals form the tailings.

3. Dolomite can be floated by soaps. It is actually depressed by metal-salt silicate and by acid dichromate.

4. Fluorspar flotation requires (a) desliming, (b) the use of soap and oelic acid as collectors, together with a dispersant and depressor, best suited for the gangue. The optimum pH is 8-9.5. Silicate gangue is depressed by sodium silicate or quebracho (gum). Calcite and barite are separated by metal salt-silicate treatment.

5. Gypsum is floated with soap type anionic reagents. e.g., oelic acid, with low alkalinity and a controlled frothing.

6. Magnesite also responds to oleic acid, but it is difficult to separate the associated calcite, dolomite, talc and serpentine.

7. Phosphates are amenable to flotation, and both froth and table flotation are employed. Fatty acids with the pH controlled between 8 and 9 with sodium silicate, soda ash, or caustic soda, are used. The frothers are resin or turpentine, in froth flotation. No frother is required in table flotation. Florida phosphate as well as Russian phosphate ore with apatite and nepheline, respond to this treatment.

Silicate minerals are also beneficiated by flotation. The following are a few important ones :

(a) Andalusite is floated with soaps; (b) Felspar is floated with soaps with cationic reagents; (c) Kyanite is floated with anionic reagents, using sodium silicate or a phosphate salt for dispersion, with a frother such as pine oil; (d) Mica is floated with either anionic (fatty acid) or cationic collectors; (e) Quartz floats with the more powerful cationic reagents; (f) Sillimanite in schist is recovered after desliming the pulp, and by the use of oleic acid and soda; (g) Zircon is also floated with oleic acid: (h) Graphite is readily floated by using pine oil. One of the draw-backs, in this method, is that graphite coated particles of gangue also float, and give rise to contamination. This can be minimised if the process is repeated or gravity concentration applied. Pine oil is used for recovering molybdenite, talc, etc.

## (G) MAGNETIC SEPARATION

The methods of magnetic separation can be broadly divided into two groups. (1) Those which are used for dealing with wet materials. (2) Those methods used for treating dry materials. In addition, these two categories can further be sub-divided, in accordances with the types of machines used (a) those using coarse or fine feeds (b) and those employing feeds with high, low or medium magnetic susceptibilities. In addition, the third method, of broadly classifying magnetic machines, depends on the manner in which the feed is acted on by the magnets. In one type of machine, magnetic material is directly picked up by the poles, and this is known as *holding type*. In the second type, magnetic particles, are picked up as the feed moves in a stream close to the poles. The material does not actually come in contact with the magnets. This type of machine is called the *pick up type*.

In the holding type machines, the feed is in contact with the collect-ing surface, but the magnetic material is removed either by the action of flowing water, or by some fluid, and then collected separately. The magnetic material can also be removed, either by the mechanical stripping of the poles and collected by gravity, or alternatively the collecting surface can be removed beyond the magnetic field, so that the material, adhering to it, is released and dropped down by gravity.

### (A) Holding Machines

#### (a) *Drum separators*

Drum separators are one of the types among the holding machines. This machine consists of a non-magnetic hollow cylinder (1), generally made of brass or bronze, inside which a cylindrical segment electromagnet (2), is located co-axially (Fig. 12.52).

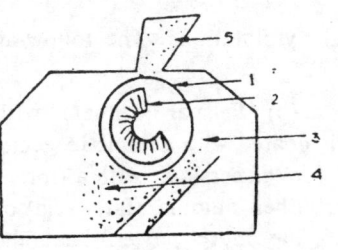

**Fig. 12.52**

The segment, of the cylindrical magnet, covers about half or 2/3 (two thirds) of the drum circumference. The feed is dropped on the cylinder through a hopper (1). When the brass drum (1) is rotated, the magnetic particles, adhering to the surface, are released, as soon as

they are transported beyond influence of the magnetic field.   The tailings
and middlings (4) drop nearer, whereas the concentrates (3)leave the drum
last.

The magnetic drum is, essentially, used in low intensity rough
separation, as in the concentration of iron ores, where clean concentrate
is made from middlings.

(b) *Other types*

(i) Magnetic pulley separators consist of two pulleys, one of which
is magnetic (1), while the other only serves as a return pulley (2).   The
two pulleys are connected by a flat, wide
rubber belt (3).   The magnetic pulley is
formed by a number of units comprising
horse-shoe and electromagnets assembled
on a common shaft.   The ore is fed on
the belt from a hopper (4).   As it passes
over the magnetic pulley, the magnetic
portion (5) is held against the belt by the
attraction of the magnet; whereas the

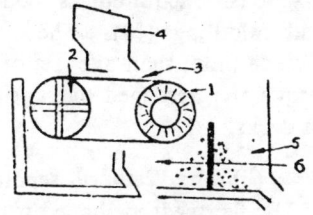

Fig. 12.53

non-magnetic portion (6), is thrown away from the moving belt, due to
acquired momentum.   Separation is thus affected (Fig. 12.53).   These
machines are generally used to remove tramp iron, as well as iron, which
is present in finished metallic products.

(ii) The induced roll separator, essentially comprises a series of high
intensity magnetic drums, placed one above the other in cascading arrangement.

(iii) The magnetic log washer is similar in design to *Akins* classifier in
which the helical screw conveyors are made of heavy copper ribbon.   The
pulp, fed into the trough, separates into the heavy portion, which sinks to
the bottom, while the light fraction is kept in the suspension, and is discharged by water currents.   The heavy portion is taken up by the helical
screw.   The machine is mainly used for the treatment of slime, coarser
than 48 mesh.   Treatment is more efficient if the feed is finer.

(iv) Frantz ferrofilter comprises a tube containing a set of steel
screens, surrounded by powerful electromagnets.   Pulp, containing magnetic materials, is passed through the screens, and the clean product is
discharged, through a pipe at the bottom or from a lip at the top.   Filters
are rated from 50 to 2000 gallons per hour, with different screen openings,
and the power consumption ranges, between 115 and 250 watts.

## B. Pick up separators

(i) Among these machines, the belt separator can be taken as a type.
This machine consists of three essential units (Fig. 12.54). (1) Feed hopper

(2) feed belt, and (3) take off belt, with electromagnets. The material is dropped on to the feed belt (2) which passes under the take off belt (3). The take off belt during rotation, picks up magnetic materials, from the feed belt under magnetic influence, and the materials are classified according to the degree of ferromagnetism acquired as concentrates (C) and middlings (M). The machines may include more than one set of these separator units, arranged one above the other, in decks.

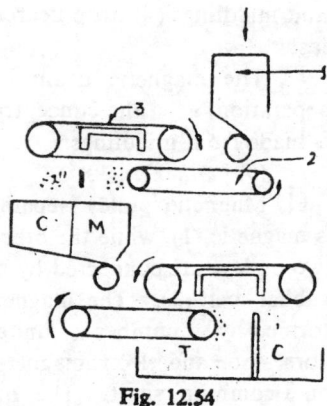

Fig. 12.54

(ii) The Wetherill separator has a long belt. Two cross belts, energesed by flat wedged shaped magnetic poles, are placed at some distance from either end of the feed belt. As the feed travels, under the magnetic poles, magnetic particles are picked up, by the cross belts and discharged, while the non-magnetic material is discharged at the return side of the feed belt. This machine is a high intensity machine, and it is used mainly in the separation of weakly magnetic materials and mixtures, where permeability differences are critical.

(iii) The rapid separator is also a high intensity machine, where powerful fields are induced by sharp wedge shaped secondary poles, placed above the primary magnets with flat poles, (Fig. 12.55). The secondary magnets (1), which are placed above the feed belt (2), are saucer shaped. The saucer shaped magnets are rotated by belts and as the feed belt, containing magnetic materials, passes underneath them, the magnetic particles are picked up by the rotating secondary poles, and discharged when the edge of the saucer leaves the field of the primary magnets, located below the belt. In most machines, there are three or more of such secondary magnets. Thus the first one is arranged to pick up the highly magnetic materials (3), the second for feebly magnetic (4), and the third for more feebly magnetic material (5).

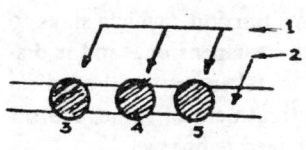

Fig. 12.55

(iv) One of the machines, used in wet magnetic separation, is Grondal drum (Fig. 12.56). This is essentially a drum type separator (1), with alternating pole electromagnets (2), mounted above a conical tank of the spitzkasten type (3), in such a manner, that the drum (1), just touches the surface of the water, in the tank. The feed (4) is kept in suspension, by the action of rising currents of water, so that all solid material comes under the influence of the magnets. As a

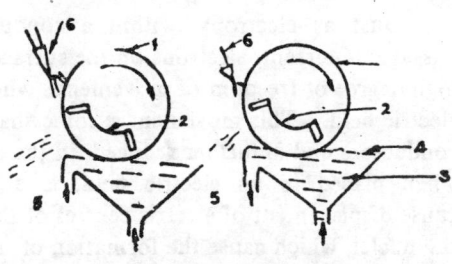

Fig. 12.56

result, the slime overflows at (5), and it is thus removed, while the concentrate adheres to the drum, until it is removed by a spray of water (6), and discharged. A series of such wet drum concentrators can also be employed, to increase the efficiency of the separation.

## (H) ELECTROSTATIC SEPARATION

### Electrostatic separators principle

From the atomic structure of matter, it is known that the positive charge, of the nucleus (proton), is balanced by the negative charge carried by the electrons. It is also known, that the electrons are mobile, and that they can be displaced, from the various orbits. In the case of static electricity, it is the electrons, in the outer shell, that are involved. If the electrons, in the outer shell are removed, the material, as a result, will acquire a positive charge. If electrons can be increased, then the substance will acquire a negative charge. Experiments to show the presence of static electricity, in the way generally exemplify what has been stated above. When two surfaces, are rubbed, e.g., fur, which is used for rubbing the ebonite rod, or silk, used for rubbing the stick of amber, the electrons configuration on both the surfaces, in contact, are changed, by mechanical force. The material, which acquires a positive charge, apparently has lost electrons, whereas the material, with the negative charge has wrenched away the electrons.

The part played by electrons, in carrying an electric current, is well known, and under the same conditions, a substance, in which the electrons are very mobile, carries higher charge. This is known as a conductor. The substance, which does not allow the passage of electric current and also inhibits electron movement, is termed an insulator, or resistor or dielectric. In between the conductors and the dielectrics, is a group of materials that can be identified, in which the electron mobility is intermediate. Electron mobility, however, is related to the strength of the electric current, so that dielectrics, for low currents, may become con-ductors, for higher currents. It is also related to the temperature.

Just as electrons, within a conductor, move freely to allow the passage of current, electrons on the surface of the material, also exhibit a high degree of freedom of movement, when the substance is placed in an electric field. This movement is somewhat restricted, in the case of semi-conductors and is further reduced, in the case of dielectrics. The substance, when placed in an electric field, is affected in such a manner, so as to cause displacement of electric centres of the electrons, as well as those of the nuclei, which cause the formation of dipoles. This process is known as polarisation. As a result of polarisation, a surface charge is induced, in the material. The extent of polarisation is represented by a constant K, for the substance, known as the dielectric constant.

Electrostatic separation utilises the effect of an electric field, together with the influence of additional forces for affecting differential movement of particles. The differences in interfacial resistance of various minerals, to the passage of electrons, together with the variations in specific gravity, size, shape, surface condition, and purity of mineral particles can be utilised for this purpose. The methods of inducing electric charges, in media, are well known. (1) Charge can be acquired by induction. When polarised particles are placed in contact with a conductor, these are found to be charged, when removed from the influence of the electric field. (2) When two spherical particles have different electron charges, a flow of electrons takes place, if a contact is established. This flow is due to conduction. (3) Pyro-electricity is a feature of natural and artificial crystals. Thus tourmaline when heated, a positive charge is developed along the C-axis. (4) Peizo-electric charge is also a property of natural and artificial crystals which lack a centre of symmetry. Such crystals are charged when sub-mitted to pressure. An example of such a mineral is quartz. Interfacial resistance is the ease with which electrons enter or leave a polarised body. It depends on the potentials of the body and conductor, at the point of contact.

Electrostatic forces are similar and analogous, in action, to perma-nent magnets. They are similar to those acting on a magnetic body, in the absence of a permanent charge. Electrostatic forces are primarily governed by the inhomogeneity and magnitude of the field intensity, and also by the inductive capacity of the body. Unlike magnetic poles, electro-static charges may move slowly over the surface of the body. If a nega-tively charged insulated sphere is brought near an uncharged particle, sus-pended from an insulated fibre, polarisation is induced and the particle is attracted towards the sphere. Since the sphere has a converging field of force, it acquires a negative charge by conduction, when the attracted particle touches the electrostatically charged sphere.

It is thus immediately repelled by the Coulombs' force When the sphere or the particle are poor conductors, then due to high interfacial resistance, a time lag, in the rate of conduction will result. Thus an initially lower potential, than the sphere, will cause the particle to be repelled, but the positive charge acquired by it, after the time lag, might cause subsequent attraction. If, however, the negatively charge particle is connected for a short length of time, to a grounded conductor, attraction can be induced. When the interfacial resistance is low enough. to drain away the earlier negative charge, loss of electrons may cause the particles to become positively charged. A prolonged contact, with the grounded conductor, may cause, even poorly conducting particles to acquire a positive charge. Thus relative conductivities of particles are of major importance, and this will apply to the mineral kingdom also. When an uncharged particle, momentarily grounded. is placed in the field of a charged electrode, it acquires an opposite polarity, in respect to that of the electrode. The time required, for the charge to build up, depends on the electrical properties of the particle. The reversibility of polarity is an important feature.

## Electrostatic Machines

Electrostatic machines are of two types. (1) In the high intensity machine, the non-conducting particles are attracted by an electrostatically charged moving roller, while the conducting particles are repelled by the same. (2) In the low intensity machine, attraction is either at a minimum or absent, while repulsion is also small. The separation is dependent more on the close regulation of the splitters, and in addition, repeated treatment improves it further. Time factor is an essential element in both types of machines. The Sutton separator (Fig. 12.57) is a high intensity machine.

The field is induced in a cylindrical electrode which is earthed (1). The electrode is also used for conveying the material. The feed is dropped on to the drum from a hopper (5). The charging electrode is made in the form of a tooth comb (2), and is placed at some distance from the material conveying electrode. A third electrode comprising a neon discharge lamp (3), is placed below the charging electrode. The entire equipment is run on D.C. through a rectified circuit

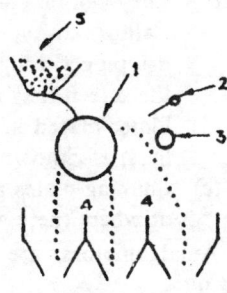

Fig. 12.57

and all the circuits are well insulated from each other. Knife edges (4), are placed at appropriate places, below the rotating material conveying electrode, separate the products into three parts viz., conducting, non-conducting, and middling.

The electrodes are so adjusted that when the neon tube electrode is switched off, the field, between the other two electrodes, i.e., the covering electrode and the charging electrode, becomes convergent. When the charging electrode is switched off a similar, though less convergent, field is induced, between the rotating electrode and the neon tube electrode. When both the charging electrode and neon tube electrode are switched on, the lines of force, which converge to a line, on the surface of the material conveying electrode, cause repulsion between the two fields. Separation can be affected by the manipulation of the three electrodes. The Johnson separator also works on a similar principle.

## MISCELLANEOUS PROCESSES

### (A)  Amalgamation

Amalgamation is the process by which gold and silver, in the native state, form an alloy with mercury inspite of the presence of water. The various equipment for the purpose are : (a) Plates, generally of copper on which a thin film of mercury has been spread. The pulp containing gold is made to flow over this plate; (b) Pocket amalgamator is fitted with riffies, and traps containing liquid mercury. The pulp is made to pass over these; (c) Pressure amalgamators force the pulp against the mercury by the use of pressure and (d) Mixing amalgamators disperse the mercury into fine droplets, thereby increasing the probability for the formation of amalgams.

Amalgamation plates are generally stationary but they can be made to oscillate as well. Amalgamation plates are employed in various situations in a mill :

(a) They can be placed inside a stamp known as battery plates, e.g., Californian type, so that gold released may get trapped due to dripping through the screen mesh. If the plates are in front of the screen they are known as splash plates.

(b) Plates placed in the launder, outside grinding mills, and classifiers, are known as sluice plates.

(c) Shaking plates are sometimes used where there is a lack of space or where there are chances of blocking due to sedimentation.

Amalgamators are generally boxed in to prevent pilferage of rich amalgam.

Plates are cleaned thoroughly with sand and fine emery paper—leaving no trace of stain, tarnish or scratches.    Fine sand impregnated with mercury and moistened with dilute cyanide solution, or a stronger ammonium chloride solution, is rubbed on to the surface, with a brush or rag. The surface is first amalgamated with silver, then mercury is rubbed directly into the surface.    Muntz metal, containing 60% copper and 40% zinc, is preferred for amalgamatian, because it has higher resistance to arsenic which causes "sickening" of the plates.    Muntz plates are easy to clean, as the mercury penetration is less.

Staining and sickening are caused by undesirable reactions, on the plate, which inhibit further amalgamation of the precious metals.    The mercury, under these conditions, fails by separating from the plate.    One of the main causes, of sickening or failure to form amalgam, is the presence of grease, in the pulp, which causes gold particles. to be coated with a fine film.    Talc, clay and graphite particles can also cause sickening when fatty acid oils are present in the pulp.    Heavy sulphide ores are unsuitable for amalgamation, as they tend to settle down, on the plate and prevent amalgamation.    If the pulp is alkaline they cause no great problem. If the ions, of heavy metals, are present in the pulp, there is a danger of the formation of scaly base metal amalgams.    Arsenic, antimony, and bismuth sulphides are the most dangerous in this respect, i.e., sickening.

Staining commonly occurs in acid pulp, and it is normally due to the formation of copper green (verdigris).    The stain is removed by the use of cyanide or ammonium chloride or with HCl or $H_2SO_4$ dil.

Pocket amalgamator is generally placed in the launder, at the discharge end of a mill.    It comprises a simple weir which holds the mercury while the pulp is kept in suspension before it flows over.

In order to prevent the loss of amalgam which peels from the amalgamation plates a mercury trap is used. This consists of a conical vessel (3) into which the pulp is dropped from the launder (1), through a pipe (2) which is placed axially in the vessel. Mercury is retained at the bottom of the vessel, while the pulp overflows from the spout (6) into launder (7) the tap (5) is for draining amalgam periodically from the trap.

A mercury trap (Fig. 12.58) is also used.    This consists of a conical vessel (3), into which the pulp is dropped from the launder (1), through a vertical pipe (2), placed axially.    Mercury is contained at the bottom of the vessel, and the pulp is thus made to come into contact with it. Tap (5) is for draining the amalgam.

Grinding amalgamators are employed where the gold is (1) coated with rusty oxides, or (2) where it is enclosed by other minerals. or (3) where agitation is essential for amalgamation, or (4) where sickening is to be prevented. The amalgamating barrel is usually a rod mill. Alkalies are added to prevent action of organic materials. Ammonium chloride is added for stopping formation of verdigris (stains). Oxidising agents $KMnSO_4$ or $K_2Cr_2O_3$ are added to prevent sickening—Plumbate mixture (containing lead) is also added for the same purpose.

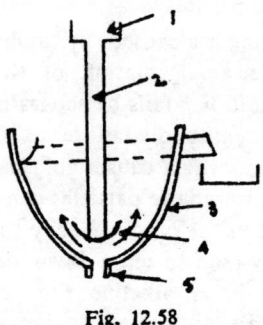

Fig. 12.58

*Pan Amalgamator* is different from the barrel type, as the pan is submitted to free oxidation, hence the choice should be made after due consi-

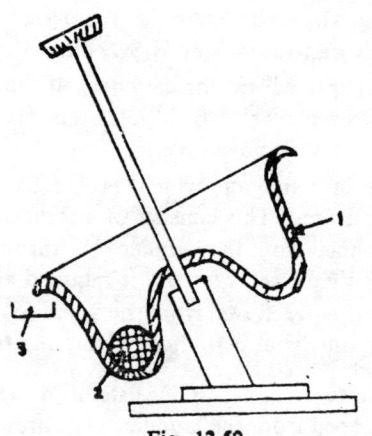

Fig. 12.59

deration. One type of pan amalgamator is shown in Fig.12.59. The pan (1) is about 3.5 ft. in diameter, and about 20 inches deep. The tilt is about 30°. One or more cast iron balls (2), are placed inside the pan, and it is loaded with sufficient mercury. The operation can be either continuous or intermittent. In continuous operation the mercury spills over into the launder (3).

**(B) Recovery of precious metals**

The process consists of (1) cleaning the amalgam, (2) filtration, (3) retorting, and (4) melting and casting the bullion.

Cleaning is done by diluting the amalgam by the addition of a further amount of mercury. The mercury is worked in by using a gold pan or in a pestle and mortar. Any extraneous material, not amalgamated thus floats up. Magnetic separation is also done at this stage. Mixing can also be carried out mechanically.

Filtration is carried out by squeezing the amalgam through a fabric. Chamois leather was used in the past for this purpose. The amalgam after manual pressing contains 30-40% Ag but where a power press is used, the gold content may be as much as 60-70%.

Retorting is carried out either in a pot or cylindrical furnace, lined with either a mixture of fire clay and graphite or chalk, clay and lime. At the opening of the retort, a water gasketted condenser is attached. The condenser discharges under water, or into a wet sack.

The gold recovered, from the retort, is spongy and is called retort sponge, as it contains about 1-1½% of Hg.

Melting is carried out in graphite crucibles, and the sponge gold is covered with borax and melted. If the presence of sulphides is suspected then silica and soda are also added.

## (B) Dewatering

Since most grinding processes involve the use of liquids, especially water and the products, recovered by flotation, are also suspension in a watery medium, it is necessary to eliminate the water, by some method. The following are some of the common methods.

Draining is the most common method. The procedure adopted is to place the wet material, on a very gently sloping floor, so that much of the water separates out and a considerable part is also lost by evaporation. especially in the hot climates. This method is adopted in the China clay mines near Chaibasa, Bihar, where the clay, from the settling tanks, is dried on cement floors. A certain amount of moisture is still retained depending on the particle size of the material. The smaller the size faction. the larger is the proportion of moisture retained. At times, the draining chamber or tank has a floor made of fabric or fibre, where fine sandy material is separated from the suspension.

Settling tanks are used for dewatering washed clays, and generally more than one tank is in operation. Coarser particles can be separated by the use of serpentine zig-zag channels, baffle tanks and washing pans.

Drainage is also affected by mechanical means. One of these is by the use of screens. This method has been described earlier under screening, and may be called filtering.

Dewatering elevator is a machine which is similar to the bucket elevator used in land and water dredges, except that the buckets are perforated to allow the liquid to drain out. Such elevators are used in the sink float machines, for separating fine sandy material. The elevator is set at such a slope that the liquid dropping out, from the higher buckets, does not fall into the lower ones. The speed of the machine is low so that sufficient time is allowed for drainage.

Scraper type dewatering machines are principally used for coarse sands, and also for separating sand from slime. They comprise a bucket wheel, in which the buckets are perforated. Here again the buckets dig up the settled sand, from the bottom of the tank (for dewatering) and empty the same on to a trough. The operation of the sand wheel, and shovel wheel are similar. Mechanical classifiers such as the drag classifier, spiral classifier, rake classifier, have been described under classifiers. These machines are also machines for dewatering.

## (C) Thickening

Thickening differs from dewatering in that the suspensions involved are very dilute, i.e., the solid content is low. The type of solid suspensions, submitted to thickening, are fine and comprise (1) slimes and (2) colloids. Slimes are very fine particles derived from the final stage of crushing and generally smaller than 0.001 mm. in size. Colloidal suspensions comprise ultra microscopic particles, and are subject to Brownian movement, associated with electrical charges, which account for their constant state of dispersion. The electrical charge will depend on the nature of the adsorbed electrolyte. The colloidal character is destroyed by the use of certain reagents known as flocculating agents, the most popular being lime. When lime is added, the colloidal particles form aggregates or flocs, and tend to settle down.

In ore dressing practice, thickening is carried out in machines called thickeners. They are large vertical cylindroconical tanks or vats, generally with a flat conical bottom, deepening towards the centre. The slime is fed at the top, while sedimentation proceeds unabated, in the deeper portion, away from the inflow. The sediment is drawn off periodically through an outlet, at the bottom of the tank. Such tanks are either intermittent or continuous in operation. The continuous tanks are discharged either by gravity or by mechanical means. In the centrifugal thickeners, the separation is affected by means of centrifugal force. These machines are essentially drums, into which the pulp is charged from one end. The drums rotate at a high speed. The solids and water are separated and are discharged at opposite ends.

In the intermittent tanks, the supernatent liquid is either siphoned out or let out through a pipe which stands above the level of the sediment found at the bottom. Even centrifugal thickeners are operated intermittently.

*Continuous thickeners* : The cone is the simplest, but it is costly and its capacity is relatively small. It is .employed where the quantities of pulp, to be treated, are small. In the larger operations, mechanical discharge type cones are normally employed. These comprise a cylindrical tank, Fig. 12.60, which is shallow as compared to its diameter. The feed

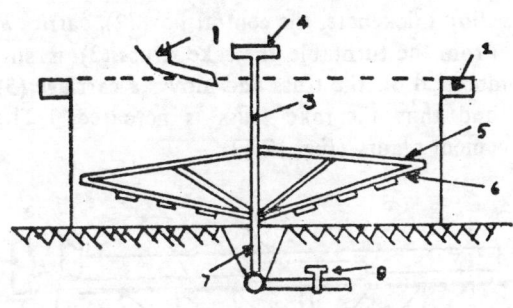

Fig. 12.60

(1) is let in at the top, but centrally, and the overflow (2) is let into a launder which surrounds the upper edge of the tank. The central shaft (3) carries the arms (5) to which rakes (6) are attached. As sedimentation takes place, the settled material is drawn into the cone (7), and discharged through the pipe (8).

The pulley, at the top of the central shaft, is used for driving the rakes. Some thickeners, especially ones in which the tank is of cement, have inclined bottoms.

In the Dorr thickener (Fig. 12.61) a motor (1) is mounted on a

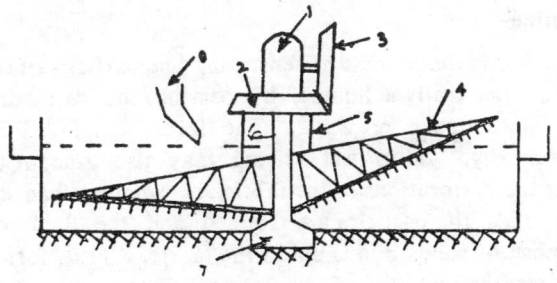

Fig. 12.61

strong central column (6), based on strong foundation (7). The motor drives a turn table (2), through the gearing (3). The truss carrying the rakes (4), is suspended from the turn table (2), by hinges. When the resistance, of the sediment, exceeds that of the suspended truss arm, the the truss acquires a tilt, as shown in the figure, as well as an inclination, which makes it trail over the sediment. This type of arrangement has been used for larger diameter thickeners.

In the traction thickeners, the central post (2), carries a free moving turntable (1). From the turntable the rake truss (3) is suspended. The motor (4) is mounted on the truss and moves a carriage (5), which runs on rails, (6), and thus the rake truss is actuated. This is used in thickeners for cement plants (Fig. 12.62).

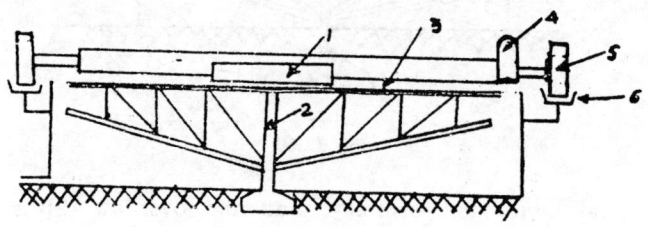

Fig. 12.62

Tray thickener comprises two or more thickening compartments arranged one above the other. It is similar to the ordinary thickener. The feed is let in centrally from a common feed box through pipes into the respective compartments. The overflow is similarly led off from the respective compartments through pipes. The thickened sediment is drawn off from the bottom valve.

(D) **Filtration**

Filtration is the process of separating fine particles of solids, suspended in a fluid (generally a liquid), by passing the suspension through a permeable fabric under pressure. The force, employed in filtration, is other than gravity. Thus wet seiving may also amount to filtration in theory. If the perforations in the fabric are smaller than the size of the suspended solids, the particles are retained and the fluid passes through. As the deposit of solids builds up on the fabric, it also forms part of the permeable membrane.

The filtering capacity of the fabric steadily falls with use, due to clogging of pores, by fine solids, and by the formation of scales of carbonates, sulphates, etc. The fabric is made of cotton, woollen, synthetic fibre, metallic mesh or perforated rubber.

The main types of filters, used in dressing plants, are (1) vacuum which may be continuous or intermittent in operation, and (2) pressure, used generally for intermittent operation.

## (1) Vacuum Filters

The most popular continuous vacuum filter is the Oliver filter (Fig. 12.63). This consists of a cylindrical drum (a) placed inside a V-shaped trough (b). At the bottom of the tank or trough is a rotary agitator (c) for stirring the suspension fed into the tank, through the pipe (d).

The drum is mounted on hollow axles or trunnions as in the case of the ball mills. Its surface is lagged by wooden laths (Fig. 12.63 (d)). Longitudinal laths (c) are fixed over these at intervals so that narrow

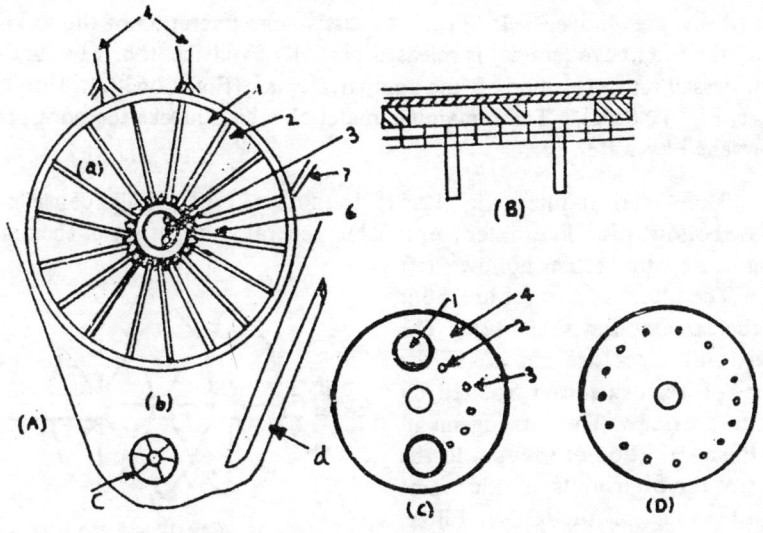

Fig. 12.63

compartments are formed (Fig. 12.63 b). On these laths a backing material (b) is fixed and on the top of this the filter fabric (a) is laid, and attached either by caulking (i.e., wedging with the wooden batons below) or by fixing with wire. The drum is rotated by the gear wheel (3).

*Principle* : The suspension is kept agitated in the trough. The drum is rotated and at the same time vacuum is applied inside the drum. The solids get attached to the filter fabric, while the liquid gets through, into the drum, from where it can be discharged. In actual practice, the vacuum is not applied to the inside of the entire drum, but only applied, to the chambers between the wooden lagging through radial pipes, Fig. 12.63 (b). The vacuum pipes are all joined to the hub of the drum (3). The hub ( ) is actually a valve comprising three plates, Fig. 12. 63(c). The upper plate contains the exit opening for the filtrate (1). Holes for pressure gauges (2) and inlet holes for compressed air (3), are also provided in this plate. The lower plate contains a ring of holes, most of which are connected to an air compressor (viz a receiver) also.

Thus, due to the applied vacuum through pipes (d), on the filter fabric, suction is produced and consequently the solids are deposited, while the liquid (filtrate) is conveyed by the tubes (d) and discharged by common openings (1) and (2). As the filter cake or layer, of solids, is built up on the fabric, a scraper plate (7) releases the blocks of sediments, and these are moved. In order to assist the operation of the scraper plate, the filter cake (solids) is released partially from the fabric by forcing compressed air, into some of the compartments, (Fig. 12.63 (e), through pipes, Fig. 12.63(d). The remaining material, which blocks the pores, can be washed by water jets.

The American filter (Fig. 12.64) is another type. This consists of a flat hollow disc like filter units (1), generally about six of them are mounted on the same hollow shaft (2). The filter unit discs are built on the same hollow shaft (2). The filter unit discs are made in the form of segments and rotated by worm gearing. The arrangement, for the application of vacuum to the discs of the filter units, is the same as in the case of the Oliver Filter. It is through pipes, that the seg-

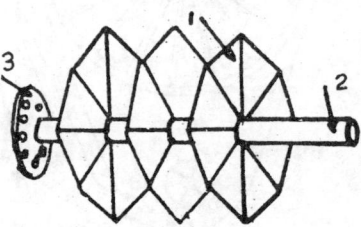

Fig. 12.64

ments are connected to the hollow shaft. The valve, for feeding compressed air and inducing vacuum, is fitted on to the hollow shaft (3). The whole arrangement (the battery of filter discs) is placed in a trough containing the suspension. It is so arranged that when a particular segment of the disc is immersed, it is subjected to vacuum. When it emerges out, it is put under pressure by feeding compressed air, so that the filter cake is loosened.

Filtering tables are also used in ore dressing plants. A typical example is the Caldecott Table, which is hollow and annular in shape

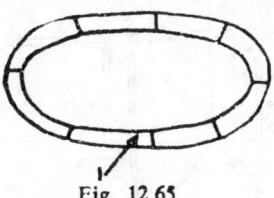

Fig. 12.65

(Fig. 12.65). It has a porous top and there are compartments (1). It moves slowly, and the suspension is fed on the top. A low vacuum is applied to the compartments and the sediment settles down. A scraper or plough is used to for removing the cake.

Filter tanks are either rectangular or cylindrical in shape. A porous fabric (4), is supported at the bottom by a grill and fibre padding (3). Over the fabric is stretched a strong wire netting (1). Vacuum is applied for filtration. The sediment is removed mechanically by shovels or clam-shell dredge from the top of the fabric, while the liquid drains through and is carried away.

Vacuum leaf filters comprise several units, as in the case of the pressure filter described earlier. Each unit consists of a flat fabric bag inside, which is a pipe frame. The cloth bag is segmented by stitching in parallel rows. The bottom pipe, of the skeleton frame, is perforated, and it is connected to a vacuum pump. All the filter units are similar. The filter units are moved by travelling cranes and lowered into a tank containing sediment in suspension and vacuum is applied. Particles of sediment adhere, and form a layer. The unit is lifted out and placed in a tank of clean water for washing before the cake is removed.

## (2) Pressure Filter

These are known as filter presses. It consists of a number of C.I. plates (6 or 8) (Fig. 12.66) which are either, perforated (1), or non-perforated (2), alternating with each other, and strung on two parallel bars. The perforated plates are ribbed while the nonperforated ones are recessed. All plates are of the same size. The perforated plates are covered over with a bag made of strong fabric. When the plates are tightened by a wheel and screw arrangement, they come into contact with each other, and the bag of fabric is now enclosed on all sides by the recessed plate except for the opening at the top (4). When it is filled with the filtered material the sediment in the fabric bag increases and expands to fill up the recesses in the nonperforate plates on either side. When the bag is full of sediment, the press is opened and the filter cake drops out. In some presses, hot air is passed into the hollow plates for drying.

## Drying

Drying is the process of removing moisture, from wet materials ob-

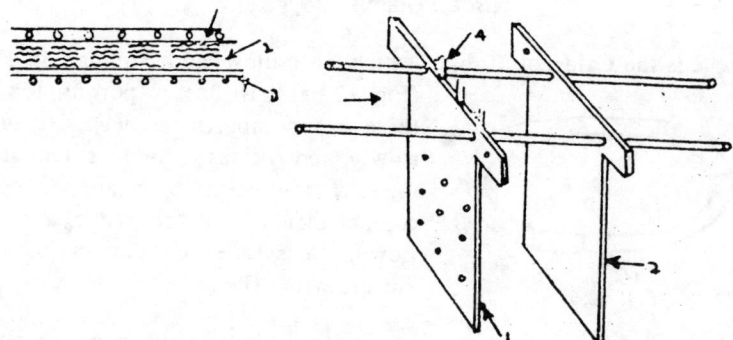

**Fig. 12.66**

tained, during various ore dressing processes such as filtering, pressing, draining, etc. Drying is undertaken with various ends in view. (1) To reduce the bulk weight, and save freight, when the material is to be shipped by steamer or rail. (2) To make the material suitable for handling or conveying. (3) To conform with terms of sale where moisture is undesirable. (4) To economise on fuel or heat in pelletising or agglomeration, and subsequent heat treatment processes. (5) To make certain ore dressing processes effective, e.g., screening, pneumatic separation, electrostatic separation.

The main types of drying equipment may be grouped under two heads. (1) Hearth type, (2) Shaft type. The other types are (3) Kiln type and (4) Spray type

Among the hearth type is the "dry" used in drying China-clay. It comprises a floor of tiles, under which is laid a number of parallel flues, through which heated gas from a furnace is made to flow. The heated floor causes the filter pressed clay to dry. In the mechanical type of hearth (Fig. 12.67), the wet material is placed over an iron belt (1) and moved slowly. Hot gas from the hearth (5) is sucked under the belt through flues (2) by a fan (6). The material on the belt is raked by mechanical rake (4).

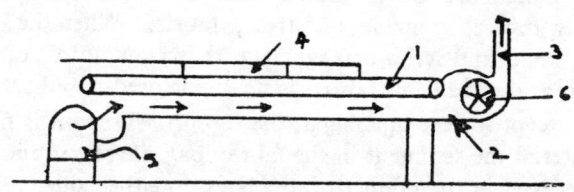

**Fig. 12.67**

In the shaft type of drying arrangement, the wet material meets the hot gases flowing in the opposite direction through a vertical cylinder, known as the tower drier. The wet material is introduced at the top and it drops downwards, being retarded in its downward passage, by baffles. Hot air enters the shaft at the bottom and travels upwards.

The main items in the rotary drying set up are (a) the furnace tube, (b) the furnace, (c) the flue with suction arrangements, and (d) the driving mechanism. The material is fed at (3), through a chute (Fig. 12.68) and the hot gases, from the furnace (4), are drawn into the kiln (1), by the fan (5). The kiln rests on roller mounting (2), which is rotated by gears (3). The dry material is delivered into a conveyor at the furnace end.

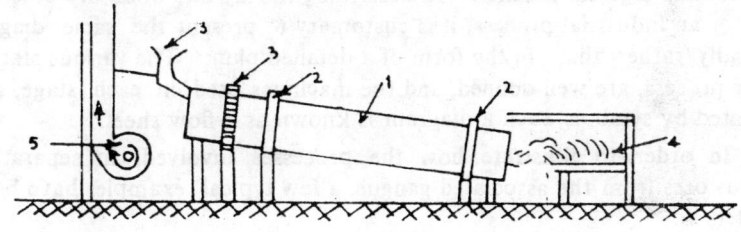

Fig. 12.68

In some driers, the furnace tube is enclosed in a refractory brick chamber so that the hot gas (air) first goes round the furnace tube and then gets into it before being discharged by the exhaust fan. These driers are more economical in fuel consumption. A further modification of this type is where the furnace tube is enclosed in another steel shell. The hot gases which travel through the annular space between the tube and outer shell, S enter the furnace tube and then are finally discharged.

Another rotary drier is the Louvre drier (Fig. 12.69). This also consists of a double shell (1) and (2) but the inside tube is segmented, as it is

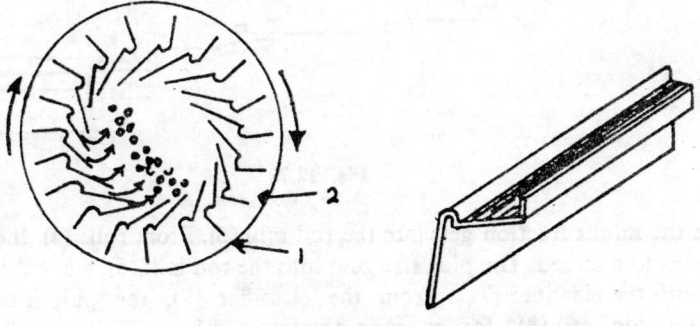

Fig. 12.69

made of a number of louvres (2). Each louvre is made from a single steel plate. The tube is so rotated that the charge drops from one louvre to another (If the rotation is opposed then the charge will get into outer tube). The hot gases entering the outer tube pass through the charge and enter the inner chamber made of louvres, thus drying is affected.

## FLOW SHEETS

After getting acquainted with the various principles of ore dressing, and the machines employed for affecting mineral separations, it is now necessary to get to know, how these can be put into use in the most effective and efficient manner. In describing the layout, of an ore dressing plant or an industrial process, it is customary to present the same diagramatically rather than in the form of a detailed plan. The various stages, of the process, are well defined, and the machines used, in each stage, are indicated by symbols, such a diagram is known as a flow sheet.

In order to illustrate how the processes involved, in separating various ores from the associated gangue, a few typical examples have been selected.

### (1) Chromite (Fig. 12.70)

Run of mine ore is passed through a grizzly (1). The plus or larger sizes go into rolls (2), while the minus or smaller size fraction goes to a trommel (3). From the trommel (3), the plus fraction is taken to rolls (4),

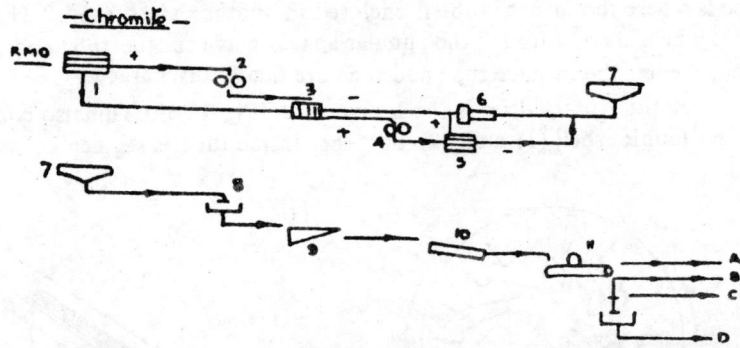

Fig. 12.70

while the minus fraction gets into the rod mill (6). From rolls (4), the material goes to a screen. The plus size goes into the rod mill (6), while the minus goes into the classifier (7). From the classifier (7), the pulp is taken to the flotation cell (8), for rougher treatment, where silicates, and quartz, are removed (A). The concentrate is transferred to a classifier (9), and then dried in a rotary drier (10) after which magnetic separation (11) is affected to remove (B) ilmenite, and magnetite, at low intensity, and (C) ferruginous chromite, at a higher intensity of magnetisation.

The concentrate is finally submitted to flotation where garnet (C), is removed and high grade chromite (E), is obtained. If zircon is present then further magnetic separation is done.

   Note : In flotation, oily collector, cresylic acid, frothing agent and creosote are employed.

## (2) Gold (Fig. 12.71)

As an example of gold dressing flow sheet the following outline may be taken.

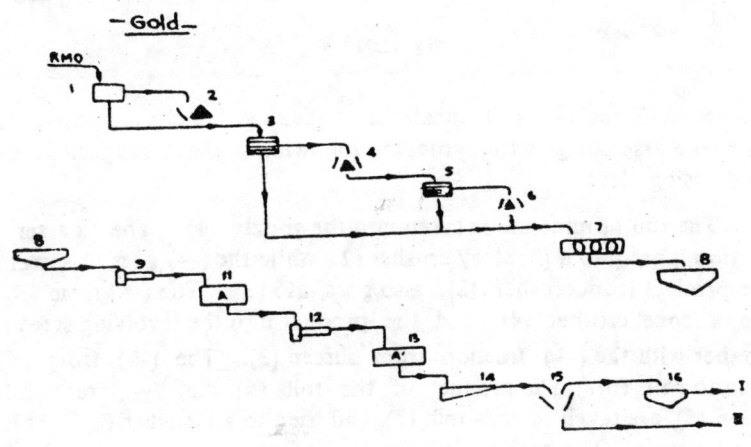

Fig. 12.71

The run of mine ore passes through a grizzly (1), the (+) goes through a jaw crusher (2), while the (—) goes straight to a screen (3). The (+) from screen (3), is transferred to a gyratory crusher (4), while the (—) goes into a stamp battery (7). From crusher (4), the discharge is taken to screen (5), and the (+) again submitted to grinding, in a gyratory machine and then (—) is led into the stamp (7). Thickening of pulp is affected in (8). From the thickener (8), the pulp is transferred to a rod mill (9), and then to amalgamating table (11), into rod mill (12), and then to amalgamating table (13), classified at (14). The next stage is (15), dewatering. The slime is led to (16), for thickening and cynadation. The sand is also treated likewise.

### (3) Copper (Fig. 12.72)

Copper is mainly found as the double sulphide chalcopyrite, while copper sulphides like bornite, chalcocite may be present in relatively smaller amounts. Pyrite and pyrrhotite are also common associates,

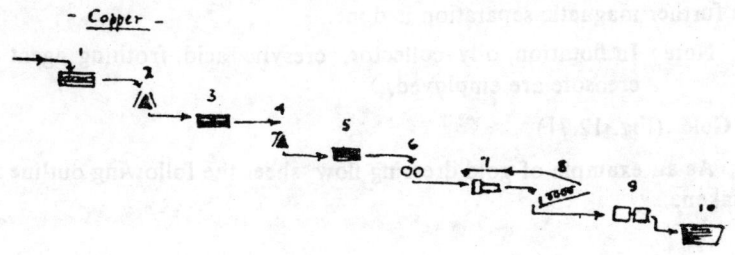

·Fig. 12.72

together with galena and sphalerite. Taking the chalcopyrite-bornite-chalcocite assemblage with pyrite, the following is the arrangement of the ore dressing plant.

The run of mine ore is taken into the grizzly (1). The coarser (+) fraction, goes into a gyratory crusher (2), while the (−) goes to screen (3). The product from crusher (2), also goes into (3). The (+) from (3) gets into a cone crusher (4), and the product into the revolving screen (5), together with the (−) fraction from screen (3). The (+) from (5), is taken to rolls (6). The product of the rolls (6), and (−) fraction from screen (5), are taken to rods mill (7), and then to a classifier (8). This is a closed circuit. The product is taken to the first battery of flotation cells (9), where pyrite is separated. From the flotation cells the material is .placed on the Wilfley table (10) where copper sulphides and chalcopyrite are recovered. *Note* : The last step may not be needed

### (4) Lead-Zinc (Fig. 12.73)

It is very common, that lead and zinc occur together, e.g. galena and sphalerite. Where the minerals are found as coarse aggregates, coarse grinding is adopted in conjunction with gravity concentration, by using jigs and/or tables, and thus a concentrate can be obtained. Flotation is adopted for separating galena from sphalerite. In many cases, however, the association of the lead and zinc minerals is so intimate, and the particle size so fine, that tube mills or balls mills are necessary for grinding while flotation is required for separation.

The flow sheet employed in Broken Hill South mines, of New South Wales, Australia, may be taken as an example.

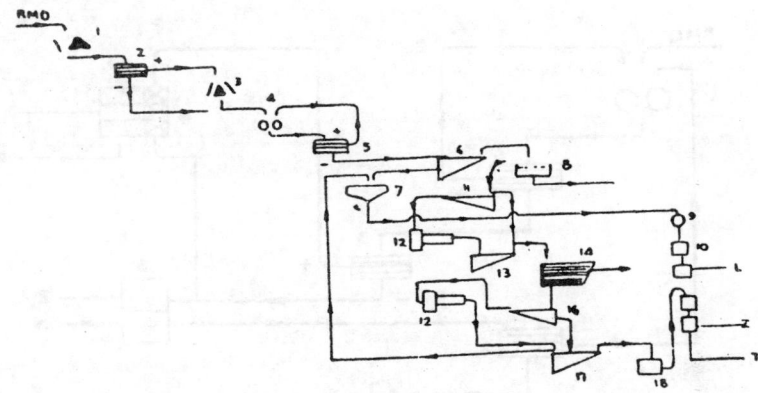

Fig. 12.73

In an alternative flow sheet the run of mine ore is fed into the primary jaw crusher (1), and then travels to the screen (2). The (+) gets into the cone crusher (3), while the (−) goes into the rolls (4). The roll discharge is fed to the screen (5). This is in a closed circuit with the rolls. The (−) from the screen (5), is taken to classifier (6), and the sand is taken to jig (8) while the overflow is taken to thickener (7). From there onwards it goes to the flotation plant comprising the conditioner (9), and cells (10). The fines from the jig (8), go into a classifier (11). The sand from classifier (11), gets into the rod mill (12), and this is a closed circuit, discharging into classifier (13). The sand from classifier (13), is tabled at (14), and lead (L) is recovered. The tailings are classified in (16), and ground in rod mill (15), and discharged into classifier (17). The sand from here is taken to a flotation cell (18), while the slime gets into the thickener (7).

Thus the products recovered as lead (L) from tabling, while lead (L) and zinc (Z) are got from flotation.

## (5) Manganese (Fig. 12.74)

The common practice in many manganese mines, is to wash the ore with a view to remove clay and laterite as well as wet material sticking to lump ore. The grade of ore can be improved, in cases, where the ore is contaminated with other gangue materials, especially carbonates, sulphates

and silicates by employing gravity methods, in combination with intermediate crushing and screening.

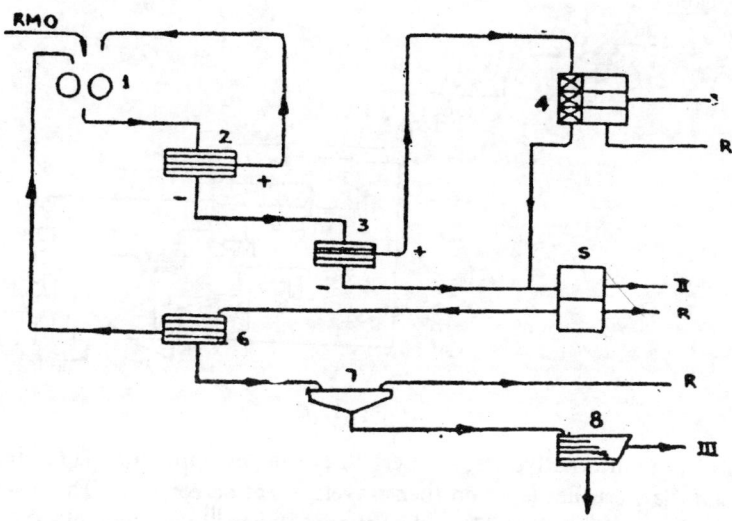

**Fig. 12.74**

In the flow sheet a simplified version of the beneficiation process, has
been depicted, giving only the broad features.. The run of mine ore is fed
into the corrugated rolls (1), the product is screened at (2). The oversize
is crushed, in the closed circuit, screened and jigged. The (—) fraction
can be floated or tabled. In practice, there may be more than one crusher
operating on a closed circuit, until finally a suitable size product is obtained. The product can be further crushed and screened on (3). The
oversize is taken to piston type jigs (4), and concentrate I is obtained.
The undersize (—) from screen (3), as well as the middlings from jigs (4),
are fed into the moving screen jig (5), and concentrate II is recovered.
The middling from jigs (8) is screened (6), classified (7), and then tabled
(8), and concentrate III is obtained.

### (6) Barite (Fig. 12.75)

Barite is generally required in the powder form by most major industries like rubber and paper, and off colour material is undesirable. Barite,
required for oil drilling, may also be coloured. Since the colour is
generally due to the associated clayey impurities, it is necessary that these

be removed. Chiefly barite occurs as lenses and seams, even as is the case with the barytes of the Vempalli limestone belt, in the Cuddapah district.

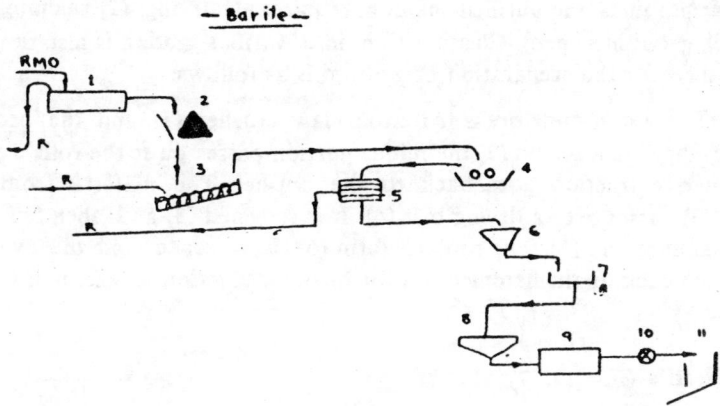

Fig. 12.75

The run of mine ore is subjected to cobbing, i.e., breaking with sledge hammers on the floor (1). The broken ore is thrown into crusher (2), and washed and in spiral or log washer where the clay is removed. The coarse sandy barite is ground, in a pan-mill (Chilean type) and screened at (5). The oversize is rejected (R), and the undersize is submitted to classification (6), and then placed in settling tanks, and treated for off colour, either with ultramarine or with potassium ferrocyanide, for imparting a slight blue colour. From the tanks the material is taken to washers (8) and then to the drying house (9). After drying the material is pulverised (10) and stored in bins (11).

(7) Gypsum (Fig. 12.76)

Gypsum often contains only clayey impurities and slight washing is likely to go a long way to improve the quality. This is true of the gypsum

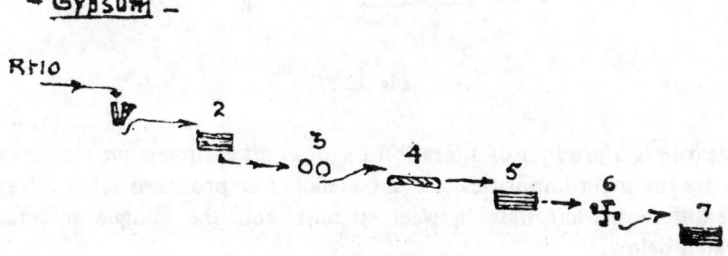

Fig. 12.76

which occurs with the Uttatur beds of Tirucherapalli district, of Tamil Nadu. Gypsum found in the lagoons of the east coast, can also be treated likewise. This treatment yields a product, sufficiently pure for the cement industry. Where gypsum is required for making plaster of Paris and similar products, the purified material requires (1) drying, (2) calcining and grinding (with screen). Classification into various grades, is also done. A flow sheet for the preparation of gypsum is as follows:

The run of mine ore is fed into a Jaw crusher (1) and the product passes through a screen (2) the minus portion passes on to the rolls 3 while the coarser fraction goes back to the crusher. The material from the polls (3) passes over a drying belt (4). It is screened (5) and then fed into the hammer mill (6). The product form (6) is screened and the oversize is taken back to the hammer mill for further reduction in size, in a closed circuit.

### (8) Steatite (Fig. 12.77)

Steatite occurs as an alteration product of ultrabasic rocks or dolomite. The impurities are silicates, tremolite, actinolite, in addition to calcite and magnetite. Quartz, as vein material, can also be found. Where

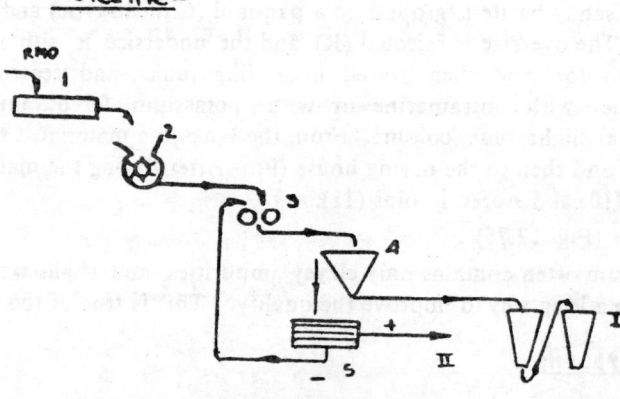

Fig. 12.77

the steatite is a product of alteration of dolomitic limestone then carbonates are the main impurities. The beneficiation processes take advantage of the differential hardness between steatite and the gangue minerals as indicated below.

The ore, which is dry, is laid on the cobbing floor (1), and the cobbed product enters the hammer mill (2). The product from (2), is taken to the rolls (3), and then to the classified (pneumatic) and then pneumatic sizing is carried out (8). This is product I. The coarser fraction from the classifier (4) is sieved. The coarser (+) is collected as a product II. The subsize (—) is returned to the grinding circuit. A magnetic separator may be included when necessary after the roll (3), for the removal of magnetic minerals, e.g., magnetite.

## (9) Clay (Fig. 12.78)

Kaolin occurs as a product of weathering of felspathic granites and pegmatites. Secondary kaolin is found in felspathic sandstones as well as

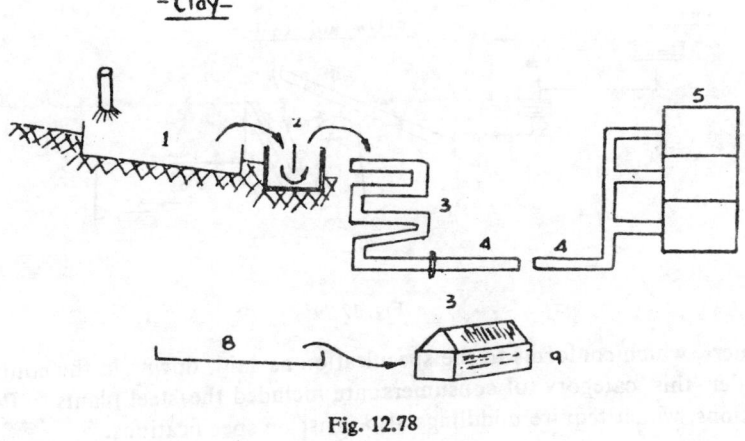

Fig. 12.78

in kaolinised sandstones. Kaolin also occurs as beds in sedimentary formations.

Kaolin of the first type occurs at Kundara in Kerala State, in Sabarmati area of Gujarat and Chaibasa area of Bihar. Sedimentary kaolin is found in Cuddapah, in the Rajmahals and in tertiary deposits. The general scheme for beneficiation aims at removing organic remains (root, leaves, twigs), sandy material (mainly quartz), and micaceous minerals (mica).

In many China clay mines, the practice is to dump the run of mine clay into a shallow trough (1), and it rake up with water, and thus most of the sand is removed. The overflow gets into a sluice tank (2), with a sluice or baffle for mainly separating floating minerals, e.g., wood, leaves etc. and some amount of sand. After this the slurry gets into a series of serpentine channels (3), (generally 2 sets). Some of the mica settles down here. In addition, a screen (3a) is also included. A long channel (4) with a screen (4a) leads the slurry into the setting tank (5) to (7). Here the clay settles down, generally on its own sometimes, a little lime or alum is added to induce flocculation. If necessary, a fluing material (ultramarine) is also mixed. The supernatent water, from the settling tank is siphoned out, and the separated clay paste is made into lumps and sun dried on floors (8) and then bagged and stocked in a hut (9).

## (10) Coal (Fig. 12.79)

Upgrading of coal is a necessity in modern economy—not only for mineral conservation, but also for supplying a uniform grade to con-

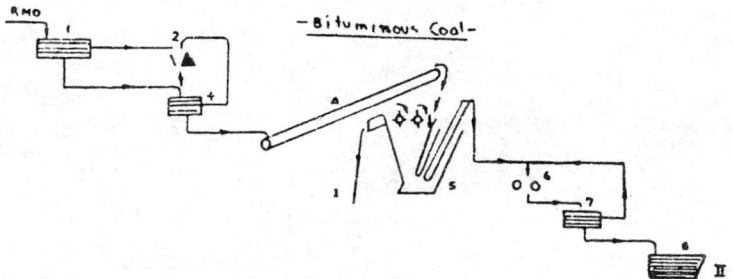

Fig. 12.79

sumers, which conforms to the specification as laid down in the contract. Under this category of consumers are included the steel plants. Power stations, which require middlings also insist on specifications.

The main characteristics, ultilised in the beneficiation process, depend essentially on (a) difference in sp. gravity of the coal and the associated stone (shaly coal and shale)—the specific gravity of the coal being less. (b) Difference in the type of fracture, and consequently on the shape and size of the fragments of coal and shale. The coal fragments are generally cubic or semi-rounded, while the fragments of shale are flat or tabular.

Methods, utilising the first property, employ sink float separation. In this method a suspension media of varying specific gravity is used for eliminating the waste. The specific gravity of the suspension depends on the nature of the products viz. floats, middlings and sinks.

The method which utilises the second principle, employs jigs for affecting separation. A flow sheet, for the usual Gondwana sub-bituminous coals, is given in Fig. 12.79.

The raw coal is screened (1), and the oversize (+), is taken to the crusher (2).   The undersize (—), is screened (3), and the whole taken to a picking belt (4) and delivered in the sinkfloat cone  (for heavy  media) (5). Clean coal is recovered  (I), while  the middlings are crushed in rolls (6), screened (7) and either tabled (8) or delivered to jigs for further concentrate recovery (II).

**(11) Tin**  (Fig. 12.80)

Tin generally occurs in the form of placers, containing mainly cassite-rite ore, associated with  gravels, and heavy minerals, e.g., magnetite.   The normal practice, where possible, is to work such deposits by  ground sluic-ing, or by dredging or hydraulicking.   If the mechanical tin saving  devices

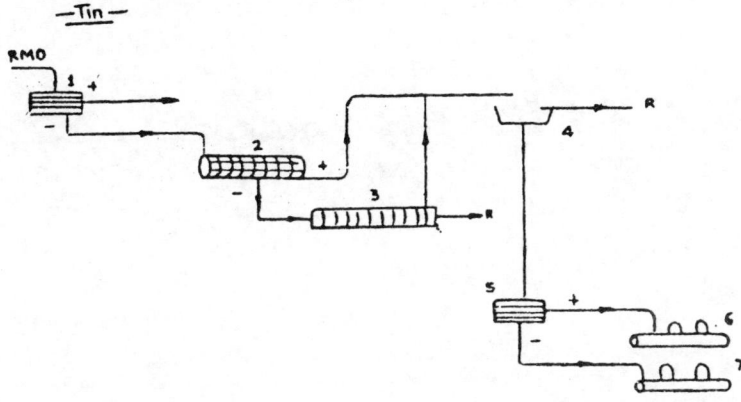

Fig. 12.80

are essential, a typical layout as employed on a dredge may be as shown in the flow sheet.

The alluvial material is fed on to a screen  (1),  the  oversize (+) is picked  out,  and the rejected  material is mostly rock fragments, etc.   The undersize (—)  is placed  on a  rotating  screen (2).   The oversize (+) is taken to jigs (4), while the  undersize (—) is led into  the  sluice box (3). The concentrate from  the  sluice  box  is also led into the jigs (4) and the tailing are discarded.   The concentrate from the jigs is screened, and sub-mitted to magnetic separation, according to  sizes, in the machines (6) and (7). The concentrate C comprises essentially of tin ore.

**(12) Iron ore.**

Fig. 11.10

# PART XIII

INDIAN EXAMPLES OF (A) GEOCHEMICAL AND (B) GEOPHYSICAL EXPLORATION (C) MINING AND ORE DRESSING PRACTICE

# INDIAN EXAMPLES OF GEOPHYSICAL EXPLORATION

1. In 1956, extensive aero-magnetic surveys were carried out by the Spartan Air Services Ltd. Canada, covering the Indo-Gangetic Plain and western Rajasthan, with the view to providing a nucleus for intensive and extensive exploration, for the Oil and Natural Gas Commission. The area covered, under this agreement, extended from the Punjab to Bihar amounting to about 133,000 sq.km. In Rajasthan roughly 44,000 sq km. was covered. From the aero-magnetic data, the thickness of the alluvium and sediments obtained was of the order of 7500 m as against 4500 m as was assumed earlier. Locally a thickness of even 15,000 m was obtained. The survey also brought to light some major tectonic features, such as the presence of large scale faulting to the north west of Moradabad, and the prominent faults in the vicinity of Lucknow and Patna.

In north west Rajasthan, to the west of Bikaner, the depth to basement was of the order of 2700 m. while west of Jaisalmer the depth was 1260 m, and again 2760 m towards the Pakistan border. In the north east part of the area, high magnetic anomalies suggesting the presence of basis intrusives associated with magnetic mineralisation were seen. These were confirmed by ground surveys.

2. *Kamptee Coalfield.* This coalfield, in Maharashtra, is partly covered by alluvium, ranging from 15-60 m in thickness. Resistivity and magnetic methods were employed in delineating the boundary of the Barakar rocks. The resistivity of the Kamptee and Barakar sandstones was only 10-40 ohm m. Whereas that of the Archean rocks was in the range of 500-1000 ohm m. This marked difference in resistivity facilitated the location of the contacts. In the western part of the coalfield, coal measure rocks are overlain by the Deccan basalts. Here the striking variation in magnetic susceptibility between the basalt and sandstone was made use of. The susceptibility of the basalt was about $2000 \times 10^6$ c.g.s. units, while that of the sandstone was only $50\text{-}200 \times 10^6$ c.g.s. units.

Further the resistivity contrast between coal and the overlying sandstones was utilised in the location of the major coal seam in the south east part of the coalfield, and also brought to light a north-south trending fault under the alluvial cover.

3. *Oil Exploration.* The presence of eocene rocks on either side of the gulf of Cambay indicated the possibility of the occurrence of oil. The occurrence of gas at Gogha, in the area, lent support to this view. Magnetic reconnaissance, over the alluvial area of Gujarat, extending from

Wadhwan in the west to Godhra in the east, and from Ahmedabad in the north to Cambay in the south, indicated the presence of two major faults; one on the western side passing through Gundhi in eastern Saurashtra, and the other along the Mahe river in the east. The thickness of sediments was estimated to be of the order of 3000-4000 m in the Cambay area due to the faulting. Gravity survey of the areas lying in between the Mahe and the Sabarmati rivers delineated a prominent closure in the Cambay area, which was also corroborated by the magnetic high.

Reflection seismic surveys, taken up in the wake of the above investigations, brought out the presence of a well defined structural high near Lunej confirming the gravity anomaly. The seismic data also showed that a broad step faulted basin was present in the alluvial tract across the Sabarmati and Mahe rivers; the deepest part of which was between 72°.28′ and 72°.38′ with the maximum thickness of sedimentaries of the order of 226 m. A test well which was drilled here, by the O.N.G.C. struck oil at a depth of about 1700 m. Success in this area opened the way to further exploration in the Cambay belt.

4. *Galena in Tamil Nadu.* E.M.S.P., magnetic, and gravity methods were used in prospecting for galena near Alangayam, North Arcot district. The major rock types, in this area, comprise, pyroxene-hornblende Gneiss, with intrusive syenite, dolerite, aplite, and quartz veins carrying baryte. The general trend of foliation is N.E-S.W. or NNE-SSW. Dipping SE, in the Kalarpatti hills and N. 70 W. in the Narsingapuram hill. Faulting was noticed in the Kalarpatti area, and lead mineralisation is associated in the shear zone.

E.M. observations made with the A.B.E.M. equipment showed prominent in-phase and out-of-phase anomalies which delineated the shear zone extending for over 7 km weak S.P. Anomalies of −20 to −30 millivolts characterised the presence of galena.

5. *Graphite, Kerala.* (Ernakulam Dt.) Very high S.P. anomalies were obtained over the graphite deposits associated with the Khondalitic rocks at Perungala and Manakad, the anomalies being of the order of 1000 mv. to 1300 mv. Test pits made at the site of the anomalies encountered the graphite lodes.

6. *Ambaji base metal deposit, Gujarat.* In the case of the Ambaji base metal deposit it has been shown that the magnitude of the S.P. anomalies alone may not be of much significance, other factors such as the depth of the ore body from the surface, and the nature of the mineralisation ore to be weighed. A drill hole, sited in the area, encountered an ore body 14 m wide at a depth of 90 m. The mineralisation being essentially galena with chalcopyrate and pyrite. The low amplitude anomaly is attributed to the depth factor as well as to the insulating effect of Galena

lead being geochemically less noble than either Cu or Fe.

7. *Manganese, Maharashtra*: Magnetic and electrical methods were used in prospecting for manganese ore in the Tirodi, Pawania, Ramtek, and other areas. The ore is associated with the sausar sediments comprising mica schist, Gneisses, calc-granulites, and limestones, the important minerals being braunite and psilomelane. The magnetic anomalies observed were of two types. 1. Irregularly distributed, isolated patches, with limited extent. 2. Narrow linear zones, with fairly strong intensities, showing regularity in strike. The first type of anomalies, coincident with electrical anomalies, were found to be due to massive ore, below soil cover or debris. Some of the second type of anomalies were associated with magnetite-quartzite or magnetite bearing schists.

8. *Diamond, Andhra Pradesh*. Resistivity and magnetic surveys were carried out in the Vajrakarur area, Anantapur district, with a view to locate the hidden diamantiferous volcanic pipes hidden by alluvium. Resistivity traverses showed a sudden drop of about 400 ohm m., in apparent resistivity values on passing the boundary between the pipe and the surrounding crystalline country rocks. The pipes so deciphered were found to be diamond bearing.

### Geochemical exploration in India

1. *Singhbhum copper belt, Bihar*: Sulphide mineralisation is found along the shear zone, immediately to the north of the Dhanjori lavas. The main rock types, in this zone, belong to the Chaibasa stage, and these are predominantly schistose e.g., chlorite and biotite schists, quartz schists, and sheared felspathised rocks known as soda granite or felspathic schists etc. Geochemical soil samples were collected along lines laid across the mineralised zones, initially at intervals of 300 m and the spacing, between the lines, was successively reduced to 120 m, to 60 m, and 15 m and samples were drawn. Where necessary closer intervals of 7 and 3 m were also adopted for sampling. The depth of sampling varied between 0.3 to 0.6 m but the samples were taken from below the 'A' zone. It was noticed that the Cu values increased progressively from a depth of 15 cm to 30 cm and decreased thereafter. Where the soil cover was deep, the maximum values were obtained at depth varying from 40 to 45 cm.

The general background value, for copper, was within the range of 50 to 100 ppm in the residual soil, and the anomalous values recorded were 4000 ppm. and even 8700 ppm. The ratio of the threshold to the anomalous values ranged between 10 to 80 in the Roam-Siddeshwar, Mahuldih, Ramachandra Pahar and other areas.

The geochemical anomalies bear a close relation to the disposition of the mineralised shear zones, and the width of the anomaly zones was found to vary from 10 m, to 167 m exploratory drilling was undertaken in the more favourable areas. In the Roam-Siddheswar area the reserves indicated by drilling were of the order of 20 million tonnes, whereas in the Rama-chandra Pahar area, a disseminated ore body was encountered. In the Mahuldih area the mineralised zone, indicated by the geochemical explora-tion, had a strike length of 600 m.

  2. *Khetri copper belt, Rajasthan.* This area is characterised by rugged topography and the height of the ridges vary from 315 m to 1000 m. The climate is of semi-arid, desert type, with an annual rainfall of about 25 cm per annum; and consequently the streams are ephemeral. The vegetation is scarce and essential xerophytic in character. Soil cover is scarce or absent except on the soft schistose bands.

  Mineralisation is associated with the meta-sediments of the Alwar and Ajabgarh belonging to the Delhi system, comprising quartzites, phylli-tes, calc-gneisses, amphibolites etc. Mineralisation is mainly confined to the schists, phyllites, quartzites with local enrichment of the basic intrusives and granites. The ore minerals are pyrite, chalcopyrite, and pyrrhotite, cobaltite and danaite occur locally.

  The average background value for copper, in the soil, is 50 ppm, while the range of the anomalous values is wide, say from about 125 ppm to 4000 ppm. over the mineralised quartzites, no significant anomaly is seen. Near Saintala-ki-dhani narrow anomalies were observed, and on dril-ling the ore body was found to be of the disseminated type.

  In the Madan Kudan area, on analysis, 50 per cent of the samples showed a copper content of 1000 ppm or more, 30 per cent showed 2000 ppm or more, 18 per cent showed 3000 ppm or more, and the rest 4000 ppm and above. The ore body in the area is found to be wide dissemina-ted type.

  3. *Malanjkhand copper deposit, Madhya Pradesh*: The deposit occu-pies a major north-south shear zone, along the southern limb of an anti-cline, in the chilpi series. The rock types include, phyllites, schistose grits, slaty shale, conglomerate etc. These have been intruded by quartz vein, granite and daibase.

  Geochemical prospecting was carried out, in an area of 0.25 sq km of the 420 samples analysed, for Cu, Pb, and Zn, 80 per cent showed more than 1000 ppm of Cu. The remaining samples showed a Cu content ranging between 200 and 900 ppm. Drilling operations revealed the presen-ce of thick ore bodies, below a leached capping 50 m to 80 m thick.

  4. *Rajpura-Dariba Zinc-lead deposit, Rajasthan*: The rock types of the area include muscovite, graphite and biotite schists, together with dolo-mite and quartz veins. The mineralised zone, marked by a band of gossan, varying in thickness from 2 m to 60 m, lies over the graphitic mica schist,

extending from south of Dariba to Rajpura. The general strike of the formations is N.E.-S.W. with steep dips to the east.

Geochemical samples were collected from depths varying from 2.0 cm to 10 cm, and about 2200 samples were analysed, for Pb, Zn, Cu, NI. The values for PB were in the range of 4000 ppm and the values for Zn were in the range of 3000-4000 ppm copper values were of local significance, while that of NI was uniformly low.

Subsurface exploration revealed the Dariba main lode under the band of gossan, extending over a strike length of over 550 m with a thickness varying from 16 m to 47 m a second lode, termed Dariba east lode, was found in the vicinity. The mineralisation occurs at the contact of the dolomitic rock and the graphitic schist. The chief sulphides seen are sphalerite, galena, pyrite and pyrrhotite.

5. *Agnigundala, Guntur Dt., Andhra Pradesh*: The area form part of the Nallamalai sequence of the Cuddapah system. The general relief of the hills is about 200 m but the high ridges are 400 m or more above MSL. Quartzites, phyllites and dolomites are the main rock types, which occur in the form of a dome, with granitic rocks forming the core. In general, mineralisation is associated with the dolomite and limestone bands. Over 4000 samples were collected from the area which on analysis showed Pb values ranging from 50 ppm to 3000 ppm, whereas the corresponding values for Cu and Zn were consistently low.

Exploratory drilling carried out in the area indicated the presence of lead-copper mineralisation, with the lead content varying from 1.0 to 5.5 per cent and the copper content varying from 0.8 to 10.61 per cent over a strike length of 2300 m.

6. *Manmandur zinc, copper deposit, N. Arcot, Tamil Nadu*: The area comprises granite gneiss interbanded with garnet-sillimanite gneiss, basic metamorphic rocks, and charnockites. Mineralisation is seen in the form of veins and segregations in the gneisses.

Geochemical soil sampling was undertaken. On analysis the samples showed the anomalous concentrations for copper, zinc and lead. The maximum values for copper were 1000 ppm whereas that for zinc were about 3000 ppm. Exploratory drilling carried out intersected an ore body assaying 0.63% Cu., 2.00% Pb., and 2.73% Zn.

## INDIAN EXAMPLES OF MINING AND ORE DRESSING PRACTICE

### 1. Iron ore

Noamundi iron ore mine of Tata Iron and Steel Co. Ltd. (TISCO) is about 80 miles from Jamshedpur, on the Tata-Gua branch line of the S.E. Railway.

The deposit comprises two ridges, extending approximately in a N-S direction. Each ridge is about 2½ miles in length, and the intervening valley is occupied by the Balijore nala.

The types and compositions of ores found in Noamundi are given in the table :

| Ore Type | Composition in % | | | | |
|---|---|---|---|---|---|
| | Fe | $SiO_2$ | $Al_2O_3$ | P | Mn |
| 1. Hard ore steel grey | 66-67 | 1-2.5 | 2-2.5 | 0-0.4 | 0.05 |
| 2. Reddish, brown grey laminated and biscuity or lump ores | 58-60 | 2-3 | 4.5-6 | 0.10-0.15 | 0.05 |
| 3. Shaly ore, pink to red | 57-60 | 2-3 | do | do | do |
| 4. Laterite ore | 53-55 | 1-1.5 | 6-9 | do | do |
| 5. Blue dust | 63-65 | 2-3 | 2-2.5 | 0.05 | 0.05 |

Mining operations were initiated in 1925, at the Balijore nala where benches were cut. The ore was transported by, 2 ft. gauge, side tilting mine cars, drawn by steam locos. From an overbridge the ore was loaded into broad guage wagons, placed at the siding.

With the increase in output, an electrically operated rotary car dumper was installed, together with a gyatory crusher and cone crusher. The 2 ft. gauge line was extended, in the eastern ridge, and crossed over to the western ridge, by a trestle bridge. Gravity haulage was used in the eastern ridge, for transporting ore from the upper benches. This was later replaced by 150 ton/hr. monocable rope way, which transported ore from the upper benches to the bin, at the railway siding, near Noamundi station. This modification affected economies in time and money. Improvement in the quality of ore was affected by putting up a washing plant on the eastern ridge. A bicable ropeway, with 250 ton/hr. capacity, was constructed, to connect up with the bin at Noamundi station, whose capacity was increased to 5000 tons. Noamundi has a daily output of 6000 tons in two shifts a day. Some of the equipment used, to deal with this increased output, are 4½ cu. yd. electric power shovel, 6 in diameter drill master for blast hole drilling, and 25 cu. yd. rear dumper trucks.

The open cast consists of two benches which are 30 ft. high Churn drills and Drill masters, with T.C bits, are used for blast hole drilling. The distance between holes is 15-18 ft., and the holes reach 6 ft. below the floor of the lower bench. Explosives employed are open cast gelignite, sunderite and liquid oxygen. Cordeau fuse and delay action detonators are used.

The flow sheet showing mining, haulage and loading at Noamundi mine is given below :

| Name of operation | Machines used | Particulars of machines |
|---|---|---|
| 1. Mining | Deep hole drilling using liquid oxygen or high explosives for blasting | Churn drills and Drill-master |
| | Excavation and loading | 4½-Cu-yd Elect. power shovel and 2½-cu.yd. power shovel |
| 2. Haulage | Haulage to crusher | 22-ton rear dumper truck |
| | Feeder | Pan type (60-in-diam.) |
| 3. Dressing | Primary crusher | 42-in. gyratory, set to 8 in. |
| | Grizzly | 5-roll |
| | Belt conveyor | 48 in. |
| | Secondary crusher | 2 in. size |
| | Belt conveyor | 48 in. |
| | Bin | 600 ton capacity and shuttle conveyor |
| 4. Haulage | Loco Haulage | 2½ miles |
| 5. 150 ton capacity bin | | 600-ton bunker (washing) conveyer |
| 6. Conveyor (24 in) | | Conveyor |
| 7. 2000 ton capacity bin | | 600 ton bin |
| 8. To railway siding | | Bicable rope way 250 bin/hr. |
| 9. Storage bin | | 5000 ton storage bin near Noamundi Station. |

*Joda East iron ore mine*: This is also an open east mine owned by Tata Iron and Steel Co. Ltd. It is located in the Keonjhar district Orissa, and lies on the railway link from Noamundi to Banspani from Dangoposi. This mine was started anew, hence mechanisation has been introduced from the initial stages unlike Noamundi, where hand mining had to be replaced at a later stage. Mining was started in two benches, cut at the crest of the hill, each bench being 50 ft. high.

The blast hole procedure adopted here is the same as that at Noamundi. Joy-Challenge compressed air drill, 4½ in diameter, equipped with T-C bits is employed. Power slovels (electric) are used for clearing the muck and loading into rear dump trucks. The ore is fed into a primary gyratory crusher with a capacity of 450 tons/hr. and then discharged into a bin with 500 ton capacity. From the bin the ore is loaded into aerial rope way buckets through electrically operated chute gates. The bicable rope way has 50 buckets each of 2-ton capacity. Keeping the bucket interval of 68 m., the capacity of the ropeway can be taken to be 250 tons/hr.

The present output of the mine is 45,000-50,000 tons per month.

The flow sheet of the Joda East iron ore mine is as follows :

Deep hole drilling—
blasting with liquid oxygen
or high explosives
|
Excavation and loading
|
Haulage to crusher
|
60-in pan feeder
|
Primary crusher (42″ gyratory set at 8 in:)
|
Table Feeder
|
Belt conveyor
|

| 500 ton storage bin | Conveyor to skip—ore bin (50 ton) |
| --- | --- |
| 250 ton/hr. aerial ropeway | Skip incline |
| 300 tons capacity storage bin at Railway siding | Skip botton bin on railway siding |

## 2. Gold—(a) *Mining*

*Kolar Gold Fields* : Popularly known as the KGF in Mysore, is about 15 miles from Bangarapet junction, on the Madras-Bangalore broad gauge line of the Southern Railway. From Bangarapet a branch line leads into the gold field and then to each of the mines.

Antiquity of mining as an 'industry', in India is undisputed and Kolar is no exception. Ancient workings, extending upto a depth of 200 ft. have been noticed, in the Nundydroog mine.

In recent times, between 1877 to 1881, small companies were floated with limited capital, for prospecting gold, in the Kolar district. John Taylor & Company, the well-known mining engineers (and managing agents), started mining tn the KGF in 1883, and in 1950, at the request of the Mysore Government, new companies were formed, and registered in India. Later on in 1955, the KGF was nationalised, by the Mysore Government, and the transfer was affected in 1956.

The gold bearing lodes occur in the well known Kolar schist belt, of the Dharwars. This belt is about 50 miles long, and extends from Srinivaspur to near Krishnagiri, in Tamil Nadu. The average width of the belt is about 3 miles.

Mineralisation is essentially within the hornblende schist, but occasionally gold is also found in the quartzites. The main lode or Champion lode has been traced for nearly 4 miles, along the strike, and it has been worked to a depth of 10,000 ft. Old workings are mainly confined to this lode. In width, the lode averages 4 ft. but at places, it is 30 ft. in width. Smaller lodes, both to the east and to the west of the Champion lode, have also been responsible for appreciable tonnages of payable ore. The western reefs also constitute an important group in the Nundydroog mine.

The Champion lode is found in the Mysore mine, and its continuity is not clearly seen. At the northern end, in the Nundydroog mine, the Champion lode is cut by the Balaghat North Fault, and no trace of the main lode is found, except for small ore shoots.

Shafts are sole means of access, to the gold mines, in the KGF. In the early days, the practice was to sink the shafts in the lode, but in recent times, the shafts have been sunk away from the lode ; so that the shaft (inclined) was driven at a desired angle, maintaining the same gradient. Both vertical and inclined shafts were sunk. In addition, auxiliary or secondary shafts were also made for communication at lower levels, beyond the depth of the main shaft. In some cases, due to the deeping of the mine workings, a tertiary system, of shafts has also become necessary.

One of the outstanding achievements, in the KGF, is the Giffords shaft, in the Champion reef mine. It was sunk in 1940, and the depth (vertical) was 6586 ft., and was reputed as the world's deepest pit. The diameter is 18 ft. and it is a brick-cum-concrete lined structure.

Direct winding is adopted, and the cages can carry 50 men or 4 nos. of $1\frac{1}{4}$ ton mine cars. Each rope weighs 20 tons, and the speed of winding is 30 m.p.h., the H.P. of the electric hoist being 4450.

The other deep shafts, in the field, are also vertical. Henry's shaft, in the Nundydroog mine, is 4088 ft. deep. The Bullen's shaft, of the Champion reef mine, is 3871 ft. in depth, and the Edgar's shaft in Mysore mine, is 3804 ft., in depth.

Development consists of drilling and blasting. Drilling is mainly done with jack hammer mounted on air legs, and T-C bits are in common use. About 20 holes, each about 3 ft. deep and $1\frac{1}{4}$ in diameter, are made in a tunnel face, approximately 7 ft. × 6 ft. The wedge cut is adopted normally. The explosive used is blasting gelatine, and either safety or electric detonators are employed, time lag ones have been especially found to be an advantage.

*Stoping* : The methods of stoping in the KGF commonly employed comprise, (*a*) underhand, (*b*) overhand flat back and (*c*) rill.

Underhand stoping is almost universal, and it is best suited, where the walls are well defined, and the dip is low. This method is also success- fully employed, even where the dips are steep, provided the walls are well defined. Overhand flat back and rill stoping are adopted where the ore body is steeply dipping, and the walls are made of flaky rocks and ill defined.

A few examples of the application of these methods of mining are given below :

In the Mysore Mine underhand stoping is normally practiced, but in the 43rd ore shoot, a modification, of the normal practice, had to be emp- loyed. The ore body dips steeply to the west at 75°-80°. Its strike length is 480 ft., while its vertical height is 800 ft. the width varying from a few inches to 10 ft.

The ore body is confined between two faults, more or less parallel to the strike, and consequently the walls, are highly brecciated. Towards the north again, it is cut off by a cross fault and southwards it thins out, and it is represented by stringers, and branch fault.

Taking into account the structure, and the nature of the walls, the ore body had to be opened up indirectly, using the stope drive method. Drives were made in all levels, parallel to the foot wall and about 75 ft. away, and connected to the ore body by cross cuts, one at either end and one in the middle.

At the next stage levels are made in the ore body, with an interval of 75-100 ft. and connected by rises driven at either end and centrally.

A long underhand stope is started (Fig. 13.1) between two rises R, from a level above—with a width of 15 ft. Granite masonry pillars 12 ft. long and about 12 ft. wide (1) are constructed between hanging and foot walls with 4-5 ft. gaps. A modified method of roof bolting is practiced. Steel straps (3) are fixed to the hanging and foot wall by bolts at 3 ft. centres. Over this a concrete mat (2) is laid and it is keyed to the walls (both hanging and foot) by bars (4) known as plug bars. The masonry is laid on this mat and the wall is extended to the stope face.

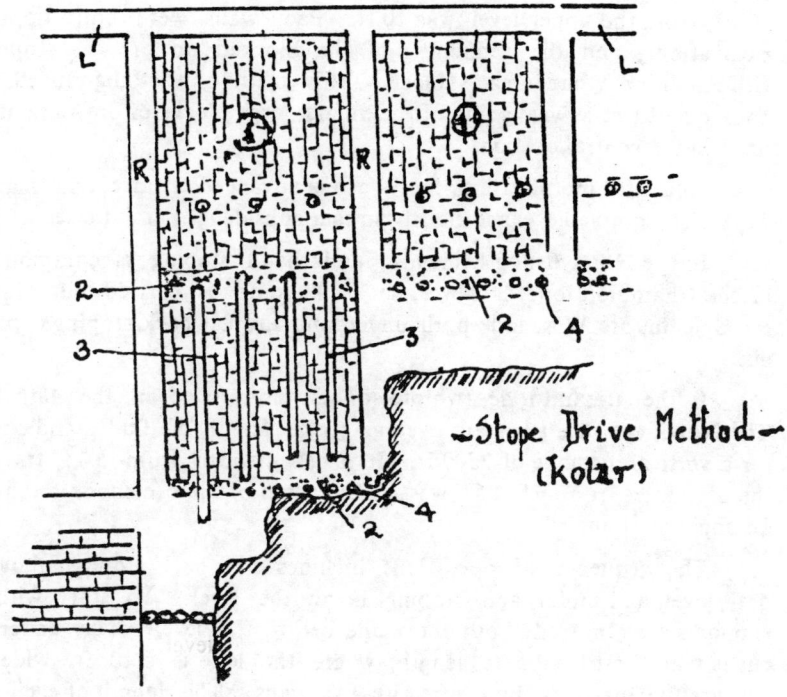

—Stope Drive Method—
(kolar)

Fig 13.1

Below the concrete mat a slice is started, in the ore body, and extended for the full width, with a height of 7 ft. Roof bolts and strapping plates are fixed to the hanging and foot wall.

If the stope is more than 10 ft., the concrete mat is laid, on the foot wall, and granite masonry pack is built up between the hanging and foot walls, leaving a passage of 3-4 ft. leading to the stope face. As the face advances, the roof bolts and strapping plates are also extended, and so also the masonry supports. Each lift, for an operation, is 16 ft. long. The process is continuously repeated, until all the ore is stoped out.

In the Champion Reef mines, underhand stoping is commonly practiced. Rill stoping has, however, been found suitable, in winning ore, from the Glens ore shoot, since the dip is steep, and the walls are ill defined—being prone to flaking and scaling. The ore body dips at 80°-85°, while the thickness averages 38 inches, giving a stoping width of 72 inches.

The earlier practice was to develop the initial cut, as a flat back stope, with a width of 20 ft. from the lower level. The bottom of the stope, from the upper level, was 10 ft. Pack walls were built up, in the excavations, and the remainder of the blocked out ore was stoped by rilling, with a V-cut in each block. This method was abandoned since consequent to the weakening of the pillars, large blocks of ore were abandoned in the central portions.

In a later modification, the V-face is advanced rapidly and masonry work also completed expeditiously so that subsidence is minimised.

In the Nundydroog Mine also underhand stoping is common, but in the Champion lode, however, flat back stoping is carried out while in the Oriental ore West lode both underhand and flat back stoping is carried out.

In the Oriental lode, two ore shoots were proved in the 48th level. The North ore shoot has an average strike length of 200 ft., and extends for a vertical distance of 2500 ft. It varies in width from 5-15 ft. while the dip varies from 75°-85° west. The pitch of the ore shoot is south, at an angle of about 50°.

The sequence of operations includes flat back stoping, above the 48th level, and underhand stoping below the level. To start with, the supports are removed, both above and below the level. The underhand cut is made first for 12 ft. length, where the lode is 6-10 ft. wide, and the granite masonry built up with 4 ft. gaps. The length of each stope is 36 ft. So also the back stope is started and supported by pig styes and the pig styes are packed with granite blocks. The gaps between supports is 7 ft.

### (b) Ore Dressing Practice

The ore obtained, from the Champion Lode, is essentially quartz with minor amounts of sulphides. Since the gold occurs as relatively coarse particles, upto 80% of the values is recovered by simple straking. It is essentially a free milling ore.

On the other hand, ore won from the west Reef is very hard, abrasive, and the mineralisation is associated with sulphides, essentially arsenopyrite and pyrrhotite. Even though the bulk of the ore is free milling. a small part is locked up in the sulphides, even after grinding to 325 mesh (Tyler). Losses on this account have risen appreciably in the Nundydroog mines.

In general, the practice, on all the KGF mines, is more or less similar, except for variations in the treatment of concentrates.

A broad outline of the flow sheet is given below. The ore from mine cars is tipped into grizzlys or trommels. The oversized gets on to picking tables and belts, and then gets into a jaw crusher, set. at $2''$-$2\frac{1}{2}''$. Material from the crusher is combined with the undersize and then fed into Californian stamp batteries (80 units in a mill). The material passing through the screen of the mill, is fed to strakes from which the free milling gold is recovered. in the case of ores from Mysore and Champion Reef mines.

After this stage, the mill product passes into classifiers, of the Caldecott cone type. The bottom discharge. from the primary and secondary classifiers, passes into tube mill, while the overflow, from the secondary classifier, enters the cyanide section, after going through thickening filtration.

In the cyanide plant, there are Brown agitators with a capacity for treating 150 tons of dry slime, at a density of 1.75. The agitation time varies from 12 to 16 hours, and then filtration is affected in Butter's type filters. Auriferous liquor, extracted. is passed through sand clarifiers, to extractor boxes, where the gold is precipitated. The filtered precipitate is treated with sulphuric acid, with the addition of manganese dioxide, where necessary, for the removal of copper. The treated precipitate is filtered, well washed, roasted and smelted in refractory lined pots and a bullion with 985 parts of gold and silver in 1000 parts is obtained.

*Flow Sheet*

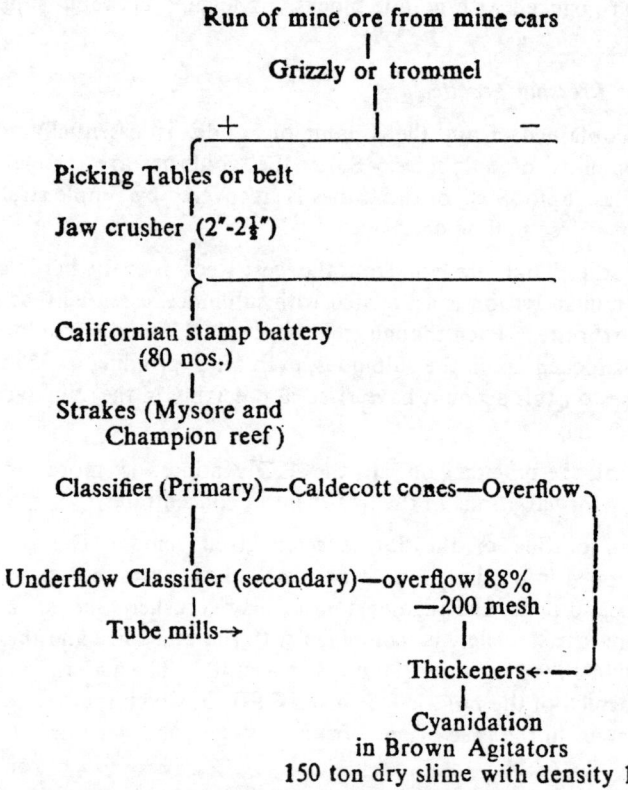

Run of mine ore from mine cars

Grizzly or trommel

\+        —

Picking Tables or belt

Jaw crusher (2″-2½″)

Californian stamp battery (80 nos.)

Strakes (Mysore and Champion reef)

Classifier (Primary)—Caldecott cones—Overflow

Underflow Classifier (secondary)—overflow 88% —200 mesh

Tube mills→

Thickeners←—

Cyanidation in Brown Agitators 150 ton dry slime with density 1.75

In the Mysore and Champion Reef mines, the blanket concentrates from the stamp, and from the tube mills are reconcentrated further by straking, as well as by tabling (James type), and the concentrates, so obtained, are smelted into bullion of high purity. The table given below shows the ore (tons) milled, ozs of bullion etc.

| Year | Tonnes milled | Gold produced (gm.) |
|------|---------------|---------------------|
| 1973/74 | 341,523 | 1801,877 |
| 1974/75 | 319,859 | 1800,406 |
| 1975/76 | 321,730 | 1747,866 |
| 1976/77 | 385,781 | 2204,160 |
| 1977/78 | 372,676 | 1941,106 |

## 3. Zawar Lead-Zinc Mines

### (a) Mining

The lead-zinc mines of Zawar are in Udaipur district, and about 17 miles due south of Udaipur City. The mine is on the national highway passing through Udaipur City and on the rail link from Udaipur to Himmatnagar, near Ahmedabad.

Mineralisation is found in the rocks of the Aravalli system, comprising conglomerates quartzites, dolomites, phyllites. The host rock is mainly dolomite traversed by parallel shear planes, which constitute the loci.

In the Mochia Mangra hill the strike is generally E-W and the dip is 75°S. The ore body varies in width from 3 to 60 m. and ore minerals are galena, sphalerite, pyrite, chalcopyrite, arsenopyrite, argentite, cassiterite.

Exploration has been carried out, in certain of the prospective portions in the lease hold, as per details given below.

|  | Surface | Underground |
|---|---|---|
| Mochia | 4.924 m. | 4790 m. |
| Balaria | 3620 m. | 1339 m. |
| Zawar Nala | 763 m. | |
| Bawa | 366 m. | |

In the central portion of Mochia Mangra hill, where the main workings are located, the total proved and probable ore is about 8 million tons with 1.8% Pb and 3.7% Zn and the reserves upto a depth of 300 m. is about 21.5 million tons. Further exploration has since been carried out.

In the eastern extension of the Mochia hill known as Balaria the values of zinc are high being 12% and the extension of the ore body is expected. The reserves upto 160 m. below valley level is about 6.37 million tons. Mochia and Balaria are contiguous areas with a aggregate strike length of 3700 m. It is estimated that the reserves will be in the neighbourhood of 50 million tons, upto a depth of 350 m.

Three mineralised zones with widths 20 m., 16 m. and 4 m., show total metal content, in respect of Pb and Zn, of 4%, 10% and 5% respectively as evaluated from intersections in the exploratory drill holes. The assessment of reserves has not been completed. In the Bawa area drilling has shown a ore bearing zone 3 m. in width with an average metal content of 8% of Pb and Zn.

Mining is confined to the Mochia Mangra hill while exploratory mining has been initiated in the Balaria area. In the Mochia Mangra area, development has proceeded to the 4th level, and the current production of the mine is 1000 ton/day.

The methods of stoping at Zawar are (a) shrinkage and (b) sub-level stoping. Shrinkage method is adopted where the width is small, and ore body is well defined. Otherwise sub-level method (Fig. 13.2) is generally adopted. The level intervals 36 m. (1). The sub-level intervals are 10 m. or more (2).

A rise is also made at one end of the pillar known as the slot rise (7), in addition to the mainway, connecting the sub-levels (8). A preparatory level (3) is made above the lower level (2). A crown pillar (6) is left at the top, while level pillars (9) are maintained at the lower part above the tramming level. Stoping starts at the preparatory level (3) by drilling long blast holes by the underhand method to form box holes (4) i.e., by drilling from the upper sub-level, and connecting to the haulage level by rises (5).

In the shrinkage stoping method adopted two raises serve as man-passes, from one level to another. It is generally carried out as a flat back stope. The wall rock being hard, massive and strong, does not require any support.

(b) *Ore Dressing*

The ore dressing practice, adopted at Zawar, is given in the flow sheet below :

Ore Bin (8″ sizes)
|
Jaw crusher (3″ size)
|
Tailor crusher (1½″ size)
|
Symons cone crusher (¼″ size)
|
Screen
|
Ore Bins
|
Ball Mill
|
Spiral classifier
|
Conditioner
|
Lead flotation battery
|          —lead concentrate—thickeners
|          —and filter
Conditioner (Zinc)
|
Zinc flotation battery
|          Zinc concentrate—to thickeners
|          —and vacuum filter
Tailings

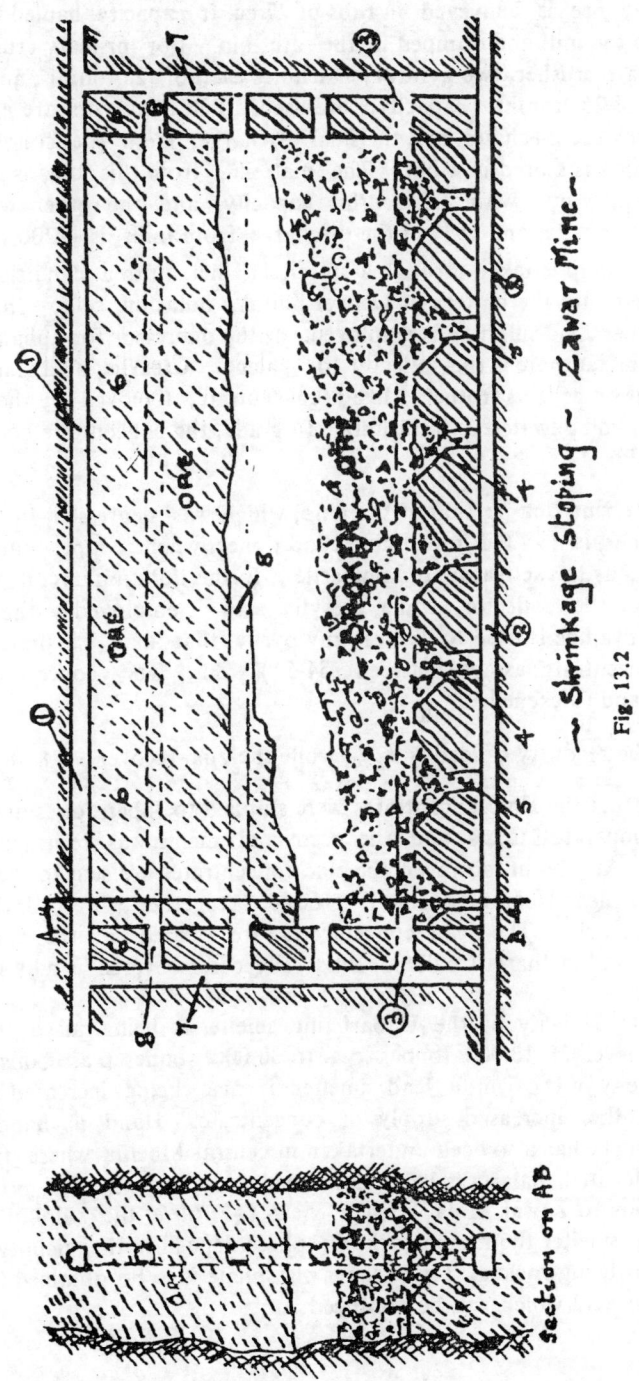

Fig. 13.2

— Shrinkage Stoping ~ Zawar Mine—

Section on AB

The ore is conveyed in tubs of 27 cu. ft capacity, hauled by diesel locos, to the mill and dumped in the ore bin. For primary crushing 40 ton/hr. jaw crusher, two gyratory machines each of 20 ton/hr , and a cone crusher of 25 ton/hr. capacity, are used. Belt conveyors are employed for linking the machines, and the final discharge, from the cone crusher, is $\frac{1}{2}''$, which is stored in a bin. The material, from this bin, is fed into 2 ball mills, one with 7.5 ton/hr. capacity, and the other with 12.5 tons/hr. capacity, and 55% of the discharge, from mills, is —200 mesh.

The pulp is taken through a spiral classifier, into a conditioner, and from there into the battery of Denver Sub-A. flotation cells. In the lead conditioner Zinc sulphide is employed, as the depressor for sphalerite and potassium xanthate is the collector for galena. Cresylic acid is added in the flotation cells as frother. Lead concentrate, removed by the froth is collected and dewatered (filtration). In grade, the concentrate contains 72 to 73% Pb.

The zinc rich portion of the pulp, which was depressed in the lead conditioner, is now fed into a zinc conditioner, where copper sulphate is employed as the activator for sphalerite. In the flotation circuit, xanthate is used as the collector, while cresylic acid is employed as the frother. Sodium cyanide is used to depress any pyrite that may be present. The zinc concentrate assays between 54-54% Zn. The concentrates are filtered and thickened.

The recovery of lead is 92%, while the zinc recovery is 85-87%.

Earlier the zinc concentrates were shipped to Japan for smelting. on international toll basis and the recovered metal was returned to the country. At present, however, the zinc concentrates are sent to the smelter at Debari near Udaipur. Zinc is produced together with cadmium and sulphuric acid. The lead concentrates are treated, at the smelter, at Tundoo, in the Jharia Coalfield, Bihar, silver is recovered as a by-product.

The capacity of the Dabari zinc smelter is being raised from the present level of 18,000 tonnes p.a. to 36,000 tonnes p.a. Consequently the capacity of the Tundu lead smelter is also being increased to cope up with the increased supply of concentrates. Hand in hand, mine development has also been undertaken in central Mochia where the main shaft is located. Balaria is being developed independently for production in addition to Zawar Mala. With a view to utilise the sulphur dioxide from the smelter the maton rock phosphate deposit in the vicinity of the smelter is being rocked. Triple super phosphate is to be obtained from the sulphuric acid which will be produced.

At present the production of zinc is 30,000 tonnes p. a. and the production of lead is 10,000 tonnes p. a These figures will be enhanced when the Vizagapatnam smelter (using imported concentrates) goes into production.

## 4. Mosabani Copper Mine

### (a) *Mining*

Mosabani is the only working copper mine in the country. It is about 150 miles WNW of Calcutta, and the nearest railhead in Ghatsila, on the Howrah-Bombay main line, of the South Eastern Railway. The mine is near the eastern end of the well known Singhbhum Copper Belt, which extends for over 100 miles, from near Dwarapuram, in the vicinity of Chakradharpur, on the west to Bharagora, in West Bengal on the east. Throughout this belt there are positive evidences of ancient workings for copper, such as deep pits, mine spoil, slag dumps, stone implements for crushing etc.

The Cordoba Copper Company exercised its option in 1924, by purchasing the mining rights, from the Cape Copper Company which had closed down earlier. Later, the Cordoba Copper Company was converted to the Indian Copper Corporation Limited. Presently the company has been taken over by the Hindustan Copper Ltd.

In the Mosabani area, the Chaibasa stage of the Iron ore series has been thrust over the Dhanjori sequence. The main rock types, of the Chaibasa stage, are mica-schist and its variants, phyllites, quartzites and quartz-kyanite rock, while those comprising the Dhanjoris are essentially quartzitic sandstones (orthoquartzites) and metamorphosed lavas (volcanics). Meta volcanics are represented by vesicular lava, epidiorite, hornblende schist, talc schist, and -biotite chlorite schist. A rock type, which is apparently granitic in composition, rich in sodic plagioclase, known as the soda granite, has been emplaced along the shear zone. The rock is mylonised and grades into quartz-chlorite schist. The contacts, between the granite and the Chaibasas, as well as that of the granite with the Dhanjoris, are gradational.

The strike of formations, in this area, is generally NNW-SSE, while the dip is 40-45° to the East. In the phyllite, mica-schist country, east of the mines, quartzites show the presence of sedimentary structures, e g., ripple marks, cross bedding besides convolutes associated with turbidites. Evidences gathered from the orientation of these structures go to show that no inversion has taken place despite of the severe deformation to which these rocks have been submitted.

Mineralisation, as far as the northern section of Mosabani is concerned, is confined to the quartz-chlorite, and quartz biotite schist. Granite is absent in this area. Three discontinuous lodes have been encountered and these are found in Chapri, Kendadih and Surda.

In the southern section, chloritic rocks are not developed. The granite is predominant and appears to have controlled mineralisation, in the main Mosabani, and Pathargora areas. Mineralisation, however, appears to diminish further south of Mosabani, in the region of Badia.

At Mosabani, two well defined lodes have been encountered, within the sheared granite, and these have been named the Main Lode, and West Lode respectively. The average dip of the lodes is about 38°, but the dip appears to ease out, at depth. In addition, the lodes also appear to converge towards each other at depth, but the Main Lode thins out at 1200 ft. and pinches out further below. In strike the Main Lode extends for more than 3000 ft. The West Lode is more extensive, and it is present in the form of an independent ore shoot, with intervening barren rock, for a distance of 2¼ miles.

As stated earlier, mineralisation is associated with the "soda granite" and adjacent rocks, of the Chaibasa and Dhanjori stages. The principal ore mineral is chalcopyrite. Pyrrhotite, pyrite, pentlandite are also found. Apatite and magnetite are common. At Mosabani the ore shoots pitch at about 60°E., while in Badia the pitch is N.30°F. and persists even to a depth of 3000 ft.

In Mosabani Mines, considerable development work has been done. Laterally, the mine has been explored for about 5 km., and vertically for about 0.7 km., i.e., 1.5 km. down the dip of the lode. The deepest part of the mine is 0.9 km. or 1.9 km. down dip.

The main shaft is an inclined shaft, reaching upto the 20th level or 670 m. below surface. Later a circular shaft (a vertical shaft) was also sunk upto the 20th level, to deal with the increase in output. The output, from below this level, is handled by inclined shafts, which feed both the Main shaft and the Circular shaft.

Compressed air drills are employed, in drilling and gelignite is the explosive, normally used in blasting.

In the payable ore shoots, the levels are interconnected by raises and winzes, from which stoping is commenced. The ore is won by breast stoping, where the stope face is advanced on true dip. Usually 1600 mm. deep blast holes are drilled, by hand held jack hammers, and blasted with 60-80% gelignite on night shift. Due to the low dip, removal of ore is effected by compressed air operated scrapers of 15.25 H.P.

Stopes are supported by pillars left *in situ*, and regular lines of timbers. Continuous lines of pillars are left at the top and bottom of stopes, for the protection of the drives   The broken ore is trammed from the stopes, in 1.4 tonne side-tipping cars, drawn by battery powered locos, to the hoisting shaft bins.   The ore is hoisted in skips (5-ton) or cars to the surface, and automatically dumped into ore bins at the top.   Here it is crushed to −3/8″ size, in the primary crushers.

Primary crushing is done in 3 stages (jaw and rotary), and the ore reduced to −¼″, while the waste and tramp are eliminated by hand picking. From the crushers the ore is conveyed, by belt to the bins at the aerial rope station. From here the ore is transported to Maubhandar on the river Subarnarekha at a distance of 6 miles, by buckets by the mono-cableropeway, with a speed of about 5 m.p.h. and a capacity of 60 ton/hr.

The aerial ropeway buckets automatically discharge into bins, at Maubhandar, and the ore is drawn from these bins for beneficiation.

(b) *Ore Dressing and Smelting*

The operation involved in producing fire-refined copper from ore, as practiced at the Indian Copper Corporation's Works, at Ghatsila cover :

1. Ore Dressing : (a) Grinding, (b) Flotation, and (c) Dewatering of concentrates.
2. Smelting : (a) Roasting, (b) Matting, (c) Converting, and (d) Refining.
3. Electrolytic refining

(1) Ore Dressing :

(a) *Grinding* : Crushed ore, from the mines, is transported by an aerial ropeway and is delivered, at the works, into ore bins.

Grinding is done in four Hardinge ball mills, the ore being fed by apron conveyors, from the ore bin, to the mill. Each mill grinds in closed circuit, with a Dorr Duplex classifier.   The classifier overflow pulp, from all mills, is combined and pumped to the primary flotation cell by Wilfley Sand Pumps.  The feed consistency is about 47%, of −200 mesh grade.

(b) *Flotation* : Flotation is carried out, in 3 cells of the pneumatic type (Callow cells) of a total length of 170 ft., and require 12,000 cu.ft. of air per minute, at a pressure of 5.5 lb/sq. inch. Pine Oil is used as frother, and potassium ethyl xanthate as collector, in the flotation process, which is carried out at a pH of 10.5, the level being maintained by the addition of lime, which also acts as a depressant for pyrite.  Calgon and soda ash are added as dispersion agents, for sliming, or for treating high grade ore.

After floating the chalcopyrite, in the primary cell, the pulp is passed through the other two cells, in series which scavenge the copper, in the tailings, to the desired degree. The middlings or scavenge concentrate, from these two cells, are returned to the grinding circuit, where some additional grinding is provided.

(c) *Dewatering of Concentrate* : The concentrate pulp is pumped to two "de-sanding" cones, which remove granular concentrate direct to the filters, whilst the overflow gravitates to a 30 ft. diam. Dorr Thickener.

The overflow water, from the thickener, is pumped back to the main water tank. The thickened concentrate pulp is fed to a battery of five Dorr Oliver drum filters, each having a surface area of 100 sq.ft.

The filter cake, with a moisture of 10-12% is discharged on to Lowden dryers, fired by pulverized coal. The dryers reduce the moisture content to 6-8%. The dried concentrate is filled into tipping cars and sent to smelter.

(2) Smelting :

(a) *Roasting* : About 1/5th of the concentrate is roasted, in eight hearths of the 20 ft. diameters. Herreschoff Roaster, to reduce the sulphur content to 4%. The concentrate is charged to the first hearth, and discharged at the bottom of the eighth hearth, into closed cars. As sufficient heat is generated, by the burning of sulphur, no additional fuel is required.

(b) *Matte Smelting* : Matte is smelted in a 47 ft. long reverberatory furnace fired with pulverized coal. Concentrate is charged, through charge holes in the roof of the reverberatory furnace and the calcine is charged through a door in back wall. The charge being self fluxing, no additional flux is required. The charge melts down to a matte containing 42% Cu, and forms an iron silicate slag. The slag is taken out, through a notch in the front wall, and is dumped. The matte is tapped, through a hole in the side wall of the skimming bay, as and when required by the converters.

(c) *Converting* : Matte, from the reverberatory furnace, is charged into 8 ft. dia. Great Falls type, converters for blowing to blister copper. Air is blown @ 4,500 cu.ft./min., at a pressure of 12-16 lbs./sq.inch. One complete cycle, of a converter, takes about $3\frac{1}{2}$ hrs. producing $2\frac{1}{2}$ tonnes of blister copper. Quartzite is used as flux and the slag formed is returned to the reverberatory furnace for further treatment.

(d) *Refining* : Blister copper from converters, containing 97% Cu, is refined in three refinery furnaces. One of these is used for making copper ingots, and the other two, which are of the tilting type, are used for delivering liquid copper to the brass making furnaces. The operations involved, in the refining process, are skimming, oxidizing by compressed air lances, and reducing the copper, to proper extent, by poling. Pulverized coal is used as fuel, in all the three furnaces.

With a view to increase the production from the Singhbhum copper belt an integrated plan of development has been initiated by the Hindustan Copper Ltd , at Mosaboni, a flash smelter has been installed as well as a plant for electrolytic refining. It proposed to increase the output from the present level of 12,000 tonnes p.a. to 15,000 t.p.a. with a view to further expansion. The Outo Kumpu smelter has a surplus capacity which will be utilised for the mines to be developed in the Rakha area closeby. The Rakha mines area will have a separate smelter later on.

## 5. Mica Mining

The method used in underground mining of mica, in Bihar, has been described as "pillar and stall", when applied to flat dipping veins, but it has been termed "drift mining", when applied to steeply dipping veins. These terms are not used correctly—as it will be realised.

The pegmatite is opened up either by sinking vertical or inclined shafts. The general pattern, is similar to breast stoping, where the dips are low, and the thickness is not excessive. In steeply dipping pegmatites, open stopes are common. However, waste filling is employed more with a view to the dispose of waste, rather than for supporting the stope.

In some mines, inclined shafts, along the veins, are used for manways, while vertical shafts, connecting to the main haulage level, are used for winding ore.

The general pattern of mining, conforms to overhand stoping, with flat back, using waste fill. Most workings are shallow and the deepest mines hardly reach the 400 ft. level.

Beneficiation of mica or dressing is a specialised art. It is carried out entirely by manual effort. After the mica arrives from the mine, the good mica is sorted out. The mica is now ready for the preparation of (a) block mica and (b) book mica. Block mica is a thick cleavage plate about 1/16″ or 1/8″ in thickness. Book mica consists of thin mica cleavage splittings, made from the same block. Both block mica and book mica are graded according to the (1) colour of the mica, (2) amount of staining, generally due to the presence of iron, (3) extent of defects inherent, such as silver (i.e., air bubbles), fracture, or ripples caused by buckling, inclusions (mainly garnet, magnetite or tourmaline), (4) any defects caused during mining and dressing, e.g., scratches, cuts etc. and (5) the area of the block or book is also an important factor, in the visual classification method.　Grading has also been covered by Indian standard specifications.

It may be mentioned here that the visual classification, even though it has been universally accepted, in the mica business, its inherent defects are evident and attempts are being made to arrive at more rational methods of classification.　A case to elucidate this point may be cited. Even though books may be visually classified under, fair stained, or badly stained, etc., electric testing for determining the dielectric strength by the Q-meter has shown that in many cases there is no difference between these grades.　It may so happen, that the badly stained grade has a greater puncture value than the fair stained.　For the present, however, the visual classification has been adopted in the Indian Standards specifications, and standard samples for comparison are obtainable.　The buyer and seller agree upon the quality, and have a sealed reference sample for comparison when the supply is made.

### 6. The Dugda Coal Washery of Hindustan Steel Jharia Coalfield

The coal washing plant at Dugda was supplied by Messrs. Mc'Nally Pittsburg International Inc., Kansas, U.S.A. with a rated capacity of 600 tonnes/hr.　The plant can work 16 hrs/day and for 300 days in the year.

Coking coal from various colleries is supplied to the plant, in 4 wheeled railway wagons, including both closed and open types.　After passing over a weighbridge, the wagon is unloaded either manually or by tipplers.　Oversizes ($+3'$) are separated by shaker screen, and after passing over picking tables, for removing waste and tramp, the coal is subjected to crushing by rolls.　The undersize material goes direct to the stock yard, and formed into two piles or stacks, each of 7500 tonnes.

From the stacks, coal is drawn through 22 feeders, at the rate of 300 tonnes/hr.　There are two such feeder systems which discharge into Baum jig washers of 300 tonnes/hr. capacity, where separation is affected at a specific gravity value ranging from 1.70-1.80, depending on the quality of the coal.　The tailings from the jig are delivered into a bunker, for disposal either as waste or sent with middlings to the Chandrapura Thermal Power Station.

Floats and middlings from the jig washers are passed through 2 double deck vibrating screens, and three sizes are obtained, $3''-1''$, $1''-\frac{1}{4}''$ and $-\frac{1}{4}''$.　The first size fraction is fed into the Tromp Washers (Heavy media). Magnetite is the material employed, in the washers for maintaining a specific gravity, ranging from 1.45 to 1.55. Similarly the second size fraction 1″-¼″, is fed into another. Tromp heavy media washer, where again magnetite is employed to maintain a specific gravity between 1.45 and 1.55.

The −¼″ size is dewatered, dried and sent to storage bunker directly. The overflow, from the settling tanks, from which the −¼″ fraction is recovered, together with the fines from the drier (−36 mesh), is taken to a thickener and then into vacuum disc filters, with 12 discs of 1½ ft. diameter. Further recovery is also made by centrifuges and the product from both the filters and centrifuges are combined. Floats and middlings from the Tromp washers are taken to separate depulping and washing screens, where any magnetite particles, adhering to the coal, are removed by water sprays. The middlings are dewatered and stored in the middlings bunker.

The clean coal is blended and stored in a bunker 2500 tonnes capacity from where it is loaded into wagons by automatically controlled chutes.

Magnetite acquired from the preparation of the heavy medium is crushed in a primary jaw crusher to −1″. This is fed into ball mills of ¼ tonnes/hr. capacity and a product −200 mesh is obtained with a counter current classifier in closed circuit. This material can now be used in the Tromp heavy media washers.

## 7. Manganese Mines

Manganese ore bodies are being mined in the Shivrajpur area, by underground methods as well as by open cast method.

In Shivrajpur Mines there are three different ore bodies, viz.,

(a) High Grade Reef     —     Average width 16 ft.
(b) Low Grade Reef     —     Average width 20 ft.
(c) Backwall Reef     —     Average width 4 ft.

The High Grade Reef produces a high proportion of good quality ore analysing above 48% Mn ; whereas the Low Grade Reef and the Backwall Reef produce Mn ore analysing 44-45% Mn.

### Underground mining

Shafts are sunk on the hanging wall of the High Grade Reef and crosscuts are driven to the ore body. After intersecting the ore body development drives are made, along the ore body, to its eastern and western limits. Since the ore body is folded, and in order to make a straight

*Flow Sheet at Dugda Coal Washery*

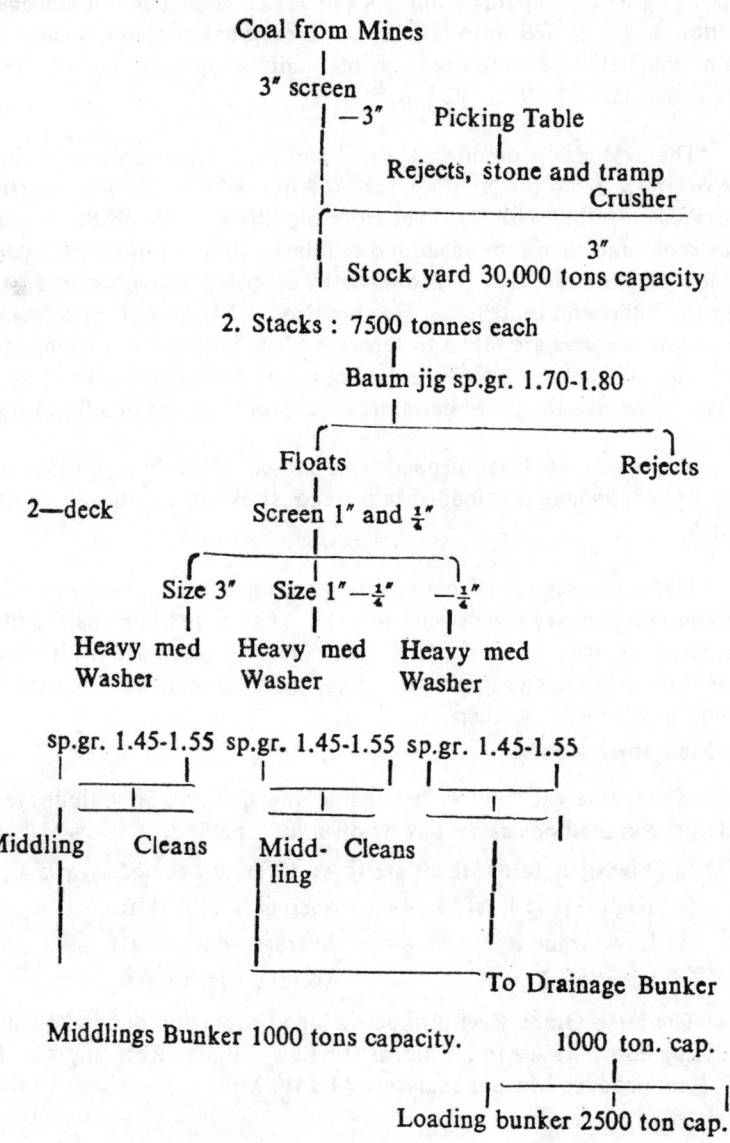

Coal from Mines

3″ screen

    —3″    Picking Table

Rejects, stone and tramp
                Crusher

                      3″
Stock yard 30,000 tons capacity

2. Stacks : 7500 tonnes each

Baum jig sp.gr. 1.70-1.80

Floats                       Rejects

2—deck     Screen 1″ and $\frac{1}{4}$″

Size 3″    Size 1″—$\frac{1}{4}$″    —$\frac{1}{4}$″

Heavy med    Heavy med    Heavy med
Washer        Washer        Washer

sp.gr. 1.45-1.55 sp.gr. 1.45-1.55 sp.gr. 1.45-1.55

Middling   Cleans    Midd-  Cleans
                     ling

                                   To Drainage Bunker

Middlings Bunker 1000 tons capacity.     1000 ton. cap.

                            Loading bunker 2500 ton cap.

*Note* : There are two identical washing circuits starting from the
2 Stacks at (2)—only one of which is indicated. The recovery
of fines is also omitted.

drive for haulage, a tramming drive is driven on the hanging wall side, approximately 40-50 ft. to the north of the ore drive, and roughly parallel to it. At regular intervals of 450-500 feet apart, crosscuts are driven from the tramming drive, to the ore drive. The ore body is blocked out, by driving winzes connecting top and bottom levels, at 500 feet intervals. Drives, in the ore body, are 8 ft. wide and later widened for manway.

In the initial stages, openings on the foot wall (inclines) are made to convey filling materials. These inclines are connected to the stopes by crosscuts, on the foot wall. At the later stages, the filling from the upper levels, is drawn into the lower levels and the working in the upper levels are allowed to cave in.

In the High Grade and Low Grade Reefs, the preset working level is at a depth of 350 feet below the surface. A new shaft, 540 feet deep, is for the development of a deeper level, approximately 150 feet below the present working level. Flat back cut and fill stoping was employed, for mining the High Grade Reef, and the Low Grade Reef. In the Backwall Reef, shrinkage stoping, with Cornish chutes was employed. In Pani mines, flat back cut and fill stoping was employed.

*Ore Beneficiation* : Mechanical screening, mechanical jigging, and hand jigging are employed for treatment of fines and waste dumps. From High Grade Reef fines analysing 33-35% Mn, concentrates analysing 48% Mn are produced by screening and jigging. From Low Grade Reef and Backwall Reef, fines and from waste dumps, feed analysing 26% to 30% Mn is treated by screening and jigging to produce a concentrate analysing 44-45% Mn. The following plant is employed for ore beneficiation : (1) Shaker Grizzly ; (2) Shaker Screen ; (3) Trommel ; (4) Gyrex Screen. (5) Hancock Jig ; (6) Hartz Jig ; and (7) Hand Jigs.

*Output*

Output from the Shivrajpur Group of Mines from 1951 to 1961 is given in Table on page 877.

Underground mining is also carried out in other manganese mines.

*Bharweli Mine* is located on the Maihar plateau, 3 miles north of Balaghat. The ore body about 10,000 ft. long, and varying between 10 ft. to 60 ft. in width, strikes N-S and dips (at an angle varying between 35° to 65°) to the west. The hill is 300 ft. in height and overburden is of considerable thickness.

The ore body is opened up through an adit 1200 ft. long, and this forms the main means of access. Due to the treacherous nature of the hanging wall, the development, of the ore body, is carried out through a foot wall haulage drive, placed at a distance of 50 ft. to 100 ft. to the east, and connected by crosscuts, driven at intervals of 200 ft. Development of the ore body is carried out by winzes, which also form the chutes for waste filling, later after the stoping is done.

*Method* : The main method is the cut and fill with flat backs, the filling being kept as close as possible to the back, with a view to avoiding the breaking up of the weak hanging. Where the dip of the ore body has flattened, a modified breasting is adopted and small pillars are left for support.

*Mansar Mine* : The method of working is similar to that in Bharweli, excepting that the adit is on the foot wall side. The ore body is itself nearly vertical.

*Open cast mining* is practiced at Tirodi, and the overburden is removed in benches each 30 ft high. Blast hole drilling is employed and power shovels are used for loading the ore into dumper trucks. Bulldozers do the levelling of the waste.

Jack hammers are used in drilling the ore, which is blasted, and hand sorted.

Waste disposal is a serious problem, as the ratio of overburden to ore is very high, especially at Shivrajpur, Kandri, Goamgaon, Mansar etc. where diesel locos are used. At Shivrajpur tractors are employed for the purpose, while at Kandri steam locos are employed.

Self acting inclines are used for the transport of ore from higher levels from hills. Electric haulage is used at Gumgaon, Kandri, Mansar for transporting ore from quarries below ground level.

The ore won from these mines is classified into the following grades: (a) above 48% (b) 46-48%Mn (c) 44-46% Mn (d) 38-44% Mn.

## 8. Coal—*Kargali Mine*

Another example of open cast working is the Kargali quarry, in Hazaribagh district, of Bihar where the 100 ft. thick Kargali coal seam is mined. The seam is worked in benches, and blast hole drilling is employed for winning the coal. Advantage is taken of the relatively shallow depths, and the railway wagons are taken down the pits where the coal is loaded directly for shipment.

## 9. Lignite—*Neyveli*

The Neyveli lignite mine, in South Arcot district, of Tamil Nadu is also another example of a large scale open cast mine. Here the lignite occurs in the immediate vicinity (just above an artesian aquifer) associated with the Cuddalore sandstones. In order to keep the peizometric surface, below level of the lignite, large scale pumping is carried out, through bore wells, drilled right round the quarry. Conventional methods are employed for excavating the top portions of the overburden, e g., bull dozers, shovels, draglines etc with increasing depth excavation is accomplished in four benches, three in the Cuddalore sandstones and the lowest in lignite, by means of specially designed equipment, such as bucket wheel excavator, land dredges, over burden conveyors, spreaders etc. The lignite is low in ash, and is suitable for use in power generation, and also as a domestic fuel. Briquetting of the lignite is carried out, at site and fed into the thermal plant. A portion of the briquettes is available for public consumption after low temperature carbonisation.

The major deposits of lignite, West Germany, in Australia, and in Neyveli, India, are worked by open cast method using specialised equipment for excavation and for overburden disposal In Germany, the excavation is by means of B.W.E. (bucket wheel excavator), and the disposal is by either overburden bridge (described earlier), or by specially designed wagons, drawn by electric locos. In Australia, where conditions of mining are very favourable, land dredges are used, for excavating the lignite, and a conveyor system for transport. At Neyveli, India, the overburden and lignite are excavated by B.W E. while the overburden is disposed of by a conveyor system in combination with spreader. The lignite is taken by a belt conveyor to the power station etc.

The B.W.E. is a continuous mining machine, whereas the other machines used in open cast mining, such as the shovels and draglines are intermittant in operation. The essential parts of a B.W.E. (Fig. 13.3)

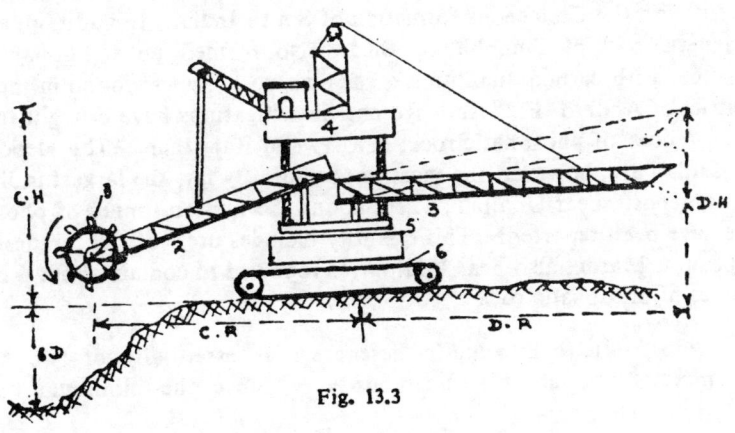

Fig. 13.3

are : (1) The bucket wheel with buckets, which is mounted on the cutting boom ; (2) Each bucket is provided with specially hardened teeth, which are so arranged that forward cutting and side cutting are made possible. Thus the machine can sump in and then make a side cut. The cutting boom also carries a conveyor. The cuttings from the bucket wheel, are transferred to this conveyor. This conveyor transfers the load to the conveyor located in the disposal boom ; and (3) Both these booms can be moved up and down, as well as laterally.

The entire mechanism rests on a turn table (5). The machine is usually crawler mounted (6). The control cabin with the primemover is located above at (4). The excavator can cut above ground level, this being referred to as the cutting height (C.H.). It can cut below ground level this being designated as cutting depth (C.D.). The extent of lateral movement is known as the cutting radius (C.R.) So also the maximum upward movement, of the disposal boom, is known as the dumping height (D.H.). The lateral movement is designated dumping radius (D R ).

A spreader, as the name implies, is used for spreading the spoil, which is to be dumped. It is essentially a belt conveyor system and comprises two conveyor units. It resembles the B.W.E., but without the excavating bucket wheel. The forward and rear conveyors can be moved up and down and also laterally. The machine is either crawler mounted (caterpiller) or rail mounted. The overburden bridge, which is used in lignite mining for disposal of spoil, has been described earlier and hence not included here.

## 10.  Phosphate Rock/Phosphorite

This is an important raw material for fertilizers.  In India upto now the known source of phosphate rock comprised phosphatic nodules associated with the Cretaceous formation of South India.   In addition, apetite-magnetite rock of Singhbhum, Bihar, also formed an additional source together with certain quantities, of similar rock types, found in the Vizag district of Andhra Pradesh.  Recent investigations have brought to light occurrences of phosphatic rock, in U.P. and Rajasthan.  The deposits in Rajasthan are, however, the most important. By far, the largest is Jhamar-kotra deposit, near Udaipur, where about 100 million tonnes of phosphatic ore have been reported.  This quantity includes ores of all grades.  The deposit at Maton, also near Udaipur, is reported to contain about 6 million tonnes of phosphatic rock.

Phosphatic rock found in these areas is essentially of two types : (a) siliceous type, and (b) calcareous type.  Since the minimum content of $P_2O_5$, required for the manufacture of ferti lisers, is in the order of 30%, all grades containing less than 30% $P_2O_5$ require beneficiation. It has been reported that grades of about 20% $P_2O_5$ could be upgraded, to the economic level, for use in the fertilizer industry.

Treatment of the siliceous phosphatic ores comprise crushing, grinding, and flotation, followed by desliming, thickening, and drying. The Maton ore is to be treated in this manner. In the case of calcareous ores, the process is rather simple. It will comprise mainly of crushing, grinding, calcination and levigation. The Jhamarkotra ore is to be treated in this manner.

## Granites with special reference to Black Granite

Granite mining has assumed considerable importance, since granite is a foreign exchange earner. Granites of all shades of colour and of varying texture are being worked. Of all the varieties of granite, one particular rock, which is known to the trade as BLACK GRANITE is the one that is most popular. The name itself is a misnomer, since the rock so called is not a granite but a dolerite. Black granite mining was started in Tamil Nadu and several deposits were developed in the South Arcot, Chengleput, and North Arcot districts. Later, the neighbouring states of Karnataka and Andhra Pradesh also followed suit. Thus, the mining of black granite became wide spread. However, with a view to standardising the quality of the materials obtained from various areas, the material obtained from the Kunnam area, in South Arcot district, was accepted by the trade as the best quality and the quality and price of any other material was estimated by coparison. For example, if the Kunnam material is rated at $1000/- and if the material under consideration is adjudged only half as good, then its price will be fixed at about $ 500/- or so.

The price of black granite has been fixed in relation to the size as well as the quality. The normal size of the blocks is one metre cube. Smaller sizes are also standardised and these are known as monument sizes. The prices for such smaller sizes are based on aggregates of one cubic metre. The presence of certain minerals also adversely affect the price. Quartz, light coloured felspars, epidote produce light coloured spots when polished. Mica, chlorite, pyrite affect the hardness and induce stains. Magnetite improves the colour but affects the hardness and has less resistance to weathering. Hair line cracks and fractures cause losses during dressing of the blocks.

Apart from the usual norms employed in evaluating the economics of a mineral deposit, the following points are of special consideration in respect of black granite deposits. 1. the width and length 2. the grain size 3. type of occurrence, whether it is above ground or below ground level. Since, petrological examination of the samples is time consuming and expensive, the detection of the presence of the deleterious elements in the black granite poses a serious problem. This difficulty can be readily got over by the field test which has been devised for the purpose. A fairly large chunk of the rock is broken from the fresh outcrop with a 5 lb. or 10 lb. sledge hammer. When clean water is poured and the surface is wetted, the various minerals, present in the rock, can be readily identified, giving an indication as to how the rock will appear when polished. In the case of hairline cracks, a very careful examination is called for, and it is very often the presence of these hairline cracks that spoil the value of the deposit. It is possible to identify the quality of the black granite by means of micro-resistivity apparatus.

Mining of black granite is done by the open cast method. If the deposit is in the form of a hill, quarry development is easy. The blocks are lowered to the ground level by inclined plane or by trolleys, and fork lifts are employed to place the blocks on the trucks for transport. If the deposit is below ground level the mobile cranes are required to lift the blocks from the pit and load them on the trucks. The exacavation of the rock is essentially by manual methods. The blocks are isolated by making deep cuts with the chisel and hammer. These cuts are made about one metre apart and parallel to each other. In most deposits, it is found that there is a parting in the depthwise direction. Wedging is done at this parting and very often the block is isolated and can be removed and lifted by the crane. Occasionally, a very small charge of gun powder is used for dislocating the block. Since, the dykes are emplaced vertically, the benching is done across the strike.

By far, the larger portion of the black granite mined is exported in the "raw" state. There are few cutting and polishing units in Tamil Nadu and also in Karnataka. Daimond impregnated discs are used for cutting and abrasives of various grades are employed to give the rock a very high polished surface.

# PART XIV

# COLLECTION OF FIELD DATA AND WRITING OF REPORTS

# COLLECTION OF FIELD DATA AND WRITING OF REPORTS

## Introduction

Not only is the field geologist required to carry out the examination of a mineral deposit, right from the prospecting stage upto the exploration stage, and assess its economic potential, but he is also required to submit a complete report embodying his observations and findings at every stage. If this is not done, then the work, however exhaustive and thorough may be, lost sight of, and serves little or no useful purpose. Often such a periodic progress report is required to be submitted, to the management or to an organisation. Frequently, the enquiry may be of a preliminary nature, and only a day or two may have to be spent by the geologist, in the field, but sometimes a complete and thorough investigation, of a large mineralised belt, may also have to be undertaken, in which case one should go prepared to stay in camp, for long periods. Systematic field work is essential in all cases.

## Observations and Materials

All observations made, during the course of the examination and mapping, must be recorded systematically, in a field note book and not trusted to memory or jotted down as notes on scraps of paper. The daily field notes should contain the co-ordinates of the area under examination, while details of the rock types and structure, i.e., faults, foliation, strike, dips, lineation etc. should be indicated on the map (scale permitting) and briefly described, in the field book. Measurement of sections, lithological or stratigraphical, from outcrops, exposures, mine workings or bore holes, is necessary for correlation and stratigraphic reconstruction. A field sketch, to scale, is a useful record. It is a good practice to collect enough specimens for petrological and petrographic examination. The location of such specimens should be indicated, on the map, and the serial numbers, with description and particulars, are to be entered, in the field note book, at a separate place, preferably at the back. As a note of caution, it may be pointed out that specimens should be numbered, in the field itself, either by marking with quick drying plastic paints, or by writing on a piece of zinc plaster tape, which is stuck to the specimen as a temporary measure. Permanent numbering with paint should be done immediately on reaching camp.

## Samples

Samples should be carefully collected, numbered, and all particulars immediately entered in the sample note book. Samples will have to be collected only from fresh exposures, trenches, pits etc. and also from underground workings and bore hole cores, when these are available. In outcrop sampling, the surface should be cleaned by brushing or by chipping. Care should be exercised to see that no powdery material is lost from the sample bag, especially in the case of cloth bags. Analysis of sulphides, coal and water samples especially should be undertaken immediately to prevent changes due to oxidation/loss of gases/volatiles etc. Immediate analysis also minimises the chances of salting, especially in the case of gold ore samples.

The techniques of sampling has been discussed elsewhere in Part XI. Care and time spent on sampling is well rewarded and sampling can make or mar a project.

### *Economic bias in Sampling/Analysis*

In connection with chemical analysis, word of caution may be added. The investigator should thoroughly understand the problem on hand. For example, if an iron ore sample is to be analysed, the deleterious elements, in iron ore, should have to be determined such as sulphur, phosphorus, besides the other undesirable constituents like $SiO_2$ and $Al_2O_3$. In addition, one should also be fully aware of the composition of the other raw materials which will be used in the charge, e.g., limestone, coke and silica. It is only then that the picture is complete.

Where a limestone is to be analysed, the purpose for which the limestone is required should be known, e.g., a limestone good enough for cement making may not be suitable for steel making. The limits of the various harmful constituents in limestones should be known. The limits of insolubles for both S.M.S. (Steel Melting Shop) grade and B.F. (blast furnace) grade as well as of MgO should be known. So also in the case of cement grade material, the upper limit for MgO should be known. This is true for several minerals required in industries, i.e., bauxite, clays, chromite, manganese ore, etc.

## Mapping

'Most economic mineral investigations necessarily involve the use of surveying methods. A thorough knowledge of the techniques, described in the sections, on surveying methods, and photogrammetry will go a long way in selecting the most suitable method for the particular problem. This will also include a working knowledge of photogeology, bore hole surveying and logging, apart from underground and stope surveys. One cannot minimise the importance of surveying in the assessment of reserves.

Normal field mapping, for preliminary work is usually carried out on $1'=1$ mile scale or $(1 : 63,360)$ or $1 : 50,000$. Where the deposit shows promise, mapping may be done on air photos on $2''=1$ mile scale (approximately). When air photos are used, full advantage should be taken of photogeological studies given in the relevant chapter. It is best to get vertical photos (stereo pairs) printed on matt paper. When choosing a scale for geological surveys and mapping, one should take into account : (a) the nature of the work, viz., preliminary or detail, (b) the time available, (c) funds available, (d) instruments available and (e) .purpose of the investigation.

It is also to be remembered that the ultimate object of geological mapping is to depict the "solid geology", i.e. the lithology and structure as they will appear after stripping the soil, alluvium or pumping out the water, where the formation occur under water. It is also good to remember that the geological map is not a "plan"; it is in reality a geological model. Consequently, from the geological map it is thus possible to prepare cross sections which are not mere topographical profiles.

With a view to achieve the above objective, every possible effort should be made. Outcrops may be few. Exposures in pits cuttings, wells, nálas should be of help. These should be located accurately on the map so that a re-examination of the area is facilitated. Further since it is necessary to prepare an "outcrop" map, this procedure will be of great help. Even though, in the field, all data in regard to lithology structure, outcrops etc. are recorded, on the same map, it is essential to prepare three separate maps depicting the outcrops, lithology (solid geology) and structure respectively. It is for this precise purpose that various techniques are to be employed, where essential pitting may be undertaken. If the problem is that of tracing a mineralised zone then the following methods may be utilised : (a) tracing float, (b) pitting and trenching, (c) geo-chemical, (d) geophysical, and (e) drilling. Sometimes photogeology can come to the rescue.

## Laboratory Work

Where metallic ores are sampled, it is often necessary to examine both thin sections as well as polished section from the samples. While thin section analysis will throw light on the gangue materials and their mode of association examination of polished section will reveal : (1) the type of ore mineral, (2) the mutual relationship between minerals, whether there is an exsolution relationship or eutectic or solid solution etc. and (3) the particle size of each of the mineral components. All these factors are of major importance in assessing the nature and extent of the deposit, and also for ore dressing studies. Hence these cannot be neglected. Suitable size chips must be collected for these purposes.

When the examination of a clay or bauxite deposit has been made, it is useful to carry out thorough laboratory testing and this may require the use of special techniques such as D.T.A. (Differential Thermal Analysis), X-ray diffraction, staining, besides involving the determination of properties such as liquid limit, plasticity limit, green strength, grain size analysis, including refractory test etc. It is only then that a full evaluation of the material can be made. The method of estimation of reserves, of such deposits has been dealt with earlier, hence not repeated here.

After the petrological and mineralogical studies are completed it may be necessary to carry out additional tests in the case of ores and industrial minerals. In the case of sulphide ores, of base metals, bench scale tests on crushing, grinding, jigging, flotation etc. may be done to find out various parameters. It may also be required to carry out pilot plant tests, so that the costing can be done.

In the case of iron ore, the physical properties, and mechanical properties, have to be determined. Beneficiation may comprise washing, as done at Naomundi or Joda, but may include grinding and magnetic separation, as proposed for Kudramukh deposit. Sometimes screening is sufficient, as done in the Bara Jamba Sector of Bihar/Orissa. Where excessive fines are produced, during mining, or where good blue dust is available, in large amounts, then even a process for pelletisation may have to be considered.

Testing of coal and lignite will comprise both the proximate analysis, as well as ultimate analysis, besides determining the calorific value. Often coal petrography, especially the polished thin section technique can help in this assessment.

## Exploration Methods

Before undertaking the more expensive exploratory programmes such as milling and mining a scheme for geochemical prospecting, may be drawn up and if needed this may be followed up by a suitable geophysical examination depending on the nature of the ore body and the controls as established by the geological and geochemical examination... For most mineral exploration work diamond drilling is inevitable. Careful planning of the programme is essential for economy.

Where drilling is indicated, the site should be selected with care. Points to be taken into account in selecting bore hole sites are : (1) Geology and structure, (2) Accessibility—so that the drill can reach safely without incurring excessive costs on road making and transport, (3) Availability of water for drilling, especially in the case of diamond drilling, and (4) Minimum drilling with maximum information should be the rule.

After the requisite quantum of drilling has been carried out further checking may be called for by exploratory mining. Based on the results of drilling such additional exploration may be undertaken with a view to : (1) verify the projections made in respect of the structure and behaviour of the ore body ; (2) compute ascertain the index of reliability of drill hole data, in respect of mineralogical/chemical composition/analysis of the ore ; (3) find out the physical and mechanical characteristics of the ore as well as the walls, i.e., the foot and hanging ; (4) drawing large samples. e g., about 5 tonnes or even more for carrying out pilot plant testing, in respect of beneficiation and smelting etc. so that the ore dressing and smelting plants can be designed. In the case of iron ore, samples may be required for testing pelletisation characteristics or sintering properties ; (5) draw up norms and parameters of cost in respect of ventilation, pumping, blasting, mine supports, etc. and (6) enable making estimates of the thickness of the overburden, in relation to the thickness of the ore body, for open cast mining. For open cast mines it is better to draw isopachs also.

Whether, it be exploratory drilling or mining, cost consciousness is of utmost importance. Exploration is a kind of capital investment. Hence the cost of exploration has to be added, totally, partially, either as lump sum payment or on deferred payment basis, to capital costs. Excessive interest charges on capital, during construction period of mining project, can act as a mill stone round its neck. A viable project can become uneconomic. On the other hand, it may also be pointed out, stinting on expenditure on exploration may prove disasterous, in the long run. It may lead to delays in commissioning the project or even abandoning the project after incurring considerable expenditure. Hence, even though it may appear contradictory, to what has been stated earlier, money spent judiciously on exploration is money well spent. Even an expenditure amounting to 2% of the total investment will seem justified, viewed from this angle.

## BROAD FRAME-WORK AND GUIDELINES FOR REPORT

The following frame-work will serve as a general guide for mustering data collected during the field examination, by the study of earlier reports and papers relating to the deposit, including reports on the geology of the adjoining areas where available, and of the mines in the neighbourhood as well as of the laboratory data.

1. Title of the report.
2. Introduction—Giving brief history of how and why the investigation was undertaken.
3. Area—its location — topography — geomorphology—drainage. Type of land and soil—fauna and flora. A study of air photos is useful.
4. Location—accessibility—roads—railway—other communication facilities.
5. Previous works—a brief review of the work carried out by earlier investigators may be given, indicating the extent and results achieved in each case. A descriptive account of old workings and a rough estimate of the extent may be included.

It is often contended that if the investigator familiarises himself with the literature pertaining to the area, before taking up the examination, he is very likely to be prejudiced. The geologist should be a person of high integrity and should not be drawn away by "mere opinions". He should base his views on facts. If this principle is adhered to then there is no danger whatever. Ignorance of available literature may entail unnecessary labour and repetition of what has been said and done.

When old workings and working mines are located within reasonable distance, from the area, no opportunity should be lost in examining the same and making a detailed study. This is more true in the case of the unstratified deposits.

6. Acknowledgements—especially when facilities and services have been rendered by an individual or an organisation. Any advice given or help received from collaboration, e.g., other geologists, drillers, chemists, geophysicists should be acknowledged and the extent of such collaboration should also be stated. It is often customary in certain organisations to include the name of the

chief also in the acknowledgements, but such formal acknowledgements for persons other than actual contributors is out of place, and should be discouraged by all concerned.

7.  Regional geology—will not only include the area examined but also the surrounding areas as well, so as to give the reader a broad idea of the geology and bring out regional structures etc. and the stratigraphy.

8.  Geology of the area—rock types encountered, with the locations on the large scale map (say 1 : 5000) or even (1 : 200). This should also include the petrography of the rock types as well as brief notes on any minera'. identifications carried out, especially to confirm zoning or wall rock alteration. Determinations using techniques, i.e., R.I. determination of felspar by staining or using the U stage, may be called for.

Structure : details regarding dip, strike, foliation lineations (a and b) etc. and descriptions, from air photos, and the presence of sedimentary structures, i.e., current bedding, ripple marks, convolutes, graded bedding, etc. should be mentioned. Folds, faults, thrusts, cross folding and drag folding should be described. Structural studies may also include both mega and microanalysis and petrofabrics.

Clay mineral studies may be carried out by X-ray, D.T.A. staining studies, infrared studies, in order to investigate wall rock alteration and the temperature of formation of the deposit. Techniques of geothermometry e.g. decrepitation, liquid inclusion etc., may be also carried out.

9.  Mineralisation—The study of zoning, wall rock alteration, controls and loci and evidences for mineralisation, minerals found are important. Gossans, type and extent should be studied. Pitting and trenching are necessary for sampling but width of samples should be noted. Cores and sludge from drill holes are to be examined and sampled—Minergraphic examination for the study of paragenesis, mutual relationship between ore minerals—for elucidating ore genesis and facilitating ore dressing operations, especially crushing and grinding/flotation is essential. Chemical analysis for the metals contents—e.g., copper, lead, zinc etc. is invariably necessary. A few samples should be analysed, spectroscopically to detect the presence of accessories, e.g., Ni, Co, Mo, etc.

10. Geochemical sampling—direction of the grid lines (generally one set is at right angles to the other), lengths—sample intervals —sampling procedure should be indicated—background values for the area and the threshold values for the particular element should be mentioned. Tabular statement showing analysis of samples and also a contour map showing the geochemical anomalies may be prepared. The anomalies should be interpreted in relation to the geology and structure.

11. A note on the geophysical survey carried out and the results achieved may contain descriptions of the methods used i.e., S.P. resistivity, gravity, etc. The anomalies obtained by the survey should be interpreted in the light of the geology of the area— whether these are related to structure or mineralisation or both etc.

12. If drilling has been carried, the following data are necessary, i.e., R.L. of bore hole, angle, and orientation. Thickness of each of the formations or bands encountered giving the depths to the top and bottom—length of core recovered—amount of sludge collected. Sections correlating lithological logs may also be drawn bringing out additional subsurface data.

13. Notes on the petrogenesis and ore genesis—as made out from field and laboratory studies—i.e., mapping, thin section studies, polished section studies and also on the origin of the rocks, igneous, metamorphic, if so, how ? Chemical and mineralogical composition of rocks—use of integrating stage and—point counter—norms—Niggli values—graphical representation of these—any evidences of metasomatism or replacement. If hydrothermal origin is postulated—evidences in favour, e.g., wall rock alteration, zoning, mineral association. Geothermometry of silicates as well as of ore minerals will be of considerable use. Use of statistical methods in petrography and the use of matrices may be made where possible for petrogenetic deductions.

14. A suggestion may also be made as to the best method of mining to be adopted in regard to the deposit examined, since the structure of the deposit, its behaviour and extension along the dip and strike direction, are all known. This information combined with the thickness variation, nature of the walls etc. should provide enough data for making the suggestion with little or no hazard.

There is no excuse, at least at this stage, for not giving

concrete suggestions for exploratory mining and exploratory drilling. The drilling programme or scheme should have a bearing on the mining and not merely prove the further extension of the deposit. It should be objective. The aim should be to develop, or aid in planning the development of the mine and at the same time prove adequate reserves.

15. A brief note on the polished section studies and any laboratory experiments carried out in regard to the beneficiation of the ores may be added. Such recommendations should be clear as to why the methods should be followed up at the pilot plant stage. A flow sheet for beneficiation may also be given, no detailed engineering being involved.

16. Description of the samples/sampling technique may be briefly stated. Account of the reserves. This may be classified as proved, probable and possible etc. depending on the status of the exploratory work. The grade and tenor of ore and its variations— if any, in depth or along the strike, whether such variations are erratic, or consistent should be indicated as well as the prospect of ore occurring at depth.

Reserves (various type) based on the grade of ore and the estimated depth should also be given based on careful chemical analysis (minerographic examination etc.).

17. A paragraph on ground water, as well as surface water resources of the area, especially when a colony is to be set up, and an ore dressing plant is contemplated, is of great importance. A brief note on rock mechanics, which is likely to play an important role in the alignment and shape and size of underground workings, should be included.

The work will not be complete unless it is followed by summary and conclusions. This portion should briefly describe the findings, and the recommendations should be made as to how the findings can be tested. If the work is considered incomplete, for some valid reason, then concrete suggestions for removing the doubts may be made.

18. A list of references is to be given. This should include :

(1) the name of the author with initials and date of publication;

(2) the title of the paper or book ;

(3) name of journal, with particulars of volume number, pages, etc.

## Some suggestions of a general nature

It should be remembered that no amount of verbose descriptions can ever substitute, sketches, diagrams, maps, and photographs, photomicrographs, etc.

Sketches, mainly field sketches, should show the scale of the structure as well as the direction of magnetic bearing or else the sketch will loose much of its value.

Camera lucid sketches are sometimes useful in bringing out details of micro textures in rocks, and are more explicit than photo micrographs.

Diagrams should include diagramatic representation of bore-hole logs. Section of exposures and profile sections correlating formations or bands encountered in bore holes help to clarify the ideas.

Map should include a location map on a small scale—large scale maps should be of suitable scales ranging from 1 : 100 to 1 : 5000 for covering the mineralised areas. Underground maps of stope, tunnels, etc. are also necessary.

It may be remembered that the thumb rule for marking structural element, on a map, (dip, strike, foliation, lineation, joint etc.) is that (1) there should be no crowding (2) priority is to be given to elements of greater significance in respect of geology (3) irrespective of the scale of the map the structural elements, where present, may be noted, keeping in a minimum spacing of 1 cm.

Photographs serve not only to beautify the report but go a long way to bring the actual representation. A good photo is necessary, a bad one may have the opposite effect. Photomicrographs are essential in the petrographic studies for the report, and are of undoubted value in explaining the features noticed in ore minerals etc.

## Economic Aspects

To conclude, it may be pointed out, that while the aim of "pure geology" is scientific research, with or without industrial implications, the aim of applied geology, and more so mining geology, which forms part of this multidisciplinary science, is not merely to satisfy scientific curiosity, but to aid in setting up viable a mineral based project or industry. Thus, a mining geologist is not only an economic geologist, with a knowledge of mineral deposits, structure, ore mineralogy and thin section petrography etc. but is also a person with a strong economic bias, capable of comprehensive techno-economic, assessment in relation to mineral based projects.

No industry can survive for long if it is not profitable. So it is with mining. Hence, the importance of knowing the fundamentals of economics, and managerial economics, cannot be over emphasised. It is not that the mining geologist is a person who has a three-in-one personality, embodying the economic geologist, the economist and the management expert. Yet the mining geologist should realise the inter-disciplinary nature of the expertise he is expected to possess. While he can, possibly, with a little arduous labour, handle matters connected with small projects, in the case of major ones he will have to resort to advice, which embraces various fields as shown in the earlier chapters.

However, even though the investigation for a project may be complete, from the technical/technological view point, it cannot constitute a feasibility study or much less a detailed project report, which is the basis on which any investment decision can be taken. If the economic aspect has not been taken into account, in the report, a technical report can be best be called a thesis or theoretical exercise. If the report has incorporated both the technical as well as the economic aspects, it can be a feasibility report. In a detailed project report, it is essential that the technical, economic and managerial aspects are brought out. The technical part should give precise specifications, drawings, plants, etc. and also, fixed costs, variable costs, prices, revenues, profitability etc. Some minerals can be marketed as mined, i.e., no further processing may be required. In this category one may include certain types of iron ore, manganese ore, coal, chromite, barite, asbestos, limestone, steatite, rock phosphate, etc. Certain minerals, like iron ore, stone and gravels, sand, asbestos, require screening, before they can be sent to the market. A large majority of minerals, however, require up-grading before they can find a market, e.g., base metal ores, gold ore, etc.

Hence, in general, the two aspects, mining and beneficiation, should be considered separately, and the techno-economic analysis for each process should be prepared on this basis. These may be combined, at a later stage, for working out the overall picture of the project.

No industrial project can survive if the infrastructure is inadequate, to meet the requirements. Without transport and communication neither raw materials can be moved inwards, nor products moved outwards. Water supply is also essential. Having settled the question of infrastructure the next is the technical aspect which will comprise, again, of two parts (a) geology and reserves, for the drawing up of which a broad outline has already been given, and (b) development and mining programme.

In keeping with the technical details, costing should also be indicated showing expenditure proposed on (1) geology and exploration and

**(2) mine development and exploitation.**

Mine development and exploitation will comprise (a) sinking of shafts either vertical or inclined (it may be that entry is by adit), (b) driving cross-cuts and levels, (c) raising and winzing and (d) stoping with a view to prove certain amount of reserves, say for lasting five years at a certain level of production. A detailed plan, of working, must be drawn up showing the layout of the proposed workings. The lengths/depths of each item of excavation, as well as the cross-sectional area (i.e. length, breadth and height) may be shown. Based on this information, an estimate of the cost can be made. An additional underground drilling should also be covered under developmental expenditure. A certain amount of expenditure may be incurred on ventilation, haulage/winding and transport. This as well as expenditure on ventilation and pumping/drainage may be incorporated. The estimates of costs are best presented in tabular form, and a summary should invariably be prepared for the same. Costing should also be carried out for the mining operations and the following tabular statement showing the break up may serve as a general guide :

Anticipated production of copper ore per annum at a cut off of 1.5% cu., the mill grade being 2.0% cu., is 200,000 tonnes.

*A. Mining Costs*

1.  Direct Costs :

   (a) Labour

   (b) Supervision including a proportionate quantum of head-office and administrative expenditure

   (c) Costs of explosives, and blasting

   (d) Cost of supports, timbering, roof bolting, pack walls etc.

   (e) Drainage and pumping, including diversion dams, sumps, making gutters, etc.

   (f) Ventilation costs, making of stoppings, air crossings, doors, ducts, bratticing etc.

   (g) Haulage and winding, including any cost of transport from the mine to the dressing plant

   (h) Maintenance

   (i) Stores and supplies not covered earlier

2.  Indirect Costs :

   (a) Royalty @ Rs. x/tonnes of ore

   (b) Head office expenditure

   (c) Other administrative expenditure not covered under (1) b

### B. Total Costs

    (a) Mining and transportation

    (b) Beneficiation and

    (c) Smelting and refining

### C. Revenue

Taking the mill efficiency to be 90%, and also assuming the smelter efficiency to be 90%, then the quantity of copper metal recovered will be

$$200,000 \times \frac{2}{100} \times \frac{90}{100} = 3600 \text{ tonnes per annum.}$$

*Note* : A certain amount of credit may be taken for ore won during development.

Thus taking the price of copper to be Rs. 12,000/- per tonne, the revenue per annum is Rs. 4,32,00000/- (This will vary with the fluctuations in metal prices).

### D. Deductions and Contributions

    (a) Depreciation and amortisation

    (b) Interest on loan capital @ 8%

    (c) Interest on working capital @ 8%

Net realisation/revenue or profit $= C - B$

### Profit/Profiitibility

The net profit can be expressed as a percentage of the equity capital (E), i.e., $\dfrac{C-B}{E} \times 100$.

It is also useful to work out the profitability of the project using the various techniques such as :

    (a) Payback method

    (b) Accounting method

    (c) Discounting cash flow method

    (d) Present value index method

Of the methods listed, the D.C.F. method is often used in evaluating the profitability. In the worked out example, given in the chapter on managerial economics, an item "cash flows" has been shown. This is really "net cash flow"—which is the difference between the "cash inflow" and "cash outflow". It may be a good practice to discount each separately.

Since it is normal that, in the case of a mineral project, the project does not earn any revenues, it is placed on capital account, say for 4 or 5 years. During this period there is only a cash outflow. After this period the project is expected to earn, and so, it is now on revenue account, showing a larger cash inflow.

Assuming a certain number of years, to be life of the project, the cash outflows are discounted and the cumulative net present worth or value computed, for the years during which the project is on capital account. Similarly, the cumulative net present worth of the cash inflows, for the remaining period of the life, when the project is remunerative or brought on to the revenue account, is also worked out. The ratio of the cumulative N.P.V. of cash inflows to the cumulative N.P.V. cash outflows as obtained is known as the profitability index. The break even analysis may also be applied.

*Capital Cost*

Since, in any sizeable industrial project, capital has to be raised either as share capital, known as equity, and loan capital, a fair approximation of the total requirements is absolutely essential. It is also to be remembered that dividend has to be declared, at a reasonable rate per cent, on equity shares, and the interest, paid on the loan, should also be at current rates. Besides these, on certain items of capital, like machinery, equipment, buildings etc. depreciation/amortisation may be allowed, in such a manner, as to redeem the investment before the life of the project.

While calculating the "internal resources" of a company or undertaking, it is the normal practice to take the depreciation/amortisation sinking fund into account. This, however, is not done in working out the profitability. It is also the practice to calculate return on capital (both equity and loan) before taxes are paid.

The following is a check list of items which are included under capital cost :

1.   Cost of land or rent.

2.   Investment on infrastructure facilities : (*a*) Roads, (*b*) water supply.

3.   Cost of overburden removal (in the case of open cast mines). This may be kept in suspense account and adjustments made after the mine comes into production, i.e., on revenue account.

4.   Other development activities, i.e., shafts, adits, cross cuts,

drilling undertaken, etc. It may be that a part or a large portion has to be capitalised depending on the purpose as to why these were undertaken.

5. Sampling and analysis of ores.

6. Expenditure incidental to development i.e. (*a*) ventilation, (*b*) drainage and pumping, (*c*) haulage, (*d*) explosives and (*e*) mine supports.

7. Power supply—sub-station—transmission.

8. Workshop.

9. Mining equipment for (*a*) ventilation, (*b*) drainage and pumping, (*c*) loading haulage and winding, (*d*) explosives drilling, (*e*) mine supports. This will include compressors, motors, other equipment like raise climbers, etc. (*f*) surveying equipment, anemometers, (*g*) underground crusher, telephone system.

10. Workshop equipment.

11. Transport vehicles—jeep/station wagon.

12. Buildings for office, store, workshop, and other structures.

13. Residential quarters of staff and workers.

14. Social requirements—club, hospital, creche, etc.

A similar statement can be prepared for the ore dressing section also and the economics worked out.

# ADDENDUMS

## Page 8

It may be pointed out that in running traverses either in rocky terrain or in areas which are slushy or covered by shrubbery, the measurement of distances by chaining is difficult, since the chain gets entangled or becomes dirty to handle. In practice it was found that a strong nylon cord 30 m, long with tag marks at every 10 m. is a very good substitute. In case off setting is needed this can be done with the metallic tape.

## Page 34

In general the value of the "instrument constant" is so small when compared to the distances to be measured, that it can be neglected. In the case of most instruments, the "stadia coefficient" is 100. Hence if the staff intercept is H, (generating number), then the distance in question is $D = H \times 100 \times Cos^2\theta$ where $\theta$ is the vertical angle, the angle obtained by the intersection of the middle cross wire on the staff. When the value of the horizontal distance so obtained is multiplied by the value of tan $\theta$, the value of the vertical height is obtained, i.e. $H \times Cos^2\theta \times \tan \theta$ = vertical height.

## Page 35

As mentioned earlier, there is no special advantage gained in using the special instruments designed for use with the plane table, such as the Beaman's Stadia Arc, or the Watt's microptic Alidade. In practice it has been found that for the usual scales adopted in geological work, the normal telescopic alidade provided with the vertical cross wire and the stadia wires meets the requirements. However, it may be mentioned that every care should be taken to level the table and note the readings only when the staff coincides with the vertical cross wire. Experience has shown that if these precautions are taken there was no difficulty in closing long traverses in rugged terrain. The computations of distances and heights are identical to that used in tacheometric traversing given earlier.

The following checks are useful in assessing the accuracy of the survey:

(a) Angular check; If there are $n$ angles in a closed traverse, then the sum of the angles should be equal to $2 \times n \times 180° - 360°$.

(b) If $f_1$ is the error in the closure and $P$ the total length of the traverse, then $f_1/P$ should be less than or equal to 1/300.

(c) The permissible closing error (angular) is ± 2 secondsx $\sqrt{n}$, where $n$ is the number of angles. The permissible closing error in distances is $\dfrac{0.04 \times P}{n}$, where $n$ is the number of sides in the traverse.

(d) If in closing a traverse there is a error in height. This error should not exceed the limit given in (c) above. If it is within the permissible limit then it is distributed in proportion to the length of the respective lines in the traverse.

Page 40

The plane table in combination with the sight vane, which can be made either of metal or of wood, requires the use of a distance measuring device such as a tape or chain. It is almost impossible to use this combination either in a rugged terrain or even in a terrain which is dissected by small ravines or deep valleys. Under these circumstances the measurement of the distances with the tape or chain is difficult and the accuracy is badly impaired. For example if the two stations are located on either side of a deep ravine which is about 100 m. wide and about 20 m. deep chaining is a problem. Under such conditions the telescopic alidade equipped with stadia wires comes in handy. In practice, it has been found that such an instrument (Indian make) is accurate even where the line of sight is 150 m. with a deep valley in between the two stations. However, in such a case, as always, care should be taken in levelling the table, maintaining the verticality of the staff intersected, and also for accurate intersection of the stations. By taking these precautions, it has been found that a plane table traverse, in a wooded area, with deep ravines and covering an aggregate distance of over 4 kms., was closed with no significant error.

In using the telescopic alidade with stadia a scientific calculator is essential for computing the distances and heights. Since the alidades available are equipped with the vertical circle also, the value of the vertical angle can also be obtained. The standard tacheometric formula for the distance (horizontal) is $H \cos^2 \theta$ (omitting the instrument constant). The formula for the vertical height is $H \cos^2 \theta \times \tan \theta$, where $H$ is the stadia intercept and $\theta$ is the value of the vertical angle.

Page 44

The instrument is set up at a convenient point, from where the line of traverse is free from all obstructions. The point so chosen need not be on the line of traverse. The temporary adjustment is carried out, as described in the earlier paragraphs. The staff,.which is held vertically at the datum point selected, is sighted, and the reading against the horizontal cross wire is recorded as "back sight"

## Page 46

When it becomes necessary to move the instrument to another location, then the last reading is recorded as a "fore sight". On shifting the instrument to the new location, the height of the plane of collimation will change. Hence this should be noted and the location also noted as "change point". The staff which is still kept at the previous location is now sighted from the new location of the instrument. This reading is recorded as a "back sight". This reading gives the height of the new plane of collimation. The same procedure is adopted in regard to the additional inter-sights until another fore sight becomes necessary.

## Page 56

When it is proposed to plot the theodolite traverse by the coordinate method, the usual practice is to measure the bearing of any one of the lines in the field. The included angles are also measured. In case the bearing was not measured then the bearing of any one of the lines is assumed. Then with reference to the actual or assumed bearing as the case may be, the bearings of all the other lines in the traverse is calculated by applying the following rule: "Add the value of the bearing, either measured or assumed, to the forward included angle of the traverse. If the sum exceed 180° deduct 180°. If the sum is less than 180° add 180° as shown in the example".

## Page 270

The pressure of the compressed air used is of the order of 7.5 to 13.5 kg/cm$^2$ and the supply of compressed air for holes 15 cms. in diameter varies from 9 to 10 m$^3$/min. The speed of rotation can be varied from 10 to 30 rpm. The slower speed is used for hard formations. The rate of drilling in hard rocks can be of the order of 10 m./hour.

## (Page 331)

## Drilling:

In the case of open cast mining, drilling for the purposes of blasting, in combination with shovel and dumper, is carried out by means of down-the-hole type of machines. The diameter of the bits used varies from about 100 mm. to about 200 mm. The common type of bit employed is called the button bit. The face of this type of bit is studded with T.C. (tungsten carbide) buttons.

These machines are in use, in limestone, iron ore, and manganese mines. In the case of coal mines, the rotary rigs using tricone bits are popular. The D.T.H. drills are very efficient in hard and dry formations. The cuttings consist of chips which are 0.25 cm. to 0.5 cm in size, and these are thrown out of the drill hole by the high pressure compressed air. In the case of the rotary drill the sludge is flushed out by the circulating water or mud suspension. The D.T.H. machines can be used for vertical drilling but the efficiency drops when the inclination increases. Another disadvantage of the D.T.H. machine is that, when drilling in jointed formations, the vibration can cause the slipping of loose blocks, and cause jamming of the rods.

Jackhammer drills are used only for secondary blasting in machanised mines In manually operated open cast mines jackhammer drills are popular for blast hole drilling.

### Terms and definitions:

There are certain terms which are frequently used in regard to explosive charges and relating to the properties of explosives and these have been given below: (Fig. 1)

1. *Booster charge*— consists of a highly concentrated explosive charge contained in cartridges, placed in the column charge for improving the velocity of detonation

2. *Bottom charge*— is a high concentration of explosive, which is placed at the bottom of the hole for obtaining the maximum breakage.

3. *Burden*— is the shortest distance from the drill hole to the free face, i.e. the surface where the breakage is anticipated.

4. *Column charge*— is a charge of lower concentration than the bottom charge, and it is located in the upper part of the drill hole.

5. *Primer*— is a cap sensitive explosive cartridge, with a high

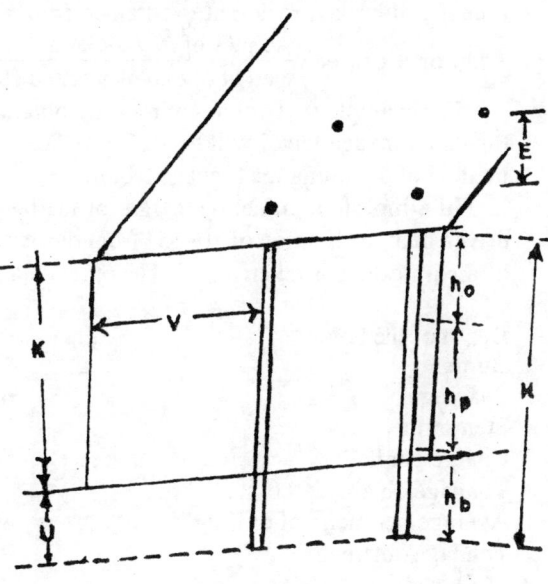

Fig. 1.

velocity of detonation which is used for initiating the explosion of non-cap sensitive explosives.

    6. *Primed cartridge*—is that cartridge in which the detonator is placed.

    7. *Rock constant*—is an estimate of the breakability of the rock and it is expressed in kg/m$^3$. The larger value are near 1.0 *m* whereas the lower ones are about 0.2, and the standard is taken to be 0.4.

    8. *Secondary blasting*—is blasting of the over-sized boulders that are got from a blast.

    9. *Spacing*—is the distance between the drill holes, arranged in a regular pattern. The pattern can be parallel or radial etc. depending on the nature and the structure.

    10. *Stemming*— is the material which is placed on the top of the explosive charge in the bore hole inorder to fill up the hole. It may consist of water or sand or clay.

    11. *Sub-drilling or sub-grade*— i that part of the drill hole which is drilled below the level of the toe of the bench. . nis is used to avoid secondary blasting of any part of the toe being left after blasting.

### Efficiency of an explosive:

    One of the standard parameter used in evalutating the efficiency of the explosive employed in blasting, is the "powder factor". Powder factor is the tonnage of broken rock that is obtained from a specific explosive charge say 1 kg. or it can be $\dfrac{\text{tonnes of rock broken}}{\text{weight of explosive used (Kg)}}$

    For example if 10 tonnes of rock are obtained by using 2 Kg. of explosive then the tonnage factor will be 10/2 = 5. The example given below shows the method of obtaining the tonnage factor:

    In a limestone mine, the height of the bench is 8.0 m. The blast hole is driven at an inclination of 10–15° from the vertical position. The bulk density of the limestone is taken as 2.5. The other data are given in the table below:

| | | |
|---|---|---|
| Depth of the hole | . . . . | 9.0 m (for avoiding toe blasting) |
| Burden | . . . . | 3.0 m |
| Spacing | . . . . | 6.0 m |
| Stemming | . . . . | 3.0 m. |
| Charge per hole | . . . . | 50.0 kgs. |
| Tonnage obtained; 9.0 | | |
| Average per metre of drilling | . . . . | 40 tonnes. |
| Tonnage obtained | . . . . | 9.0 |
| Powder factor | . . . . | $\dfrac{2.5 \times 9.0 \times 3.0 \times 6.0}{50} = 8.1$ |

Even though the value got is about 8, for practical purposes it is taken as varying from 6 to 7.

The example given above can be used as a guide in working out the powder factor in other cases also with slight modifications as necessary, to allow for the rock factor and bulk density.

With the view to surmount some of the difficulties due to the number of variables to be accounted in evaluating the tonnage factor etc., the makers of the explosives such as Du Pont, Imperial Chemical Industries, Nobel Explosives etc, have from experience and experimentation, with the various types of explosives and in various rock conditions, provided guidelines for determining the drilling parameters:
$D$ = diameter of the bit. (mm), $K$ = height of the bench, $H$ = depth of the hole, $V_{max}$ = maximum burden, $V$ = practical burden, $E$ = spacing of the holes, (m), $U$ = sub-grade, (m), $B_b$ = concentration of the bottom charge (Kg/m), $H_b$ = height of the bottom charge (m), $Q_b$ = weight of bottom charge (Kg), $1_p$ = concentration of the column charge (Kg/m), $H_b$ = height of column charge (m), $Q_p$ = weight of column charge (Kg), $Q$ = the total charge (kg/hole), $H_o$ = the height of the empty hole (m) used for stemming, $Q$ = specific charge (Kg/solid Cu.m) $G$ = specific drilling (meters drilled/soild $m^3$). q = equivalent to powder factor, $G$ = tonnes of rock per meter of drilling (Fig. 1).

### The value of $V_{max}$ is dependent on:

(a) the diameter of the bottom portion of the hole, (b) strength (weight) of the explosive used in the bottom charge (s), (c) on the degree of packing of the bottom charge (P), (d) on the rock constant (c), (e) the angle of inclination of the hole from the vertical (C), (f) on the ratio E/V.

Inorder to simplify the computations which involve all the above listed parameters, a standard case has been made out which can be applied any other situation by introducing modifications as necessary.

Assuming that the explosive used has a weight strength equal to about 78% of blasting gelatine, then the bottom charge carefully loaded into the hole will be P = 1.25 Kg/d $m^3$. In this type of explosive the charge required to loosen the rock or rock factor (e) = 0.40 Kg/$m^3$. If the inclination of the hole is 3:1 then a factor (f) = 1.0 is to be applied. If E/V = 1.25, then K = 2$V_{max}$ $V_{max}$ = $\dfrac{45\,d}{1000}$ (m).

Page 342

*Marginal/sub-marginal grade* : These terms are often used in the industry to signify the grade of ore which is just below the cut off value/mill grade.

Hence the reserves of such material are calculated separately, and such materials are also stacked separately during the course of mining so that these can be used if the need arises.

### Page 352

A single seam or several seams of coal may extend in an area of several square meters. This area is termed, ''a mining field'' or in this case ''coal field''. A property acquired within this field, for mining, is a lease. The excavation carried out will be known as a quarry or pit. The cover or overburden comprising sandstone or shale, as also the coal occurring below, within the area of the lease, is extracted systematically, in horizontal slices, beginning from the top. The slicing is so made such that the lower slice extends beyond the limit of the one above. The arrangement gives rise to a step-like appearance. The steps are known as benches. The arrangement of the benches can be from one side or from all the sides depending on the local geology, structure as well as the plan of mining, and in accordance with the Mining Regulations. Benches are generally of two types: 1. working benches and 2. non-working benches. A working bench is that on which active mining is done. A non-working bench is one from which no more extraction is possible, either of the mineral or of the overburden. Such benches are located on the flanks or boundary of the lease.

There are certain terms used to name parts of a bench. The various parts have been named and shown in the Fig. 2.

Apart from the normal berm as shown in the Fig. 2, there is another type of berm which is sometimes used called the ''Safety berm''. This kind of berm is generally about 5.0 m wide, and about 0.2 m. in height. Such berms are interspaced among the benches at regular intervals so as to give a protection to workman at the lower levels, against any loose material which may roll down from the top.

The height of the bench is determined by the type of working which can be either manual or machanised. In the case of manual open cast workings the bench height is normally 1.5 m to 2.0 m, whereas in the mechanised

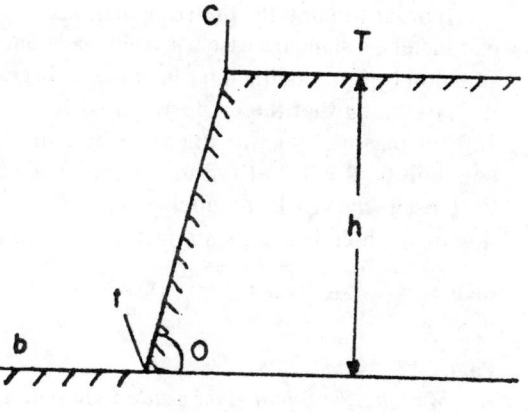

Fig. 2.

mines the height is related to the boom height of the machine excavator such as shovel or front end loader etc. Each bench is worked out in parallel strips known as cuts. The width of the bench, in the case of mechanised mining, can be of the order of 30 or 40 m. for accommodating the shovel and dumper, and the width of each cut can about 5 to 6 m. In case the length of the bench is excessive then it divided longitudinally into blocks. The overall slope of the working benches, in a pit, should be less than 45° from the horizontal and that of the non-working benches can be 60°. The floors of the berms are made to slope at a small angle so that water does not stand and the face is kept dry (Fig. 3).

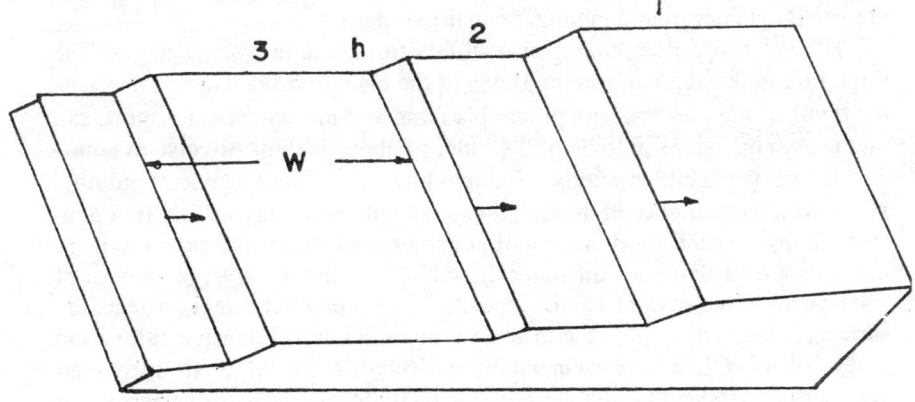

Fig. 3.

**Open cast mining :**

*Conditions favourable* : Opencast mining or openpit mining is the earliest method of extracting minerals including petroleum and ground water. The ideal conditions, for the use of this method, are when the mineral occurs at a shallow depth, say, of the order of 30 m. from the surface, and where the ratio of the thickness of the mineral body to that of the overburden of the order of 1 : 3 for minerals like limestone, or clay or moulding sand etc. This ratio is not fixed. It will vary with the value of the mineral to be extracted, namely, on the demand and supply position. This ratio will also be modified in relation to the unit value of the mineral. In consideration of this factor, the ratio of iron ore will be different to that for base metals such as copper or lead or zinc. This ratio can even be 1 : 5 or more.

Another consideration which favours open cast mining is that this method is generally cheaper than underground mining. It is also safer since no artificial lighting is needed or special arrangements for mine ventilation. Generally the out put is larger since heavy earth moving machinery can be employed, and this factor is what makes it applicable to low grade deposits such as porphyry copper deposits. Some of the other advantages, which favour the application of open cast mining, are that the environment in which the workers are deployed is cleaner and it is less polluted either by gaseous or liquid effluents. As compared to under-ground mining, mine planning is much simpler, and selective mining is easier, and hence the percentage of extraction is higher. More often than not, it is less capital intensive, and so also the time involved in gestation and development, as compared to an underground mine of the same size. It is because of these advantages that open cast mining is preferred to underground mining if conditions permit.

However, opencast mining also suffers from certain disadvantages. The limitation in the depth of mining is one of the main disadvantages. The limiting depth at the present, with proper planning and mechanised transport, can on the average be taken to be of the order of about 100 m. Adverse overburden to ore (mineral) is another factor which can inhibit opencast mining. Areas which experience high rainfall or snowfall are not favourable for opencast mining. In such areas the cost of pumping can be high and certain type of minerals are liable to be contaminated, which can involve a large amount of wastage as in the case of kaolin deposits. Since opencast mining affects the surface, it may not be possible to apply it in areas where intensive cultivation is carried on. Often land reclamation is difficult since the waste recovered may not be sufficient to fill the pit, and even with care the loss of top soil can affect the fertility. During mining operations, the dumps occupy large areas, and they can be a serious cause of contamination and pollution. The dust from the dumps and from blasting can also cause problems to the environment. It may also be mentioned that land reclamation is a costly operation.

### Stages in opencast mining:

Opencast Mining can be considered as comprising two distinct operations:

1. The first of these operations is stripping. Stripping or the removal of the overburden is done so as to expose the mineral deposit, in case it is not exposed. At the same time the material stripped is also to be dumped in an area which is free from mineralisation. At times where this is not possible, the dump may be kept temporarily over a mineralised part, and then reclaimed for back-filling, as soon as a block of mineral is mined out.

2. The second part is the exploitation or winning of the mineral and transporting it out of the pit either for temporary stacking or for further processing as the case may be.

In India, opencast mining is classified as a mechanised and b. non-mechanised or manual. Mechanised mines are those in which blast hole drills are used in conjunction with heavy earth moving machinery. Even the use of the front-end loader or wheel loader makes the operation "mechanised", non-mechanised mines are those where manual working is adopted. Men are employed for mining as well as for loading the mineral extracted either into trucks or tippers. Drilling is done either by the use of jumper bars or by jack-hammer drills.

Irrespective of the type of mine, whether it be opencast or underground, the classification for statutory purposes is two-fold, namely, first class and second class. Hence depending on the number of persons employed in the mine the mine's manager should hold either the "first class certificate of competency" or the "second class certificate of competency" awarded by the Department of Mines Safety. The holder of the first class certificate can be appointed as manager of a mine where more than 150 men are below ground, the total number of workmen being 400. In a mine employing more than 75 men below ground, but the total number is 400 the holder of the second class certificate can be the manager. Again there is a third distinction the certificate issued by the Department of Mines Safety can be either "restricted" or "unristricted". The holder of an unristricted certificate can manage both-under-ground and opencast mines, whereas the holder of the restricted certificate can a manage only opencast mines.

The term "Mine" as defined in the Mines Act is very comprehensive. It includes not only the actual workings but also, all prospecting operations, in addition to other ancillary operations related to mining. It is essential that every mining geologist be fully oware of the stipulations and the implications of not only the "Mines Act but also those of the Regulations made under the Act as applicable to Coal Mines and to Metalliferous Mines".

**Stripping :**

More often than not, opencast mining involves either the removal of the overburden or the removal of both overburden and side burden, depending on the nature of occurrence. For example, a coal seam, with a horizontal disposition, which is overlain by sandstone and shale, will need only stripping, of the over burden. On the other hand, in the case of a steeply dipping ore body with a width of 50 m. or more, not only the overburden comprising the hanging wall has to be removed but also the footwall comprising the side burden for maintaining desirable pit slopes.

Since the extent of stripping is critical in determining the overall economics, the methods, adopted in calculating the stripping ratio, require consideration.

1. Commercial stripping ratio is defined as the ratio of the barren rock extracted per $m^3$ of mineral won.

2. Average stripping ratio is the ratio of the volume of stripping involved to the volume of the mineral extracted, when the pit has reached its ultimate limits, aerially and in depth.

3. Contour stripping ratio is the ratio of the barren rock to the reserves of mineral occurring within a specific horizon or in a slice.

4. Current stripping ratio is the volume of overburden to the volume of mineral won, within a specific period of time.

5. Boundary stripping ratio is the index obtained by analysing the relative economics of working the deposit by open cast method, to the economics if the exploitation were to be done by the underground method.

Certain methods, of estimating the depth of the quarry and also the over-all angle of the quarry wall on the hanging well side ($Q_h$), and that of the wall on the foot wall side ($Q_f$), employ the graphical representation or the use of the formula $D = (W_t - W_b) \, Sin\theta_h \, Sin\theta / Sin(\theta_h - \theta_f)$.

Where D = the depth in m., $W_t$ = the width of the pit, either at the top or at a particular R.L., and $W_b$ = the width of the pit at the bottom.

The method mentioned above is generally applicable in the case of deposits which have a regular shape such as lensoid or tabular (Fig. 4).

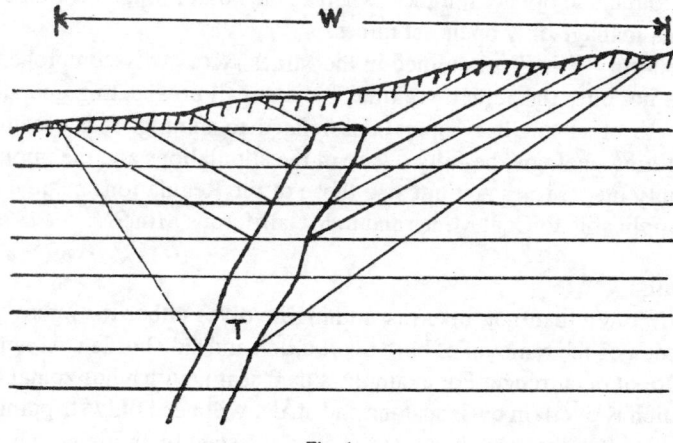

Fig. 4.

An approximate method of arriving at the ultimate depth of the pit, is also used and this employs the current stripping ratio as the criterion. The principle involved in the use of this method is depicted in (Fig. 4).

The ore body, with a thickness of T m. is cut into horizontal slices keeping the height of the slice equal to that of the bench. From the bottom of each slice the cut in the footwall is drawn at an angle of $\theta_f$, and the cut on the hanging wall side is drawn at an angle of $\theta_h$. Both these angles are drawn from the horizontal. The same procedure is continued depthwise until the width of the excavation as noticed at a particular R.L. is equal to W m. When the value of W/T is equal to that of the boundary stripping ratio, that depth is taken to be the limit of economic working.

### Entry to the Deposit :

In the case of underground workings, shafts, either vertical or inclined, are made in order to gain entry to the deposit. More often than not mineral deposits occur under a cover of either soil or rock, known as overburden. It is hence necessary to make a trench in the overburden which will form the approach to the deposit. These trenches are often horizontal, but when the situation warrants the trench can be inclined at low angles, the gradient being of the order of about 1 : 20 or 1 : 18 so as to facilitate transportation by dumpers. Where belt conveyors are proposed to be used, then the gradients can be of the order of 1 : 5 or 1 : 4. The width of the trench can be about 15 to 20 m, depending on the size and the type of machine.

Making of trenches is expensive, and at the same time it involves locking up of capital, hence it is essential that the volume of excavation be kept to the minimum. The minimum depth is calculated by the use of the formula H = L i, where L is the length of the trench in m. and i = $\tan\theta$, and $\theta$ is the angle of the slope. In case the overburden is excessive then more than one trench is needed for carrying out the stripping operations. Consequently, the trench will also have two or more benches, depending on the thickness of the overburden. It can so happen that separate trenches are needed for entry and exit. It also possible that both the trenches can be used either for entry or for exit.

### Methods of trenching:

There are two common methods used in the making of trenches: a) trench cutting by using a continuous face and adopting bottom loading; b) trench cutting by using a continuous face while adopting top loading. Of these two methods trench cutting by bottom loading is more commonly practised, than the top loading method.

### Bottom Loading Method: (Fig. 5)

In this method the trench is made in such a manner so as to expose the top of the mineral deposit, assuming the ideal case where the deposit is horizontal. By this a working face is made in the overburden. The face of the trench is drilled using the blasting pattern best suited for the type of formation com-

prising the overburden.
Blast hole drills are commonly employed. The
rock is blasted and
benches are formed start-
ing from the top. The
blasted rock falls to the
bottom of the trench. The
loading is commonly

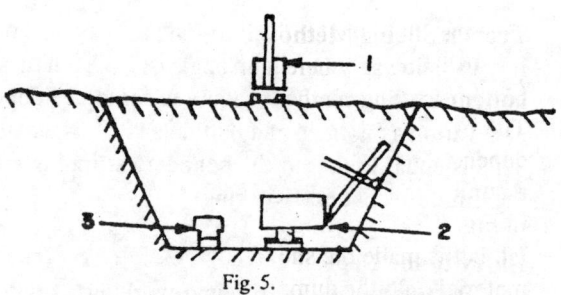

Fig. 5.

done by shovels and the transport is done by dumpers. The Fig. 4 shows the
trench as well as the arrangement of the machines. The advantages of the
method are that it is possible to extend the trench for the entire length, with
the use of power shovels for excavation as well as for loading. Some of the
disadvantages of this method are that only a restricted area is available for the
loading by the shovel, and this causes loss of time, in the interval between
loading, unloading and the return of either the dumpers of transporting
machines. This causes losses in the output of the shovel.

**Top loading method:** (Fig. 6) & (Fig. 6a)

In a situation where the
overburden is fairly soft comprising either sandy clay or
semi-consolidated sandstone,
the continuous top loading
method is generally preferred.
As shown in Fig. 5 the shovel
stands at the floor of the
trench, and it functions as
both the excavator as well as
the loader. The shovel digs
into the formation and loads
the material onto the dumper
which is positioned at the top
of the trench. It will be seen
that in this method, the depth

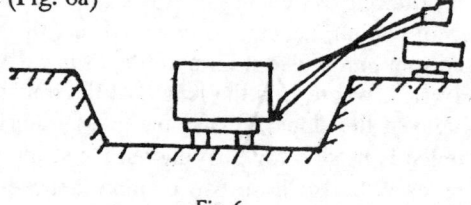

Fig. 6.

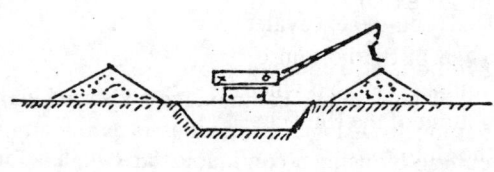

Fig. 6a.

of the trench is limited by the maximum dumping height of the shovel, which
is equal to the depth of the trench + the height of the dumper. The advantages
of this method of stripping are that the planning is relatively simple, and
quicker transport by cutting the delays. One of the disadvantages in addition
to the limit in the depth of trench, is that the costs of excavation exceed those
of the bottom loading method by 20 to 30%.

## Trench Slicing Method: (Fig. 7)

In the case where the thickness of the overburden is excessive and the bottom loading method is not applicable, trench slicing method is adopted. The procedure involves making a number of sequential cuts systematically depending on the depth of the trench. This method requires the deployment of a number of excavators (shovels). The arrangement of the machines is shown in Fig. 7. Considering the case as shown in the Fig. 7. It is seen that when the Ist. cut is made the shovel operates at the bottom of the Ist. cut and loads the material onto the dumper stationed at the top of the trench. When making the

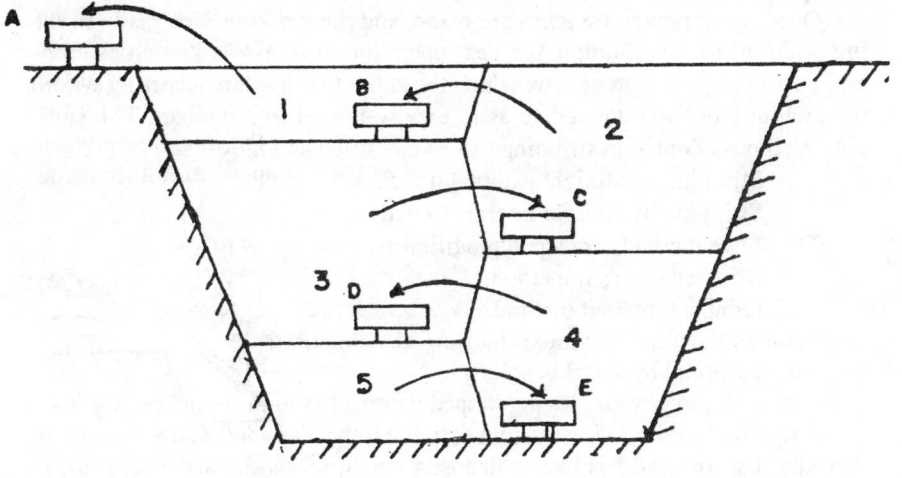

Fig. 7.

IInd. cut, the excavated material is loaded onto the dumper which is now located at the bottom of the Ist. cut, and transported along the Ist. cut to be dumped. So also when the IIIrd. cut is made the dumper is stationed at the bottom of the IInd. cut and the material is loaded onto it and transported along the IInd. cut for disposal. When the IVth. cut is made the dumper now stands on the floor of the IIIrd. cut, and when loaded it moves along the IIIrd. cut for dumping. In the same way when the Vth. cut is made the dumper is positioned on the floor of the IVth. cut and when loaded moves at the same level for dumping.

This method of trenching has the advantage of employing a large number of conventional machines, in each bench, of the trench, for speeding up the operations it also has certain disadvantages. The time and energy spent in the making of roads for the movement of the dumper are appreciable. Since there is only limited space for the movement of the shovels, the efficiency is affected adversely. Further, the trench cannot be commissioned until the final slice or the Vth. slice in the present case, is removed. It may also be mentioned that, where the overburden is soft or sandy, bucket wheel excavators known as land dredges can be used with greater efficiency.

**Opencast Mining Practice :**
The common methods employed in opencast mining can be classified as:

**Stripping method :**
Once the approach trenches are made, and the working face made in the full width of the overburden, the next operation to follow is known as stripping. Stripping is the process by which the entire overburden occurring within the mining block is removed so as to expose the mineral deposit. The common methods adopted in stripping can be classified as follows:
    1. Stripping by straight external trenches.
    2. Stripping by straight internal trenches.
The above methods are again modified as
    a. Grouped external method.
    b. Grouped internal method.
Some of the other methods which are also practised are
    a. Stripping by spiral benching.
    b. Stripping by employing looped approaches to the benches.
Stripping by employing straight external trenches is best suited to mining deposits that are flat dipping or deposits which are horizontal occurring at shallow depths. It is also applicable to deposits that occur in hilly terrain. All the above-mentioned types of occurrences are amenable to stripping either by individual, group or main external trenches. The trenches are made on the flanks and not at the center of the deposit.

If individual external trenches are used then the ore can be hauled independently from each horizon or bench. Since with increase in the depth, of the workings, the trench also requires deepening and at the same time it is not able to cope with the increase in the output, additional trenches are needed in order to open up two or three benches (Fig. 8). These additional trenches can be either of the external type or of in the external type (grouped). It is seen that when a group of two trenches are is used, about 5 to 6 benches can be opened, and it is more economical than having separate trenches.

Stripping by using straight internal trenches is practised in opening up thick deposits which occur at shallow, but not at great a depth. Individual trenches permit independent haulage from each horizon or bench. When the trenches are developed within the boundary of the open cast, and either inter-

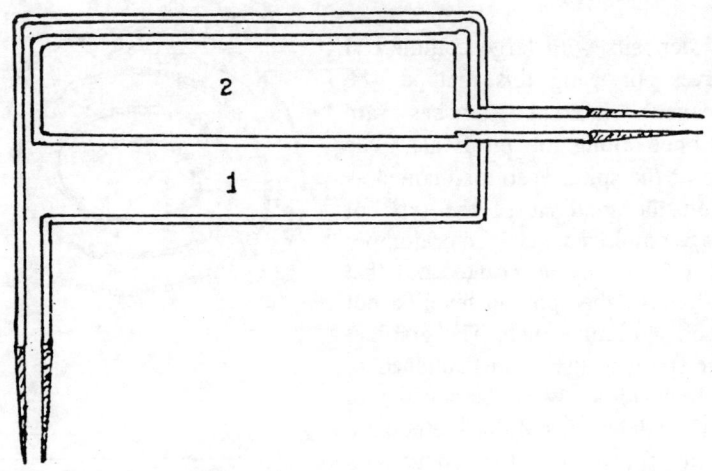

Fig. 8.

nal grouped or main trenches are made use of, the output is lowered. Hence in the case of deep pits (300 to 400 m) the practise is to use main internal straight trenches (Fig. 9).

Stripping by the use of spiral approaches is advantageous in the case of

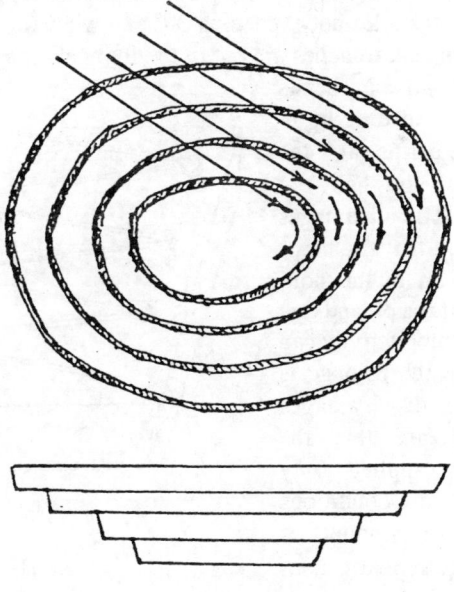

Fig. 9.

deep deposits with large commercial reserves. In using this method, the permanent internal trenches are developed along the pit flanks. The slope of the spiral is so maintained as to suit the gradient of the type of haulage employed, say loco or dumper (Fig. 10). It may be pointed out that the slope of the spiral it levelled out at each working bench. The gradient of the spiral is again re-established so that the level below can be developed.

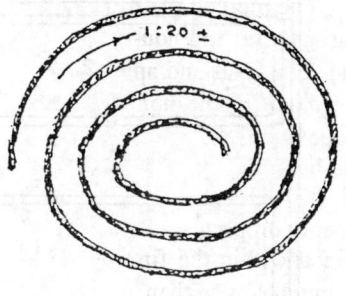

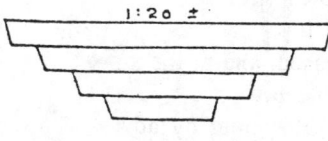

Fig. 10.

The advantages of this method are that it allows for the continuous transport of material, and since no by-pass is needed for the incoming and out-going traffic the productivity is increased. The main disadvantage of the method is that it is applicable to areas only where the walls are strong enough to ensure the stability of the roads. Further the pit should be sufficiently wide. In addition the bulk of the over burden has to be excavated at the very start.

Stripping by the method of loop approaches is applicable when the deposit dips at low angles not exceeding 30° and when the footwall rocks are competent. Permanent trenches are cut in the footwall in a zigzag manner, in the manner of hair pin bends made in the roads leading up to hill stations. The berms carry the turning point of the hair pin bends (Fig. 11). The advantages of the method are that the loops can be made even in the non-working flanks of the pit and that a higher level of productivity can be attained since the haulage is continuous. The disadvantages of the method are that the method can be applied only under certain specific conditions and further more amount of stripping is required for constructing the berms.

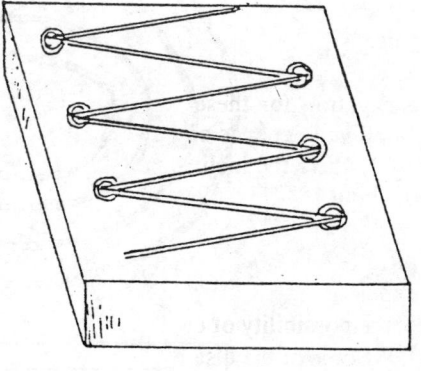

Fig. 11.

The method of stripping by dead end approaches is best suited for working metallic and non-metallic mineral deposits, dipping at steep angles (Fig. 11). The dead end approaches comprise a system of internal trenches which are made in the non-working flank of the pit. The trenches have a gentle slope but reverse at the dead end, where they become horizontal in level with the berm. The dead ends enable the reversing of the transporting machines (rail). The use of the dead end approaches is an advantage in the case of steeply dipping, and elongate deposits, since the successive trenches are contained within the first trench, and so the volume of rock extracted (barren) is much lower than in the case of the spiral method. This method is less dependent on the nature of occurrence of the deposit, and it is possible to make all the trenches on a single flank. The disadvantages are that the first approach is excessive in length, and further on account of the dead ends and the horizontal berms at the reversal points, the volume of stripping is increased, and at the same time the average speed of the transport (trains) is reduced.

Stripping by advancing approaches is preferred when stripping is required for steeply dipping deposits as well as for deposits which have considerable thickness. In this method an inclined approach is made diagonally splitting the bench into halves, and this facilitates easy extraction. The height of the bench can be varied. Since this method permits the opening of the mine from the center of the deposit, it helps to quickly develop the individual horizons. The advantages of this method are the saving in the time taken for constructing, with possibly, steady rate in the stripping. The disadvantages are that where tracks are used these require constant shifting, and the cost involved in making the diagonal approaches.

Stripping by means of steep trenches is useful in working deposits which persist in depth. In this method steep trenches or trench is made in the non-working flank of the pit, or in the footwall as the case may be. The volume of excavation for these trenches is comparatively small. The type of haulage employed depends on the gradient of the trenches. Skip haulage (winding) is used where the inclination permits at lower gradients when the inclination is of about 18° (Fig. 12) shows the arrangement for skip winding.

The advantages of trench stripping are: 1. time saved in developing the pit; 2. speed with which material can be lifted from deep horizons; 3. lower expenditure on transport from the pit; 4. less excavation needed for the trench; 5. the possibility of employing more than one skip.

Some of the disadvantages of the method are that it is difficult to develop additional horizons with one permanent trench. Loading cannot be done at all the levels, and may be difficult to handle the loads concentrated at the loading level, which is generally the bottom level.

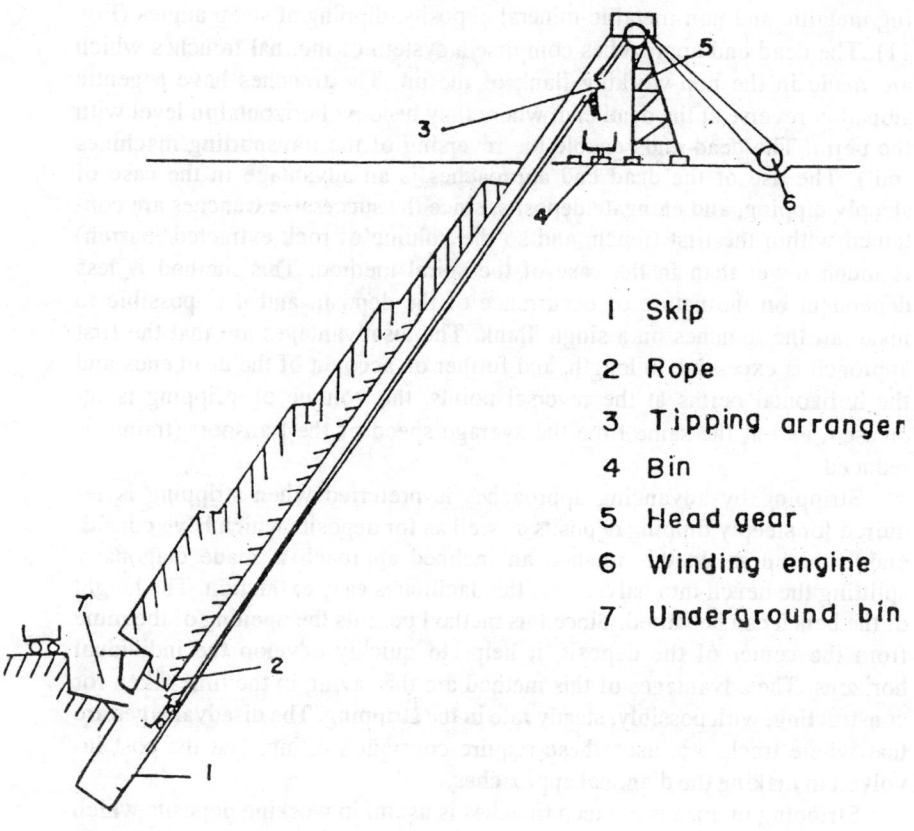

1 Skip
2 Rope
3 Tipping arranger
4 Bin
5 Head gear
6 Winding engine
7 Underground bin

Fig. 12.

Stripping by underground access such as adits, shafts or drifts is a modification of the glory hole method of working. This is shown in Fig. 13.

Page 357

## Utilisation of Heavy Earth Moving Machinery :

Heavy earth moving machines are essential for large scale opencast mining operations. At the same time these machines are expensive. Hence the extent to which these machines are utilised to the maximum possible capacity, within the time available for working, is the measure of the efficiency of the operations taken as a whole. Thus in the planning of large scale open pit mining, it is necessary to assess the nature and the quantum of the excavations

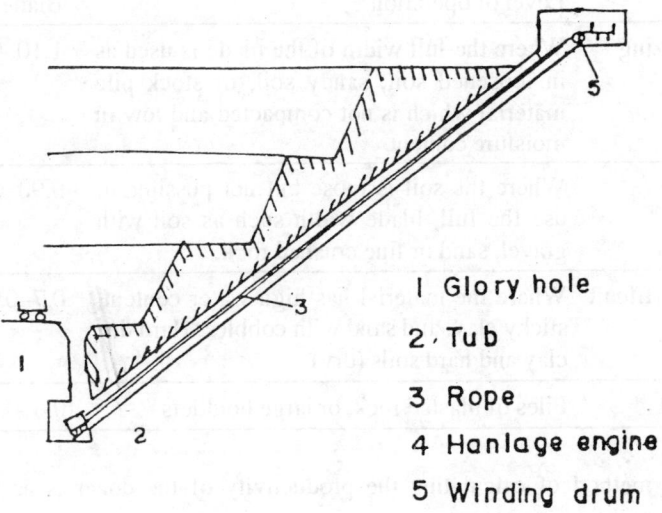

1 Glory hole

2 Tub

3 Rope

4 Hanlage engine

5 Winding drum

Fig. 13.

involved, so that the type and the capacity of the machines can be decided. The factors that influence the selection of the machinery, are the type of the earth or rock to be worked, the rate of planned output, capital available, and also the availably of the machines, whether indigenous or imported.

The machines which is most popular in India, for forward operation, i.e. for the removal of top soil, boulder, etc. is the bull dozer, with the straight cutting blade. For purpose of blast hole drilling (opencast), the D.T.H. type machine is preferred. For loading, in the larger mines, the power shovel is popular, while the wheel loader is sometimes used in the smaller mines. One popular method of transport of the broken materials is by the dumper, belt conveyor, and the aerial ropeway are used to a lesser extent. With the three operations going hand-in-hand it becomes necessary to see that no machine is idle. Hence it is necessary to have a comprehensive understanding of machine operation for making a matching schedule.

*Bull dozer* : The most common type of dozer in use is the machine with the straight blade. The efficiency of the dozer varies with such factors as: i. whether the machine is of the push type; ii. whether the material is clayey, or sandy or rocky etc. Hence to accommodate such variables a factor called the "blade factor" is employed as shown in the table below:

**Blade Factor:**

| Type | Level of operation | Blade factor |
|---|---|---|
| Easy dozing | Where the full width of the blade is used as in loosened soil, sandy soil, or stock pile material which is not compacted and low in moisture content. | 1.10–0.50 |
| Average | Where the soil is loose but not possible to use the full, blade width such as soil with gravel, sand or fine crushed rock. | 0.90–0.70 |
| Rather difficult | Where the material has high water content, sticky clay, and sand with cobbles. Hard dry clay and hard soils (dry). | 0.7–0.6 |
| Difficult | Piles of blasted rock, or large boulders | 0.6–0.4 |

The method of calculating the productivity of the dozer is as shown below:

The production/hour $(Q) = \dfrac{q \times 60 \times E}{cm}$

Where Q = hourly production $(m^3/h)$, q = production per cycle $(m^3)$, cm = cycle time (min.), and E = Job efficiency.
The production per cycle (q) in $m^3 = L \times H^2 \times a$.
where L = blade width (m), a = blade factor, H = blade height (m)

Cycle time (cm) in minutes $= \dfrac{D}{F} + \dfrac{D}{R} + Z$

Where D = Haul distance (m), F = Forward speed (m/min), R = Reverse speed (m/min) and Z = Time required for gear shifting (min).

In calculating the standard productivity of the dozer, the value of q (production per cycle) is taken to be blade width × (blade height)$^2$, since the productivity depends on the conditions of the soil, the blade factor is also taken into consideration.

*Blast hole drills* : This machine uses button bits ranging from 100 mm to 200 mm in diameter. The capacity of the drill is to be matched with the capacity of the excavator, i.e., shovel which again will have to be matched with that of the dumper. Thus, for example, the shovel to match a 100 mm. diameter D.T.H. machine, should have a bucket capacity of about 3.0 $m^3$ to 4.0 $m^3$. With larger drills, the bucket capacity can even of the order of 5.0 $m^3$ or 6.0 $m^3$. The other factors to be taken into account are: 1. rate of penetration of the drill, 2. frequency of blasting, 3. spacing and the burden pattern, 4.

explosive or powder factor, 5. the number of machine employed.

1. Rate of penetration depends not only on the hardness of the rock but also on the pressure used. High pressure not only improves the rate of cutting, but also the efficiency of flushing the bore hole of the cutting. The normal pressure that is used is about 100 psi. The rate of penetration, for example, in hard limestone is about 5.0 m/hr. A higher rate can be achieved by employing pressure of 250 psi.

2. Frequency of drilling is dependent on the schedule of production and on the local conditions. It can be once a week, but then the material blasted should be sufficient to meet the requirements for the entire week. In such an eventuality it may become necessary to employ more than one drill.

3. Spacing and burden are factors which depend on the type of fragmentation desired. The larger blocks should not be a hindrance to loading, since bridging by large blocks will foul the bucket of the shovel. If the spacing is reduced then the total depth of drilling as well as the time consumed will also increase together with the costs.

4. Explosive factor is an index of the efficiency of blasting and also of the drilling pattern.

5. Number of drills can be decided only after a few trials since there are too many unknowns to be taken into account. However, as a first approximation the number can be calculated by a straightforward method as shown below:

Page 337

Taking the example of a limestone quarry feeding a one million tonne cement plant, one million tonnes of cement will requires about 1.60 million tonnes of limestone per annum. Taking 300 working days/year, after allowing for stoppages, for maintenance, mines flooding during rains, etc. the quantity of limestone to be produced per day will be $\dfrac{1,600,000}{300}$ = 5,300 or say 5400 tonnes.

The bench height is taken to be 8.0 m. and for drilling 9.0 m is taken the extra 1.0 m. had been allowed for avoiding the toe and secondary blasting. The burden is 3.0 m and the spacing is 6.0 m. with 3.0 m of stemming. The charge per hole is 50 Kg. and the bulk density of the limestone is taken as 2.50. The powder factor is $\dfrac{9 \times 3 \times 6 \times 2.5}{50}$ = 8. This, for practical purpose, is taken as 7. The tonnage obtained from one blast is $9 \times 3 \times 6 \times 2.5 = 405$ tonnes. Hence the number of blasts needed for getting 5400 tonnes will be $\frac{5400}{405}$ = 13.3 or 14. The quantum of drilling required to produce 5400 tonnes/day will be of the order of $14 \times 9 = 126$ m. Assuming that the rate of drilling is 3.0

m/hr. The meterage drilled per day in 2 shifts of 8 hours each, will be $16 \times 3 =$ 48. Hence to drill 126 m it will take $\dfrac{126}{48} = 2.65$ days. Since in practice it is observed that the rate of drilling in hard limestone is of the order of 8.0 m or even 10 m. the meterage drilled/day by 2 shifts will be about 140 to 150. Hence one drill will be sufficient with one for standby to meet contingencies such as break downs or maintenance.

*Power shovels*: In computing the hourly production of any heavy earth moving machinery the common formula used is $Q = q \times N \times E \dfrac{60}{q} \times cm \times E$, where Q = prouduction/hr. (m³/hr), q = production/cycle (m³) of loose and excavated material, N = the number of cycles/hr. or $\dfrac{60}{cm}$, cm = cycle time (min), E = Job Efficiency. Since when the in situ rock is broken by blasting, the volume is increased by a certain ratio, in respect of the original volume, and this increase in the volume has to be taken into account in handling. Hence a factor known as the ''swell factor'' has to be included in handling broken or loose rock.

The swell factor to be applied to various situations is given below, and this can be employed to yield compatible results:

| Material | | Type of material handled | | |
|---|---|---|---|---|
| Natural | Condition | In situ | Loose | Compacted |
| sand | (A) | 1.00 | 1.11 | 0.95 |
| | (B) | 0.90 | 1.00 | 0.86 |
| | (C) | 1.05 | 1.17 | 1.00 |
| Sandy clay | (A) | 1.00 | 1.25 | 0.90 |
| | (B) | 0.80 | 1.00 | 0.72 |
| | (C) | 1.11 | 1.59 | 1.00 |
| Gravelly soil | (A) | 1.00 | 1.18 | 1.08 |
| | (B) | 0.85 | 1.00 | 0.91 |
| | (C) | 0.93 | 1.09 | 1.00 |
| Gravels | (A) | 1.00 | 1.13 | 1.03 |
| | (B) | 0.88 | 1.00 | 0.91 |
| | (C) | 0.97 | 1.10 | 1.00 |
| Solid/rigged | (A) | 1.00 | 1.42 | 1.29 |
| Gravels | (B) | 0.70 | 1.00 | 0.91 |
| | (C) | 0.77 | 1.10 | 1.00 |

| Broken limestone/ | (A) | 1.00 | 1.65 | 1.22 |
| sandstone, or soft | (B) | 0.61 | 1.00 | 0.74 |
| rocks | (C) | 0.82 | 1.35 | 1.00 |
| Broken granite/ | (A) | 1.00 | 1.70 | 1.31 |
| basalt, etc. | (B) | 0.59 | 1.00 | 0.77 |
| | (C) | 0.76 | 1.30 | 1.00 |
| Broken rocks | (A) | 1.00 | 1.75 | 1.40 |
| | (B) | 0.57 | 1.00 | 0.80 |
| | (C) | 0.71 | 1.21 | 1.00 |
| Blasted bulky | (A) | 1.00 | 1.80 | 1.30 |
| rocks | (B) | 0.56 | 1.00 | 0.72 |
| | (C) | 0.77 | 1.38 | 1.00 |

A = In situ          B = Loose          C = Compacted

In addition to the swell factor, the productivity of all earth moving machine is affected by the condition of the machine, which again depends on the upkeep. Further the skill of the operator is also to be taken into account, in evaluating the overall productivity. Hence a factor known as the "Job efficiency" is introduced in computing the productivity. In India, the factor is taken to be of the order of 80–90 per cent, of the normal standards. A lower value of the job efficiency, is not allowable in view of the very high capital investment on the machines.

*Output of excavators* : The operating capacity of shovels is obtained by the formula $Q = \dfrac{q \times 3600 \times E}{cm}$, where Q = production/hr. (m$^3$/hr.), q = production per cycle (m$^3$), cm = cycle time (sec), E = job efficiency. Production per cycle $q = q_1 \times k$, where $q_1$ is the capacity taking account into the swell factor and k = bucket factor. The bucket factor is influenced by the height of the bench, nature of fragmentation, efficiency of the bucket machanism, as well as the operators, skill. The loading influencing the value of the bucket factor and the relative values are given in the Table: Cycle time is the time taken in excavation, swinging, dumping, swinging back to the excavation (crowding) position. This time is taken to vary from 25–40 seconds cycle time (cm) = Excavating or crowding time + swinging time = 2 × dumping time. Excavating time is dependent on the digging depth and condition of material as shown below:

| Digging depth | Digging conditions | Easy | Average | Rather difficult | Difficult |
|---|---|---|---|---|---|
| 0–2 m | A | 6 | 9 | 15 | 26 |
| 0–4 m | B | 7 | 11 | 17 | 28 |
| 4 m–above | C | 8 | 13 | 19 | 30 |

A = Easy     B = Average     C = Difficult

Swing time is dependent on the arc of swinging, the speed, and the skill of the operator. The table below, shows the relation of the arc of the swing to time:

| Arc of swing | Time in seconds |
|---|---|
| 45–90% | 4–7 |
| 90–100% | 5–8 |

**Bucket factors:**

| Type | Description | Factor |
|---|---|---|
| Easy | Digging and loading from stock piles or materials cut or broken by some other excavator, which do not require effort in digging, and can be readily filled into the bucket. | 1.0–0.8 |
| Average | Digging and loading from loose stock piles of the soil which is more difficult to penetrate, and scoop up, but yet the material is capable of filling the bucket to the full capacity. These include dry sand and sandy soil, clayey soil, clay, unscreened gravels, etc. as well as soft gravels from hill slopes. | 0.8–0.6 |
| Rather difficult | Digging and loading finely crushed rock, hard clay, gravelly sand/sandy soil, clay etc. with high moisture content which have been stockpiled by excavators. | 0.6–0.5 |
| Difficult | Bulky or irregular-shaped blocks of rocks forming a loose heap, such as blasted bouldry materials, sand mixed with boulders, or clay etc. These cannot be easily scooped up by the bucket. | 0.5–0.4 |

Dumping time or unloading time, which is adopted under Indian conditions, should be based on proper spotting or positioning of the dumper, as well as the skills of both the "spotter" and the dumper operators. As a general case the values given in the table may be utilised:

|  | Time in seconds |
| --- | --- |
| Dumping in a fixed location by the dumper | Dumping in an area which is not fixed |
| 5–8 | 3–6 |

The dumping time as related to the operating conditions is taken as:

| Operating conditions | $(t_1)$ in minutes |
| --- | --- |
| Favourable | 0.5 to 0.7 |
| Average | 1.0 to 1.3 |
| Unfavourable | 1.5 to 2.0 |

Taking into account the factors that influence the cycle time of the dumper, the following formula can be used to obtain a good approximation of its value:

$$\text{Cycle time (Cmt)} = n \cdot Cms + \frac{D}{V_1} + t_1 + \frac{D}{V_2} + t_2$$

Where nCmt is the loading time, $D/V_1$ is the hauling time, $t_1$ is the dumping time, $D/V_2$ is the returning time, and $t_2$ is the spotting and the returning time. All the times are in minutes.

n = the number of cycles required for the loader to fill the dumper $n = c_1/q_1 \times k$ where $c_1$ rated capacity of the dumper $(m^3)$: $q_1$ = Bucket capacity of the shovel, k = bucket factor of the shovel; Cms = cycle time of the shovel (min); D = the hauling distance of the dumper (m); $V_1$ = Average speed of the loaded dumper (m/min). $V_2$ = The average speed of the empty dumper (m/min); $t_1$ = the time required for dumping + the time required for waiting, until the dumping can start (min.).

**Estimation of the number of dumpers required:**

The number of dumpers that are needed for operating in combination with the shovel, which is working at the maximum job efficiency, can be obtained as shown below:

$$\text{Number of dumpers (m)} = \text{Cycle time of the dumper} = \frac{Cmt}{n \cdot Cms}$$

Where n is the number of cycles required for the shovel to fill the dumper; Cms is the cycle time of the shovel (min); Cmt is the cycle time of the dumper (min); M is the number of dumpers in operation.

## Method of estimating the productivity of the dumpers:

The production of a fleet of dumpers per hour, where the fleet is employed in doing the same job can be estimated in the manner shown below:

$$\text{Integrated production of the fleet } (P) = \frac{C \times 60 \times Et \times M}{Cmt}$$

Where M = Number of dumpers in operation; P = the hourly production; C = Production per cycle i.e. C = n × q × k. Et = job efficiency of dumper; and Cmt = the cycle time of the dumper (min).

P 436

It is also a good practice to work out the "standard error of estimate" which is a measure of the variation or scatter of the value, about the line of regression. This is expressed by either

$$S_y = \sqrt{\frac{(y - y_c)^2}{N}} \text{ and } S_x = \sqrt{\frac{(X - X_c)^2}{N}}$$

Page 524

## Other methods of Computing Reserves—Types of Reserves :

A. *Cross-sectional Method:* This method is most commonly used and it is applicable to any type of deposit irrespective of its shape, size etc. In the application of this method, the preliminary step is to establish the shape as well as the extent of the deposit, in question, by detailed exploration. After this is done, then the deposit is subdivided into blocks, with respect to the regularity or uniformity, in dimension, as shown in Fig. 14. The blocks are designated as A,B,C,D, etc. Within each of the blocks cross sections are measured, 1,2,3,4, etc. at equal distances starting from and ending at P. The areas of two adjacent cross section are measured and averaged and the mean area is multiplied by the distance between the two sections to obtain the volume of the material contained between the two cross-sections i.e. Block 1 = section 1 + 2/2 × distance 1–2 = volume. The method is adopted to get the volumes of the other blocks. The volumes of all the blocks are added up so as to obtain the volume of the mineral deposit.

Since the cross sections of most ore bodies are irregular the "Simpson's Rule" is used in the computation. The rule is stated in the following terms:

a , b , c , d are drill hole locations

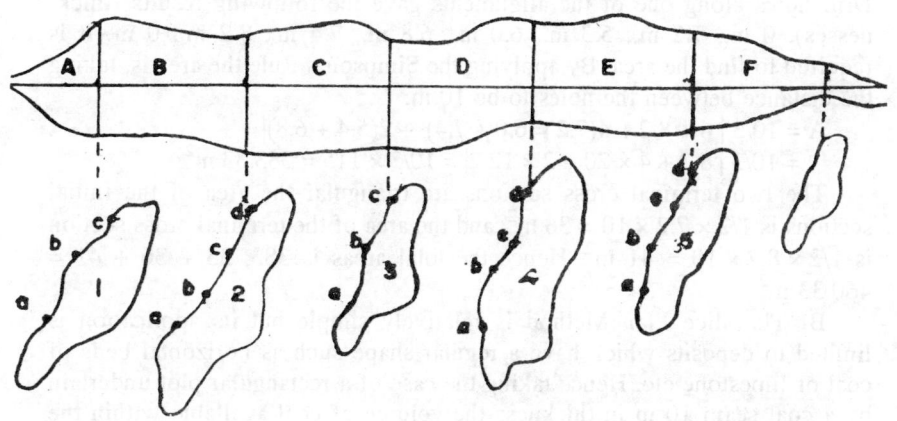

Fig. 14.

"To the sum of the first and the last ordinates, add twice the sum of the remaining odd ordinates, and four times the sum of all the even ordinates. Multiply the total sum by one third the total distance between the ordinates to obtain the area". Hence for example if, $O_0$, $O_1$, $O_2$, $O_3$, $O_4$, $O_5$, $O_6$ are the ordinates and D is the distance between adjacent ordinate, the area is

$$A = d/3 \,|\, 0_1 + 0_6 + 4(0_1 + 0_3 + 0_5) + 2(0_2 + 0_4 + 0_6)|$$

In case there are odd number of divisions and even number of ordinates, the area of the last two divisions have to be calculated separately and then added to the rest. In case the terminal ordinates are zero then again these are measured separately. Generally these will be triangular and hence they can be taken to be triangles.

*Example*: A lensoid ore body was drilled in the transverse direction at regular intervals of 30 m. The thicknesses obtained along the alignment are 5.9 m., 12.4 m., 16.5 m., 15.3 m., 18.4 m., 20.9 m., 24.2 m., 21.8 m., and 19.2 m. It is required to calculate the area of the cross section.

By applying the Simpson's Rule
$A = |5.9 + 19.2 + 4(12.4 + 15.3 + 20.9 + 21.8) + 2(16.5 + 18.4 + 24.2)| \times 30/3$
$= 10(25.1 + 4 \times 70.4 + 2 \times 59) = 10(25.1 + 218.6 + 118.2)$
$$= 4549 \text{ m}^2$$

*Note*: That the above area in only 1/2 the area of the lenticular cross section. Hence the area of the lens is 2 × area obtained.

*Example* : A lenticular ore body was explored in detail by drilling on an alignment in the transverse direction so as to get the cross sectional area. Drill holes along one of the alignments gave the following results (thicknesses). 0 m., 7.2 m., 5.3 m., 6.0 m., 6.8 m., 7.4 m., 8.2 m., 0 m. It is required to find the area. By applying the Simpson's Rule the area is, taking the distance between the holes to be 10 m.

A = 10/3 | 6 + 8.2 + 4(7.2 + 6.0 + 7.4) + 2(5.4 + 6.8)|
    = 10/3 | 8.2 + 4 × 20 + 2 × 12.2| = 10/3 × 115 = 383.33 m².

The two terminal cross sections are triangular the area of the initial sections is 1/2 × 7.2 × 10 = 36 m², and the area of the terminal cross section is 1/2 × 8.2 × 10 = 41 m². Hence the total areas is 383, 33 + 36 + 41 = 460.33 m².

B. The slice Plan Method is relatively simple but its application is limited to deposits which have a regular shape such as horizontal beds of coal or limestone etc. Hence taking the case of a rectangular plot underlain by a coal seam 10 m in thickness the volume of coal available within the plot is Area of plot × 10 m. If, however, the seam has a dip and if the true dip is parallel to one of the sides on which it outcrops, and its depth at the opposite side 100 m. away, is 20 m. then tan of the dip = 20/100 = 0.2 or dip is about 11°. From Fig. 15 it is seen that 20/x = sin dip, 20/x = 0.191 or x = about 105 m. In case the direction of true dip is not coincident with the side then the apparent dip in the direction of the side is taken in the calculation. In calculating the reserves of a mineral deposit the term in common use is "tonnage factor", in the FPS system. It is defined as the weight of 1 c.ft. of the in situ mineral in pounds. In the CGS system, which is now accepted in India, "bulk density" is used. Bulk density is defined as the weight of 1 m³ of the in situ mineral in tonnes.

In the estimation of the reserves of a mineral certain terms are used such as "geological reserves" and "mineable reverves". The geological reserves of a mineral are those present within the entire area under mining lease. The geological reserves are got by multiplying the volume of the deposit by the bulk density. The volume is calculated by using the methods mentioned earlier.

The mineable reverves are those which can be extracted during the mining operations. This quantity is got after making deductions for mineral locked up and also that which is lost during mining. The mineral locked up is that which is under the barrier and that which is under the benches, in the case of opencast mines. Mining loss comprises mineral lost in powdery form. These reserves which cannot be recovered during mining operations are to be deducted from the geological reserves and what remains is the mineable reserves. It can so happen that the mineable reserves are limited by geological factors such as the structural or petrological, competence of the ore body, or the associated rocks. Ground water conditions can also prove to be one of such factor. Further the reserves falling in the above mentioned two categories can be again classified as either "proved", "probable", or "possible". These classes of reserves are defined taking into consideration the geological and structural features, as well as the amount of exploration (probing) that has been carried out. It is evident that a deposit which is uniform in nature with little or no structural disturbance will require less probing than one which is disturbed, or erratic in respect of the intensity of mineralisation. The criterion employed in these classifications has been indicated earlier under the terms "measured", "indicated", and "inferred" reserves.

Page 530

The norms that are used in defining the various categories of reserves, such as proved, indicated or probable, and inferred or possible, in respect of bedded deposits, such as coal, lignite, certain types of sedimentary limestone, or similar deposits, cannot be applied straightaway, in general, to metalliferous ore bodies. Generally, in the case of bedded deposits, there is a certain amount of regularity or uniformity, in respect of certain parameters, such as thickness, dip, or grade. On the other hand, on the same score, metalliferous ore bodies are highly unpredictable. As a consequence the estimation of reserves, either tonnage-wise or grade-wise, is fraught with grave uncertainties requiring considerable caution, even in the case of a deposit where exploration has been undertaken.

In general, systematic exploration is initiated by pinpointing a deposit from satellite imageries. This study is followed by identifying and demarcating the deposits, on air photographs, on the scale 1:5000. Such preliminaries having been completed, the first operation is to carry out detailed geological mapping. This is successively followed by geochemical exploration, geophysical exploration, exploratory and structural drilling, and finally by exploratory mining. Sampling and analysis is done at every stage of the exploratory programme. During geological mapping, surface sampling is done, from outcrops, and exposures, in cuttings etc. Samples are also collected, by trenching and pitting, where the ore body is not exposed. Cores, got from exploratory drilling, are sampled and analysed. Exploratory mining is undertaken to collect bulk samples for pilot plant tests as well as for assessing the nature of the ore body and its physical and mechanical properties as well as those of the wall rocks.

Hence in defining the type of reserves, in a base metal deposit it is essential to take into account not only the quantum but also the type exploration, as listed above, that has been carried out. Thus, for example, if the reserves are to be considered as falling with in the "proved" category, then it is expected that sufficient quantum of probing has been carried out either by or together with pitting, trenching or drilling, or exploratory mining, to such an extent, that it is possible to assess the extent of the variability of the parameters such as the thickness, (volume) grade, dip etc. So also, in computing probable reserves judgement should be based on the extent and nature of the exploration carried out. The reserves, falling within the probable or indicated category, can be taken as those, in a part of the deposit where the geological, geochemical and geophysical explorations have been completed, together with some amount of probing by core drilling by means of which the geologist is able to surmise the probable grade and extent of reserves. Reserves, that fall in the inferred or possible category, are those where geological exploration including, geochemical, geophysical exploration have positively indicated the presence or geological/structural continuity of the ore body (see p. 522).

Page 668

The granite trade which started with the export of the so-called "black granite (dolerite) has since undergone considerable change. Presently the stock piling of black granite by the foreign buyers has caused a substantial drop in the prices. As a consequence, exporters have started diversification and granites of various kinds have come into the market. Most of the so-called granites exported are gneisses and migmatitic rocks, which have prominent banding and pleasing colour. Such rocks are marketed under fancy trade names. It may be pointed out that the mining of "black granite" is often very unsystematic. Hence, the recovery is not more than 15%. Consequently areas of granite mining are littered with unsightly dumps with little chance of land reclamation. In mining dykes the benching of the side burden, comprising the country rock (granitic) is often not done, and the dyke is extracted straight away. Blocks are cut manually by chisling and sometimes a little explosive is used to ease out the block that has been isolated. The blocks are generally one cubic metre in size. Sometimes smaller dimensions are also accepted as "monumental" blocks or "tile" blocks.

When the workings are below the ground level, diesel operated cranes are used for lifting. The block is lashed by chains which form a "cradle" and then the chain is attached to the hook of the crane. When the block is lifted out it is directly loaded on to the truck which is already in position. When the workings are at ground level or above ground level, lifting forks are employed for loading onto the trucks. Sometimes where the terrain permits a trench is made sufficiently deep such that the depth at the back is equal to the height of the chassis of the truck. In such a case the blocks are loaded by levering the blocks with wooden poles.

It is often seen that the blocks of "black granite" when extracted from the ground and left exposed to the sun are affected by spalling and cracking.

The reason for this is not very clear. It is felt that the spalling can be due to one of the two reasons or both: 1. since the dyke was intruded into possibly a joint plane in the country rock it was in a confined state and in a state of compression. Hence when the block is extracted there is a relief of the stresses causing expansion with resultant spalling 2. it is a well known fact that dolerite dykes are characterised by breaking up into more or less cubical blocks, which are prone to spheroidal weathering.

Granite inclusive of "black granite" is considered to be 'minor mineral" under the Mineral Concession Rules. Hence it is out of the purview of the Indian Bureau of Mines which is the Department charged with the responsibility of the conservation aspect in regard to mineral deposits and scruting of mining plan prior to working.

## Mining Plan—Its Preparation

With the revision of the Mineral Conservation and Development Rules in 1987, the Indian Bureau of Mines has been endowed with statutory powers, and at the same time it has been empowered to oversee mining practices, adopted by the exploiting agencies, in the context of mineral conservation. As a corollary to the above mentioned responsibility, the Indian Bureau of Mines is also empowered to examine the effects of mining and attendant operations on the environment. With the view to facilitating the preparation of the mining plan in accordance with the statutory requirements, the Indian Bureau of Mines has come out with a note entitled "Guidelines in Respect of the Outline for the "Mining Plan" including "Environmental Management Plan." The Indian Bureau of Mines has also prepared another note entitled "Outline of Mining Plan for Small Opencast Mines, Excluding Mines with Mineral Beneficiation Plants" (see Annexures I). The qualification of the persones authorised to prepare mining plans have been given in Chapter II of the Mineral Conservation and Development Rules, 1978. See Rule 4(2) and Rule 21(4).

# INDIAN BUREAU OF MINES

# OUTLINE OF MINING PLAN

**1.0 INTRODUCTION**

**2.0 GENERAL**
- 2.1 Name and address of the applicant
- 2.2 Status of the applicant
- 2.3 Mineral or minerals which the applicant intends to mine
- 2.4 Name, address and registration number of the recognised person who prepared the Mining Plan
- 2.5 Name and address of the prospecting agency
- 2.6 Details of the 'Area'
- 2.7 Period for which the mining lease is required
- 2.8 Infrastructure

**3.0 GEOLOGY AND RESERVES**
- 3.1 Physiography
- 3.2 Geology
- 3.3 Details of exploration
  - (i) already carried out in the area
  - (ii) proposed to be carried out
- 3.4 Method of estimation of reserves
- 3.5 Geological reserves and grade

**4.0 MINING**
- 4.1 Year-wise development for the first five years
- 4.2 Year-wise production for the first five years
- 4.3 Proposed rate of production when the mine is fully developed
- 4.4 Mineable reserves and anticipated life of the mine
- 4.5 Proposed method of mining

4.5.1 Open cast working:
Mine layout to be illustrated with plans and sections
4.5.2 Underground working:
(a) Mode of entry
(b) System of winding/hoisting
(c) Underground layout (illustrated with plan and longitudinal sect.)
(d) Method of stoping
(e) Mine ventilation
4.6 Extent of mechanisation
4.6.1 Extent of manual and/or machine mining
4.6.2 Drilling
4.6.3 Loading
4.6.4 Hauling/Transport
4.6.5 Miscellaneous operations

5.0 **BLASTING**
5.1 Broad blasting parameters
5.2 Type of explosives to be used
5.3 Storage of explosives

6.0 **MINE DRAINAGE**

7.0 **DISPOSAL OF WASTE**
7.1 Nature of waste
7.2 Selection of dumping site
7.3 Maximum height and spread of dumps
7.4 Stacking of subgrade minerals
7.5 Selection of site for stacking
7.6 Height and spread of stacks

8.0 **USE OF MINERAL**
8.1 If for captive use, the location of the plant and distance from the mine.
8.2 If intended for sale—Industry in India or abroad for whom intended

9.0 **MINERAL BENEFICIATION**
9.1 Average grade and tonnage of feed
9.2 Specifications of beneficiation products
9.3 Laboratory/pilot plant test data if any on the sample from the area

9.4     Method of beneficiation
9.5     Expected recovery and grade
9.6     Physical and chemical composition of tailings
9.7     Design, size and capacity of tailings pond
9.8     Quality of water before final discharge

## 10.0  SURFACE TRANSPORT

10.1    Mode of transport of minerals to the despatch point

## 11.0  SITE SERVICES

## 12.0  EMPLOYMENT POTENTIAL

12.1    Management and supervisory personnel
12.2    Labour—skilled, semi-skilled, unskilled

## 13.0  ENVIRONMENT MANAGEMENT PLAN

13.1    Base Line information
          (i)    Existing land use pattern
         (ii)    Water regime
        (iii)    Flora and fauna
         (iv)    Climatic conditions
          (v)    Human settlements
         (vi)    Public buildings, places and monuments
        (vii)    Quality of air and water
       (viii)    Whether the area falls under notified area under Water Act, 1974
13.2    Environmental impact assessment statement
          (i)    Impact of mining and beneficiation on environment
13.3    Management Plan
          (i)    Storage & preservation of top soil
         (ii)    Proposal for reclamation of land affected by mining activities—during and at the end of mining
        (iii)    In case of forest—programme of phased compensatory afforestation
         (iv)    Measures for dust suppression
          (v)    Measures to minimise vibrations due to blasting and check noise pollution
         (vi)    Stabilisation and vegetation of dumps
        (vii)    Treatment and disposal of water from mine and beneficiation plant
       (viii)    Measures for minimising adverse effects on water regime

(ix) Afforestation of tailing ponds

(x) Preparation of dumping ground for stacking toxic mineral substance

## 14.0 ANY OTHER RELEVANT INFORMATION

## Outline of Mining Plan for Small Open Cast Mines, Excluding Mines with Mineral Beneficiation Plants (to be read with guidelines already circulated)

## 1.0 INTRODUCTION

## 2.0 GENERAL

2.1 Name and address of the applicant

2.2 Status of the applicant

2.3 Mineral or minerals which the applicant intends to mine

2.4 Name, address and reg. No. of the recognised person who prepared the mining plan

2.5 Name and address of the agency

2.6 Details of the area

2.7 Period for which mining lease is required

2.8 Infrastructure

## 3.0 GEOLOGY AND RESERVES

3.1 Physiography

3.2 Geological report of the areas with necessary plans and sections

3.3. Details of exploration;

(1) already carried out

(2) proposed to be carried out

3.4 Method of estimation of reserves

3.5 Geological reserves and grade

3.6 Mineable reserves and anticipated life of the mine

## 4.0 MINING

4.1 Year-wise development for the first five years

4.2 Year-wise production for the first five years

4.3 Proposed rate of production when the mine is fully developed

4.4 Proposed method of mining

4.4.1 Open cast workings: Mine layout to be illustrated with plans and sections

4.5.1 Extent of manual mining
4.5.2 Drilling
4.5.3 Loading
4.5.4 Hauling/transport
4.5.5 Miscellaneous operations

## 5.0 BLASTING

5.1 Broad blasting parameters
5.2 Type of explosives to be used
5.3 Storage of explosives

## 6.0 MINE DRAINAGE

## 7.0 STACKING OF MINERAL REJECTS AND DISPOSAL OF WASTE

7.1 Nature of waste
7.2 Selection of dumping site
7.3 Maximum height and area of dumps
7.4 Stacking of sub-grade minerals
7.5 Selection of the site of for stacking
7.6 Height and spread of stacks

## 8.0 USE OF MINERAL

8.1 If for captive use, location of the plant and distance from mine
8.2 If intended for sale—industry in India or abroad whom intended

## 9.0 SURFACE TRANSPORT

9.1 Mode of transport of mineral to the despatch point

## 10.0 SITE SERVICES

## 11.0 EMPLOYMENT POTENTIAL

11.1 Management and advisory personal
11.2 Labour—skilled, semi-skilled and unskilled

## 12.0 ENVIRONMENT MANAGEMENT PLAN

12.1 Base line information
    1. Existing land use pattern
    2. Water regime
    3. Human settlements
    4. Public buildings, places and monuments

## DETAILS OF ESTIMATION OF RESERVES

| Cross Section with Reference to geol. plan | Area in sq.m. | Strike length considered in m. (based on influence of section) | Volume cu.m. | Bulk density | Total insitu reserves (ton) | Percentage recovery of unsaleable/saleable material | Effective reserves Quantity (ton) | Grade | Category of reserves | Remarks |
|---|---|---|---|---|---|---|---|---|---|---|
| 1 | 2 | 3 | 4 | 5 | 6 | 7 | 8 | 9 | 10 | 11 |

## YEARWISE PRODUCTION SCHEDULE

| Year | Slice (as corresponding to each bench height) | Production | | Generation of waste material in tonnes | | Generation of mineral rejects | | Remarks |
|---|---|---|---|---|---|---|---|---|
| | | Quantity | Grade | overburden & side burden | interstitial waste material if any | Quantity, tonnes | Grade | |
| 1 | 2 | 3 | 4 | 5 | 6 | 7 | 8 | 9 |

# INDEX TO THE ADDENDUMS

# SUBJECT INDEX